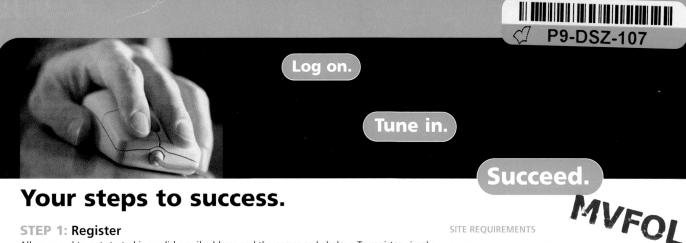

Log on.

Tune in.

Succeed.

MVFOL

Your steps to success.

STEP 1: Register

All you need to get started is a valid email address and the access code below. To register, simply:

1. Go to **www.physicsplace.com**
2. Click the appropriate book cover.
   Cover must match the textbook edition being used for your class.
3. Click "**Register**" under "**First-Time User?**"
4. Leave "**No, I Am a New User**" selected.
5. Using a coin, scratch off the silver coating below to reveal your access code.
   Do not use a knife or other sharp object, which can damage the code.
6. Enter your access code in lowercase or uppercase, without the dashes.
7. Follow the on-screen instructions to complete registration.
   During registration, you will establish a personal login name and password to use for logging into the website. You will also be sent a registration confirmation email that contains your login name and password.

Your Access Code is:

Note: If there is no silver foil covering the access code, it may already have been redeemed, and therefore may no longer be valid. In that case, you can purchase access online using a major credit card. To do so, go to **www.physicsplace.com**, click the cover of your textbook, click "**Buy Now**", and follow the on-screen instructions.

STEP 2: Log in

1. Go to **www.physicsplace.com** and click the appropriate book cover.
2. Under "**Established User?**" enter the login name and password that you created during registration. *If unsure of this information, refer to your registration confirmation email.*
3. Click "**Log In**".

STEP 3: (Optional) Join a class

Instructors have the option of creating an online class for you to use with this website. If your instructor decides to do this, you'll need to complete the following steps using the Class ID your instructor provides you. By "joining a class," you enable your instructor to view the scored results of your work on the website in his or her online gradebook.

To join a class:

1. Log into the website. For instructions, see "STEP 2: Log in."
2. Click "**Join a Class**" near the top right.
3. Enter your instructor's "**Class ID**" and then click "**Next**".
4. At the Confirm Class page you will see your instructor's name and class information. If this information is correct, click "**Next**".
5. Click "**Enter Class Now**" from the Class Confirmation page.
- *To confirm your enrollment in the class, check for your instructor and class name at the top right of the page. You will be sent a class enrollment confirmation email.*
- *As you complete activities on the website from now through the class end date, your results will post to your instructor's gradebook, in addition to appearing in your personal view of the Results Reporter.*

To log into the class later, follow the instructions under "STEP 2: Log in."

Got technical questions?

Customer Technical Support: To obtain support, please visit us online anytime at http://247.aw.com where you can search our knowledgebase for common solutions, view product alerts, and review all options for additional assistance.

SITE REQUIREMENTS

For the latest updates on Site Requirements, go to **www.physicsplace.com**, choose your text cover, and click Site Reqs.

WINDOWS
OS: Windows 2000, XP
Resolution: 1024 x 768
Plugins: Latest Version of Flash/QuickTime/Shockwave (as needed)
Browsers: Internet Explorer 6.0; Firefox 1.0

MACINTOSH
OS: 10.2.4, 10.3.2
Resolution: 1024 x 768
Plugins: Latest Version of Flash/QuickTime/Shockwave (as needed)
Browsers: Firefox 1.0; Safari 1.3

Internet Connection: 56k modem minimum

Register and log in

Join a class

Important: Please read the Subscription and End-User License agreement, accessible from the book website's login page, before using The Physics Place website. By using the website, you indicate that you have read, understood, and accepted the terms of this agreement.

Unfold to explore how **The Physics Place website** makes learning easy and fun!

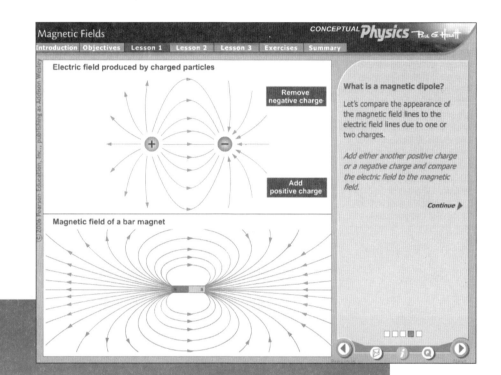

This media grid provides you with an at-a-glance, chapter-by-chapter overview of how the text and media work together to illustrate key Physics concepts. The grid outlines the extensive library of Interactive Figures™, Tutorials, and Video Demonstrations available for use with *Conceptual Physics Fundamentals*. In addition to these resources, The Physics Place provides self-assessment quizzes, a wealth of ActivPhysics™ simulations and animations, and other self-study tools for each chapter.

CONCEPTUAL

Physics Fundamentals

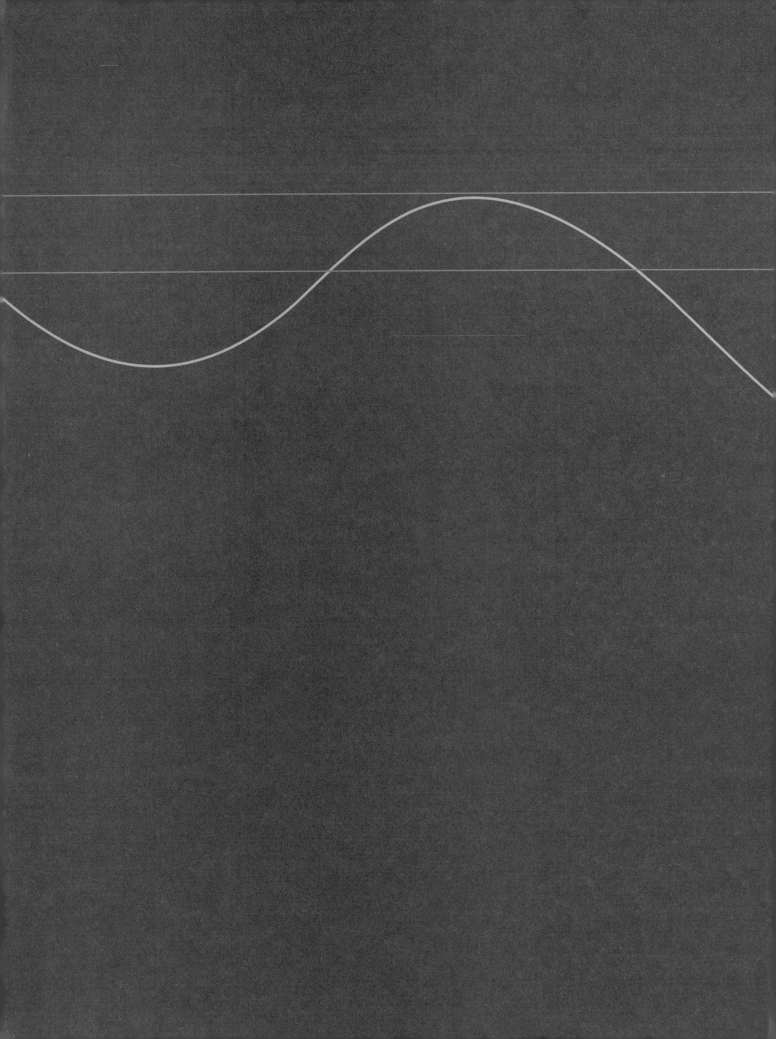

CONCEPTUAL
Physics Fundamentals

written and illustrated by

Paul G. Hewitt

City College of San Francisco

with contributions from

Phillip R. Wolf

Mt. San Antonio College

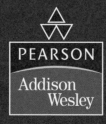

PEARSON

Addison
Wesley

San Francisco Boston New York
Capetown Hong Kong London Madrid Mexico City
Montreal Munich Paris Singapore Sydney Tokyo Toronto

Library of Congress Cataloging-in-Publication Data

Hewitt, Paul G.
  Conceptual physics fundamentals / written and illustrated by Paul G. Hewitt; with contributions from Phillip R. Wolf.
      p. cm.
  Includes bibliographical references and index.
    ISBN 0-321-50136-5 (alk. paper)
1. Physics—Textbooks.   I. Wolf, Phillip R.   II. Title.
  QC23.2.H495 2008
530—dc22
2007015739

ISBN **0-321-50136-5 / 978-0-321-50136-3**
(Student Edition)
ISBN **0-321-51697-4 / 978-0-321-51697-8**
(Professional Copy)

Publisher: Jim Smith
Project Editor: Chandrika Madhavan
Development Manager: Michael Gillespie
Editorial Manager: Laura Kenney
Managing Editor: Corinne Benson
Production Supervisor: Lori Newman
Production Management: Progressive Publishing Alternatives
Production Editor: David Bailey
Compositor: Progressive Information Technologies
Interior Designer: Seventeenth Street Studios
Cover Designer: Riezebos Design
Illustrators: Dartmouth Publishing, Inc.
Photo Researcher: Elaine Soares, Image Resource Center
Director, Image Resource Center: Melinda Patelli
Image Rights and Permissions Manager: Zina Arabia
Image Permissions Coordinator: Elaine Soares, Image Resource Center
Manufacturing Buyer: Pam Augspurger
Executive Marketing Manager: Scott Dustan
Text Printer: Quebecor World Versailles
Cover Printer: Phoenix Color Corp./Hagerstown
Cover Photo Credit: Wave on water surface, Getty Images—Photonica Amana America, Inc.

PEARSON
Addison
Wesley   www.aw-bc.com

8 9 10—QGV—14 13 12

To my wife Lillian,
to
wonderful parents Wai Tsan and Siu Bik Lee,
to
students who value physics in their education,
and to
instructors who nurture their love of science

CONTENTS IN BRIEF

CONTENTS IN DETAIL

This is a personalized book, as noted in the many photographs of family and friends that grace its pages. Foremost to providing suggestions and feedback to this and various editions of *Conceptual Physics* is Kenneth Ford, former CEO of the American Institute of Physics, to whom the eighth edition of *Conceptual Physics* and the third edition of the high-school version were dedicated. Ken's hobby of flying is appropriately shown on page 257. Assisting in the production of all my books in recent years is my wife, Lillian Lee Hewitt. She is shown on pages 1, 178, and 251. We both demonstrate Newton's third law on page 66. Lil holds our colorful pet conure, Sneezlee, on page 281. She is also seen with friend Sushi Shah in a British supermarket on page 216. Her dad Wai Tsan Lee showing iron nails is on page 229, her mom Siu Bik Lee using solar energy is on page 192, and her niece and nephew, Allison and Erik Wong, demonstrating thermodynamics on page 189.

Part openers with the cartoon style blurbs begin on page 29 with my great nephew Evan Suchocki holding a pet chickie while sitting on my lap. Part 2 opens on page 161 with nephew Terrence Jones. Part 3, page 197, is my granddaughter Megan Abrams. Part 4 opens with my grandson Alexander Hewitt on page 245. Alexander and Grace Hewitt open Part 5 on page 319.

City College of San Francisco Will Maynez shows the air track he designed and built on page 86, and he burns a peanut on page 175. Former Chicago high-school teacher and close friend, Marshall Ellenstein, walks barefoot on broken glass on page 155. Over many years Marshall has been and still is a contributor to *Conceptual Physics*. He had more recently converted videos of my CCSF lectures in 1982 to a 3-disc DVD set, *Conceptual Physics Alive!—The San Francisco Years* (which predate the DVDs of 34 lectures in Hawaii). Former teacher and friend Howard Brand, from college days, demonstrates momentum on page 78. Page 135 shows San Mateo high-school physics teacher Paul Robinson, who risks his body for science sandwiched between beds of nails. Paul is shown again on page 280. Paul's wife Ellyn, author of *Biotechnology: Science for the New Millennium*, EMC-Paradigm Publishing, 2006, is on page 160.

Family photos begin with the touching photo on page 75 of my brother Steve with his daughter Gretchen at their coffee farm in Costa Rica. My son Paul is shown on pages 165 and 180. His lovely wife, Ludmila, holds crossed Polaroids on page 312. The endearing girl on page 18 is my daughter Leslie, earth science co-author of the *Conceptual Physical Science* textbooks. Leslie and husband Bob's children, Megan and Emily, along with son Paul's children (Alex and Grace), make up the colorful set of photos on page 280. A photo of my late son James is on page 308. He left me my first grandson, Manuel, seen on page 269. Manuel's grandmom, my late wife Millie, before passing away in early 2004, bravely holds her hand above the active pressure cooker on page 180. Brother Dave and his wife Barbara demonstrate atmospheric pressure on page 145. Their son David is on page 217. Sister Marjorie Hewitt Suchocki, an author and emeritus theologian at Claremont School of Theology, illustrates reflection on page 295. Marjorie's son, John Suchocki, author of *Conceptual Chemistry*, Benjamin Cummings, third edition, and my chemistry co-author of the *Conceptual Physical Science* textbooks, walks fearlessly across hot coals on page 177. Nephew John, a talented vocalist and guitarist known as John Andrew in his popular CDs, strums his guitar on page 235. The group listening to music on page 265 are part of John and Tracy's wedding party; from left to right, Butch Orr, my niece Cathy Candler, bride and groom, niece Joan Lucas, sister Marjorie, Tracy's parents Sharon and David Hopwood, teacher friends Kellie Dippel and Mark Werkmeister, and myself.

Physics professor friends include Tsing Bardin illustrating liquid pressure on page 136, Bob Greenler showing a colorful giant bubble on page 289, Ron Hipschman freezing water on page 190, Peter Hopkinson with his zany mirror on page 294, David Housden with an impressive circuit display on page 198, David Kagan with his wave model on page 320, Darlene Librero and Paul Doherty illustrating Newton's third law on page 52, Chelcie Liu with his novel race tracks on page 30, Jennie McKelvie making waves on page 271, Fred Myers showing the force produced by a small magnet on page 225, Sheron Snyder generating light on page 240, Jim Stith with his impressive Wimshurst generator on page 206, Neil deGrasse Tyson extolling the wonders of gravity on page 105, Roy Unruh with an electric car on page 218, Lynda Williams singing her heart out on page 262, Per Olof and Johan Zetterberg demonstrating Magdeburg hemispheres on page 133, and Dean Zollman making nuclear measurements on page 346.

Personal friends who were my former students begin with Tenny Lim, a "rocket scientist" at Jet Propulsion Lab in Pasadena, drawing her bow on page 89. On page 184 is another "rocket-scientist" friend and former teaching assistant, Helen Yan, who develops satellites for Lockheed Martin in Sunnyvale in addition to teaching physics part-time at CCSF. The photo of the karate gal on page 82 is former CCSF student Cassy Cosme.

Ernie Brown, a lifelong friend, cartoonist, and designer of the cover logos for all my conceptual books, is seen on page 383 as a cartoon. My friend and early influence toward science from sign painting days, Burl Grey, is on page 37. Other cherished friends are Tim Gardner, applying Bernoulli's principle on page 150 and induction on page 243. Lori Patterson in an energizing pose on page 204, her son Ryan on page 255. Then there's Paul Ryan, dragging his finger through molten lead on page 192, and Suzanne Lyons, co-author of *Conceptual Integrated Science*, posing with her children Tristan and Simone on page 292. Physics buddy John Hubisz appears in the entropy photo on page 166. Charlie Spiegel shows refraction on page 276. Friends Larry and Tammy Tunison wear radiation badges on page 350 and their dogs are on page 187. Phil Wolf, a contributor to this book and co-author of the *Problem Solving in Conceptual Physics* book that accompanies *Conceptual Physics*, tenth edition (and a recommended supplement to this book), is shown on page 321. Helping create that book is teacher Diane Riendeau, shown on page 246.

My dear Hawaii friends include Walter Steiger, page 356, Jean and George Curtis, pages 237 and 322, and Chiu Man Wu on page 186, and his daughter, Andrea, on page 102.

The inclusion of these people who are so dear to me makes *Conceptual Physics Fundamentals* all the more my labor of love.

You know you can't enjoy a game unless you know its rules; whether it's a ball game, a computer game, or simply a party game. Likewise, you can't fully appreciate your surroundings until you understand the rules of nature. Physics is the study of these rules, which show how everything in nature is beautifully connected. So the main reason to study physics is to enhance the way you see the physical world. You'll see the mathematical structure of physics in frequent equations, but more than being recipes for computation, you'll see the equations as guides to thinking.

I enjoy physics, and you will too — because you'll understand it. If you get hooked and take a follow-up course, then you can dig deeper into mathematical problems. Go for comprehension of concepts now, and if more computation follows, it will be with understanding.
Enjoy your physics!

Paul G. Hewitt

Unlike the editions of *Conceptual Physics*, immediately after the introductory chapter on science we begin with atoms in this book. This serves as needed background for the study of fluids, heat, and electricity.

Part One, "Mechanics," begins with the chapter on mechanical equilibrium (a departure from beginning with kinematics, which too often gets bogged down with the distinction between velocity and acceleration—and their units). Your students get their first taste of physics via a very comprehensible treatment of parallel force vectors. They enter a comfortable part of physics before being introduced to kinematics, which is covered briefly in the latter part of the chapter—just enough to prepare for the next chapter on Newton's laws. I strongly advise going quickly through kinematics. After all, it pre-dates Newton and is devoid of physics laws. I advise making the distinction between velocity and acceleration without great fanfare, and continuing to Newton's laws.

Chapter 5 treats momentum before energy, with the thought that students find mv much simpler and easier to grasp than $\frac{1}{2}mv^2$. Another reason for treating momentum first is that the vectors of the previous chapter are employed with momentum but not with energy. Momentum, work, energy, and power also get generous coverage in the *Practicing Physics* book.

For brevity, rotational motion is briefly covered in Appendix B. Gravity and projectile motion are combined in Chapter 6, extending to satellite motion. Students are fascinated to learn that any projectile moving fast enough can become an Earth satellite. Moving faster, it can become a satellite of the Sun. Gravity, projectile motion, and satellite motion belong together. Mechanics concludes with fluids.

Part 2 treats heat, followed by electricity and magnetism in Part 3. Waves, sound, and light make up Part 4. The book concludes with an overview of quantum physics, leading to nuclear processes.

Boxed features on technology and pseudoscience are scattered throughout this book. To the person who works in the science field, who knows about the care, checking, and cross-checking that go into understanding something, pseudoscience fads and misconceptions are laughable. But to those who don't work in the science arena, including even your best students, pseudoscience can seem compelling when purveyors clothe their wares in the language of science while skillfully sidestepping the tenets of

science. It is my hope that these boxes may help to stem this rising tide.

Featured in this book are "One-Step Calculations," sets of straightforward "plug-and-chug" problems requiring only single-step solutions as the title indicates. They appear in more equation-oriented chapters. Students become familiar with the equations by substitution of given numerical values. More math-physics challenges are found in the problem sets. These are preceded by qualitative exercises.

Many of the physics problems have two parts: part a expressed in symbols rather than numbers for conceptual understanding, then part b where magnitudes are given. If you can get your students to get the hang of problem solving without numbers, you'll be treating physics conceptually. More problems are available in the student supplement *Problem Solving in Conceptual Physics,* co-authored with Phil Wolf. It is a supplement to *Conceptual Physics*, tenth edition and can complement any introductory physics course. Phil and I think that many professors will enjoy the options offered by this student supplement. Problem solutions are posted on the website in the Instructor's Resource area for this text.

Supporting this edition is the *Instructor's Manual,* in editable Word format with suggested lectures, demonstrations, and answers to all chapter-end material. The *Next-Time Questions* book has a greater number of insightful questions, which are in a horizontal format to make them more compatible with computer monitors and PowerPoint® displays. As the title suggests, I strongly recommend a long wait time before a student sees the question and its answer. Although the printed book is still in black and white, the electronic version is in color. The student supplement *Practicing Physics* includes answers to the odd-numbered exercises and problems herein, and is also available electronically. A *Test Bank* (both in print and in computerized format) is also available. Perhaps most important, however, is the ambitious range of teaching and learning media developed to support this book.

Another supplement to help with your in-class presentations is *The Conceptual Physics Fundamentals Lecture Launcher*. This CD-ROM provides a wealth of presentation tools to help support your fun and dynamic lectures. It includes more than 100 clips from my favorite video demonstrations, more than 100 interactive applets developed specifically to illustrate particularly tricky concepts, and chapter-by-chapter weekly in-class quizzes for use with Classroom Response Systems. These easy-to-use wireless polling systems allow you to pose questions in class, see how students "vote," and then display and discuss the results in real time. This CD-ROM also provides all the line images from the book (in high resolution).

For out-of-class help for your students, a critically acclaimed website, which can be found at http://www.physicsplace.com, provides even more study resources. The Physics Place is the most educationally advanced, most highly rated by students, and most widely used website available for students taking this course. The enhanced website now provides more of the students' favorite interactive online tutorials (covering many instructor-requested topics), and a new library of Interactive Figures (key figures from each chapter in the book that are better understood through interactive experimentation owing to reasons of scale, geometry, time evolution,

or multiple representation). Quizzes, flash cards, and a wealth of other chapter-specific study aids are also provided.

All of these innovative, targeted, and effective online learning media are easily integrated in your course using a new online gradebook that allows you to "assign" the tutorials, quizzes, and other activities as out-of-class homework or projects that are automatically graded and recorded. Simple icons throughout the text highlight for you and your students key tutorials, Interactive Figures, and other online resources. A new Online Resources section at the Physics Place summarizes the media available to you and your students, chapter by chapter, week by week.

For more information on the support ancillaries, check out http://www.aw-bc.com/physics or contact your Addison Wesley representative.

For general questions contact me: pghewitt@ aol.com.

ACKNOWLEDGMENTS

I am enormously grateful to Ken Ford for his many insightful suggestions. One of Ken's admirable books, *Basic Physics,* many years ago inspired me to write *Conceptual Physics.* In more recent years I find myself honored that he has devoted so much of his time and energy to assist me in refining my explanations of physics. Errors invariably crop up after manuscript is submitted, so I take full responsibility for any errors that have survived his scrutiny.

For extensive feedback, I'm thankful to Diane Riendeau and Phil Wolf. Phil is the architect of many of the physics problems in the chapter end-matter. For valued suggestions, I thank my friends Dean Baird, Tsing Bardin, Howie Brand, George Curtis, Marshall Ellenstein, Mona El Tawil-Nassar, Herb Gottlieb, Jim Hicks, David Housden, John Hubisz, Dan Johnson, Evan Jones, Iain McInnes, Fred Myers, Kenn Sherey, Chuck Stone, John Suchocki, Paul Robinson, and David Williamson. I'm grateful for suggestions from Michael Crivello, Mike Diamond, Matthew Griffiths, Paul Hammer, Kevin Hope, Francisco Izaguirre, Serhii Klaynovs'kyi, Homer Neal, Mary Page Ouzts, Rex Paris, Ethel Petrou, Les Sawyer, Stan Schiocchio, Dan Sulke, Richard W. Tarara, Lawrence Weinstein, David Williamson, and Dean Zollman. I am grateful to the resourcefulness of my Exploratorium friends and colleagues: Judith Brand, Paul Doherty, Ron Hipschman, and Modesto Tamez. For photographs, I thank my son-in-law Bob Abrams, Keith Bardin, Mark Clark, Burl Grey, my brother Dave Hewitt, my wife Lillian, my son Paul, Will Maynez, Fred Myers, Jay Pasachoff, and the late Milo Patterson.

Also, my thanks to the following reviewers:
Christopher Roddy, *Broward Community College*
Rex Ramsier, *University of Akron*
Tom McCaffrey, *SUNY Oswego*
Sulakshana Plumley, *CC Allegheny*
Wayne Hayes, *Greenville Tech*
Jennifer Leigh Burris, *Aims University*
Mary Paige Ouzts, *Lander University*
Patrick Hecking, *Thiel University*

Serhii Klaynovs'kyi, *Columbia Union College*
Ethel Petrou, *Erie Community College South*
Homer Neal, *Yale University*
Kevin M. Hope, *University of Montevallo*
Charles W. Rogers, *Southwestern Oklahoma State University*
Renee Lathrop, *Dutchess Community College*
Ethan Bourkoff, *Baruch College, CUNY*
John Hauptman, *Iowa State University*
John Hopkins, *Penn State University*
Rex Paris, *Grossmont College*
Michael Crivello, *San Diego Mesa College*

I remain grateful to the authors of books that initially served as influences and references many years ago: Theodore Ashford, *From Atoms to Stars;* Albert Baez, *The New College Physics: A Spiral Approach;* John N. Cooper and Alpheus W. Smith, *Elements of Physics;* Richard P. Feynman, *The Feynman Lectures on Physics;* Kenneth Ford, *Basic Physics;* Eric Rogers, *Physics for the Inquiring Mind;* Alexander Taffel, *Physics: Its Methods and Meanings;* UNESCO, *700 Science Experiments for Everyone;* and Harvey E. White, *Descriptive College Physics.* I'm thankful to Bob Park, whose book *Voodoo Science* motivated me to include the boxes on pseudoscience.

I am most grateful to my resourceful wife, Lillian, for her assistance in all phases of book- and ancillary preparation.

For their dedication to this edition, I am grateful to the staff at Addison Wesley in San Francisco. I am especially thankful to Ashley Anderson Taylor, Adam Black, Lothlorien Homet, and Chandrika Madhavan. A note of appreciation is due Claire Masson for the cyberspace components for this and my other books. For his insightful tutorials I thank David Vasquez, my dear friend of many years. And I thank Crystal Clifton, Marsha Hall, and the production team at Progressive Publishing Alternatives for their patience with my last-minute changes. Thanks to you all!

Paul G. Hewitt
St. Petersburg, Florida

About Science

The circular spots of light surrounding Lillian are images of the Sun, cast through small openings between the leaves in the tree above. During a partial eclipse, the spots are crescents.

First of all, science is the body of knowledge that describes the order within nature and the causes of that order. Second, science is an ongoing human activity that represents the collective efforts, findings, and wisdom of the human race, an activity that is dedicated to gathering knowledge about the world and organizing and condensing it into testable laws and theories. Science had its beginnings before recorded history, when people first discovered regularities and relationships in nature, such as star patterns in the night sky and weather patterns—when the rainy season started or when the days grew longer. From these regularities, people learned to make predictions that gave them some control over their surroundings.

Science made great headway in Greece in the fourth and third centuries BC. It spread throughout the Mediterranean world. Scientific advance came to a near halt in Europe when the Roman Empire fell in the fifth century AD.

Barbarian hordes destroyed almost everything in their paths as they overran Europe. Reason gave way to religion, which ushered in what came to be known as the Dark Ages. During this time, the Chinese and Polynesians were charting the stars and the planets. Before the rise of religion in the Islamic world, Arab nations developed mathematics and learned about the production of glass, paper, metals, and various chemicals. Greek science was reintroduced to Europe by Islamic influences that penetrated into Spain during the tenth, eleventh, and twelfth centuries. Universities emerged in Europe in the thirteenth century, and the introduction of gunpowder changed the social and political structure of Europe in the fourteenth century. The fifteenth century saw art and science beautifully blended by Leonardo da Vinci. Scientific thought was furthered in the sixteenth century with the advent of the printing press.

The sixteenth-century Polish astronomer Nicolaus Copernicus caused great controversy when he published a book proposing that the Sun is stationary and that Earth revolves around the Sun. These ideas conflicted with the popular view that Earth was the center of the universe. They also conflicted with Church teachings and were banned for 200 years. The Italian physicist Galileo Galilei was arrested for popularizing the Copernican theory and for his other contributions to scientific thought. Yet, a century later, those who advocated Copernican ideas were accepted.

These cycles occur age after age. In the early 1800s, geologists met with violent condemnation because they differed with the account of creation in the book of Genesis. Later in the same century, geology was accepted, but theories of evolution were condemned and the teaching of them was forbidden. Every age has its intellectual rebels who are scoffed at, condemned, and sometimes even persecuted during their lifetime, but who later seem beneficial and often essential to the elevation of human conditions. "At every crossway on the road that leads to the future, each progressive spirit is opposed by a thousand men appointed to guard the past."*

1.1 Mathematics—The Language of Science

Science and human conditions advanced dramatically after science and mathematics became integrated some four centuries ago. When the ideas of science are expressed in mathematical terms, they are unambiguous. The equations of science provide compact expressions of relationships between concepts. They don't have the multiple meanings that so often confuse the discussion of ideas expressed in common language. When findings in nature are expressed mathematically, they are easier to verify or to disprove by experiment. The mathematical structure of physics will be evident in the many equations you will encounter throughout this book. The equations are your guides to thinking that show the connections between concepts in nature. The methods of mathematics and experimentation led to enormous success in science.**

* From Count Maurice Maeterlinck's "Our Social Duty."

Physicists have a deep-seated need to know *why* and *what if*. Mathematics is foremost in the tool kits they develop to tackle these questions.

1.2 Scientific Measurements

Measurements are a hallmark of good science. How much you know about something is often related to how well you can measure it. This was well put in the nineteenth century by the famous physicist Lord Kelvin: "I often say that when you can measure something and express it in numbers, you know something about it. When you cannot measure it, when you cannot express it in numbers, your knowledge is of a meager and unsatisfactory kind. It may be the beginning of knowledge, but you have scarcely in your thoughts advanced to the stage of science, whatever it may be." Scientific measurements are not something new but go back to ancient times.

FIGURE 1.1

The round spot of light cast by the pinhole is an image of the Sun. Its *diameter/distance* ratio is the same as the *Sun's diameter/Sun's distance* ratio, 1/110. The Sun's diameter is 1/110 its distance from Earth.

$$\frac{d}{h} = \frac{D}{150,000,000 \text{ Km}} = \frac{1}{110}$$

** We distinguish between the mathematical structure of physics and the practice of mathematical problem solving—the focus of most nonconceptual courses. Note that there are fewer mathematical problems than there are exercises at the ends of chapters of this book. The focus is on comprehension before computation.

A simple and intriguing scientific measurement is that of the Sun. Have you noticed that the spots of sunlight you see on the ground beneath trees are perfectly round when the Sun is overhead, and spread into ellipses when the Sun is low in the sky? These are pinhole images of the Sun, where light shines through openings in the leaves that are small compared with the distance to the ground below. A round spot 5 centimeters in diameter, for example, is cast by an opening that is 110 × 5 centimeters above ground. Tall trees make large images; short trees make small images. With a bit of elementary geometry, you can measure the diameter of the Sun. This is done in the *Practicing Physics* book.

Interestingly, at the time of a partial solar eclipse, the images are crescents (Figure 1.3).

FIGURE 1.3

The crescent-shaped spots of sunlight are images of the Sun when it is partially eclipsed.

FIGURE 1.2

Renoir accurately painted the spots of sunlight on his subjects' clothing and surroundings—images of the Sun cast by relatively small openings in the leaves above.

1.3 Scientific Methods

There is no *one* scientific method. But there are common features in the way scientists do their work. This all dates back to the Italian physicist Galileo Galilei (1564–1642) and the English philosopher Francis Bacon (1561–1626). Theybroke free from the methods of the Greeks, who worked "upward or downward," depending on the circumstances, reaching conclusions about the physical world by reasoning from arbitrary assumptions (axioms). The modern scientist works "upward," first examining the way the world actually works and then building a structure to explain findings.

Although no cookbook description of the scientific method is really adequate, some or all of the following steps are likely to be found in the way most scientists carry out their work.

1. Observe: Closely observe the physical world about you. Recognize a question or a puzzle—such as an unexplained observation.

2. Question: Make an educated guess—a **hypothesis**—that might resolve the puzzle.

3. Predict: Predict consequences of the hypothesis.

4. Test predictions: Do experiments or make calculations to test the predictions.

5. Draw a conclusion: Formulate the simplest general rule that organizes the three main ingredients: hypothesis, predicted effects, and experimental findings.

Although these steps are appealing, much progress in science has come from trial and error, experimentation without hypotheses, or just plain accidental discovery by a well-prepared mind. The success of science rests more on an attitude common to scientists than on a particular method. This attitude is one of inquiry, experimentation, and humility—that is, a willingness to admit error.

1.4 The Scientific Attitude

t is common to think of a fact as something that is unchanging and absolute. But, in science, a fact is generally a close agreement by competent observers who make a series of observations about the same phenomenon. For example, where it was once a fact that the universe is unchanging and permanent, today it is a fact that the universe is expanding and evolving. A scientific hypothesis, on the other hand, is an educated guess that is only presumed to be factual until supported by experiment. When a hypothesis has been tested over and over again and has not been contradicted, it may become known as a law or *principle*.

If a scientist finds evidence that contradicts a hypothesis, law, or principle, the scientific spirit requires that the hypothesis must be changed or abandoned (unless the contradicting evidence, upon testing, turns out to be wrong—which sometimes happens). For example, the greatly respected Greek philosopher Aristotle (384–322 BC) claimed that an object falls at a speed proportional to its weight. This idea was held to be true for nearly 2000 years because of Aristotle's compelling authority. Galileo allegedly showed the falseness of Aristotle's claim with one experiment— demonstrating that heavy and light objects dropped from the Leaning Tower of Pisa fell at nearly equal speeds. In the scientific spirit, a single verifiable experiment to the contrary outweighs any authority, regardless of reputation or the number of followers or advocates. In modern science, argument by appeal to authority has little value.*

> **Experiment, not philosophical discussion, decides what is correct in science.**

Scientists must accept their experimental findings even when they would like them to be different. They must strive to distinguish between what they see and what they wish to see, for scientists,

* But appeal to *beauty* has value in science. More than one experimental result in modern times has contradicted a lovely theory, which, upon further investigation, proved to be wrong. This has bolstered scientists' faith that the ultimately correct description of nature involves conciseness of expression and economy of concepts—a combination that deserves to be called beautiful.

like most people, have a vast capacity for fooling themselves.** People have always tended to adopt general rules, beliefs, creeds, ideas, and hypotheses without thoroughly questioning their validity and to retain them long after they have been shown to be meaningless, false, or at least questionable. The most widespread assumptions are often the least questioned. Most often, when an idea is adopted, particular attention is given to cases that seem to support it, whereas cases that seem to refute it are distorted, belittled, or ignored.

Scientists use the word *theory* in a way that differs from its usage in everyday speech. In everyday speech, a theory is no different from a hypothesis—a supposition that has not been verified. A scientific theory, on the other hand, is a synthesis of a large body of information that encompasses well-tested and verified hypotheses about certain aspects of the natural world. Physicists, for example, speak of the quark theory of the atomic nucleus, chemists speak of the theory of metallic bonding in metals, and biologists speak of the cell theory.

The theories of science are not fixed; rather, they undergo change. Scientific theories evolve as they go through stages of redefinition and refinement. During the past hundred years, for example, the theory of the atom has been repeatedly refined as new evidence on atomic behavior has been gathered. Models of the atom progress as new information is found. Similarly, chemists have refined their view of the way molecules bond together. Astronomers speak of the theory of the Big Bang to account for the observation that galaxies are moving away from one another. Biologists have refined the cell theory and have made enormous inroads toward understanding life. The refinement of theories is a strength of science, not a weakness. Many people feel that it is a sign of weakness to change their minds. Competent scientists must be experts at changing their minds. They change their

** In your education it is not enough to be aware that other people may try to fool you; it is more important to be aware of your own tendency to fool yourself.

minds, however, only when confronted with solid experimental evidence or when a conceptually simpler hypothesis forces them to a new point of view. More important than defending beliefs is improving them. Better hypotheses are made by those who are honest in the face of experimental evidence.

Away from their profession, scientists are inherently no more honest or ethical than most other people. But in their profession they work in an arena that places a high premium on honesty. The cardinal rule in science is that all hypotheses must be testable— they must be susceptible, at least in principle, to being shown to be *wrong*. In science, it is more important that there be a means of proving an idea wrong than that there be a means of proving it right. This is a major factor that distinguishes science from non-science. At first this may seem strange, for when we wonder about most things, we concern ourselves with ways of finding out whether they are true. Scientific hypotheses are different. In fact, if you want to distinguish whether a hypothesis is scientific or not, look to see if there is a test for proving it wrong. If there is no test for its possible wrongness, then the hypothesis is not scientific. Albert Einstein put it well when he stated, "No number of experiments can prove me right; a single experiment can prove me wrong."

Consider the biologist Charles Darwin's hypothesis that life forms evolve from simpler to more complex forms. This could be proved wrong if paleontologists discover more complex forms of life appeared before their simpler counterparts. Einstein hypothesized that light is bent by gravity. This might be proved wrong if starlight that grazed the Sun and could be seen during a solar eclipse were undeflected from its normal path. As it turns out, less complex life forms are found to precede their more complex counterparts and starlight is found to bend as it passes close to the Sun, which support the claims. If and when a hypothesis or scientific claim is confirmed, it is regarded as useful and as a stepping-stone to additional knowledge.

Consider the hypothesis "The alignment of planets in the sky determines the best time for making decisions." Many people believe it, but this hypothesis is not scientific. It cannot be proven wrong, nor can it be proven right. It is *speculation*. Likewise, the hypothesis "Intelligent life exists on other planets somewhere in the universe" is not sci-

entific. Although it can be proven correct by the verification of a single instance of intelligent life existing elsewhere in the universe, there is no way to prove it wrong if no intelligent life is ever found. If we searched the far reaches of the universe for eons and found no life, then that would not prove that it doesn't exist "around the next corner." A hypothesis that is capable of being proved right but not capable of being proved wrong is not a scientific hypothesis. Many such statements are quite reasonable and useful, but they lie outside the domain of science.

Before a theory is accepted, it must be tested by experiment and make one or more new predictions —different from those made by previous theories.

None of us has the time, energy, or resources to test every idea, so most of the time we take somebody's word. How do we know whose word to accept? To reduce the likelihood of error, scientists accept only the word of those whose ideas, theories, and findings are testable—if not in practice, at least in principle. Speculations that cannot be tested are regarded as "unscientific." This has the long-run effect of compelling honesty—findings widely publicized among fellow scientists are generally subjected to further testing. Sooner or later, mistakes (and deception) are found out; wishful thinking is exposed. A discredited scientist does not get a second chance in the community of scientists. The penalty for fraud is professional excommunication. Honesty, so important to the progress of science, thus becomes a matter of self-interest to scientists. There is relatively little bluffing in a game in which all bets are called. In fields of study where right and wrong are not so easily established, the pressure to be honest is considerably less.

The essence of science is expressed in two questions: How would we know? What evidence would prove this idea wrong? Assertions without evidence are unscientific, and can be dismissed without evidence.

Which of these statements is a scientific hypothesis?

a. Atoms are the smallest particles of matter that exist.

b. Space is permeated with an essence that is undetectable.

c. Albert Einstein was the greatest physicist of the twentieth century.

CHECK YOUR ANSWER

*Think about the Stop and Check Yourself questions throughout this book **before** reading the answers. When you first formulate your own answers, you'll find yourself learning more—much more!*

Only *a* is scientific, because there is a test for falseness. The statement is not only *capable* of being proved wrong, but it in fact *has* been proved wrong. Statement *b* has no test for possible wrongness and is therefore unscientific. Likewise, any principle or concept for which there is no means, procedure, or test whereby it can be shown to be wrong (if it is wrong) is unscientific. Some pseudoscientists and other pretenders of knowledge will not even consider a test for the possible wrongness of their statements. Statement *c* is an assertion that has no test for possible wrongness. If Einstein was not the greatest physicist, how could we know? It is important to note that, because the name Einstein is generally held in high esteem, it is a favorite of pseudoscientists. So we should not be surprised that the name of Einstein, like that of Jesus or of any other highly respected person, is cited often by charlatans who wish to bring respect to themselves and their points of view. In all fields, it is prudent to be skeptical of those who wish to credit themselves by calling upon the authority of others.

The ideas and concepts most important to our everyday life are often unscientific; their correctness or incorrectness cannot be determined in the laboratory. Interestingly enough, it seems that people honestly believe their own ideas about things to be correct, and almost everyone is acquainted with people who hold completely opposite views—so the ideas of some (or all) must be incorrect. How do you know whether or not *you* are one of those possessing erroneous beliefs? There is a test. Before you can be reasonably convinced that you are right about a particular idea, you should be certain that you understand the objections and the positions of your most articulate antagonists. You should find out whether your views are supported by sound knowledge of opposing ideas or by your *misconceptions* of opposing ideas. You make this distinction by seeing whether or not you can state the objections and positions of your opposition to *their* satisfaction. Even if you can successfully do this, you cannot be absolutely certain of being right about your own ideas, but the probability of being right is considerably higher if you pass this test.

Suppose that, in a disagreement between two people, A and B, you note that person A states and restates only one point of view, whereas person B clearly states both her own position and that of person A. Who is more likely to be correct? *(Think before you read the answer below!)*

CHECK YOUR ANSWER

Who knows for sure? Person B may have the cleverness of a lawyer who can state various points of view and still be incorrect. We can't be sure about the "other guy." The test for correctness or incorrectness suggested here is not a test of others, but of and for *you*. It can aid your personal development. As you attempt to articulate the ideas of your antagonists, be prepared, like scientists who are prepared to change their minds, to discover evidence contrary to your own ideas—evidence that may alter your views. Intellectual growth often develops in this way.

> **We each need a knowledge filter to tell the difference between what is valid and what only pretends to be valid. The best knowledge filter ever invented is science.**

Although the notion of being familiar with counter points of view seems reasonable to most thinking people, just the opposite—shielding ourselves and others from opposing ideas—has been more widely practiced. We have been taught to discredit unpopular ideas without understanding them in proper context. With the 20/20 vision of hindsight, we can see that many of the "deep truths" that were the cornerstones of whole civilizations were shallow reflections of the prevailing ignorance of the time. Many of the problems that plagued societies stemmed from this ignorance and the resulting misconceptions;

much of what was held to be true simply wasn't true. This is not confined to the past. Every scientific advance is by necessity incomplete and partly inaccurate, for the discoverer sees with the blinders of the day, and can only discard a part of that blockage.

1.5 Science, Art, and Religion

The search for order and meaning in the world around us has taken different forms: One is science, another is art, and the other is religion. Although the roots of all three go back thousands of years, the traditions of science are relatively recent. More important, the domains of science, art, and religion are different, although they often overlap. Science is principally engaged with discovering and recording natural phenomena, the arts are concerned with personal interpretation and creative expression, and religion addresses the source, purpose, and meaning of it all.

Science and the arts are comparable. In the art of literature, we discover what is possible in human experience. We can learn about emotions ranging from anguish to love, even if we haven't experienced them. The arts do not necessarily give us those experiences, but they describe them to us and suggest what may be possible for us. Science tells us what is possible in nature. Scientific knowledge helps us to predict possibilities in nature even before those possibilities have been experienced. It provides us with a way of connecting things, of seeing relationships between and among them, and of making sense of the great variety of natural events around us. Science broadens our perspective of nature. A knowledge of both the arts and the sciences makes for a wholeness that affects the way we view the world and the decisions we make about it and ourselves. A truly educated person is knowledgeable in both the arts and the sciences.

Science and religion have similarities also, but they are basically different—principally because their domains are different. The domain of science is natural order; the domain of religion is nature's purpose. Religious beliefs and practices usually involve faith in, and worship of, a supreme being. By its very nature, religion deals

with those parts of human experience that are not subject to controlled experiments. In this respect, science and religion are as different as apples and oranges: They are two different yet complementary fields of human activity.

When we study the nature of light later in this book, we will treat light first as a wave and then as a particle. To the person who knows a little bit about science, waves and particles are contradictory; light can be only one or the other, and we have to choose between them. But to the enlightened person, waves and particles complement each other and provide a deeper understanding of light. In a similar way, it is mainly people who are either uninformed or misinformed about the deeper natures of both science and religion who feel that they must choose between believing in religion and believing in science. Unless one has a shallow understanding of either or both, there is no contradiction in being religious and being scientific in one's thinking.*

Many people are troubled about not knowing the answers to religious and philosophical questions. Some avoid uncertainty by uncritically accepting almost any comforting answer. An important message in science, however, is that uncertainty is acceptable. For example, in Chapter 15 you'll learn that it is not possible to know with certainty both the momentum and position of an electron in an atom. The more you know about one, the less you can know about the other. Uncertainty is a part of the scientific process. It's okay not to know the answers to fundamental questions. Why are apples gravitationally attracted to Earth? Why do electrons repel one another? Why do magnets interact with other magnets? Why does energy have mass? At the deepest level, scientists don't know the answers to these questions—at least not yet. We know a lot about where we are, but nothing really about *why* we are. It's okay not to know the answers to such religious questions. Given a choice between a closed mind with comforting answers, and an open and exploring mind without answers, most scientists choose the latter. Scientists in general are comfortable with not knowing.

* Of course, this doesn't apply to certain religious extremists who steadfastly assert that one cannot embrace both their brand of religion and science.

SCIENCE AND SOCIETY

■ PSEUDOSCIENCE

In prescientific times, any attempt to harness nature meant forcing nature against her will. Nature had to be subjugated, usually with some form of magic or by means that were above nature—that is, supernatural. Science does just the opposite, and it works within nature's laws. The methods of science have largely displaced reliance on the supernatural—but not entirely. The old ways persist, full force in primitive cultures, and they survive in technologically advanced cultures too, sometimes disguised as science. This is fake science—**pseudoscience**. The hallmark of a pseudoscience is that it lacks the key ingredients of evidence and having a test for wrongness. In the realm of pseudoscience, skepticism and tests for possible wrongness are downplayed or flatly ignored.

There are various ways to view cause-and-effect relations in the universe. Mysticism is one view, appropriate perhaps in religion but not applicable to science. Astrology is an ancient belief system that supposes there is a mystical correspondence between individuals and the universe as a whole—that human affairs are influenced by the positions and movements of planets and other celestial bodies. This nonscientific view can be quite appealing. However insignificant we may feel at times, astrologers assure us that we are intimately connected to the workings of the cosmos, which has been created for humans—particularly the humans belonging to one's own tribe, community, or religious group. Astrology as ancient magic is one thing, but astrology in the guise of science is another. When it poses as a science related to astronomy, then it becomes pseudoscience. Some astrologers present their craft in a scientific guise. When they use up-to-date astronomical information and computers that chart the movements of heavenly bodies, astrologers are operating in the realm of science. But when they use these data to concoct astrological revelations, they have crossed over into full-fledged pseudoscience.

Pseudoscience, like science, makes predictions. The predictions of a dowser, who locates underground water with a dowsing stick, have a very high rate of success—nearly 100%. Whenever the dowser goes through his or her ritual and points to a spot on the ground, the well digger is sure to find water. Dowsing works. Of course, the dowser can hardly miss, because there is groundwater within 100 meters of the surface at nearly every spot on Earth. (The real test of a dowser would be finding a place where water wouldn't be found!)

A shaman who studies the oscillations of a pendulum suspended over the abdomen of a pregnant woman can predict the sex of the fetus with an accuracy of 50%. This means that, if he tries his magic many times on many fetuses, half his predictions will be right and half will be wrong—the predictability of ordinary guessing. In comparison, determining the sex of unborns by scientific means gives a 95% success rate via sonograms, and 100% by amniocentesis. The best that can be said for the shaman is that the 50% success rate is a lot better than that of astrologers, palm readers, or other pseudoscientists who predict the future.

An example of a pseudoscience that has zero success is provided by energy-multiplying machines. These machines, which are alleged to deliver more energy than they take in, are, we are told, "still on the drawing boards and needing funds for development." They are touted by quacks who sell shares to an ignorant public who succumb to the pie-in-the-sky promises of success. This is junk science. Pseudoscientists are everywhere, are usually successful in recruiting apprentices for money or labor, and can be very convincing even to seemingly reasonable people. Their books have widespread respect and greatly outnumber books on science in bookstores. Junk science is thriving.

Four centuries ago, most humans were dominated by superstition, devils, demons, disease, and magic in their short and difficult lives. Only through enormous effort did they gain scientific knowledge and overthrow superstition. We have come a long way in comprehending nature and freeing ourselves from ignorance. We should rejoice in what we've learned. We no longer have to die whenever an infectious disease strikes or to live in fear of demons. Life was cruel in medieval times. Today we have no need to pretend that superstition is anything but superstition, or that junk notions are anything but junk notions— whether voiced by street-corner quacks, by loose thinkers who write promise-heavy health books, by hucksters who sell magnetic therapy, or by demagogues who inflict fear.

Yet there is cause for alarm when the superstitions that people once fought to erase are back in force, enchanting a growing number of people.

James Randi reports in his book *Flim-Flam!* that more than twenty thousand practicing astrologers in the United States serve millions of credulous believers. Science writer Martin Gardner reports that a greater percentage of Americans today believe in astrology and occult phenomena than did citizens of medieval Europe. Few newspapers print a daily science column, but nearly all provide daily horoscopes. Although goods and medicines around us have improved with scientific advances, much human thinking has not.

Many believe that the human condition is slipping backward because of growing technology. More likely, however, we'll slip backward because science and technology will bow to the irrationality, superstitions, and demagoguery of the past. "Equal time" will be allotted to irrationality in our classrooms. Watch out for the spokespeople of irrationality. Pseudoscience is a huge and lucrative business.

SCIENCE AND SOCIETY

■ RISK ASSESSMENT

The numerous benefits of technology are paired with risks. X rays, for example, continue to be used for medical diagnosis despite their potential for causing cancer. But when the risks of a technology are perceived to outweigh its benefits, it should be used very sparingly or not at all.

Risk can vary for different groups. Aspirin is useful for adults, but for young children it can cause a potentially lethal condition known as *Reye's syndrome*. Dumping raw sewage into the local river may pose little risk for a town located upstream, but for towns downstream the untreated sewage is a health hazard. Similarly, storing radioactive wastes underground may pose little risk for us today, but for future generations the risks of such storage are greater if there is leakage into groundwater. Technologies involving different risks for different people, as well as differing benefits, raise questions that are often hotly debated. Which medications should be sold to the general public over the counter and how should they be labeled? Should food be irradiated in order to put an end to food poisoning, which kills more than 5000 Americans each year? The risks to all members of society need consideration when public policies are decided.

The risks of technology are not always immediately apparent. No one fully realized the dangers of combustion products when petroleum was selected as the fuel of choice for automobiles early in the last century. From the hindsight of 20/20 vision, alcohols from biomass would have been a superior choice environmentally, but they were banned by the prohibition movements of the day.

Because we are now more aware of the environmental costs of fossil-fuel combustion, biomass fuels are making a comeback. More environmentally friendly power sources are being developed. An awareness of both the short-term risks and the long-term risks of a technology is crucial.

People seem to have a difficult time accepting the impossibility of zero risk. Airplanes cannot be made perfectly safe. Processed foods cannot be rendered completely free of toxicity, for all foods are toxic to some degree. You cannot go to the beach without risking skin cancer, no matter how much sunscreen you apply. You cannot avoid radioactivity, for it's in the air you breathe and the foods you eat, and it has been that way since before humans first walked on Earth. Even the cleanest rain contains radioactive carbon-14, as do our bodies. Between each heartbeat in each human body, there have always been about 10,000 naturally occurring radioactive decays. You might hide yourself in the hills, eat the most natural of foods, practice obsessive hygiene, and still die from cancer caused by the radioactivity emanating from your own body. The probability of eventual death is 100%. Nobody is exempt.

Science helps to determine the most probable. As the tools of science improve, then assessment of the most probable gets closer to being on target. Acceptance of risk, on the other hand, is a societal issue. Placing zero risk as a societal goal is not only impractical but selfish. Any society striving toward a policy of zero risk would consume its present and future economic resources. Isn't it more noble to accept nonzero risk and to minimize risk as much as possible within the limits of practicality? A society that accepts no risks receives no benefits.

1.6 Science and Technology

Science and technology are also different from each other. Science is concerned with gathering knowledge and organizing it. Technology lets humans use that knowledge for practical purposes, and it provides the tools needed by scientists in their further explorations.

Technology is a double-edged sword that can be both helpful and harmful. We have the technology, for example, to extract fossil fuels from the ground and then to burn the fossil fuels for the production of energy. Energy production from fossil fuels has benefited our society in countless ways. On the flip side, the burning of fossil fuels endangers the environment. It is tempting to blame technology itself for problems such as pollution, resource depletion, and even overpopulation. These problems, however, are not the fault of technology any more than a shotgun wound is the fault of the shotgun. It is humans who use the technology, and humans who are responsible for how it is used.

Remarkably, we already possess the technology to solve many environmental problems. This twenty-first century will likely see a switch from fossil fuels to more sustainable energy sources, such as photovoltaics and solar thermal electric generation. The greatest obstacle to solving today's problems lies more with social inertia than with a lack of technology. Technology is our tool. What we do with this tool is up to us. The promise of technology is a cleaner and healthier world. Wise applications of technology *can* lead to a better world.

1.7 Physics—The Basic Science

Science, once called *natural philosophy,* encompasses the study of living things and nonliving things, the life sciences and the physical sciences. The life sciences include biology, zoology, and botany. The physical sciences include geology, astronomy, chemistry, and physics.

Physics is more than a part of the physical sciences. It is the *basic* science. It's about the nature of basic things such as motion, forces, energy, matter, heat, sound, light, and the structure of atoms. Chemistry is about how matter is put together, how atoms combine to form molecules, and how the molecules combine to make up the many kinds of matter around us. Biology is more complex and

involves matter that is alive. So underneath biology is chemistry, and underneath chemistry is physics. The concepts of physics reach up to these more complicated sciences. That's why physics is the most basic science.

An understanding of science begins with an understanding of physics. The following chapters present physics conceptually so that you can enjoy understanding it.

STOP AND CHECK YOURSELF

Which of the following activities involves the utmost human expression of passion, talent, and intelligence?

a. painting and sculpture

b. literature

c. music

d. religion

e. science

CHECK YOUR ANSWER

All of them! The human value of science, however, is the least understood by most individuals in our society. The reasons are varied, ranging from the common notion that science is incomprehensible to people of average ability to the extreme view that science is a dehumanizing force in our society. Most of the misconceptions about science probably stem from the confusion between the *abuses* of science and science itself.

Science is an enchanting human activity shared by a wide variety of people who, with present-day tools and know-how, are reaching further and finding out more about themselves and their environment than people in the past were ever able to do. The more you know about science, the more passionate you feel toward your surroundings. There is physics in everything you see, hear, smell, taste, and touch!

1.8 In Perspective

Only a few centuries ago the most talented and most skilled artists, architects, and artisans of the world directed their genius and effort to the construction of the great cathedrals, synagogues, temples, and mosques. Some of these architectural structures took centuries to build, which means that nobody witnessed both the beginning and the end of construction. Even the architects and early builders who lived to a ripe old age never saw the finished results of their labors. Entire lifetimes were spent in the shadows of construction that must have seemed without beginning or end. This enormous focus of human energy was inspired by a vision that went beyond worldly concerns—a vision of the cosmos. To the people of that time, the structures they erected were their "spaceships of faith," firmly anchored but pointing to the cosmos.

Today the efforts of many of our most skilled scientists, engineers, artists, and artisans are directed to building the spaceships that already orbit Earth and others that will voyage beyond. The time required to build these spaceships is extremely brief compared with the time spent building the stone and marble structures of the past. Many people working on today's spaceships were alive before the first jetliner carried passengers. Where will younger lives lead in a comparable time?

We seem to be at the dawn of a major change in human growth, for as little Evan suggests in the photo on page 29, we may be like the hatching chicken who has exhausted the resources of its inner-egg environment and is about to break through to a whole new range of possibilities. Earth is our cradle and has served us well. But cradles, however comfortable, are one day outgrown. So with the inspiration that in many ways is similar to the inspiration of those who built the early cathedrals, synagogues, temples, and mosques, we aim for the cosmos.

We live at a challenging and exciting time!

> Science is a way to teach how something gets to be known, what is not known, to what extent things are known (for nothing is known absolutely), how to handle doubt and uncertainty, what the rules of evidence are, how to think about things so that judgments can be made, and how to distinguish truth from fraud and from show.
> — *Richard Feynman*

SUMMARY OF TERMS

Scientific method Principles and procedures for the systematic pursuit of knowledge involving the recognition and formulation of a problem, the collection of data through observation and experiment, and the formulation and testing of hypotheses.

Hypothesis An educated guess; a reasonable explanation of an observation or experimental result that is not fully accepted as factual until tested over and over again by experiment.

Fact A phenomenon about which competent observers who have made a series of observations are in agreement.

Law A general hypothesis or statement about the relationship of natural quantities that has been tested over and over again and has not been contradicted. Also known as a *principle*.

Theory A synthesis of a large body of information that encompasses well-tested and verified hypotheses about certain aspects of the natural world.

Pseudoscience Fake science that pretends to be real science.

SUGGESTED READING

Bodanis, David. *E = mc²: A Biography of the World's Most Famous Equation*. New York: Berkeley Publishing Group, 2002.

Bryson, Bill. *A Short History of Nearly Everything*. New York: Broadway Books, 2003.

Cole, K. C. *First You Build a Cloud*. New York: Morrow, 1999.

Feynman, Richard P. *Surely You're Joking, Mr. Feynman*. New York: Norton, 1986.

Gleick, James. *Genius—The Life and Science of Richard Feynman*. New York: Pantheon Books, 1992.

Sagan, Carl. *The Demon-Haunted World*. New York: Random House, 1995.

REVIEW QUESTIONS

1.1 Mathematics—The Language of Science

1. What is the role of equations in this course?

1.2 Scientific Measurements

2. What are the circular spots of light seen on the ground beneath a tree on a sunny day?

1.3 Scientific Methods

3. Outline the steps of the classic scientific method.

1.4 The Scientific Attitude

4. Distinguish between a scientific fact and a scientific law.
5. How is a scientific theory different from a theory, as used in everyday speech?
6. In daily life, people are often praised for maintaining some particular point of view, for the "courage of their convictions." A change of mind is seen as a sign of weakness. How is this different in science?
7. What is the test for whether a hypothesis is scientific or not?
8. In daily life, we see many cases of people who are caught misrepresenting things and who soon thereafter are excused and accepted by their contemporaries. How is this different in science?
9. What test can you perform to increase the chance in your own mind that you are right about a particular idea?

1.5 Science, Art, and Religion

10. Cite a reason for encouraging students of the arts to learn about science, and for encouraging science students to learn about the arts.
11. Why do many people believe they must choose between science and religion?
12. Psychological comfort is a benefit of having solid answers to religious questions. What benefit accompanies a position of not knowing the answers?

1.6 Science and Technology

13. Clearly distinguish between science and technology.

1.7 Physics—The Basic Science

14. Why is physics considered to be the basic science?

1.8 In Perspective

15. Look ahead to page 29. Little Evan poses a question to the author. What is the message of this question?

EXERCISES

1. What is the penalty for scientific fraud in the science community?
2. Which of the following are scientific hypotheses?
   (a) Chlorophyll makes grass green.
   (b) Earth rotates about its axis because living things need an alternation of light and darkness.
   (c) Tides are caused by the Moon.
3. In answer to the question, when a plant grows, where does the material come from? Aristotle hypothesized by logic that all material came from the soil. Do you consider his hypothesis to be correct, incorrect, or partially correct? What experiments do you propose to support your choice?
4. The great philosopher and mathematician Bertrand Russell (1872–1970) wrote about ideas in the early part of his life that he rejected in the latter part of his life. Do you see this as a sign of weakness or as a sign of strength in Bertrand Russell? (Do you speculate that your present ideas about the world around you will change as you learn and experience more, or do you speculate that further knowledge and experience will solidify your present understanding?)

5. Bertrand Russell wrote, "I think we must retain the belief that scientific knowledge is one of the glories of man. I will not maintain that knowledge can never do harm. I think such general propositions can almost always be refuted by well-chosen examples. What I will maintain— and maintain vigorously—is that knowledge is very much more often useful than harmful and that fear of knowledge is very much more often harmful than useful." Think of examples to support this statement.
6. When you step from the shade into the sunlight, the Sun's heat is as evident as the heat from hot coals in a fireplace in an otherwise cold room. You feel the Sun's heat not because of its high temperature (higher temperatures can be found in some welder's torches), but because the Sun is big. Which do you estimate is larger, the Sun's radius or the distance between the Moon and Earth? Check your answer in the list of physical data on the inside front cover of this book.
7. What is probably being misunderstood by a person who says, "But that's only a scientific theory"?
8. Scientists call a theory that unites many ideas in a simple way beautiful. Are unity and simplicity among the criteria of beauty outside of science? Support your answer.

CHAPTER 1 ONLINE RESOURCES

The Physics Place

Quiz

Flashcards

Links

Atoms

Extraordinary twentieth-century physicist Richard Feynman contributed enormously to our understanding of atoms and physics in general.

I f, in some cataclysm, all of scientific knowledge were to be destroyed, and only one sentence could be passed on to the next generation of creatures, what statement would contain the most information in the fewest words? American physicist Richard Feynman's answer to this question is: *"All things are made of atoms— little particles that move around in perpetual motion, attracting each other when they are a little distance apart, but repelling upon being squeezed into one another."* All matter—shoes, ships, sealing wax, cabbages, and kings—any material we can think of—is made of atoms. We will see that an **atom** is the smallest particle of an element that has all of the element's chemical properties.

2.1 The Atomic Hypothesis

T he idea that matter is composed of atoms goes back to the Greeks in the fifth century BC. Investigators of nature back then wondered whether matter was continuous or not. We can break a rock into pebbles, and the pebbles into fine gravel. The gravel can be broken into fine sand, which then can be pulverized into powder. To the fifth-century Greeks, there was a smallest bit of rock, an "atom," that could not be divided any further.

Aristotle, the most famous of the early Greek philosophers, didn't agree with the idea of atoms. In the fourth century BC, he taught that all matter is composed of various combinations of four elements—earth, air, fire, and water. This view seemed reasonable because, in the world around us, matter is seen in only four forms: solids (earth), gases (air), liquids (water), and the state of flames (fire). The Greeks viewed fire as the element of change, since fire was observed to produce changes on substances that burned. Aristotle's ideas

The Physics Place
Evidence for Atoms

An early model of the atom, with a central nucleus and orbiting electrons, much like a solar system with orbiting planets.

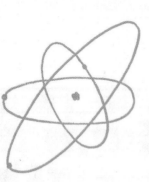

We can't "see" atoms because they're too small. We can't see the farthest star either. There's much that we can't see. But that doesn't prevent investigation of such things or even collecting indirect evidence.

about the nature of matter persisted for more than 2000 years.

The atomic idea was revived in the early 1800s by an English meteorologist and schoolteacher, John Dalton. He successfully explained the nature of chemical reactions by proposing that all matter is made of atoms. He and others of the time, however, had no direct evidence for their existence. Then, in 1827, a Scottish botanist named Robert Brown noticed something very unusual in his microscope. He was studying grains of pollen suspended in water, and he saw that the grains were continually moving and jumping about. At first he thought the grains were some kind of moving life forms, but later he found that dust particles and grains of soot suspended in water moved in the same way. This perpetual and haphazard jiggling of tiny particles—now called **Brownian motion**—results from collisions between visible particles and invisible atoms. The atoms are invisible because they're so small. Although he couldn't see the atoms, Brown could see the effect they had on particles he *could* see. It's like a supergiant beach ball being bounced around by a crowd of people at a football game. From a high-flying airplane, you wouldn't see the people because they are small relative to the enormous ball, which you would be able to see. The pollen grains that Brown observed moved because they were constantly being jostled by the atoms (actually, by the atomic combinations referred to as molecules) that made up the water surrounding them.

Brownian motion was explained in 1905 by Albert Einstein, the same year that he announced the theory of special relativity. Until Einstein's explanation—which made it possible to find the masses of atoms—many prominent physicists remained skeptical about the existence of atoms. So we see that the reality of the atom was not firmly established until the early twentieth century.

All matter, however solid it appears, is made up of tiny building blocks, which themselves are mostly empty space. These are atoms.

2.2 Characteristics of Atoms

Atoms are *incredibly tiny*. An atom is as many times smaller than you as an average star is larger than you. A nice way to say this is that we stand between the atoms and the stars. Or another way of stating the smallness of atoms is that the diameter of an atom is to the diameter of an apple as the diameter of an apple is to the diameter of Earth. So, to imagine an apple full of atoms, think of Earth solid-packed with apples. Both have about the same number.

Atoms Are Numerous

There are about 100,000,000,000,000,000,000,000 atoms in a gram of water (a thimbleful). In scientific notation, that's 10^{23} atoms. The number 10^{23} is an enormous number, more than the number of drops of water in all the lakes and rivers of the world. So there are more atoms in a thimbleful of water than there are drops of water in the world's lakes and rivers. In the atmosphere, there are about 10^{22} atoms in a liter of air. Interestingly, the volume of the atmosphere contains about 10^{22} liters of air. That's an incredibly large number of atoms, and the same incredibly large number of liters of atmosphere. Atoms are so small and so numerous that there are about as many atoms in the air in your lungs at any moment as there are breathfuls of air in Earth's atmosphere.

fyi

How long would it take you to count to one million? If each count takes one second, counting nonstop to a million would take 11.6 days. To count to a billion (10^9) would take 31.7 years. To count to a trillion (10^{12}) would take 31,700 years. Counting to 10^{22} would take about ten thousand times the age of the universe!

Life is not measured by the number of breaths we take, but by the moments that take our breath away.
— *George Carlin*

Atoms Are Perpetually Moving

In solids, they vibrate in place; in liquids, they migrate from one location to another; and in gases, the rate of migration is even greater. Drops of food coloring in a glass of water, for example, soon spread to color the entire glass of water. The same would be true of a cupful of food coloring thrown into an ocean: It would spread around and later be found in every part of the world's oceans.

Atoms and molecules in the atmosphere zip around at speeds up to ten times the speed of sound. They spread rapidly, so oxygen that surrounds you today may have been halfway across the country a few days ago. Taking Figure 2.2 further, your exhaled breaths of air quite soon mix with other atoms in the atmosphere. After the few years it takes for your breath to mix uniformly in the atmosphere, anyone, anywhere on Earth, who inhales a breath of air will take in, on the average, one of the atoms in that exhaled breath of yours. But you exhale many, many breaths, so other people breathe in many, many atoms that were once in your lungs—that were once a part of you; and, of course, the reverse is true. Believe it or not, with each breath you take in, you breathe atoms that were once a part of everyone who ever lived! Considering that the atoms we exhale were part of our bodies (the nose of a dog has no trouble discerning this), it can be truly said that we are literally breathing one another.

Atoms Are Ageless

Many atoms in your body are nearly as old as the universe itself. When you breathe, for example, only some of the atoms that you inhale are exhaled in your next breath. The remaining atoms are taken into your body to become part of you, and they later leave your body by various means. You don't "own" the atoms that make up your body; you borrow them. We all share from the same atom pool, as atoms forever migrate around, within, and among us. Atoms cycle from person to person as we breathe and as our perspiration is vaporized. We recycle atoms on a grand scale.

The Physics Place
Atoms Are Recyclable

The origin of the lightest atoms goes back to the origin of the universe, and most heavier atoms are older than the Sun and Earth. There are atoms in your body that have existed since the first moments of time, recycling throughout the universe among innumerable forms, both nonliving and living. You're the present caretaker of the atoms in your body. There will be many who will follow you.

2.3 Atomic Imagery

Atoms are too small to be seen with visible light. Because of diffraction, you can discern details no smaller than the wavelength of light you use to look with. This can be understood by an analogy with water waves. A ship is much larger than the water waves that roll on by it. As Figure 2.3 shows, water waves can reveal features of the ship. The waves *diffract* as they pass the ship. But diffraction is nil for waves that pass by the anchor chain, revealing little or nothing about it. Similarly, waves of visible light are too coarse compared with the size of an atom to show details of atomic size and shape.

Yet here in Figure 2.4 we see a picture of atoms—the historic 1970 image of chains of individual thorium atoms. The picture is not a photograph but

FIGURE 2.2
There are as many atoms in a normal breath of air as there are breathfuls of air in the atmosphere of Earth.

STOP AND
CHECK YOURSELF

1. Which are older, the atoms in the body of an elderly person or those in the body of a baby?

2. World population grows each year. Does this mean that the mass of Earth increases each year?

3. Are there really atoms that were once a part of Albert Einstein incorporated in the brains of all the members of your family?

CHECK YOUR ANSWERS

1. The age of the atoms in both is the same. Most of the atoms were manufactured in stars that exploded before the solar system came into existence.

2. The greater number of people increases the mass of Earth by zero. The atoms that make up our bodies are the same atoms that were here before we were born—we are but dust, and unto dust we shall return. The material that makes up human cells is a rearrangement of material already present. The atoms that make up a baby forming in its mother's womb are supplied by the food the mother eats. And those atoms originated in stars—some of them in far-away galaxies. (Interestingly, the mass of Earth *does* increase by the incidence of roughly 40,000 tons of interplanetary dust each year, but not by the birth and survival of more people.)

3. Quite so, and of Oprah Winfrey, too. However, these atoms are combined differently than they were previously. If you experience one of those days when you feel like you'll never amount to anything, take comfort in the thought that many of the atoms that now constitute your body will live forever in the bodies of all the people on Earth who are yet to be. Our atoms are immortal.

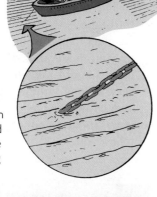

FIGURE 2.3

Information about the ship is revealed by passing waves because the distance between wave crests is small compared with the size of the ship. The passing waves reveal nothing about the chain.

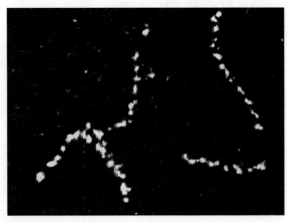

FIGURE 2.4

The strings of dots are chains of thorium atoms imaged with a scanning electron microscope. This historic photograph of individual atoms was taken in 1970 by researchers at the University of Chicago's Enrico Fermi Institute.

an electron micrograph—it was not made with light but with a thin electron beam in a scanning electron microscope (SEM) developed by Albert Crewe at the University of Chicago's Enrico Fermi Institute. An electron beam, such as the one that sprays a picture on an early television screen, is a stream of particles that have wave properties. The wavelength of an electron beam is smaller than the wavelengths of visible light, and atoms are larger than the tiny wavelengths of an electron beam. Crewe's electron micrograph is the first high-resolution image of individual atoms.

In the mid-1980s, researchers developed a new kind of microscope—the scanning tunneling microscope (STM). It employs a sharp tip that is scanned over a surface at a distance of a few atomic diameters in a point-by-point and line-by-line fashion. At each point, a tiny electric current, called a tunneling current, is measured between the tip and the surface. Variations in the current reveal the surface topology. The image of Figure 2.5 beautifully shows the position of a ring of atoms. The ripples shown in the ring of atoms reveal the wave nature of matter. This image, among many others, underscores the delightful interplay of art and science.

Sometimes a model is useful even when it's incorrect. Scotsman James Watt constructed a workable steam engine in the eighteenth century based on a model of heat that turned out to be quite incorrect.

FIGURE **2.5**

An image of 48 iron atoms positioned into a circular ring that "corrals" electrons on a copper crystal surface; taken with a scanning tunneling microscope at the IBM Almaden Laboratory in San Jose, California.

Because we can't see inside an atom, we construct models. A model is an abstraction that helps us to visualize what we can't see, and, very important, it enables us to make predictions about unseen portions of the natural world. An early model of the atom (and the one most familiar to the general public) is akin to that of the solar system. As with the solar system, most of an atom's volume is empty space. At the center is a tiny and very dense nucleus in which most of the mass is concentrated. Surrounding the nucleus are "shells" of orbiting particles. These are **electrons**, electrically charged basic units of matter (the same electrons that constitute the electric current in your calculator). Although electrons electrically repel other electrons, they are electrically attracted to the nucleus, which has a net positive charge. As the size and charge of the nuclei

increase, electrons are pulled closer to the nucleus, and the shells become smaller. Interestingly, the uranium atom, with its 92 electrons, is not appreciably larger in diameter than the lightest atom, hydrogen. This model was first proposed in the early twentieth century, and it reflects a rather simplified understanding of the atom. It was soon discovered, for example, that electrons don't orbit the atom's center like planets orbit the Sun. Like most early models, however, the planetary atomic model served as a useful stepping stone to further understanding and more accurate models. Any atomic model, no matter how refined, is nothing more than a symbolic representation of the atom and not a physical picture of the actual atom.

> Artists and scientists both look for patterns in nature, finding connections that have always been there yet have missed the eye.

2.4 Atomic Structure

Nearly all the mass of an atom is concentrated in the **atomic nucleus**, which occupies only a few quadrillionths of its volume. The nucleus, therefore, is extremely dense. If bare atomic nuclei could be packed against each other into a lump that is 1 centimeter in diameter (about the size of a large pea), the lump would weigh 133,000,000 tons! Huge electrical forces of repulsion prevent such close packing of atomic nuclei because each nucleus is electrically charged and repels all other nuclei. Only under special circumstances are the nuclei of two or more atoms squashed into contact. When this happens, a violent kind of nuclear reaction may occur. Such a reaction, a *thermonuclear fusion reaction*, occurs in the centers of stars and ultimately makes them shine. (We'll discuss these nuclear reactions in Chapter 16.)

The principal building block of the nucleus is the *nucleon*, which is in turn composed of fundamental particles called *quarks*. When a nucleon is in an electrically neutral state, it is a *neutron;* when it is in a positively charged state, it is a **proton**. All protons are identical; they are copies of one another. Likewise with neutrons: Each neutron is like every other neutron. The lighter nuclei have roughly equal numbers of protons and neutrons; more massive nuclei have more neutrons than protons. Protons

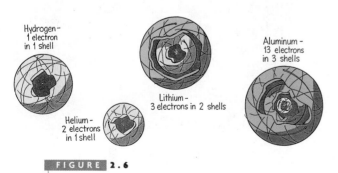

Hydrogen – 1 electron in 1 shell

Helium – 2 electrons in 1 shell

Lithium – 3 electrons in 2 shells

Aluminum – 13 electrons in 3 shells

FIGURE **2.6**

The classical model of the atom consists of a tiny nucleus surrounded by electrons that orbit within spherical shells. As the charges of nuclei increase, electrons are pulled closer, and the shells become smaller.

have a positive electric charge that repels other positive charges but attracts negative charges. So like kinds of electrical charges repel one another, and unlike charges attract one another. It is the positive protons in the nucleus that attract a surrounding cloud of negatively charged electrons to constitute an atom. (The strong nuclear force, which binds the protons to neutrons and to each other within the nucleus, is described in Chapter 16.)

2.5 The Elements

When a substance is composed of only one kind of atom, we call that substance an **element**. A pure 24-carat gold ring, for example, is composed only of gold atoms. A gold ring with a lower carat value is composed of gold and other elements, such as nickel. The silvery liquid in a barometer or thermometer is the element mercury. The entire liquid consists of only mercury atoms. An atom of a particular element is the smallest sample of that element. Although *atom* and *element* are often used interchangeably, *element* refers to a type of substance (one containing only one type of atom), whereas atom refers to the individual particles that make up that substance. For example, we speak of isolating a mercury *atom* from a flask of the *element* mercury.

The lightest element of all is hydrogen. In the universe at large, it is the most abundant element—more than 90% of the atoms in the known universe are hydrogen atoms. Helium, the second-lightest element, provides most of the remaining atoms in the universe. Heavier atoms in our surroundings were manufactured by the fusion of light elements in the hot, high-pressure cauldrons deep within the interiors of stars. The heaviest elements form when huge stars implode and then explode—supernovas. Nearly all the elements on Earth are remnants of stars that exploded long before the solar system came into existence.

To date, some 115 elements are known. Of these, about 90 occur in nature. The others are produced in the laboratory with high-energy atomic accelerators and nuclear reactors. These laboratory-produced elements are too unstable (radioactive) to

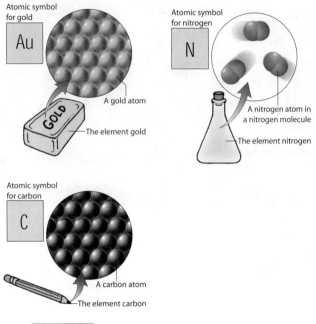

FIGURE 2.7

Any element consists only of one kind of atom. Gold consists only of gold atoms, a flask of gaseous nitrogen consists only of nitrogen atoms, and the carbon of a graphite pencil is composed only of carbon atoms.

occur naturally in appreciable quantities. From a pantry containing less than 100 elements, we have the atoms that constitute almost every simple, complex, living, or nonliving substance in the known universe. More than 99% of the material on Earth

FIGURE 2.8

Both you and Leslie are made of stardust—in the sense that the carbon, oxygen, nitrogen, and other atoms that make up your body originated in the deep interiors of ancient stars that have long since exploded.

Group

Legend:
- Metal
- Metalloid
- Nonmetal

Group →	1	2	3	4	5	6	7	8	9	10	11	12	13	14	15	16	17	18
Period 1	1 H Hydrogen 1.0079																	2 He Helium 4.003
Period 2	3 Li Lithium 6.941	4 Be Beryllium 9.012											5 B Boron 10.811	6 C Carbon 12.011	7 N Nitrogen 14.007	8 O Oxygen 15.999	9 F Fluorine 18.998	10 Ne Neon 20.180
Period 3	11 Na Sodium 22.990	12 Mg Magnesium 24.305											13 Al Aluminum 26.982	14 Si Silicon 28.086	15 P Phosphorus 30.974	16 S Sulfur 32.066	17 Cl Chlorine 35.453	18 Ar Argon 39.948
Period 4	19 K Potassium 39.098	20 Ca Calcium 40.078	21 Sc Scadium 44.956	22 Ti Titanium 47.88	23 V Vanadium 50.942	24 Cr Chromium 51.996	25 Mn Manganese 54.938	26 Fe Iron 55.845	27 Co Cobalt 58.933	28 Ni Nickel 58.69	29 Cu Copper 63.546	30 Zn Zinc 65.39	31 Ga Gallium 69.723	32 Ge Germanium 72.61	33 As Arsenic 74.922	34 Se Selenium 78.96	35 Br Bromine 79.904	36 Kr Krypton 83.8
Period 5	37 Rb Rubidium 85.468	38 Sr Strontium 87.62	39 Y Yttrium 88.906	40 Zr Zirconium 91.224	41 Nb Niobium 92.906	42 Mo Molybdenum 95.94	43 Tc Technetium 98	44 Ru Ruthenium 101.07	45 Rh Rhodium 102.906	46 Pd Palladium 106.42	47 Ag Silver 107.868	48 Cd Cadmium 112.411	49 In Indium 114.82	50 Sn Tin 118.71	51 Sb Antimony 121.76	52 Te Tellurium 127.60	53 I Iodine 126.905	54 Xe Xenon 131.29
Period 6	55 Cs Cesium 132.905	56 Ba Barium 137.327	57 La Lanthanum 138.906	72 Hf Hafnium 178.49	73 Ta Tantalum 180.948	74 W Tungsten 183.84	75 Re Rhenium 186.207	76 Os Osmium 190.23	77 Ir Iridium 192.22	78 Pt Platinum 195.08	79 Au Gold 196.967	80 Hg Mercury 200.59	81 Tl Thallium 204.383	82 Pb Lead 207.2	83 Bi Bismuth 208.980	84 Po Polonium 209	85 At Astatine 210	86 Rn Radon 222
Period 7	87 Fr Francium 223	88 Ra Radium 226.025	89 Ac Actinium 227.028	104 Rf Rutherfordium (261)	105 Db Dubnium (262)	106 Sg Seaborgium (266)	107 Bh Bohrium (264)	108 Hs Hassium (269)	109 Mt Meitnerium (268)	110 Ds Darmstadtium (269)	111 Ro Roentgenium (272)	112 Uub (285)		114 Uuq (289)		116 Uuh (292)		

Lanthanides:

58 Ce Cerium 140.115	59 Pr Praseodymium 140.908	60 Nd Neodymium 144.24	61 Pm Promethium 145	62 Sm Samarium 150.36	63 Eu Europium 151.964	64 Gd Gadolinium 157.25	65 Tb Terbium 158.925	66 Dy Dysprosium 162.5	67 Ho Holmium 164.93	68 Er 68 Erbium 167.26	69 Tm Thulium 168.934	70 Yb Ytterbium 173.04	71 Lu Lutetium 174.967

Actinides:

90 Th Thorium 232.038	91 Pa Protactinium 231.036	92 U Uranium 238.029	93 Np Neptunium (237)	94 Pu Plutonium (244)	95 Am Americium (243)	96 Cm Curium (247)	97 Bk Berkelium (247)	98 Cf Californium (251)	99 Es Einsteinium (252)	100 Fm Fermium (257)	101 Md Mendelevium (258)	102 No Nobelium (259)	103 Lr Lawrencium (262)

FIGURE 2.9

The periodic table of the elements. The number above the chemical symbol is the *atomic number,* and the number below is the *atomic mass* averaged by isotopic abundance in Earth's surface and expressed in atomic mass units (amu). Atomic masses for radioactive elements shown in parentheses are the whole number nearest the most stable isotope of that element.

is formed from only about a dozen of the elements. The other elements are relatively rare. Living things are composed primarily of five elements: oxygen (O), carbon (C), hydrogen (H), nitrogen (N), and calcium (Ca). The letters in parentheses represent the chemical symbols for these elements.

2.6 Periodic Table of the Elements

Elements are classified by the number of protons in their nucleus—their **atomic number**. Hydrogen, containing one proton per atom, has atomic number 1; helium, containing two protons per atom, has atomic number 2; and so on in sequence to the heaviest naturally occurring element, uranium, with atomic number 92. The numbers continue beyond atomic number 92 through the artificially produced transuranic (beyond uranium) elements. The arrangement of elements by their atomic numbers makes up the **periodic table of the elements** (Figure 2.9).

The periodic table is a chart that lists atoms by their atomic number and also by their electrical arrangements. Like the rows of a calendar that list the days of the week, each element, from left to right, has one more proton and electron than the preceding element. Reading down the table, each element has one more shell than the one above. The inner shells are filled to their capacities, and the outer shell may or may not be, depending on the element. Only the elements at the far right of the table, like the column of Saturdays on a calendar, have their outer shells filled to capacity. These are the *noble gases*—helium, neon, argon, krypton, xenon, and radon. The periodic table is the chemist's road map—and much more. Most scientists consider the periodic table to be the most elegant organizational chart ever devised. The enormous human effort and

fyi Most of the elements of the periodic table are found in interstellar gases.

ingenuity that went into finding its regularities makes a fascinating atomic detective story.*

Elements may have up to seven shells, and each shell can hold some maximum number of electrons. The first and innermost shell has a capacity for two electrons, while the second shell has a capacity for eight electrons. The arrangement of electrons in the shells dictates such properties of substances as their melting and freezing temperatures; their electrical conductivity; and their taste, texture, appearance, and color. The arrangements of electrons quite literally give life and color to the world.

Models of the atom evolve with new findings. The classical model of the atom has given way to a model that views the electron as a standing wave—altogether different from the idea of an orbiting particle. This is the quantum mechanical model, introduced in the 1920s, which is a theory of the small-scale world that includes predicted wave properties of matter. It deals with "lumps" occurring at the subatomic level—lumps of matter or lumps of such things as energy and angular momentum. (More about the quantum in Chapter 15.)

2.7 Relative Sizes of Atoms

The diameters of the outer electron shells of atoms are determined by the amount of electrical charge in the nucleus. For example, the positive proton in the hydrogen atom holds one electron in an orbit at a certain radius. If we double the positive charge in the nucleus, the orbiting electron will be pulled into a tighter orbit with half its former radius, since the electrical attraction is doubled. This occurs for helium, which has a doubly charged nucleus when it attracts a single electron. Interestingly, an added second electron isn't pulled in as far because the first electron partially offsets the attraction of the doubly charged nucleus. This is a neutral helium atom, which is somewhat smaller than a hydrogen atom.

An atom with unbalanced electrical charge—for example, a nucleus with more positive charge than the surrounding negative charges—is called an **ion**.

---

* The creation of the periodic table is credited to Russian chemistry professor Dmitri Mendeleev (1834–1907). By using his table, Mendeleev predicted the existence of elements not then known. Mendeleev was a much-loved and devoted teacher whose lecture halls were filled with students who were eager to hear him speak. He was both a great teacher and a great scientist. Element 101 is named in his honor.

An ion is a charged atom. A helium atom that holds only one electron, for example, is a helium ion. We say it is a positive ion because it has more positive charge than negative charge.

Two electrons around a doubly charged nucleus assume a configuration characteristic of helium. Two electrons surrounding a helium nucleus constitute a neutral atom. A third proton in an atomic nucleus can pull two electrons into an even closer orbit and, furthermore, hold a third electron in a somewhat larger orbit. This is the lithium atom, atomic number 3. We can continue with this process, increasing the positive charge of the nucleus and adding successively more electrons and more orbits all the way up to atomic numbers above 100, to the "synthetic" radioactive elements.**

We find that, as the nuclear charge increases and additional electrons are added in outer orbits, the inner orbits shrink in size because of the stronger nuclear attraction. This means that the heavier elements are not much larger in diameter than the lighter elements. The diameter of the xenon atom, for example, is only about four helium diameters, even though it is nearly 33 times as massive. The relative sizes of atoms in Figure 2.10 are approximately to the same scale.

STOP AND CHECK YOURSELF

What fundamental force dictates the size of an atom?

CHECK YOUR ANSWER

The electrical force.

2.8 Isotopes

Whereas the number of protons in a nucleus exactly matches the number of electrons around the nucleus in a neutral atom, the number of protons in the nucleus need not match the number of neutrons there. For example, most iron nuclei with

---

**Each orbit will hold only so many electrons. A rule of *quantum mechanics* is that an orbit is filled when it contains a number of electrons given by $2n^2$, where n is 1 for the first orbit, 2 for the second orbit, 3 for the third orbit, and so on. For $n = -1$, there are 2 electrons; for $n = -2$, there are $2(2^2)$ or 8 electrons; for $n = 3$, there are a maximum of $2(3^2)$ or 18 electrons, etc. The number n is called the *principal quantum number*.

GROUPS

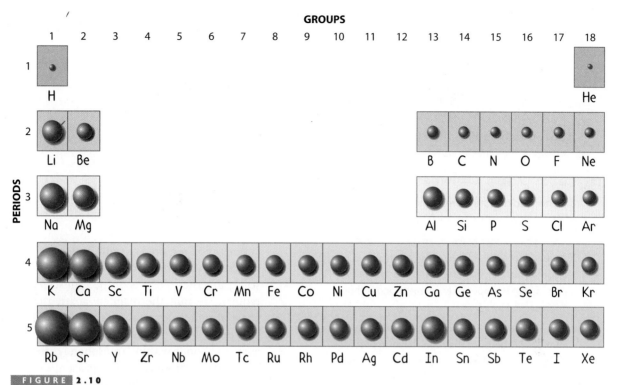

FIGURE 2.10

The sizes of atoms gradually decrease from left to right across the periodic table (only the first 5 periods are shown here).

26 protons contain 30 neutrons, while a small percentage contain 29 neutrons. Atoms of the same element that contain different numbers of neutrons are **isotopes** of the element. The various isotopes of an element all have the same number of electrons, and so, for the most part, they behave identically. We shall return to isotopes in Chapter 16.

We identify isotopes by their *mass number*, which is the total number of protons and neutrons (in other words, the number of nucleons) in the nucleus. A hydrogen isotope with one proton and no neutrons, for example, has a mass number of 1, and is referred to as hydrogen-1. Likewise, an iron atom with 26 protons and 30 neutrons has a mass number of 56 and is referred to as iron-56. An iron atom with 26 protons and only 29 neutrons would be called iron-55.

> **Don't confuse an isotope with an *ion*, which is an atom that is electrically charged owing to an excess or a deficiency of electrons.**

The total mass of an atom is called its *atomic mass*. This is the sum of the masses of all the atom's components (electrons, protons, and neutrons). Because electrons are so much less massive than protons and neutrons, their contribution to atomic mass is negligible. Atoms are so small that expressing their masses in units of grams or kilograms is not practical. Instead, scientists use a specially defined unit of mass known as the **atomic mass unit** or **amu**. A nucleon has a mass of about 1 amu. An atom with 12 nucleons, such as carbon-12, therefore, has a mass of about 12 amu. The periodic table lists atomic masses in units of amu.

Most elements have a variety of isotopes. The atomic mass for each element listed in the periodic table is the weighted average of the masses of these isotopes based on the occurrence of each isotope on Earth. For example, carbon with 6 protons and 6 neutrons has an atomic mass of 12.000 amu. About 1% of all carbon atoms, however, contain 7 neutrons. The heavier isotope raises the average atomic mass of carbon from 12.000 amu to 12.011 amu.

> **Don't confuse mass number with the atomic mass. *Mass number* is an integer that specifies an isotope and has no units—it's simply equal to the number of nucleons in a nucleus. *Atomic mass* is an average of the isotopes of a given element, with units of kilograms. An *atomic mass unit* is expressed in units of amu.**

STOP AND
CHECK YOURSELF

1. Which contributes more to an atom's mass, electrons or protons? Which contributes more to an atom's volume (its size)?

2. Which is represented by a whole number—the mass number or the atomic mass?

3. Do two isotopes of iron have the same *atomic number*? The same *atomic mass number*?

CHECK YOUR ANSWERS

1. Protons contribute more to an atom's mass; electrons contribute more to its size.

2. The mass number is always given as a whole number, such as hydrogen-1 or carbon-12. Atomic mass, by contrast, is the average mass of the various isotopes of an element and is thus represented by a fractional number.

3. The two isotopes of iron have the same atomic number, 26, because they each have 26 protons in the nucleus. They have different atomic mass numbers if they have different numbers of neutrons in the nucleus.

2.9 Molecules

A molecule consists of two or more atoms that bond together by sharing electrons. (We say such atoms are *covalently bonded*.) A molecule may be as simple as the two-atom combination of oxygen, O_2, or the two-atom combination of nitrogen, N_2, which are the elements that make up most of the air we breathe. Two atoms of hydrogen combine with a single atom of oxygen to form a water molecule, H_2O. Changing a molecule by one atom can make a big difference. Replacing the oxygen atom with a sulfur atom produces hydrogen sulfide, H_2S, a strong-smelling toxic gas.

fyi Organic molecules, complex and simple, are found in interstellar gases.

STOP AND
CHECK YOURSELF

How many atomic nuclei are in a single oxygen atom? In a single oxygen molecule?

CHECK YOUR ANSWER

There is one nucleus in an oxygen atom, O, and two in the combination of two oxygen atoms—an oxygen molecule, O_2.

Energy is required to pull molecules apart. We can understand this by considering a pair of magnets stuck together. Just as some "muscle energy" is required to pull the magnets apart, the breaking apart of molecules requires energy. During photosynthesis, plants use the energy of sunlight to break apart the bonds within atmospheric carbon dioxide and water to produce oxygen gas and carbohydrate molecules. These carbohydrate molecules retain this solar energy until the process is reversed—the plant is oxidized, either slowly by rotting or quickly by burning. Then the energy that came from the sunlight is released back into the environment. So the slow warmth of decaying compost or the quick warmth of a campfire is really the warmth of stored sunlight!

More things can burn besides those that contain carbon and hydrogen. Iron "burns" (oxidizes) too. That's what rusting is—the slow combination of oxygen atoms with iron atoms, releasing energy. When the rusting of iron is speeded up, it makes nice hand-warmer packs for skiers and winter hikers. Any process in which atoms rearrange to form different molecules is called a *chemical reaction*.

Our sense of smell is sensitive to exceedingly small quantities of molecules. Our olfactory organs easily detect small concentrations of such noxious gases as hydrogen sulfide (the stuff that smells like rotten eggs), ammonia, and ether. The smell of perfume is the result of molecules that rapidly evaporate and diffuse haphazardly in the air until some of them get close enough to your nose to be inhaled. They are just a few of the billions of jostling molecules that, in their aimless wanderings, happen to end up in the nose. You can get an idea of the speed of molecular diffusion in the air when you are in your bedroom and smell

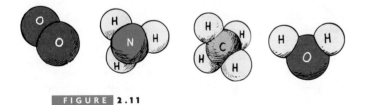

FIGURE 2.11

Models of simple molecules. The atoms in a molecule are not just mixed together but are joined in a well-defined way.

■ THE PLACEBO EFFECT

People have always sought healers for help with physical pain and fear. As treatment, traditional healers often administer herbs, or chant, or wave their hands over a patient's body. And improvement, more often than not, actually occurs! This is the *placebo effect*. A placebo may be a healing practice or a substance (pill) containing elements or molecules that have no medical value. But, remarkably, the placebo effect does have a biological basis. It so happens that, when you are fearful or in pain, your brain response is *not* to mobilize your body's healing mechanism—it instead prepares your body for some external threat. It's an evolutionary adaptation that assigns highest priority to preventing additional injury. Stress hormones released into the bloodstream increase respiration, blood pressure, and heart rate—changes that usually *impede* recovery. The brain prepares your body for action; recovery can wait.

That's why the first objective of a good healer or physician is to relieve stress. Most of us begin feeling better even before leaving the healer's (or doctor's) office. Prior to 1940, most medicine was based on the placebo effect, when about the only medicines doctors had in their bags were laxatives, aspirin, and sugar pills. In about half the cases, a sugar pill is as effective in stopping pain as an aspirin. Here's why: Pain is a signal to the brain that something is wrong and needs attention. The signal is induced at the site of inflammation by prostaglandins released by white blood cells. Aspirin blocks the production of

prostaglandins and therefore relieves the pain. The mechanism for pain relief by a placebo is altogether different. The placebo fools the brain into thinking that whatever is wrong is being cared for. Then the pain signal is lowered by the release of endorphins, opiate-like proteins found naturally in the brain. So, instead of blocking the *production* of prostaglandins, the endorphins block their *effect*. With pain alleviated, the body can focus on healing.

The placebo effect has always been employed (and still is!) by healers and others who claim to have wondrous cures that lie outside modern medicine. These healers benefit from the public's tendency to believe that, if *B* follows *A*, then *B* is *caused* by *A*. The cure could be due to the healer, but it could also merely be the body repairing itself. Although the placebo effect can certainly influence the perception of pain, it has not been shown to influence the body's ability to fight infection or repair injury.

Is the placebo effect at work for those who believe that better health is bequeathed to those who wear crystals, magnets, or certain metal bracelets? If so, is there any harm in thinking so— even if there is no scientific evidence for it? Harboring positive beliefs is usually quite harmless—but not always. For serious problems requiring modern medical treatment, reliance on these aids can be disastrous if used to the exclusion of modern medical treatment. The placebo effect has real limitations.

*Adapted from *Voodoo Science: The Road from Foolishness to Fraud*, by Robert L. Park, Oxford University Press, New York, 2000.

---

food very soon after the oven door has been opened in the kitchen.

Although H_2O is the major greenhouse gas in the atmosphere, the molecule CO_2, the second most abundant greenhouse gas, is notorious because its numbers are rapidly increasing. Unhappily, further warming by CO_2 can trigger more H_2O as well. So present concern is focused on the combination of growing amounts of both these molecules in the atmosphere.

2.10 Antimatter

Whereas matter is composed of atoms with positively charged nuclei and negatively charged electrons, **antimatter** is composed of atoms with negative nuclei and positive electrons, or *positrons*.

Positrons were first discovered in 1932, in cosmic rays bombarding Earth's atmosphere. Today, antiparticles of all types are regularly produced in laboratories using large nuclear accelerators. A positron has the same mass as an electron and the same magnitude of charge but the opposite sign.

FIGURE **2.12**

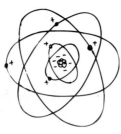

An atom of antimatter has a negatively charged nucleus surrounded by positrons.

Antiprotons have the same mass as protons but are negatively charged. The first complete artificial anti-atom, a positron orbiting an antiproton, was constructed in 1995. Every charged particle has an antiparticle of the same mass and opposite charge. Neutral particles (such as the neutron) also have antiparticles, which are alike in mass and in some other properties but opposite in certain other properties. Every particle has an antiparticle. There are even antiquarks.

Gravitational force does not distinguish between matter and antimatter—each attracts the other. Also, there is no way to indicate whether something is made of matter or antimatter by the light it emits. Only through much subtler, hard-to-measure nuclear effects could we determine whether a distant galaxy is made of matter or antimatter. But, if an antistar were to meet a star, it would be a different story. They would mutually annihilate each other, with most of the matter converting to radiant energy (this is what happened to the anti-atom created in 1995, when it encountered normal matter and rapidly annihilated in a puff of energy). This process, more so than any other known, results in the maximum energy output per gram of substance—$E = mc^2$, with a 100% mass conversion.* (Nuclear fission and fusion, by contrast, convert less than 1% of the matter involved.)

There cannot be both matter and antimatter in our immediate environment, at least not in appreciable amounts or for appreciable times. That's because something made of antimatter would be completely transformed to radiant energy as soon as it touched matter, consuming an equal amount of normal matter in the process. If the Moon were made of antimatter, for example, a flash of energetic radiation would result as soon as one of our spaceships touched it. Both the spaceship and an equal amount of the anti-

*Some physicists speculate that, right after the Big Bang, the early universe had billions of times more particles than it has now, and that a near total extinction of matter and antimatter caused by their mutual annihilation left only the relatively small amount of matter now present in the universe.

matter Moon would disappear in a burst of radiant energy. We know the Moon is not antimatter because this didn't happen during the Moon missions. (Actually, astronauts weren't in this kind of danger, for previous evidence showed that the Moon is made of ordinary matter.) But what about other galaxies? There is strong reason to believe that in the part of the universe we know (the "observable universe"), galaxies are made only of normal matter—apart from the occasional transitory antiparticle. But what of the universe beyond? Or other universes? We don't know.

STOP AND CHECK YOURSELF

If a 1-gram body of antimatter meets a 10-gram body of matter, what mass survives?

CHECK YOUR ANSWER

Nine grams of matter survive (the other 2 grams are transformed into radiant energy).

2.11 Dark Matter

We know that the elements in the periodic table are not confined to our planet. From studies of radiation coming from other parts of the universe, we find that stars and other objects "out there" are composed of the same particles we have on Earth. Stars emit light with the same "fingerprints" (*atomic spectra*, Chapter 15) as the elements in the periodic table. How wonderful to find that the laws that govern matter on Earth extend throughout the observable universe. Yet there remains one troubling detail. Gravitational forces within galaxies are measured to be far greater than visible matter can account for.

Astrophysicists talk of the **dark matter**—matter we can't see that tugs on stars and galaxies that we *can* see. In the closing years of the twentieth century, astrophysicists confirmed that some 23% of matter in the universe is composed of the unseen dark matter. Whatever dark matter is, most or all of it is likely to be "exotic" matter—very different from the elements that make up the periodic table, and different from any extension of the present list of elements. Much of the rest of the universe is *dark energy*, which pushes outward on the expanding universe. Both dark matter and dark energy make up some 90% of the universe. Only in this twenty-first century has this type of energy been apparent. At this writing, neither has been identified. Speculations abound about dark matter and dark energy, but we don't know what they are.

Richard Feynman often used to shake his head and say he didn't know anything. When he and other top physicists say they don't know anything, they mean that what they *do* know is closer to nothing than to what they *can* know. Scientists know enough to realize that they have a relatively small handle on an enormous universe still full of myster-ies. From a looking-backward point of view, today's scientists know enormously more than their fore-bears a century ago, and scientists then knew much more than *their* forebears. But, from our present vantage point, looking forward, there is so much yet to be learned. Physicist John A. Wheeler, Feynman's graduate school adviser, envisions the next level of physics as going beyond *how* to *why*—to meaning. We have scarcely scratched the surface.

> Finding the nature of the dark matter and the nature of the vacuum energy are high-priority quests in these times. What we will have learned by mid-century will likely dwarf all that we have ever known.

> I can live with doubt and uncertainty and not knowing. I think it is much more interesting to live not knowing than to have answers that might be wrong.
>
> *— Richard Feynman*

SUMMARY OF TERMS

Atom The smallest particle of an element that has all of the element's chemical properties.

Brownian motion The haphazard movement of tiny particles suspended in a gas or liquid resulting from their bombardment by the fast-moving atoms or molecules of the gas or liquid.

Electron Negatively charged particle in the shell of an atom.

Atomic nucleus The core of an atom, consisting of two basic subatomic particles—protons and neutrons.

Proton Positively charged particle in the nucleus of an atom.

Element A pure substance consisting of only one kind of atom.

Atomic number The number that designates the identity of an element, which is the number of protons in the nucleus of an atom; in a neutral atom, the atomic number is also the number of electrons in the atom.

Periodic table of the elements A chart that lists the elements in horizontal rows by their atomic number and in vertical columns by their similar electron arrangements and chemical properties. (See Figure 2.9.)

Ion An electrically charged atom; an atom with an excess or deficiency of electrons.

Isotope An atom of the same element that contains a different number of neutrons.

Atomic mass unit (amu) The standard unit of atomic mass, which is equal to one-twelfth the mass of the common atom of carbon, arbitrarily given the value of exactly 12. One amu has a mass of 1.661×10^{-24} grams.

Molecule Two or more atoms that bond together by a sharing of electrons. Atoms combine to form molecules.

Antimatter A "complementary" form of matter composed of atoms with negative nuclei and positive electrons.

Dark matter Unseen and unidentified matter that is evident by its gravitational pull on stars in the galaxies. Dark matter, along with dark energy, constitutes perhaps 90% of the stuff of the universe.

SUGGESTED READINGS

Feynman, R. P., R. B. Leighton, and M. Sands. *The Feynman Lectures on Physics,* vol. 1, chap. 1. Reading, MA: Addison-Wesley, 1963.

Rigden, John S. *Hydrogen: The Essential Element.* Cambridge, MA: Harvard University Press, 2002.

Suchocki, J. *Conceptual Chemistry,* 3rd ed., chap. 5. San Francisco, CA: Benjamin Cummings, 2006. Contains an excellent treatment of the periodic table.

REVIEW QUESTIONS

2.1 The Atomic Hypothesis

1. What causes dust particles and tiny grains of soot to move with Brownian motion?
2. Who first explained Brownian motion and made a convincing case for the existence of atoms?
3. According to Richard Feynman, when do atoms attract each other, and when do they repel?

2.2 Characteristics of Atoms

4. How does the approximate number of atoms in the air in your lungs compare with the number of breaths of air in the atmosphere of Earth?
5. Are most of the atoms around us younger or older than the Sun?

2.3 Atomic Imagery

6. Why can atoms not be seen with a powerful optical microscope?
7. Why can atoms be seen with an electron beam?
8. What is the purpose of a model in science?

2.4 Atomic Structure

9. How does the mass of an atomic nucleus compare with the mass of an atom as a whole?
10. What is a nucleon?
11. How do the mass and electric charge of a proton compare with the mass and charge of an electron?
12. Since atoms are mostly empty space, why don't we fall through a floor we stand on?

2.5 The Elements

13. What is the lightest of the elements?
14. What is the most abundant element in the known universe?
15. How were elements that are heavier than hydrogen formed?

16. Where did the heaviest elements originate?
17. What are the five most common elements in living things?

2.6 Periodic Table of the Elements

18. What does the atomic number of an element tell you about the element?
19. What is characteristic of the columns in the periodic table?

2.7 Relative Sizes of Atoms

20. What kind of basic force pulls electrons close to the atomic nucleus?
21. Why are heavier elements not much larger than lighter elements?

2.8 Isotopes

22. How does an isotope differ from a normal atom?
23. Distinguish between *mass number* and *atomic mass*.

2.9 Molecules

24. How does a molecule differ from an atom?
25. Compared to the energy it takes to separate oxygen and hydrogen from water, how much energy is released when they recombine? (In Chapter 5 you'll study the conservation of energy.)

2.10 Antimatter

26. How do matter and antimatter differ?
27. What occurs when a particle of matter and a particle of antimatter meet?

2.11 Dark Matter

28. What is the evidence that dark matter exists?

ACTIVE EXPLORATIONS

1. A candle will burn only if oxygen is present. Will a candle burn twice as long in an inverted liter jar as it will in an inverted half-liter jar? Try it and see.

2. Write a letter to Grandma or Grandpa and describe how long the atoms that make up their bodies have been around. Describe how long they will continue to be around.

EXERCISES

1. How many types of atoms can you expect to find in a pure sample of any element?
2. How many individual atoms are in a water molecule?
3. When a container of gas is heated, what happens to the average speed of its molecules?

4. The average speed of a perfume-vapor molecule at room temperature may be about 300 m/s, but you'll find the speed at which the scent travels across the room is much less. Why?
5. A cat strolls across your backyard. An hour later, a dog with his nose to the ground follows the trail of the cat.

Explain this occurrence from a molecular point of view.

6. If no molecules in a body could escape, would the body have any odor?

7. Where were the atoms that make up a newborn infant "manufactured"?

8. Which of the following is not an element: hydrogen, carbon, oxygen, water?

9. Which of the following are pure elements: H_2, H_2O, He, Na, NaCl, H_2SO_4, U?

10. Your friend says that what makes one element distinct from another is the number of electrons about the atomic nucleus. Do you agree wholeheartedly, partially, or not at all? Explain.

11. What is the cause of the Brownian motion of dust particles? Why aren't larger objects, such as baseballs, similarly affected?

12. Why don't equal masses of golf balls and Ping-Pong balls contain the same number of balls?

13. Why don't equal masses of carbon atoms and oxygen atoms contain the same number of particles?

14. Which contains more atoms: 1 kg of lead or 1 kg of aluminum?

15. How many atoms are in a molecule of ethanol, C_2H_6O?

16. The atomic masses of two isotopes of cobalt are 59 and 60.
    a. What is the number of protons and neutrons in each?
    b. What is the number of orbiting electrons in each when the isotopes are electrically neutral?

17. A particular atom contains 29 electrons, 34 neutrons, and 29 protons. What is the identity of this element and its atomic number?

18. If two protons and two neutrons are removed from the nucleus of an oxygen atom, what nucleus remains?

19. What element results if you add a pair of protons to the nucleus of mercury? (See the periodic table.)

20. What element results if two protons and two neutrons are ejected from a radium nucleus?

21. To become a negative ion, does an atom lose or gain an electron?

22. To become a positive ion, does an atom lose or gain an electron?

23. You could swallow a capsule of germanium without ill effects. But, if a proton were added to each of the germanium nuclei, you would not want to swallow the capsule. Why? (Consult the periodic table.)

24. Helium is an inert gas, meaning that it doesn't readily combine with other elements. What five other elements would you also expect to be inert gases? (See the periodic table.)

25. Which of the following elements would you predict to have properties most like those of silicon (Si): aluminum (Al), phosphorus (P), or germanium (Ge)? (Consult the periodic table.)

26. Carbon, with a half-full outer shell of electrons (four in a shell that can hold eight), readily shares its electrons with other atoms and forms a vast number of molecules, many of which are the organic molecules that form the bulk of living matter. Looking at the periodic table, what other element do you think might play a role like carbon in life forms on some other planet?

27. Which contributes more to an atom's mass: electrons or protons? Which contributes more to an atom's size?

28. A hydrogen atom and a carbon atom move at the same speed. Which has the greater kinetic energy?

29. In a gaseous mixture of hydrogen and oxygen gas, both with the same average kinetic energy, which molecules move faster on average?

30. The atoms that constitute your body are mostly empty space, and structures such as the chair you're sitting on are composed of atoms that are also mostly empty space. So why don't you fall through the chair?

31. In what sense can you truthfully say that you are a part of every person in history? In what sense can you say that you will tangibly contribute to every person on Earth who will follow?

32. What are the chances that at least one of the atoms exhaled by your very first breath will be in your next breath?

33. Hydrogen and oxygen always react in a 1:8 ratio by mass to form water. Early investigators thought this meant that oxygen was eight times more massive than hydrogen. What chemical formula did these investigators assume for water?

34. Somebody told your friend that, if an antimatter alien ever set foot on Earth, the whole world would explode into pure radiant energy. Your friend looks to you for verification or refutation of this claim. What do you say?

35. Make up a multiple-choice question that will test your classmates on the distinction between any two terms in the Summary of Terms list.

PROBLEMS

● BEGINNER ■ INTERMEDIATE ◆ EXPERT

1. ■ Show that there are 16 grams of oxygen in 18 g of water.
2. ■ Show that there are 4 grams of hydrogen in 16 g of methane gas. (The chemical formula for methane is CH_4.)
3. ■ Gas A is composed of diatomic molecules (two atoms to a molecule) of a pure element. Gas B is composed of monatomic molecules (one atom to a "molecule") of another pure element. Gas A has three times the mass of an equal volume of gas B at the same temperature and pressure. How do the atomic masses of elements A and B compare?

4. ■ A teaspoon of an organic oil dropped on the surface of a quiet pond spreads out to cover almost an acre. The oil film has a thickness equal to the size of a molecule. In the lab, when you drop 0.001 milliliter (10^{-9} m^3) of the organic oil on the still surface of water, you find that it spreads to cover an area of 1.0 m^2. If the layer is one molecule thick, show how

the size of a single molecule is 10^{-9} m (about 10 atomic diameters).

5. ■ There are approximately 10^{23} H_2O molecules in a thimbleful of water and 10^{46} H_2O molecules in Earth's oceans. Suppose that Columbus threw a thimbleful of water into the ocean and that those water molecules have by now mixed uniformly with all the water molecules in the oceans. Show that, if you dip a sample thimbleful of water from anywhere in the ocean, you'll probably scoop up at least one of the molecules that was in Columbus's thimble. (Hint: The ratio of the number of molecules in a thimble to the number of molecules in the ocean will equal the ratio of the number of molecules in question to the number of molecules that the thimble can hold.)

6. ■ There are approximately 10^{22} molecules in a single medium-size breath of air and approximately 10^{44} molecules in the atmosphere of the whole world. The number 10^{22} squared is equal to 10^{44}. So how many breaths of air are there in the world's atmosphere? How does this number compare with the number of molecules in a single breath? If all the molecules from Julius Caesar's last dying breath are now thoroughly mixed in the atmosphere, how many of these, on the average, do we inhale with each single breath?

7. ■ Assume that the present world population of about 6×10^9 people is about 1/20 the number of people who ever lived on Earth. How does the number of people who ever lived compare to the number of air molecules in a single breath?

CHAPTER 2 ONLINE RESOURCES

The Physics Place

Videos
Evidence for Atoms
Atoms Are Recyclable

Quiz

Flashcards

Links

PART ONE

Mechanics

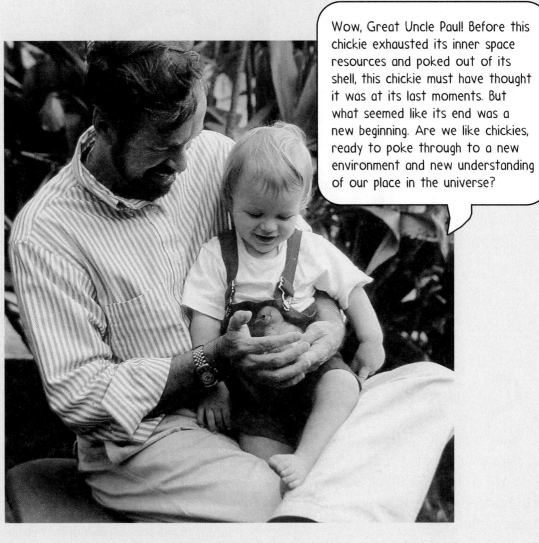

Wow, Great Uncle Paul! Before this chickie exhausted its inner space resources and poked out of its shell, this chickie must have thought it was at its last moments. But what seemed like its end was a new beginning. Are we like chickies, ready to poke through to a new environment and new understanding of our place in the universe?

Equilibrium and Linear Motion

Chelcie Liu asks his students to check their neighbors and predict which ball will reach the end of the equal-length tracks first.

Some two thousand years ago Greek scientists understood some of the physics we understand today. They had a good grasp of the physics of floating objects and some of the properties of light. But they were confused about motion. One of the first to study motion seriously was Aristotle, the most outstanding philosopher-scientist in ancient Greece. Aristotle attempted to clarify motion by classification.

> **fyi**
>
> Rather than reading chapters in this book slowly, try reading quickly, and more than once. You'll better learn physics by going over the same material several times. With each time, it makes more sense. Don't worry if you don't understand things right away — just keep on reading.

3.1 Aristotle on Motion

Aristotle classified motion into two kinds: natural motion and violent motion. We shall briefly consider each, not as study material, but as a background to modern ideas about motion.

In Aristotle's view, *natural motion* proceeds from the "nature" of an object. He believed all objects were some combination of four elements—earth, water, air, and fire—and asserted that motion depended on the combination of elements that an object contains. He taught that every object in the universe has a proper place, determined by this "nature"; any object not in its proper place will "strive" to get there. For example, an unsupported lump of clay, being of the earth, properly falls to the ground; an unimpeded puff of smoke, being of the air, properly rises; a feather properly falls to the ground, since it is a mixture of air and earth, but not as rapidly as a lump of clay would fall. Aristotle stated that heavier objects would strive harder and fall faster than lighter ones would.

Natural motion was understood to be either straight up or straight down for all things on Earth.

■ ARISTOTLE (384–322 BC)

Aristotle was the foremost philosopher, scientist, and educator of his time. Born in Greece, he was the son of a physician who personally served the king of Macedonia. At the age of 17, he entered the Academy of Plato, where he worked and studied for 20 years until Plato's death. He then became the tutor of young Alexander the Great. Eight years later, he formed his own school. Aristotle's aim was to systematize existing knowledge, just as Euclid had systematized geometry. Aristotle made critical observations, collected specimens, and gathered, summarized, and classified almost all of existing knowledge of the physical world. His systematic approach became the method from which Western science later arose. After his death, his voluminous notebooks were preserved in caves near his home and were later sold to the library at Alexandria. Scholarly activity ceased in most of Europe through the Dark Ages, and the works of Aristotle were forgotten and lost in the scholarship that continued in the Byzantine and Islamic empires. Several of his texts were reintroduced to Europe during the eleventh and twelfth centuries and were translated into Latin. The Church, the dominant political and cultural force in Western Europe, at first prohibited the works of Aristotle and then accepted and incorporated them into Christian doctrine.

Natural motion beyond Earth, such as the motion of celestial objects, was circular. Both the Sun and the Moon continually circle Earth in paths without beginning or end. Aristotle taught that different rules apply in the heavens, and that celestial bodies are perfect spheres made of a perfect and unchanging substance, which he called *quintessence.**

Violent motion, Aristotle's other class of motion, is produced by pushes or pulls. Violent motion is imposed motion. A person pushing a cart or lifting a heavy boulder imposes motion, as does someone hurling a stone or winning a tug-of-war. The wind imposes motion on ships. Floodwaters impose it on boulders and tree trunks. Violent motion is externally caused and is imparted to objects; they move not of themselves, not by their "nature," but because of impressed *forces*—pushes or pulls.

The concept of violent motion had its difficulties, for the forces responsible for it were not always evident. For example, a bowstring moved an arrow until the arrow left the bow; after that, further explanation of the arrow's motion seemed to require some other pushing agent. Aristotle imagined, therefore, that a parting of the air by the moving arrow resulted in a squeezing effect on the rear of the arrow as the air rushed back to prevent a vacuum from forming. The arrow was propelled through the air as a bar of soap is propelled in the bathtub when you squeeze one end of the bar.

To summarize, Aristotle taught that all motions are due to the nature of the moving object, or due to a sustained push or pull. Provided that an object is in its proper place, it would not move unless subjected to a force. Except for celestial objects, the normal state was one of rest.

FIGURE 3.1

Does a force keep the cannonball moving after it leaves the cannon?

* Quintessence is the *fifth* essence, the other four being earth, water, air, and fire.

STOP AND
CHECK YOURSELF

Isn't it common sense to think of Earth as in its proper place, and that a force to move it is inconceivable, as Aristotle held, and that Earth *is* at rest in this universe?

CHECK YOUR ANSWER

Common sense is relative to one's time and place. Aristotle's views were logical and consistent with everyday observations. So unless you become familiar with the physics to follow in this book, Aristotle's views about motion *do* make common sense (and are held by many uneducated people today). But as you acquire new information about nature's rules, you'll likely find your common sense progressing beyond Aristotelian thinking.

3.2 Galileo's Concept of Inertia

Aristotle's ideas were accepted as fact for nearly 2000 years. Then, in the early 1500s, the Italian scientist Galileo demolished Aristotle's belief that heavy objects fall faster than light ones. According to legend, Galileo dropped both heavy and light objects from the Leaning Tower of Pisa. He showed that, except for the effects of air resistance, objects of different weights fell to the ground at the same time.

📖 **The Physics Place**
Newton's Law of Inertia
The Old Tablecloth Trick
Toilet Paper Roll
Inertia of a Cylinder
Inertia of an Anvil

Galileo made another huge discovery. He showed that Aristotle was wrong about forces being necessary to keep objects in motion. In the simplest sense, a **force** is a push or a pull. Although a force is needed to start an object moving, Galileo showed that once moving, no force is needed to keep it moving—except for the force needed to overcome friction (more about friction later in this chapter). When friction is absent, a moving object needs no force to keep it moving.

Galileo tested his revolutionary idea by *experiment*. This was the beginning of modern science. He rolled balls down inclined planes and observed and recorded the gain in speed as rolling continued (Figure 3.3). On downward-sloping planes the force of gravity increases a ball's speed. On an upward slope the force of gravity decreases the ball's speed. What about a ball rolling on a level surface? While rolling level, the ball doesn't roll with or against the vertical force of gravity—it neither speeds up nor slows down. The rolling ball maintains a constant

FIGURE 3.2
Galileo's famous demonstration.

Slope downward—
Speed increases

Slope upward—
Speed decreases

No slope—
Does speed change?

FIGURE 3.3
Motion of balls on various planes.

speed. Galileo reasoned that a ball moving horizontally would move forever if friction were entirely absent. A ball would move of itself—without being pushed or pulled.

> Inertia isn't a kind of force; it's a property of all matter to resist changes in motion.

Galileo noted that moving objects tend to remain moving, without the need of an imposed force. Objects at rest tend to remain at rest. This property of objects to maintain their state of motion is called **inertia**.

STOP AND CHECK YOURSELF

A ball rolling along a level surface slowly comes to a stop. How would Aristotle explain this behavior? How would Galileo explain it? How would you explain it?

CHECK YOUR ANSWERS

Aristotle would probably say that the ball stops because it seeks its natural state of rest. Galileo would probably say that friction overcomes the ball's natural tendency to continue rolling—that friction overcomes the ball's *inertia*, and brings it to a stop. Only you can answer the last question!

FIGURE 3.4
A ball rolling down an incline on the left tends to roll up to its initial height on the right. The ball must roll a greater distance as the angle of incline on the right is reduced.

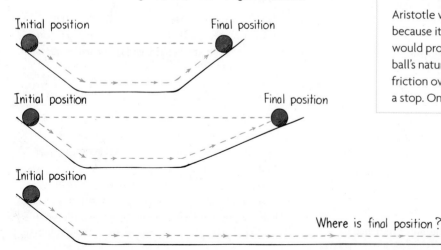

Initial position   Final position

Initial position   Final position

Initial position

Where is final position?

HISTORY OF SCIENCE

▪ GALILEO GALILEI (1564–1642)

Galileo was born in Pisa, Italy, in the same year Shakespeare was born and Michelangelo died. He studied medicine at the University of Pisa and then changed to mathematics. He developed an early interest in motion and was soon at odds with others around him, who held to Aristotelian ideas on falling bodies. He left Pisa to teach at the University of Padua and became an advocate of the new theory of the solar system advanced by the Polish astronomer Copernicus. Galileo was one of the first to build a telescope, and the first to direct it to the night-time sky and discover mountains on the Moon and the moons of Jupiter. Because he published his findings in Italian instead of in Latin, the language that such a reputable scholar would ordinarily use, and because of the recent invention of the printing press, his ideas reached many people. He soon ran afoul of the Church and was warned not to teach and not to hold to Copernican views. He restrained himself publicly for nearly 15 years. Then he defiantly published his observations and conclusions, which were counter to Church doctrine. The outcome was a trial in which he was found guilty, and he was forced to renounce his discoveries. By then an old man broken in health and spirit, he was sentenced to perpetual house arrest. Nevertheless, he completed his studies on motion, and his writings were smuggled out of Italy and published in Holland. His eyes had been damaged earlier by viewing the Sun through a telescope, which led to blindness at the age of 74. He died 4 years later.

3.3 Mass—A Measure of Inertia

When an object changes its state of motion—by speeding up, slowing down, or changing course—we say it undergoes *acceleration*. How much acceleration it will undergo depends on the forces applied to it and on the inertia of the object—how much it resists changes in motion. The amount of inertia an object possesses depends on the amount of matter in the object—the more matter, the more inertia. In speaking of how much matter something has, we use the term *mass*: the greater the mass of an object, the greater its inertia. **Mass** is a measure of the inertia of a material object.

The Physics Place
Definition of a Newton

Mass corresponds to our intuitive notion of **weight**. We casually say that something contains a lot of matter if it weighs a lot. But there is a difference between mass and weight. We can define each as follows:

Mass: The quantity of matter in an object. It is also the measure of the inertia or sluggishness that an object exhibits in response to any effort made to start it, stop it, or change its state of motion in any way.

Weight: The force upon an object due to gravity.

The standard unit of mass is the **kilogram**, abbreviated kg. Weight is measured in units of force. The scientific unit of force is the **newton**, abbreviated N, which we'll use in this book. The abbreviation is written with a capital letter because the unit is named after a person. Mass and weight are directly proportional to each other.* If the mass of an object is doubled, its weight is also doubled; if the mass is halved, the weight is halved. Because of this, mass and weight are often interchanged. Also, mass and weight are sometimes confused because it is customary to measure the quantity of matter in things (their mass) by their gravitational attraction to Earth (their weight). But mass doesn't depend on gravity. Gravity on the Moon, for example, is much less than it is on Earth. Whereas your weight on

> Mass (quantity of matter) and weight (force due to gravity) are directly proportional to each other.

FIGURE 3.5

An anvil in outer space—beyond the Sun, for example—may be weightless, but it still has mass.

* *Directly proportional* means directly related. If you change one, the other changes proportionally. The constant of proportionality is *g*, the acceleration due to gravity. As we shall soon see, weight = *mg* (or mass × acceleration due to gravity), so $9.8\,\text{N} = (1\,\text{kg})(9.8\,\text{m/s}^2)$. In Chapter 6, we'll extend our definition of weight to be the gravitational force of a body pressing against a support (for example, against a weighing scale).

the surface of the Moon would be much less than it is on Earth, your mass would be the same in both locations. Mass is a fundamental quantity that completely escapes the notice of most people.

You can sense how much mass is in an object by sensing its inertia. When you shake an object back and forth, you can feel its inertia. If it has a lot of mass, it's difficult to change the object's direction. If it has a small mass, shaking the object is easier. To-and-fro shaking requires the same force even in regions where gravity is different—on the Moon, for example. An object's inertia, or mass, is a property of the object itself and not its location.

If an object has a large mass, it may or may not have a large volume. Do not confuse mass and volume. Volume is a measure of space, measured in units such as cubic centimeters, cubic meters, or liters. How many kilograms of matter an object contains and how much space the object occupies are two different things. Mass is different from volume.

A nice demonstration that distinguishes mass from weight is the massive ball suspended on the string, shown in Figure 3.7. The top string breaks when the lower string is pulled with a gradual increase in force, but the bottom string breaks when the string is jerked. Which of these cases illustrates the weight of the ball, and which illustrates the mass of the ball? Note that only the top string bears the weight of the ball. So when the lower string is gradually pulled, the tension supplied by the pull is transmitted to the top string. So total tension in the top string is pull plus the weight of the ball. The top string breaks when the breaking

FIGURE 3.6
The astronaut in space finds it is just as difficult to shake the "weightless" anvil as it would be on Earth. If the anvil is more massive than the astronaut, which shakes more—the anvil or the astronaut?

FIGURE 3.7
Why will a slow continuous increase in downward force break the string above the massive ball, whereas a sudden increase in downward force breaks the lower string?

FIGURE 3.8
Why does the blow of the hammer not harm her?

point is reached. But when the bottom string is jerked, the mass of the ball—its tendency to remain at rest—is responsible for breakage of the bottom string.

STOP AND CHECK YOURSELF

1. Does a 2-kilogram bar of gold have twice as much *inertia* as a 1-kilogram bar of gold? Twice as much *mass*? Twice as much *volume*? Twice as much *weight* when weighed in the same location?

2. Does a 2-kilogram bar of gold have twice as much *inertia* as a 1-kilogram bunch of bananas? Twice as much *mass*? Twice as much *volume*? Twice as much *weight* when weighed in the same location?

3. How does the mass of a bar of gold vary with location?

CHECK YOUR ANSWERS

1. The answer is yes to all questions. A 2-kilogram bar of gold has twice as many gold atoms, and therefore twice the amount of matter, mass, and weight. The bars consist of the same material, so the 2-kilogram bar also has twice the volume.

2. Two kilograms of *anything* has twice the inertia and twice the mass of one kilogram of anything else. Since mass and weight are proportional in the same location, two kilograms of anything will weigh twice as much as one kilogram of anything. Except for volume, the answer to all the questions is yes. Volume and mass are proportional only when the materials are identical—when they have the same *density*. (Density is mass/volume, as we'll discuss in Chapter 7). Gold is much more dense than bananas, so two kilograms of gold must occupy less volume than one kilogram of bananas.

3. Not at all! It consists of the same number of atoms no matter what the location. Although its weight may vary with location, it has the same mass everywhere. This is why mass is preferred to weight in scientific studies.

A pillow is bigger than an auto battery, but which has more matter? Which has more inertia? Which has more mass?

FIGURE **3.9**

One kilogram of nails weighs 9.8 newtons, which is equal to 2.2 pounds.

One Kilogram Weighs 9.8 Newtons

A 1-kilogram bag of any material at Earth's surface has a weight of 9.8 newtons. Away from Earth's surface where the force of gravity is less (on the Moon for example), the bag would weigh less.

Except in cases where precision is needed, we will round off 9.8 and call it 10. So 1 kilogram of something on Earth's surface weighs about 10 newtons. If you know the mass in kilograms and want weight in newtons, multiply the number of kilograms by 10. Or, if you know the weight in newtons, divide by 10 and you'll have the mass in kilograms. As previously mentioned, weight and mass are proportional to each other.

The relationship between kilograms and pounds is that 1 kg weighs 2.2 lb at Earth's surface. (That means 1 lb is the same as 4.45 N.)

3.4 Net Force

In simplest terms, a force is a push or a pull. Objects don't speed up, slow down, or change direction unless a force acts. When we say "force" we imply the total force, or *net* force, acting on an object. Often more than one force acts. For example, when you throw a baseball, the force of gravity, air friction, and the pushing force you apply with your muscles all act on the ball. The **net force** on the ball is the combination of all these forces. It is the net force that changes an object's state of motion.

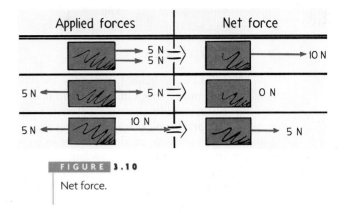

Applied forces		Net force

FIGURE **3.10**

Net force.

For example, suppose you pull on a box with a force of 5 N (slightly more than 1 pound). If your friend also pulls with 5 N in the same direction, the net force on the box is 10 N. If your friend pulls on the box with the same magnitude of force as you in the opposite direction, the net force on it is zero. Now if you increase your pull to 10 N and your friend pulls oppositely with 5 N, the net force is 5 N in the direction of your pull. This is shown in Figure 3.10.

The forces in Figure 3.10 are shown by arrows. Forces are vector quantities. A **vector quantity** has both magnitude (how much) and direction (which way). When an arrow represents a vector quantity, the arrow's length represents magnitude and its direction shows the direction of the quantity. Such an arrow is called a **vector.** (You'll find more on vectors in the next chapter, in Appendix C, and in the *Practice Book for Conceptual Physics Fundamentals*.)

A zero net force on an object doesn't mean the object must be at rest, but that its state of motion remains unchanged. It can be at rest or moving uniformly in a straight line.

3.5 The Equilibrium Rule

If you tie a string around a 2-pound bag of flour and suspend it on a weighing scale (Figure 3.11 on p. 9), a spring in the scale stretches until the scale reads 2 pounds. The stretched spring is under a "stretching force" called *tension*. A scale in a science lab is likely calibrated to read the same force as 9 newtons. Both pounds and newtons are units of weight, which in

SCIENCE AND SOCIETY

■ PERSONAL ESSAY

When I was in high school, my counselor advised me not to enroll in science and math classes but instead focus on what seemed to be my gift for art. I took this advice. I was then interested in drawing comic strips and in boxing, neither of which earned me much success. After a stint in the army, I tried my luck at sign painting, and the cold Boston winters drove me south to Miami, Florida. There, at age 26, I got a job painting billboards and met an intellectual friend, Burl Grey. Like me, Burl had never studied physics in high school. But he was passionate about science in general, and shared his passion with many questions as we painted together.

I remember Burl asking me about the tensions in the ropes that held up the scaffold we were on. The scaffold was simply a heavy horizontal plank suspended by a pair of ropes. Burl twanged the rope nearest his end of the scaffold and asked me to do the same with mine. He was comparing the tensions in both ropes—to determine which was greater. Burl was heavier than I was, and he guessed the tension in his rope was greater. Like a more tightly stretched guitar string, the rope with greater tension twangs at a higher pitch. The finding that Burl's rope had a higher pitch seemed reasonable because his rope supported more of the load.

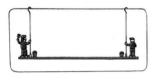

When I walked toward Burl to borrow one of his brushes, he asked if tensions in the ropes changed. Did tension in his rope increase as I moved closer? We agreed that it should have, because even more of the load was then supported by Burl's rope. How about my rope? Would its tension decrease? We agreed that it would, for it would be supporting less of the total load. I was unaware at the time that I was discussing physics.

Burl and I used exaggeration to bolster our reasoning (just as physicists do). If we both stood at an extreme end of the scaffold and leaned outward, it was easy to imagine the opposite end of the scaffold rising like the end of a seesaw, with the opposite rope going limp. Then there would be no tension in that rope. We then reasoned the tension in my rope would gradually decrease as I walked toward Burl. It was fun posing such questions and seeing if we could answer them.

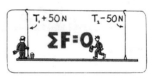

A question that we couldn't answer was whether or not the decrease in tension in my rope when I walked away from it would be *exactly* compensated by a tension increase in Burl's rope. For example, if my rope underwent a decrease of 50 newtons, would Burl's rope gain 50 newtons? (We talked pounds back then, but here we use the scientific unit of force, the *newton*—abbreviated N.) Would the gain be *exactly* 50 N? And if so, would this be a grand coincidence? I didn't know the answer until more than a year later, when Burl's stimulation resulted in my leaving full-time painting and going to college to learn more about science.*

In college I learned that any object at rest, such as the sign-painting scaffold I worked on with Burl, is said to be in equilibrium. That is, all the forces that act on it balance to zero. So the sum of the upward forces supplied by the supporting ropes indeed do add up to our weights plus the weight of the scaffold. A 50 N loss in one would be accompanied by a 50 N gain in the other.

I tell this true story to make the point that one's thinking is very different when there is a rule to guide it. Now, when I look at any motionless object, I know immediately that all the forces acting on it cancel out. We see nature differently when we know its rules. It makes nature simpler and easier to understand. Without the rules of

(continued)

physics, we tend to be superstitious and see magic where there is none. Quite wonderfully, everything is beautifully connected to everything else by a surprisingly small number of rules. Physics is the study of nature's rules.

*I am forever indebted to Burl Grey for the stimulation he provided, for when I continued with formal education, it was with enthusiasm. I lost contact with Burl for 40 years. A student in my class at the Exploratorium in San Francisco, Jayson Wechter, who was a private detective, located him in 1998 and put us back in contact. Friendship renewed, we continue in our spirited conversations.

turn are units of *force*. The bag of flour is attracted to Earth with a gravitational force of 2 pounds—or, equivalently, 9 newtons. Suspend twice as much flour from the scale and the reading will be 18 newtons.

There are two forces acting on the bag of flour— tension force acting upward and weight acting downward. The two forces on the bag are equal in magnitude and opposite in direction, and they cancel to zero. Hence the bag remains at rest.

Everything that isn't undergoing a change in motion is in mechanical equilibrium. That's because $\Sigma F = 0$.

When the net force on something is zero, we say that something is in *mechanical equilibrium*.* In mathematical notation, the **equilibrium rule** is

$$\Sigma F = 0$$

The symbol Σ stands for "the vector sum of" and F stands for "forces" (we put F in bold to indicate that it's a vector quantity). For a suspended object at rest, like the bag of flour, the rule states that the forces acting upward on the body must be balanced by other forces acting downward to make the vector sum equal zero. (Vector quantities take direction into account, so, if upward forces are positive, downward ones are negative, and the resulting sum is equal to zero.)

In Figure 3.12 we see the forces of interest to Burl and Paul on their sign-painting scaffold. The sum of the upward tensions is equal to the sum of their weights plus the weight of the scaffold. Note how the magnitudes of the two upward vectors equal the magnitude of the three downward vectors. Net force on the scaffold is zero, so we say it is in mechanical equilibrium.

Can you see evidence of $\Sigma F = 0$. in bridges and other structures around you?

FIGURE **3.11**

Burl Grey, who first taught the author about tension forces, suspends a 1-kg bag of flour from a spring scale, showing its weight and the tension in the string of nearly 10 newtons.

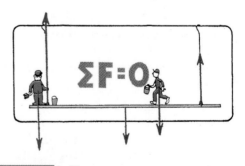

FIGURE **3.12**

The sum of the upward vectors equals the sum of the downward vectors. $\Sigma F = 0$, and the scaffold is in equilibrium.

* We'll see in Appendix B that another condition for mechanical equilibrium is that the net ***torque*** equals zero.

PROBLEM SOLVING

Problems

1. When Burl stands alone in the exact middle of his scaffold, the left scale reads 500 N. Fill in the reading on the right scale. The total weight of Burl and the scaffold must be _____N.

2. Burl stands farther from the left. Fill in the reading on the right scale.

3. In a silly mood, Burl dangles from the right end. Fill in the reading on the right scale.

Solutions

Do your answers illustrate the equilibrium rule?

1. *The total weight is* **1000 N**. The right rope must be under 500 N of tension because Burl is in the middle, and both ropes support his weight equally. Since the sum of upward tensions is 1000 N, the total weight of Burl and the scaffold must be **1000 N**. Let's call the upward tension forces +1000 N. Then the downward weights are −1000 N. What happens when you add +1000 N and −1000 N? The answer is they equal zero. So we see that $\Sigma \boldsymbol{F} = 0$.

2. *Did you get the correct answer of* **830 N**? Reasoning: We know from question 1 that the sum of the rope tensions equals 1000 N, and since the left rope has a tension of 170 N, the other rope must make up the difference—that 1000 N − 170 N = 830 N. Get it? If so, great. If not, discuss it with your friends until you do. Then read further.

3. *The answer is* **1000 N**. Do you see that this illustrates $\Sigma \boldsymbol{F} = 0$?

STOP AND CHECK YOURSELF

Consider Nellie Newton suspended from the rings.

1. If she hangs with her weight evenly distributed between the two rings, how would scale readings in both supporting ropes compare with her weight?

2. Suppose she hangs with slightly more of her weight supported by the left ring. How would a scale on the right read?

CHECK YOUR ANSWERS

1. The reading on each scale will be *half her weight*. The sum of the readings on both scales then equals her weight.

2. When more of her weight is supported by the left ring, the reading on the right is *less than half her weight*. No matter how she hangs, the sum of the scale readings equals her weight. For example, if one scale reads two-thirds her weight, the other scale will read one-third her weight. Get it?

| 3.6 | **Support Force** |

Consider a book lying at rest on a table. It is in equilibrium. What forces act on the book? One force is that due to gravity—the *weight* of the book. Since the book is in equilibrium, there must be another force acting on it to produce a net force of zero—an upward force opposite to the force of gravity. The table exerts this upward force, called the **support force.** This upward support force, often called the *normal force*, must equal the weight of the book.* If we designate the upward force positive, then the downward force (weight) is negative, and the sum of the two is zero. The net force on the book is zero. Stating it another way, $\Sigma F = 0$.

To better understand that the table pushes up on the book, compare the case of compressing a spring (Figure 3.13). If you push the spring down, you can feel the spring pushing up on your hand. Similarly, the book lying on the table compresses atoms in the table, which behave like microscopic springs. The weight of the book squeezes downward on the atoms, and they squeeze upward on the book. In this way, the compressed atoms produce the support force.

When you step on a bathroom scale, two forces act on the scale. One is the downward pull of gravity, your weight, and the other is the upward support force of the floor. These forces compress a spring that is calibrated to show your weight (Figure 3.14). In effect, the scale shows the support

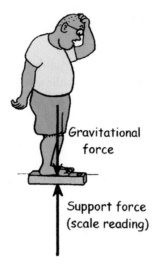

FIGURE 3.14

The upward support is as much as the downward gravitational force.

Gravitational force

Support force (scale reading)

force. When you weigh yourself on a bathroom scale at rest, the support force and your weight have the same magnitude.

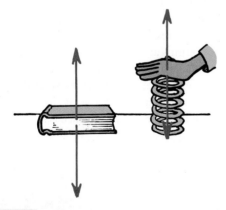

FIGURE 3.13

The table pushes up on the book with as much force as the downward force of gravity on the book. The spring pushes up on your hand with as much force as you exert to push down on the spring.

---

* This force acts at right angles to the surface. When we say "normal to," we are saying "at right angles to," which is why this force is called a normal force.

STOP AND CHECK YOURSELF

1. What is the net force on a bathroom scale when a 150-pound person stands on it?

2. Suppose you stand on two bathroom scales with your weight evenly distributed between the two scales. What is the reading on each of the scales? What happens when you stand with more of your weight on one foot than the other?

CHECK YOUR ANSWERS

1. Zero, for the scale remains at rest. The scale reads *support force* (which has the same magnitude as weight), not the net force.

2. The reading on each scale is half your weight, because the sum of the scale readings must balance your weight, so the net force on you will be zero. If you lean more on one scale than on the other, more than half your weight will be read on that scale but less on the other, so they will still add up to equal your weight. Like the example of Nellie hanging by the rings, if one scale reads two-thirds your weight, the other scale will read one-third your weight.

3.7 Equilibrium of Moving Things

When an object isn't moving, it's in equilibrium. The forces on it add up to zero. But the state of rest is only one form of equilibrium. An object moving at constant speed in a straight-line path is also in equilibrium. Once in motion, if there is no net force to change the state of motion, it's in equilibrium. Whether at rest or steadily rolling in a straight-line path, $\Sigma F = 0$.

Equilibrium is a state of *no change*. A hockey puck sliding along slippery ice or a bowling ball rolling at constant velocity is in equilibrium—until either experiences a nonzero net force. Whether at rest or steadily moving in a straight-line path, the sum of the forces on both is zero: $\Sigma F = 0$.

Interestingly, an object under the influence of only one force cannot be in equilibrium. Net force couldn't be zero. Only when there is no force at all, or when two or more forces combine to zero, can an object be in equilibrium. We can test whether or not something is in equilibrium by noting whether or not it undergoes changes in motion.

Consider a refrigerator being pushed horizontally across a kitchen floor. If it moves steadily at constant speed, without change in its motion, it is in equilibrium. This tells us that more than one horizontal force acts on the refrigerator—likely the force of friction between the refrigerator and the floor. The fact that the net force on it equals zero means that the force of friction must be equal in magnitude and act opposite to our pushing force.

We say objects at rest are in *static* equilibrium, and objects moving at constant speed in a straight-line path are in *dynamic* equilibrium. Both of these situations are examples of mechanical equilibrium. In Chapter 8 we'll talk about thermal equilibrium, and in Appendix B we'll discuss rotational equilibrium.

FIGURE 3.15

When the push on the refrigerator is as great as the force of friction between it and the floor, the net force on the refrigerator is zero and it slides at an unchanging speed.

3.8 The Force of Friction

Friction occurs when one object rubs against something else.* Friction occurs for solids, liquids, and gases. An important rule of friction is that it always acts in a direction to oppose motion. If you push a solid block along a floor to the right, the force of friction on the block will be to the left. A boat propelled to the east by its motor experiences water friction to the west. When an object falls downward through the air, the force of friction, **air resistance,** acts upward. Again, for emphasis, let's state that friction always acts in a direction to oppose motion.

The Physics Place
Friction

STOP AND CHECK YOURSELF

You push on a piece of furniture and it slides at constant speed across the kitchen floor. In other words, it is in equilibrium. Two horizontal forces act on it. One is your push and the other is the force of friction that acts in the opposite direction. Which force is greater?

CHECK YOUR ANSWER

Neither, for both forces have the same magnitude. If you call your push positive, then the friction is negative. Since the pushed furniture is in equilibrium, can you see that the two forces combine to equal zero?

The amount of friction between two surfaces depends on the kinds of material and how much they are pressed together. Friction is due to tiny surface bumps and also to the "stickiness" of the atoms on the surfaces of the two materials (Figure 3.16). Friction between a crate and a smooth wooden floor is less than friction between the same crate and a rough floor. And if the surface is inclined, friction is

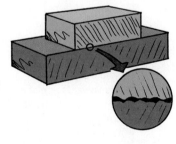

FIGURE 3.16

Friction results from the mutual contact of irregularities in the surfaces of sliding objects. Even surfaces that appear to be smooth have irregular surfaces when viewed at the microscopic level.

* Friction is a very complicated phenomenon. The findings are empirical (gained from a wide range of experiments) and the predictions are approximate (also based on experiment).

less because the crate doesn't press as much on the inclined surface.

When you push horizontally on a crate and it slides across a factory floor, both your force and the opposite force of friction affect the motion. When you push hard enough on the crate to match the friction, the net force on the crate is zero, and it slides at constant velocity. Notice that we are talking about what we recently learned—that no change in motion occurs when $\Sigma F = 0$.

FIGURE 3.17

A cheetah can maintain a very high speed, but only for a short time.

STOP AND CHECK YOURSELF

1. Suppose you exert a 100-N horizontal force on a heavy crate of computer supplies resting motionless on an office floor. The fact that it remains at rest indicates that 100 N isn't great enough to make it slide. How does the force of friction between the crate and floor compare with your push?

2. You push harder—say, 110 N—and the crate still doesn't slide. How much friction acts on the crate?

3. You push still harder and the crate moves. Once in motion, you push with 115 N, which is just sufficient to keep it sliding at constant velocity. How much friction acts on the crate?

4. What net force does a sliding crate experience when you exert a force of 125 N and friction between the crate and the floor is 115 N?

CHECK YOUR ANSWERS

1. 100 N in the opposite direction. Friction opposes the motion that would occur other-wise. The fact that the crate is at rest is evidence that $\Sigma F = 0$.

2. Friction increases to 110 N, again $\Sigma F = 0$.

3. 115 N, because when moving at constant velocity, $\Sigma F = 0$.

4. 10 N, because $\Sigma F = 125 \, N - 115 \, N$. In this case the crate picks up speed—it accelerates.

3.9 Speed and Velocity

Speed

Before the time of Galileo, people described moving things as simply "slow" or "fast." Such descriptions were vague. Galileo was the first to measure speed by comparing the distance covered with the *time* it

The Physics Place
Definition of Speed
Average Speed
Velocity
Changing Velocity

takes to move that distance. He defined **speed** as the distance covered per amount of travel time.

$$\text{Speed} = \frac{\text{distance covered}}{\text{travel time}}$$

For example if a bicyclist covers 20 kilometers in 1 hour, her speed is 20 km/h. Or, if she runs 6 meters in 1 second, her speed is 6 m/s.

Any combination of units for distance and time can be used for speed—kilometers per hour (km/h), centimeters per day (the speed of a sick snail), or whatever is useful and convenient. The slash symbol (/) is read as "per," and means "divided by." In physics the preferred unit of speed is meters per second (m/s). Table 3.1 compares some speeds in different units.

Instantaneous Speed

Moving things often have variations in speed. A car, for example, may travel along a street at 50 km/h, slow to 0 km/h at a red light, and speed up to only 30 km/h because of traffic. At any instant you can tell the speed of the car by looking at its

If you're cited for speeding, which does the police officer write on your ticket, your *instantaneous speed* or your *average speed?*

TABLE 3.1

Approximate Speeds in Different Units

12 mi/h = 20 km/h = 6 m/s (bowling ball)
25 mi/h = 40 km/h = 11 m/s (very good sprinter)
37 mi/h = 60 km/h = 17 m/s (sprinting rabbit)
50 mi/h = 80 km/h = 22 m/s (tsunami)
62 mi/h = 100 km/h = 28 m/s (sprinting cheetah)
75 mi/h = 120 km/h = 33 m/s (batted softball)
100 mi/h = 160 km/h = 44 m/s (batted baseball)

FIGURE 3.18

A common automobile speedometer. Note that speed is shown in units km/h and mph.

speedometer. The speed at any instant is the *instantaneous speed*.

Average Speed

In planning a trip by car, the driver often wants to know the travel time. The driver is concerned with the *average speed* for the trip. How is average speed defined?

$$\text{Average speed} = \frac{\text{total distance covered}}{\text{travel time}}$$

Average speed can be calculated rather easily. For example, if you drive a distance of 80 kilometers in 1 hour, your average speed is 80 kilometers per

hour. Likewise, if you travel 320 kilometers in 4 hours,

$$\text{Average speed} = \frac{\text{total distance covered}}{\text{travel time}}$$

$$= \frac{320 \text{ km}}{4 \text{ h}} = 80 \text{ km/h}$$

Note that, when a distance in kilometers (km) is divided by a time in hours (h), the answer is in kilometers per hour (km/h).

Since average speed is the entire distance covered divided by the total time of travel, it doesn't indicate the various instantaneous speeds that may have occurred along the way. On most trips, the instantaneous speed is often different from the average speed.

If we know average speed and travel time, distance traveled is easy to find. A simple rearrangement of the definition above gives

Total distance covered = average speed × travel time

For example, if your average speed on a 4-hour trip is 80 kilometers per hour, then you cover a total distance of 320 kilometers.

STOP AND
CHECK YOURSELF

1. What is the average speed of a cheetah that sprints 100 m in 4 seconds? How about if it sprints 50 m in 2 s?

2. If a car travels with an average speed of 60 km/h for an hour, it will cover a distance of 60 km.
   a. How far would it travel if it moved at this rate for 4 h?
   b. For 10 h?

3. In addition to the speedometer on the dashboard of every car, there is an odometer, which records the distance traveled. If the initial reading is set at zero at the beginning of a trip and the reading is 40 km one-half hour later, what was the car's average speed?

4. Would it be possible to attain this average speed and never go faster than 80 km/h?

CHECK YOUR ANSWERS

(Are you reading this before you have formed a reasoned answer in your mind? As mentioned earlier, **think** *before you read answers. You'll not only learn more; you'll enjoy learning more.)*

1. In both cases the answer is 25 m/s:

$$\text{Average speed} = \frac{\text{total distance covered}}{\text{travel time}}$$

$$= \frac{100 \text{ meters}}{4 \text{ seconds}} = \frac{50 \text{ meters}}{2 \text{ seconds}} = 25 \text{ m/s}$$

2. The distance traveled is the average speed × time of travel, so
   a. Distance = 60 km/h × 4 h = 240 km
   b. Distance = 60 km/h × 10 h = 600 km

3. $$\text{Average speed} = \frac{\text{total distance covered}}{\text{travel time}}$$

$$= \frac{40 \text{ km}}{0.5 \text{ h}} = 80 \text{ km/h}$$

4. No, not if the trip starts from rest and ends at rest. During the trip, there are times when the instantaneous speeds are less than 80 km/h, so the driver must at some time drive faster than 80 km/h in order to average 80 km/h. In practice, average speeds are usually much lower than high instantaneous speeds.

FIGURE 3.19

Although the car can maintain a constant speed along the circular track, it cannot maintain a constant velocity. Why?

FIGURE 3.20

Although you may be at rest relative to Earth's surface, you're moving about 100,000 km/h relative to the Sun.

Velocity

When we know both the speed and direction of an object, we know its **velocity.** For example, if a vehicle travels at 60 km/h, we know its speed. But, if we say it moves at 60 km/h to the north, we specify its *velocity.* Speed is a description of how fast; velocity is a description of how fast *and* in what direction. As previously mentioned, a quantity such as velocity that specifies direction as well as magnitude is called a **vector quantity.** Velocity is a vector quantity. (Vectors are treated in the next chapter and in Appendix D, and are nicely developed in the *Practice Book for Conceptual Physics Fundamentals*).

Constant speed means steady speed, neither speeding up nor slowing down. Constant velocity, on the other hand, means both constant speed *and* constant direction. Constant direction is a straight line—the object's path doesn't curve. So constant velocity means motion in a straight line at constant speed—motion with no acceleration.

| Velocity is "directed" speed. |

STOP AND CHECK YOURSELF

"She moves at a constant speed in a constant direction." Say the same sentence in fewer words.

CHECK YOUR ANSWER

"She moves at constant velocity."

Motion Is Relative

Everything is always moving. Even when you think you're standing still, you're actually speeding through space. You're moving relative to the Sun and stars—though you are at rest relative to Earth. At this moment, your speed relative to the Sun is about 100,000 kilometers per hour, and that speed is even faster relative to the center of our galaxy.

When we discuss the speed or velocity of something, we mean speed or velocity relative to something else. For example, when we say a space shuttle travels at 30,000 kilometers per hour, we mean relative to Earth below. Or when we say a racing car reaches a speed of 300 kilometers per hour, we mean relative to the track. Unless stated otherwise, all speeds discussed in this book are relative to the surface of Earth. Motion is relative.

STOP AND CHECK YOURSELF

A hungry mosquito sees you resting in a hammock in a 3-m/s breeze. How fast and in what direction should the mosquito fly in order to hover above you for lunch?

CHECK YOUR ANSWER

The mosquito should fly toward you into the breeze. When barely above you it should fly at 3 m/s in order to hover at rest. Unless its grip on your skin is strong enough after landing, it must continue flying at 3 m/s to keep from being blown off. That's why a breeze is an effective deterrent to mosquito bites.

3.10 Acceleration

Most moving things undergo variations in their motion. We say they undergo *acceleration.* The first to formulate the concept of acceleration was Galileo, who developed the concept in his experiments with inclined planes. He found that balls rolling down inclines rolled faster and faster. Their velocity changed as they rolled. Further, the balls gained the same amount of velocity in equal time intervals.

The Physics Place
Definition of Acceleration
Force Causes Acceleration
Numerical Example of Acceleration

The Physics Place
Parachuting and Newton's Second Law

Galileo defined the rate of change of velocity as **acceleration.***

$$\text{Acceleration} = \frac{\text{change of velocity}}{\text{time interval}}$$

Acceleration is experienced when you're in a moving car or bus. When the driver steps on the gas pedal, the vehicle gains speed. We say that the bus accelerates. We can see why the gas pedal is called the "accelerator"! When the brakes are applied, the vehicle slows. This is also acceleration, because the velocity of the vehicle is changing. When something slows down, we often call this *deceleration*.

Can you see that a car has three controls that change velocity—the gas pedal (accelerator), the brakes, and the steering wheel?

Consider driving in a car that steadily increases in speed. Suppose that in 1 second, you steadily increase your velocity from 30 kilometers per hour to 35 kilometers per hour. In the next second, you go from 35 kilometers per hour to 40 kilometers per hour, and so on. You change your velocity by 5 kilometers per hour each second. We see that

$$\text{Acceleration} = \frac{\text{change of velocity}}{\text{time interval}}$$
$$= \frac{5 \text{ km/h}}{1 \text{ s}} = 5 \text{ km/h} \cdot \text{s}$$

In this example the acceleration is 5 kilometers per hour-second (abbreviated as 5 km/h·s).** Note that the unit for time enters twice: once for the unit of velocity and again for the interval of time in which the velocity is changing. Also note that acceleration is not just the change in velocity; it is the *change in velocity per second*. If either speed or

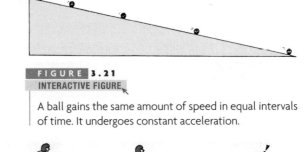

FIGURE 3.21
INTERACTIVE FIGURE

A ball gains the same amount of speed in equal intervals of time. It undergoes constant acceleration.

FIGURE 3.22

We say that a body undergoes acceleration when there is a *change* in its state of motion.

direction changes, or if both change, then velocity changes.

When a car makes a turn, even if its speed does not change, it is accelerating. Can you see why? Acceleration occurs because the car's direction is changing. Acceleration refers to a change in velocity. So acceleration involves a change in speed, a change in direction, or a change in both speed *and* direction. Figure 3.22 illustrates this.

Hold a stone above your head (but off to the side a bit) and drop it. It accelerates during its fall. When the only force that acts on a falling object is that due to gravity, when air resistance doesn't affect its motion, we say the object is in *free fall*. All freely falling objects in the same vicinity have the same acceleration. At Earth's surface an object in free fall gains speed at the rate of 10 m/s each second, as shown in Table 3.2.

$$\text{Acceleration} = \frac{\text{change in speed}}{\text{time interval}} = \frac{10 \text{ m/s}}{1 \text{ s}}$$
$$= 10 \text{ m/s} \cdot \text{s} = 10 \text{ m/s}^2$$

* The Greek letter Δ (delta) is often used as a symbol for "change in" or "difference in." In "delta" notation, $a = \frac{\Delta v}{\Delta t}$, where Δv is the change in velocity and Δt is the change in time (the time interval). From this we see that $v = at$. See further development of linear motion in Appendix B.

** When we divide $\frac{\text{km}}{\text{h}}$ by s $\left(\frac{\text{km}}{\text{h}} \div \text{s}\right)$, we can express this as $\frac{\text{km}}{\text{h}} \times \frac{1}{\text{s}} = \frac{\text{km}}{\text{h} \cdot \text{s}}$ (some textbooks express this as km/h/s). Or when we divide $\frac{\text{m}}{\text{s}}$ by s $\left(\frac{\text{m}}{\text{s}} \div \text{s}\right)$, we can express this as $\frac{\text{m}}{\text{s}} \times \frac{1}{\text{s}} = \frac{\text{m}}{\text{s} \cdot \text{s}} = \frac{\text{m}}{\text{s}^2}$ (which can also be written as (m/s)/s or ms^{-2}).

TABLE 3.2

Free-Fall Velocity Acquired and Distance Fallen

Time of Fall (s)	Velocity Acquired (m/s)	Distance Fallen (m)
0	0	0
1	10	5
2	20	20
3	30	45
4	40	80
5	50	125

We read the acceleration of free fall as 10 meters per second squared. (More precisely, 9.8 m/s².) This is the same as saying that acceleration is 10 meters per second per second. Note again that the unit of time, the second, appears twice. It appears once for the unit of velocity and again for the time during which the velocity changes.

When you're over the hill, that's when you pick up speed.
—Quincy Jones

STOP AND CHECK YOURSELF

In 2.0 seconds a car increases its speed from 60 km/h to 65 km/h while a bicycle goes from rest to 5 km/h. Which has the greater acceleration?

CHECK YOUR ANSWER

Both have the same acceleration since both gain the same amount of speed in the same time. Both accelerate at 2.5 km/h · s.

In Figure 3.23, we imagine a freely falling boulder with a speedometer attached to it. As the boulder falls, the speedometer shows that the boulder goes 10 m/s faster each second. This 10 m/s gain each second is the boulder's acceleration. Velocity acquired and distance fallen* are shown in Table 3.2. (The acceleration of free fall is further developed in Appendix B and in the *Practice Book for Conceptual Physics Fundamentals*.) We see that the distance of free fall from rest is directly proportional to the square of the time of fall. In equation form,

$$d = \frac{1}{2}gt^2$$

Up-and-down motion is shown in Figure 3.24. The ball leaves the thrower's hand at 20 m/s. Call this the initial velocity. The figure uses the convention

* Distance fallen from rest: d = average velocity × time

$$d = \frac{\text{initial velocity} + \text{final velocity}}{2} \times \text{time}$$

$$d = \frac{0 + gt}{2} \times t$$

$$d = \frac{1}{2}gt^2 \text{ (See Appendix B for further explanation.)}$$

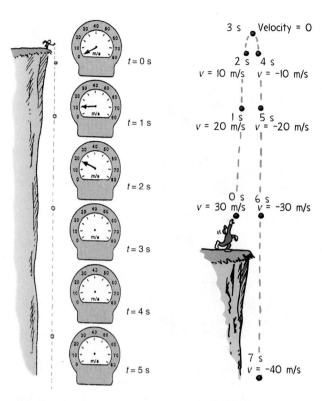

FIGURE 3.23

Imagine that a falling boulder is equipped with a speedometer. In each succeeding second of fall, you'd find the boulder's speed increasing by the same amount: 10 m/s. Sketch in the missing speedometer needle at $t = 3$ s, $t = 4$ s, and $t = 5$ s.

FIGURE 3.24
INTERACTIVE FIGURE

The rate at which velocity changes each second is the same.

of up being + and down being −. Notice that the 1-second interval positions correspond to 10-m/s velocity changes.

Aristotle used logic to establish his ideas of motion, whereas Galileo used experiment. Galileo showed that experiments are superior to logic in testing knowledge. Galileo was concerned with *how* things move rather than *why* they move. The path was paved for Isaac Newton to make further connections of concepts of motion.

Why do all freely falling objects fall with equal acceleration? The answer to this question awaits you in Chapter 4.

■ HANG TIME

Some athletes and dancers have great jumping ability. Leaping straight up, they seem to "hang in the air," apparently defying gravity. Ask your friends to estimate the "hang time" of the great jumpers—the time a jumper is airborne with his or her feet off the ground. They may estimate two or three seconds. But, surprisingly, the hang time of the greatest jumpers is almost always less than 1 second. The perception of a longer time is one of many illusions we have about nature.

People often have a related illusion about the vertical height a human can jump. Most of your classmates probably cannot jump higher than one-half meter. They can easily step over a 0.5 meter fence, but in doing so, their bodies rise only slightly. The height of the barrier is different from the height a jumper's "center of gravity" rises. Many people can leap over a 1-meter fence, but only rarely does anybody raise the "center of gravity" of their body by 1 meter. Even basketball star Michael Jordan in a standing jump during his prime couldn't raise his body 1.25 meters high, although he could easily reach considerably above the basket, which is more than 3 meters high.

Jumping ability is best measured by a standing vertical jump. Stand facing a wall with feet flat on the floor and arms extended upward. Make a mark on the wall at the top of your reach. Then make your jump, and, at the point you are able to reach, make another mark. The distance between these two marks measures your vertical leap. If it's more than 0.6 meters (2 feet), you're exceptional.

Here's the physics. When you leap upward, jumping force is applied only while your feet are still making contact with the ground. The greater the force, the greater your launch speed and the higher your jump. When your feet leave the ground, your upward speed immediately decreases at the steady rate of g, which is 10 m/s². At the top of your jump, your upward speed decreases to zero. Then you begin to fall, gaining speed at exactly the same rate, g. If you land as you took off, upright with legs extended, then your time rising equals your time falling; hang time is time up plus time down. While you are airborne, no amount of leg or arm pumping or other bodily motions can change your hang time.

As will be shown in Appendix B, the relationship between time up or down and vertical height is given by

$$d = \frac{1}{2} g t^2$$

If the vertical height d is known, we can rearrange this expression to read

$$t = \sqrt{\frac{2d}{g}}$$

A world-record vertical standing jump is 1.25 meters. Let's use this jumping height of 1.25 meters for d, and the more precise value of 9.8 m/s² for g. Solving for t, half the hang time (one way), we get

$$t = \sqrt{\frac{2d}{g}} = \sqrt{\frac{2(1.25 \text{ m})}{9.8 \text{ m/s}^2}} = 0.50 \text{ s}$$

Double this amount (because this is the time for one way of an up-and-down round trip) and we see that the record-breaking hang time is 1 second.

We're discussing vertical motion here. How about running jumps? We'll see in Chapter 8 that hang time depends only on the jumper's vertical speed at launch. While airborne, the jumper's horizontal speed remains constant while the vertical speed undergoes acceleration. Intriguing physics!

SUMMARY OF TERMS

Force Simply stated, a push or a pull.

Inertia The property of things to resist changes in motion.

Mass The quantity of matter in an object. More specifically, it is the measure of the inertia or sluggishness that an object exhibits in response to any effort made to start it, stop it, deflect it, or change in any way its state of motion.

Weight Simply stated, the force due to gravity on an object. More specifically, the gravitational force with which a body presses against a supporting surface.

Kilogram The unit of mass. One kilogram (symbol kg) is the mass of 1 liter (l) of water at 4°C.

Newton The scientific unit of force.

Net force The combination of all forces that act on an object.

Vector quantity A quantity whose description requires both magnitude and direction.

Support force The force that supports an object against gravity, often called the *normal force*.

Equilibrium rule The vector sum of forces acting on a nonaccelerating object equals zero: $\Sigma F = 0$.

Friction The resistive force that opposes the motion or attempted motion of an object past another with which it is in contact, or through a fluid.

Air resistance The force of friction acting on an object due to its motion in air.

Speed The distance traveled per time.

Velocity The speed of an object and specification of its direction of motion.

Acceleration The rate at which velocity changes with time; the change in velocity may be in magnitude or direction or both, usually measured in units m/s^2.

REVIEW QUESTIONS

*Each chapter in this book concludes with a set of review questions and exercises, and for some chapters, there are problems. In some chapters there is a set of single-step numerical problems that are meant for acquainting you with equations in the chapter—**One-Step Calculations**. The **Review Questions** are designed to help you comprehend ideas and catch the essentials of the chapter material. You'll notice that answers to the questions can be found within the chapters. The **Exercises** stress thinking rather than mere recall of information and call for an understanding of the definitions, principles, and relationships of the chapter material. In many cases the intention of particular exercises is to help you apply the ideas of physics to familiar situations. Unless you cover only a few chapters in your course, you will likely be expected to tackle only a few exercises for each chapter. Answers should be in complete sentences, with an explanation or sketches when applicable. The large number of exercises is to allow your instructor a wide choice of assignments. **Problems** go further than One-Step Calculations and feature concepts that are more clearly understood with more challenging computations. Additional problems can be found in the supplement to the 10th edition of* Conceptual Physics: Problem Solving in Conceptual Physics.

3.1 Aristotle on Motion

1. What were the two main classifications of motion in Aristotle's view of science?
2. Did Aristotle believe that forces are necessary to keep moving objects moving, or did he believe that once moving, they'd move of themselves?

3.2 Galileo's Concept of Inertia

3. What two main ideas of Aristotle did Galileo discredit?
4. Which dominated Galileo's way of extending knowledge, philosophical discussion or experiment?
5. What name is given to the property of objects to maintain their states of motion?

3.3 Mass—A Measure of Inertia

6. Which depends on location, weight or mass?
7. Where is your weight greater, on Earth or on the Moon? How about your mass?
8. What are the units of measurement for weight and for mass?
9. One kilogram weighs 9.8 newtons on Earth. Would it weigh more or less on the Moon?

3.4 Net Force

10. What is the net force on a box pushed to the right with 50 N of force, while being pushed to the left with 20 N of force?
11. What two quantities are necessary for a vector quantity?

3.5 The Equilibrium Rule

12. Name the force that occurs in a rope when both ends are pulled in opposite directions.
13. How much tension is there in a vertical rope that holds a 20-N bag of apples at rest?
14. What does $\Sigma F = 0$ mean?

3.6 Support Force

15. Why is the support force on an object often called the normal force?
16. When you weigh yourself, how does the support force of the scale acting on you compare with the gravitational force between you and Earth?

3.7 Equilibrium of Moving Things

17. A bowling ball sits at rest. Another ball rolls down a lane at constant speed. Which, if either, is in equilibrium? Defend your answer.
18. If we push a crate at constant velocity, how do we know how much friction acts on the crate compared to our pushing force?

3.8 The Force of Friction

19. How does the direction of a friction force compare with the velocity of a sliding object?
20. If you push to the right on a heavy crate and it slides, what is the direction of friction on the crate?
21. Suppose you push to the right on a heavy crate, but not hard enough to make it slide. Does a friction force act on the crate?

3.9 Speed and Velocity

22. Distinguish between speed and velocity.
23. Why do we say velocity is a vector and speed is not?
24. Does the speedometer on a vehicle show average speed or instantaneous speed?
25. How can you be both at rest and also moving at 100,000 km/h at the same time?

3.10 Acceleration

26. Distinguish between velocity and acceleration.
27. What is the acceleration of an object that moves at constant velocity? What is the net force on the object in this case?
28. What is the acceleration of an object in free fall at Earth's surface?

ACTIVE EXPLORATIONS

1. Grandma is interested in your educational progress. Like many grandmothers, she may have little science background and may not be familiar with advanced mathematics. Write a letter to grandma, without using equations, and explain to her the difference between velocity and acceleration. Tell her why some of your classmates confuse the two, and state some examples that clear the confusion. Also consider a letter to grandpa.

2. Stand flatfooted next to a wall and make a mark at the highest point you can reach. Then jump vertically and mark this highest point. The distance between the marks is your vertical jumping distance. Use this to calculate your hang time.
3. Using any method you choose, determine both your walking speed and your running speed.
4. Go further than the previous activity, and try walking across a room with a constant acceleration. (Not so easy!)

ONE-STEP CALCULATIONS

These are "plug-in-the-number" type activities to familiarize you with the equations that link the concepts of physics. They are mainly one-step substitutions and are less challenging than the Problems.

$$\text{Speed} = \frac{\text{distance}}{\text{time}}$$

1. Calculate your walking speed when you step 1 meter in 0.5 second.
2. Calculate the speed of a bowling ball that travels 4 meters in 2 seconds.

$$\text{Average speed} = \frac{\text{total distance covered}}{\text{time interval}}$$

3. Calculate your average speed if you run 50 meters in 10 seconds.
4. Calculate the average speed of a tennis ball that travels the full length of the court, 24 meters, in 0.5 second.
5. Calculate the average speed of a cheetah that runs 140 meters in 5 seconds.
6. Calculate the average speed (in km/h) of Larry, who runs to the store 4 kilometers away in 30 minutes.

$$\text{Distance} = \text{average speed} \times \text{time}$$

7. Calculate the distance (in km) that Larry runs if he maintains an average speed of 8 km/h for 1 hour.
8. Calculate the distance you will travel if you maintain an average speed of 10 m/s for 40 seconds.
9. Calculate the distance you will travel if you maintain an average speed of 10 km/h for one-half hour.

$$\text{Acceleration} = \frac{\text{change of velocity}}{\text{time interval}}$$

10. Calculate the acceleration of a car (in km/h · s) that can go from rest to 100 km/h in 10 s.
11. Calculate the acceleration of a bus that goes from 10 km/h to a speed of 50 km/h in 10 seconds.
12. Calculate the acceleration of a ball that starts from rest and rolls down a ramp and gains a speed of 25 m/s in 5 seconds.
13. On a distant planet, a freely falling object gains speed at a steady rate of 20 m/s during each second of fall. Calculate its acceleration.

Instantaneous speed = acceleration × time

14. Calculate the instantaneous speed (in m/s) at the 10-second mark for a car that accelerates at 2 m/s² from a position of rest.
15. Calculate the speed (in m/s) of a skateboarder who accelerates from rest for 3 seconds down a ramp at an acceleration of 5 m/s².

Velocity acquired in free fall, from rest:
$v = gt$ (where $g = 10$ m/s²)

16. Calculate the instantaneous speed of an apple that falls freely from a rest position and accelerates at 10 m/s² for 1.5 seconds.

17. An object is dropped from rest and falls freely. After 7 seconds, calculate its instantaneous speed.
18. A skydiver steps from a high-flying helicopter. In the absence of air resistance, how fast would she be falling at the end of a 12-second jump?

Distance fallen in free fall, from rest: $d = 1/2\, gt^2$

19. A coconut drops from a tree and hits the ground in 1.5 seconds. Calculate how far it falls.
20. Calculate the vertical distance that a coconut dropped from rest covers in 12 seconds of free fall.

EXERCISES

Please do not be intimidated by the large number of exercises in this book. If your course work is to cover many chapters, your instructor will likely assign only a few exercises from each.

1. A bowling ball rolling along a lane gradually slows as it rolls. How would Aristotle interpret this observation? How would Galileo interpret it?
2. What Aristotelian idea did Galileo discredit in his fabled Leaning Tower of Pisa experiment? With his inclined plane experiments?
3. When a ball rolls down an inclined plane, it gains speed because of gravity. When rolling up, it loses speed because of gravity. Why doesn't gravity play a role when it rolls on a horizontal surface?
4. What physical quantity is a measure of how much inertia an object has?
5. Which has more mass, a 2-kg fluffy pillow or a 3-kg small piece of iron? More volume? Why are your answers different?
6. Does a dieting person more accurately lose mass or lose weight?
7. A favorite class demonstration by the author is lying on his back with a blacksmith's anvil placed on his chest. When an assistant whacks the anvil with a strong sledge-hammer blow, Hewitt is not hurt. How is the physics here similar to that illustrated in Figure 3.8?
8. What is your own mass in kilograms? Your weight in newtons?
9. Gravitational force on the Moon is only 1/6 that of the gravitational force on Earth. What would be the weight of a 10-kg object on the Moon and on Earth? What would its mass be on the Moon and on Earth?
10. Consider a pair of forces, one with a magnitude of 25 N and the other with a magnitude of 15 N. What maximum net force is possible for these two forces? What minimum net force is possible?

11. The sketch shows a painter's scaffold in mechanical equilibrium. The person in the middle weighs 250 N, and the tensions in each rope are 200 N. What is the weight of the scaffold?
12. A different scaffold that weighs 300 N supports two painters, one weighing 250 N and the other weighing 300 N. The reading in the left scale is 400 N. What should the reading in the right scale be?

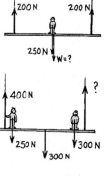

13. Nellie Newton hangs at rest from the ends of the rope as shown. How does the reading on the scale compare to her weight?
14. Harry the painter swings year after year from his bosun's chair. His weight is 500 N, and the rope, unknown to him, has a breaking point of 300 N. Why doesn't the rope break when he is supported as shown below? One day, Harry was painting near a flag-pole, and, for a change, he tied the free end of the rope to the flagpole instead of to his chair, as shown to the right. Why did Harry end up taking his vacation early?

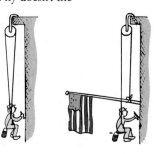

15. A hockey puck slides across the ice at a constant velocity. Is it in mechanical equilibrium? Why or why not?
16. If you push horizontally on a crate and it slides across the floor, slightly gaining speed, how does the friction acting on the crate compare with your push?

17. When you place a heavy book on a table, the table pushes up on the book. Why doesn't this upward push cause the book to rise from the table?

18. An empty jug of weight W rests on a table. What is the support force exerted on the jug by the table? What is the support force when water of weight w is poured into the jug?

19. In order to slide a heavy cabinet across the floor at constant speed, you exert a horizontal force of 600 N. Is the force of friction between the cabinet and the floor greater than, less than, or equal to 600 N? Defend your answer.

20. Consider a crate at rest on a factory floor. As a pair of workers starts to lift it, does the support force on the crate provided by the floor increase, decrease, or remain unchanged? What happens to the support force on the workers' feet?

21. Correct your friend who says, "The dragster rounded the curve at a constant velocity of 100 km/h."

22. What is the impact speed when a car moving at 100 km/h bumps into the rear of another car traveling in the same direction at 98 km/h?

23. Harry Hotshot can paddle a canoe in still water at 8 km/h. How successful will he be at canoeing upstream in a river that flows at 8 km/h?

24. A destination 120 miles away is posted on a highway sign, and the speed limit is 60 miles/hour. If you drive at the posted speed, will you reach the destination in 2 hours? Or will you reach it in more than 2 hours?

25. Suppose that a freely falling object were somehow equipped with a speedometer. By how much would its speed reading increase with each second of fall?

26. Suppose that the freely falling object in the preceding exercise were also equipped with an odometer. Would the readings of distance fallen each second indicate equal or unequal distances of fall for successive seconds? Explain.

27. When a ballplayer throws a ball straight up, by how much does the speed of the ball decrease each second while moving upward? In the absence of air resistance, by how much does its speed increase each second while it is moving downward? How much time is required while moving

upward? How much time is required while moving downward?

28. Someone standing at the edge of a cliff (as in Figure 3.24) throws a ball straight up at a certain speed and another ball straight down with the same initial speed. If air resistance is negligible, which ball has the greater speed when it strikes the ground below?

29. What is the acceleration of a car that moves at a steady velocity of 100 km/h for 100 seconds? Explain your answer, and state why this question is an exercise in careful reading as well as in physics.

30. For a freely falling object dropped from rest, what is its acceleration at the end of the fifth second of fall? At the end of the tenth second? Defend your answers (and distinguish between velocity and acceleration).

31. Two balls, A and B, are released simultaneously from rest at the left end of the equal-length tracks A and B, as shown. Which ball will reach the end of its track first?

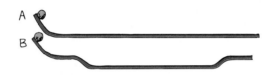

32. Refer to the tracks above.
   a. Does ball B roll faster along the lower part of track B than ball A rolls along track A?
   b. Is the speed gained by ball B going down the extra dip the same as the speed it loses going up near the right-hand end—and doesn't this mean that the speed of balls A and B will be the same at the ends of both tracks?
   c. On track B, won't the average speed dipping down and up be greater than the average speed of ball A during the same time?
   d. So, overall, does ball A or ball B have the greater average speed? (Do you wish to change your answer to the previous exercise?)

PROBLEMS

● **BEGINNER** ■ **INTERMEDIATE** ◆ **EXPERT**

1. ● Find the net force produced by a 30-N force and a 20-N force in each of the following cases:
   a. Both forces act in the same direction.
   b. The two forces act in opposite directions.

2. ● A horizontal force of 100 N is required to push a box across a floor at a constant velocity.
   a. What is the net force acting on the box?
   b. How much is the friction force that acts on the box?

3. ● A firefighter with a mass of 100 kg slides down a vertical pole at a constant speed. What is the force of friction provided by the pole?

4. ● The ocean's level is currently rising at about 1.5 mm per year. If this rate remained the same, not increasing, show that it would take 2000 years for the sea level to be 3 meters higher than it is now.

5. ● A vehicle changes its velocity from 100 km/h to a dead stop in 10 s. Show that the acceleration in stopping is -10 km/h · s.

6. ● Extend Table 3.2 (which gives values of from 0 to 5 s) to 10 s, assuming no air resistance.

7. ● A ball is thrown straight up with an initial speed of 30 m/s.
   a. Show that the time it takes to reach the top of its trajectory will be 3 s.
   b. Show that it will reach a height of 45 m (neglecting air resistance).

8. ■ A ball is thrown straight up with enough speed so that it is in the air for several seconds.
   a. What is the velocity of the ball when it reaches its highest point?

b. What is its velocity 1 s before it reaches its highest point?

c. What is the change in its velocity during this 1-s interval?

d. What is its velocity 1 s after it reaches its highest point?

e. What is the change in velocity during this 1-s interval?

f. What is the change in velocity during the 2-s interval from 1 s before the highest point to 1 s after the highest point? (Caution: we are asking for velocity, not speed.)

g. What is the acceleration of the ball during any of these time intervals and at the moment the ball has zero velocity?

9. ◆ Starting from rest, the change in an object's velocity $= at$. That is, $v_f - v_0 = at$, or

$$t = \frac{v_f - v_0}{a}.$$

The distance traveled by an object is given by $d = v_{ave}t$, where

$$v_{ave} = \frac{v_f + v_0}{2}.$$

Begin with $d = v_{ave}t$ and with appropriate substitutions show that

$$d = \frac{v_f^2 - v_0^2}{2a}.$$

Note that this equation does not include time, so it's a good one to use when time is not given in a problem!

10. ◆ An electrically charged particle accelerates uniformly from rest to speed v while traveling a distance x.

a. Show that the acceleration of the particle is $a = \frac{v^2}{2x}$.

b. If the particle starts from rest and reaches a speed of 1.8×10^7 m/s over a distance of 0.10 m, show that its acceleration is 1.6×10^{15} m/s^2.

CHAPTER 3 ONLINE RESOURCES

The Physics Place

Interactive Figures
3.21, 3.23, 3.24

Tutorials
Parachuting and Newton's Second Law

Videos
Newton's Law of Inertia
The Old Tablecloth Trick
Toilet Paper Roll
Inertia of a Cylinder
Inertia of an Anvil
Definition of a Newton

Friction
Definition of Speed
Average Speed
Velocity
Changing Velocity
Definition of Acceleration
Force Causes Acceleration
Numerical Example of Acceleration

Quiz

Flashcards

Links

CHAPTER 4

Newton's Laws of Motion

Darlene Librero pulls with one finger; Paul Doherty pulls with both hands. This is the question they ask their Exploratorium class: "Who exerts more force on the scale?"

Galileo's work set the stage for Isaac Newton, who was born shortly after Galileo's death in 1642. By the time Newton was 23, he had developed his famous three laws of motion that completed the overthrow of Aristotelian physics. These three laws first appeared in one of the most famous books of all time, Newton's *Philosophiae Naturalis Principia Mathematica*,* often simply known as the *Principia*. The first law is a restatement of Galileo's concept of inertia; the second law relates acceleration to its cause—force; and the third is the law of action and reaction.

* Translated from the Latin: "Mathematical Principles of Natural Philosophy." See Newton's biography on page 71.

4.1 Newton's First Law of Motion

Newton's first law, usually called the **law of inertia**, is a restatement of Galileo's idea.

Every object continues in a state of rest or of uniform speed in a straight line unless acted on by a nonzero force.

> You can think of inertia as another word for *laziness* (or resistance to change).

The key word in this law is *continues*: an object *continues* to do whatever it happens to be doing unless a force is exerted upon it. If the object is at rest, it *continues* in a state of rest. This is nicely demonstrated when a tablecloth is skillfully whipped from beneath dishes sitting on a tabletop, leaving the dishes in their initial state of rest.** On the

** Close inspection shows that brief friction between the dishes and fast-moving tablecloth starts the dishes moving, but then friction between the dishes and table stops the dishes before they slide very far. If you try this, use unbreakable dishes!

FIGURE 4.1
INTERACTIVE FIGURE

Inertia in action.

other hand, if an object is moving, it *continues* to move without changing its speed or direction, as evidenced by a sliding puck on a laboratory air track, or space probes that continually move in outer space. This property of objects to resist changes in motion is called **inertia**.

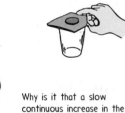

Why will the coin drop into the glass when a force accelerates the card?

Why is it that a slow continuous increase in the downward force breaks the string above the massive ball, but a sudden increase breaks the lower string?

Why does the downward motion and sudden stop of the hammer tighten the hammerhead?

FIGURE 4.2

Examples of inertia.

FIGURE 4.3

Rapid deceleration is sensed by the driver who lurches forward—inertia in action!

STOP AND CHECK YOURSELF

When a space shuttle travels in a circular orbit around Earth, is a force required to maintain its high speed? If suddenly the force of gravity were cut off, what type of path would the shuttle follow?

CHECK YOUR ANSWERS

No force in the direction of the shuttle's motion exists. The shuttle "coasts" by its own inertia. The only force acting on it is the force of gravity, which acts at right angles to its motion (toward Earth's center). We'll see later that this right-angled force holds the shuttle in a circular path. If it were cut off, the shuttle would move in a straight-line path at constant speed (constant velocity).

The Moving Earth

In 1543, the Polish astronomer Copernicus caused great controversy when he published a book proposing that Earth revolves around the Sun.* This idea conflicted with the popular view that Earth was the center of the universe. Copernicus's concept of a Sun-centered solar system was the result of years of studying the motion of the planets. He had kept his theory from the public—for two reasons. The first reason was that he feared persecution; a theory so completely different from common opinion would surely be taken as an attack on established order. The second reason was reservations about it himself; he could not reconcile the idea of a moving Earth with the prevailing ideas of motion. The concept of inertia was unknown to him and others of his time. In the final days of his life, at the urging of close friends, he sent his manuscript, *De Revolutionibus Orbium Coelestium,*** to the printer. The first copy

* Copernicus was certainly not the first to think of a Sun-centered solar system. In the fifth century, for example, the Indian astronomer Aryabhatta taught that Earth circles the Sun, not the other way around (as most of the world believed).

** Translated from the Latin: "On the Revolutions of Heavenly Spheres."

of his famous exposition reached him on the day he died—May 24, 1543.

The idea of a moving Earth was much debated. Europeans thought as Aristotle had, and the existence of a force big enough to keep Earth moving was beyond their imagination. They had no idea of the concept of inertia. One of the arguments against a moving Earth was the following:

Consider a bird sitting at rest atop a branch on a tall tree. On the ground below is a fat, juicy worm. The bird sees the worm and drops vertically below and catches it. It was argued that this would be impossible if Earth were moving. A moving Earth would have to travel at an enormous speed to circle the Sun in one year. While the bird descends from its branch to the ground below, the worm would be swept far away along with the moving Earth. It seemed that catching a worm on a moving Earth would be an impossible task. The fact that birds do catch worms from tree branches seemed to be clear evidence that Earth must be at rest.

> Force *changes* motion, it doesn't *cause* motion.

Can you see the mistake in this argument? You can if you use the concept of inertia. You see, not only is Earth moving at a great speed, but so are the tree, the branch of the tree, the bird that sits on it, the worm below, and even the air in between. Things in motion remain in motion if no unbalanced forces are acting upon them. So when the bird drops from the branch, its initial sideways motion remains unchanged. It catches the worm quite unaffected by the motion of its total environment.

We live on a moving Earth. If you stand next to a wall and jump up so that your feet are no longer in contact with the floor, does the moving wall slam into you? Why not? It doesn't because you are also traveling at the same speed as Earth, before, during, and after your jump. The speed of Earth relative to the Sun is not the speed of the wall relative to you.

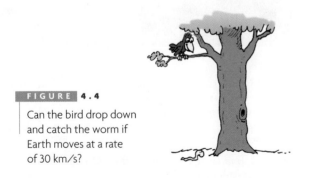

Can the bird drop down and catch the worm if Earth moves at a rate of 30 km/s?

When you flip a coin in a high-speed airplane, it behaves as if the airplane were at rest. The coin keeps up with you—inertia in action!

Four hundred years ago, people had difficulty with ideas such as these. One reason is that they didn't yet travel in high-speed vehicles. Rather, they experienced slow, bumpy rides in horse-drawn carts. People were less aware of the effects of inertia. Today we flip a coin in a high-speed car, bus, or plane and catch the vertically moving coin as we would if the vehicle were at rest. We see evidence for the law of inertia when the horizontal motion of the coin prior, during, and after the catch is the same. The coin always keeps up with us.

4.2 Newton's Second Law of Motion

Isaac Newton was the first to realize the connection between force and mass in producing acceleration, which is one of the most central rules of nature. He expressed it in his *second law of motion*. **Newton's second law** states:

The acceleration produced by a net force on an object is directly proportional to the net force, is in the same direction as the net force, and is inversely proportional to the mass of the object.

Or, in shorter notation,

$$\text{Acceleration} \sim \frac{\text{net force}}{\text{mass}}$$

By using consistent units such as newtons (N) for force, kilograms (kg) for mass, and meters per second squared (m/s^2) for acceleration, we produce the exact equation:

$$\text{Acceleration} = \frac{\text{net force}}{\text{mass}}$$

Here's directly proportional.

Here's inversely proportional.

In briefest form, where a is acceleration, F_{net} is net force, and m is mass:

$$a = \frac{F_{net}}{m}$$

When we use the symbol F to mean net force, we can shorten this to:

$$a = \frac{F}{m}$$

Acceleration equals the net force divided by the mass. If the net force acting on an object is doubled, the object's acceleration will be doubled. Suppose instead that the mass is doubled. Then the

FIGURE 4.6
INTERACTIVE FIGURE

Acceleration depends on both the amount of push and the mass being pushed.

Force of hand accelerates the brick

Twice as much force produces twice as much acceleration

Twice the force on twice the mass gives the same acceleration

FIGURE 4.7

Acceleration is directly proportional to force.

acceleration will be halved. If both the net force and the mass are doubled, then the acceleration will be unchanged. (These relations are nicely developed in the *Practice Book for Conceptual Physics Fundamentals*.)

STOP AND
CHECK YOURSELF

1. In the previous chapter we defined acceleration to be the time rate of change of velocity; that is, $a = \dfrac{\text{change in } v}{\text{time}}$. Are we now saying that acceleration is instead the ratio of force to mass—that is, $a = \dfrac{F}{m}$? Which is it?

2. A jumbo jet cruises at constant velocity of 1000 km/h when the thrusting force of its engines is a constant 100,000 N. What is the acceleration of the jet? What is the force of air resistance on the jet?

3. Suppose you apply the same amount of force to two separate carts, one cart with a mass of 1 kg, and the other with a mass of 2 kg. Which cart will accelerate more, and how much greater will the acceleration be?

CHECK YOUR ANSWERS

1. Both. Acceleration is *defined* as the time rate of change of velocity and is *produced by* a force. How much force/mass (usually the cause) determines the rate change in velocity/time (usually the effect). So we must first define acceleration and then define the terms that produce acceleration.

2. The acceleration is zero, as evidenced by the constant velocity. Because the acceleration is zero, it follows from Newton's second law that the net force is zero, which means that the force of air resistance must just equal the thrusting force of 100,000 N and act in the opposite direction. So the air resistance on the jet is 100,000 N. This is in accord with $\Sigma F = 0$. (Note that we don't need to know the velocity of the jet to answer this question, but only that it is constant—our clue that acceleration, and therefore net force, is zero.)

3. The 1-kg cart will have more acceleration—twice as much, in fact—because it has half as much mass—which means half as much resistance to a change in motion.

FIGURE 4.8

Acceleration is inversely proportional to mass.

Force of hand accelerates the brick

The same force accelerates 2 bricks ¹/₂ as much

3 bricks, ¹/₃ as much acceleration

FIGURE 4.9

When you accelerate in the direction of your velocity, you speed up; against your velocity, you slow down; at an angle to your velocity, your direction changes.

When Acceleration Is *g*—Free Fall

Although Galileo founded both the concepts of inertia and acceleration, and was the first to measure the acceleration of falling objects, he was unable to explain why objects of various masses fall with equal accelerations. Newton's second law provides the explanation.

The Physics Place

Parachuting and Newton's Second Law

The Physics Place

Newton's Second Law
Free-Fall Acceleration Explained
Free Fall: How Fast?
Free Fall: How Far?
V = *gt*
Air Resistance and Falling Objects
Falling and Air Resistance

We know that a falling object accelerates toward Earth because of the gravitational force of attraction between the object and Earth. As mentioned earlier, when the force of gravity is the only force—that is, when air resistance is negligible—we say that the object is in a state of **free fall**. An object in free fall accelerates toward Earth at 10 m/s² (or, more precisely, at 9.8 m/s²).

The greater the mass of an object, the stronger is the gravitational pull between it and Earth. The double brick in Figure 4.10, for example, has twice the gravitational attraction as the single brick. Why,

FIGURE 4.10

The ratio of weight (*F*) to mass (*m*) is the same for all objects in the same locality; hence, their accelerations are the same in the absence of air resistance.

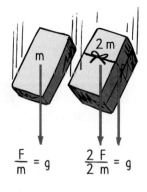

$$\frac{F}{m} = g \qquad \frac{2\,F}{2\,m} = g$$

then, doesn't the double brick fall twice as fast (as Aristotle supposed it would)? The answer is evident in Newton's second law: the acceleration of an object depends not only on the force (weight, in this case), but on the object's resistance to motion—its inertia. Whereas a force produces an acceleration, inertia is a *resistance* to acceleration. So twice the force exerted on twice the inertia produces the same acceleration as half the force exerted on half the inertia. Both accelerate equally. The acceleration due to gravity is symbolized by *g*. We use the symbol *g*, rather than *a*, to denote that acceleration is due to gravity alone.

When Galileo tried to explain why all objects fall with equal accelerations, wouldn't he have loved to know the rule *a* = *F/m*?

The ratio of weight to mass for freely falling objects equals the constant *g*. This is similar to the constant ratio of circumference to diameter for circles, which equals the constant π. The ratio of weight to mass is identical for both heavy and light objects, just as the ratio of circumference to diameter is the same for both large and small circles (Figure 4.11).

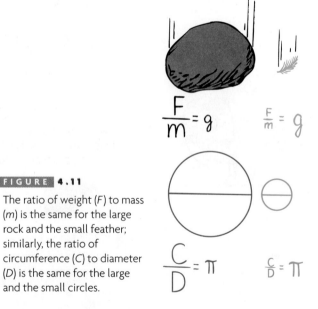

FIGURE 4.11

The ratio of weight (*F*) to mass (*m*) is the same for the large rock and the small feather; similarly, the ratio of circumference (*C*) to diameter (*D*) is the same for the large and the small circles.

$$\frac{F}{m} = g \qquad \frac{F}{m} = g$$

$$\frac{C}{D} = \pi \qquad \frac{C}{D} = \pi$$

PROBLEM SOLVING

■ SAMPLE PROBLEM SOLVING

Almost everywhere in physics you see equations, which foremost are guides to thinking. Equations are also useful in solving physics problems. When solving problems, you should think first in terms of the concepts and their symbols. The symbols in equations guide the way. Your solution is a reexpression of these and other symbols. Often you'll want to express an answer in numbers. Through numbers you learn about units and magnitudes and uncertainties, and you get a feel for what is "large" and what is "small." So after you come to a solution expressed in symbols, bring in the numbers. Consider this pair of sample problems.

Problems

1. A force F acts in the forward direction on a cart of mass m. A friction force f opposes this motion.

a. Use Newton's second law and show that the acceleration of the cart is $\dfrac{F - f}{m}$.

b. If the cart's mass is 4.0 kg, the applied force is 12.0 N, and the friction force is 6.0 N, show that the cart's acceleration is 1.5 m/s².

2. A rock band's tour bus, mass M, is accelerating from a stop sign at rate a when a chunk of heavy metal, mass $M/5$, falls off of the bus. The applied force acting on the bus remains unchanged.

a. Show that the bus's acceleration is then $\tfrac{5}{4}a$.

b. If the initial acceleration of the bus is 1.2 m/s², show that the acceleration of the bus will be 1.5 m/s² when it no longer carries the heavy metal.

Solutions

1. a. We're asked to find the acceleration. From Newton's second law we know that $a = \dfrac{F_{net}}{m}$. Here the net force is $F - f$. So the solution is $a = \dfrac{F - f}{m}$ (where all quantities represented are known values). Notice that this answer applies to

all situations in which a steady applied force is opposed by a steady frictional force. It covers many possibilities.

b. Here we simply substitute the numerical values given:

$$a = \frac{F - f}{m} = \frac{12.0\,\text{N} - 6.0\,\text{N}}{4.0\,\text{kg}}$$
$$= 1.5\frac{\text{N}}{\text{kg}} = 1.5\ \text{m/s}^2.$$

(The units N/kg are equivalent to m/s².) Note that the answer, about 15 percent of g, is reasonable.

2. a. Again, we're asked to find the acceleration. From Newton's second law we know that $a = \dfrac{F_{net}}{m}$, so $F_{net} = ma$. Before the chunk of metal falls off, the mass of the bus was M, so the net force was Ma, the mass of the bus multiplied by its acceleration. We're told that this same force acts after the metal falls off, so the final acceleration is

$$a = \frac{\text{the same force}}{\text{the new mass}} = \frac{Ma}{M - M/5}$$
$$= \frac{Ma}{\left(\dfrac{5M - M}{5}\right)} = \frac{Ma}{\left(\dfrac{4M}{5}\right)} = \frac{5Ma}{4M} = \frac{5}{4}a.$$

It makes sense that the acceleration after the metal falls off is greater than it was initially.

b. Again, here we simply substitute the numerical values given:

$$\text{New acceleration} = \frac{5}{4}a = \frac{5}{4}1.2\ \text{m/s}^2$$
$$= 1.5\ \text{m/s}^2,$$

again a reasonable answer. All the physics in both problems occurs in part (a). The focus is on concepts and reasoning, not on numbers. In part (b) the answer is found by substituting numerical values in the solution to part (a). Judgment about units of measurement and significant figures, and the reasonableness of an answer, can be employed in part (b). For information on units of measurement and significant figures, see Appendix A.

We now understand that the acceleration of free fall is independent of an object's mass. A boulder 100 times more massive than a pebble falls at the same acceleration as the pebble because although the force on the boulder (its weight) is 100 times

greater than the force (or weight) on the pebble, its resistance to a change in motion (mass) is 100 times that of the pebble. The greater force offsets the correspondingly greater mass.

FIGURE 4.12

In a vacuum, a feather and a coin fall at an equal acceleration.

STOP AND
CHECK YOURSELF

In a vacuum, a coin and a feather fall equally, side by side. Would it be correct to say that *equal forces of gravity* act on both the coin and the feather when in a vacuum?

CHECK YOUR ANSWER

No, no, no—a thousand times no! These objects accelerate equally not because the forces of gravity on them are equal, but because the *ratios* of their weights to masses are equal. Although air resistance is not present in a vacuum, gravity is. (You'd know this if you placed your hand into a vacuum chamber and a cement truck rolled over it!) If you answered yes to this question, let this be a signal to be more careful when you think physics!

The speed of a vertically thrown ball at the top of its path is zero. Is the acceleration there zero also? (Answer begins with an N.)

When Acceleration of Fall Is Less Than *g*—Non-Free Fall

Most often, air resistance is not negligible for falling objects. Then the acceleration of fall is less. Air resistance depends primarily on two things: speed and surface area. When a skydiver steps from a high-flying plane, the air resistance on the skydiver's body builds up as the falling speed increases. The result is reduced acceleration. The acceleration can be reduced further by increasing

surface area. A skydiver does this by orienting his or her body so more air is encountered—by spreading out like a flying squirrel. So air resistance depends on speed and the frontal area encountered by the air.

In free fall, only a single force acts—the force of gravity. Whenever the force of air resistance also occurs, the falling object is not in free fall.

For free fall, the downward net force is weight—only weight. But when air is present, the downward net force = weight − air resistance. Can you see that the presence of air resistance reduces net force? And that less net force means less acceleration? So as a skydiver falls faster and faster, the acceleration of fall becomes less and less.* What happens to the net force if air resistance builds up to equal the weight of the diver? The answer is that net force becomes zero. Here we see $\Sigma F = 0$ again! Then acceleration becomes zero. Does this mean the skydiver comes to a stop? No! What it means is the skydiver no longer gains speed. Acceleration terminates—it no longer occurs. We say the skydiver has reached **terminal speed**. If we are concerned with direction—down, for falling objects—we say the diver has reached **terminal velocity**.

Terminal speed for a human skydiver varies from about 150 to 200 km/h, depending on weight, size, and orientation of the body. A heavier person has to fall faster for air resistance to balance weight.** The greater weight is more effective in "plowing through" air, resulting in more terminal speed for a heavier person. Increasing frontal area reduces terminal speed. That's where a parachute is useful. A parachute increases frontal area, which greatly

* In mathematical notation,

$$a = \frac{F_{\text{net}}}{m} = \frac{mg - R}{m}$$

where *mg* is the weight and *R* is the air resistance. Note that when $R = mg$, $a = 0$; then, with no acceleration, the object falls at constant velocity. With elementary algebra we proceed to another step and get

$$a = \frac{F_{\text{net}}}{m} = \frac{mg - R}{m} = g - \frac{R}{m}$$

We see that the acceleration *a* will always be less than *g* if air resistance *R* impedes falling. Only when $R = 0$ does $a = g$.

** A skydiver's air resistance is proportional to speed squared.

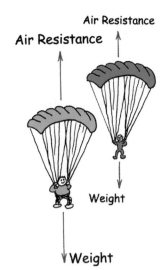

A skydiver jumps from a high-flying helicopter. As she falls faster and faster through the air, does her acceleration increase, decrease, or remain the same?

CHECK YOUR ANSWER

Acceleration decreases because the net force on her decreases. Net force is equal to her weight minus her air resistance, and since air resistance increases with increasing speed, net force and hence acceleration decrease. By Newton's second law,

$$a = \frac{F_{net}}{m} = \frac{mg - R}{m}$$

where mg is her weight and R is the air resistance she encounters. As R increases, both net force and a decrease. Note that if she falls fast enough so that $R = mg$, $a = 0$, then, with no acceleration, she falls at constant speed.

FIGURE 4.13

The heavier parachutist must fall faster than the lighter parachutist for air resistance to cancel her greater weight.

FIGURE 4.14

A flying squirrel increases its frontal area when it jumps.

increases air resistance, reducing the terminal speed to a safe 15 to 25 km/h.

Consider the interesting demonstration of the falling coin and feather in the glass tube (Figure 4.12). When air is inside, we see that the feather falls more slowly due to air resistance. The feather's weight is very small, so it reaches terminal speed very quickly. Can you see that it doesn't have to fall

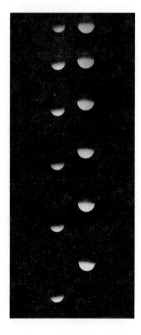

FIGURE 4.15

A stroboscopic study of a golf ball (*left*) and a Styrofoam ball (*right*) falling in air. The air resistance is negligible for the heavier golf ball, and its acceleration is nearly equal to *g*. Air resistance is not negligible for the lighter Styrofoam ball, which reaches its terminal velocity sooner.

very far or fast before air resistance builds up to equal its small weight? The coin, on the other hand, doesn't have enough time to fall fast enough for air resistance to build up to equal its weight. Interestingly, if you drop a coin from a very high location, such as off the top of a tall building, its terminal speed would be reached when the speed of the coin would be about 200 km/h. This is a much, much higher terminal speed than that of a falling feather!

STOP AND CHECK YOURSELF

Consider two parachutists, a heavy person and a light person, who jump from the same altitude with parachutes of the same size.

1. Which person reaches terminal speed first?
2. Which person has the greater terminal speed?
3. Which person reaches the ground first?
4. If there were no air resistance, like on the Moon, how would your answers to these questions differ?

CHECK YOUR ANSWERS

To answer these questions think of a coin and a feather falling in air.

1. Just as a feather reaches terminal speed very quickly, the lighter person reaches terminal speed first.
2. Just as a coin falls faster than a feather through air, the heavy person falls faster and reaches a terminal speed that is faster.
3. Just like the race between a falling coin and feather, the heavier person falls faster and will reach the ground first.
4. If there were no air resistance there would be no terminal speed at all. Both would be in free fall and hit the ground at the same time.

fyi Depending on the size and weight of packages dropped from airplanes, 160 km/h (100 miles per hour) is a typical terminal speed. That's about how fast a pitched baseball travels, or almost as fast as a tennis ball is served. Objects such as bags of rice and flour can survive this terminal speed, so parachutes are seldom used. In fact, parachutes are not used when dropping food supplies to citizens in the midst of an army whose troops would confiscate the supplies.

When Galileo allegedly dropped objects of different weights from the Leaning Tower of Pisa, they didn't actually hit at the same time. They almost did, but because of air resistance, the heavier one hit a split second before the other. But this contradicted the much longer time difference expected by the followers of Aristotle. The behavior of falling objects was never really understood until Newton announced his second law of motion.

4.3 Forces and Interactions

So far, we've treated force in its simplest sense—as a push or pull. In a broader sense, a force is not a thing in itself but makes up an **interaction** between one thing and another. If you push on a wall with your fingers, more is happening than you just pushing on the wall. While you're interacting with the wall, the wall also pushes on you. The fact that your fingers and the wall push on each other is evident in your bent fingers (Figure 4.16). These two forces are equal in magnitude (amount) and opposite in direction. This **force pair** makes up a single interaction. In fact, you can't push on the wall unless the wall pushes back. There is a pair of forces involved: your push on the wall and the wall's push back on you.*

The Physics Place
Forces and Interactions

In Figure 4.17, we see a boxer's fist hitting a massive punching bag. The fist hits the bag (and dents it) while the bag hits back on the fist (and stops its motion). This force pair is fairly large. But what if the boxer were hitting a piece of tissue paper? The

FIGURE 4.16
INTERACTIVE FIGURE

When you lean against a wall, you exert a force on the wall. The wall simultaneously exerts an equal and opposite force on you. Hence you don't topple over.

* We tend to think of only living things pushing and pulling. But inanimate things can do likewise. So please don't be troubled about the idea of the inanimate wall pushing on you. It does, just as another person leaning against you would.

FIGURE 4.17

The boxer can hit the massive bag with considerable force. But with the same punch he can exert only a tiny force on the tissue paper in midair.

boxer's fist can exert only as much force on the tissue paper as the tissue paper can exert on the boxer's fist. Furthermore, the fist can't exert any force at all unless what is being hit exerts the same amount of reaction force. An interaction requires a *pair* of forces acting on *two* objects.

Can a boxer hurt his hand when punching a piece of tissue paper?

When a hammer hits a stake and drives it into the ground, the stake exerts an equal amount of force on the hammer that brings it to an abrupt halt. And when you pull on a cart and it accelerates, the cart pulls back on you, as evidenced perhaps by the tightening of the rope wrapped around your hand. One thing interacts with another; the hammer interacts with the stake, and you interact with the cart.

Which exerts the force and which receives the force? Isaac Newton's answer to this was that neither force has to be identified as "exerter" or "receiver," and concluded both objects must be treated equally. For example, when the hammer exerts a force on the stake, it is brought to a halt by the force the stake exerts on the hammer. Both forces are equal and oppositely directed. When you pull the cart, the cart simultaneously pulls on you. This pair of forces, your pull on the cart and the cart's pull on you, makes up the single interaction between you and the cart. Such observations led Newton to his third law of motion.

4.4 Newton's Third Law of Motion

Newton's third law states:

Whenever one object exerts a force on a second object, the second object exerts an equal and opposite force on the first.

We can call one force the *action force*, and the other the *reaction force*. Then we can express Newton's third law in the form:

To every action there is always an opposed equal reaction.

It doesn't matter which force we call *action* and which we call *reaction*. The important thing is that they are co-parts of a single interaction and that neither force exists without the other. Action and reaction forces are equal in strength and opposite in direction. They occur in pairs and make up one interaction between two things.

When walking, you interact with the floor. Your push against the floor is coupled to the floor's push against you. The pair of forces occurs simultaneously. Likewise, the tires of a car push against the road while the road pushes back on the tires—the tires and road push against each other. In swimming, you interact with the water that you push backward, while the water pushes you forward—you and the water push against each other. The reaction forces are what account for our motion in these cases. These forces depend on friction; a person or car on ice, for example, may not be able to exert the action force to produce the needed reaction force. Neither force exists without the other.

The Physics Place
Newton's Third Law

The Physics Place
Action and Reaction on Different Masses
Action and Reaction on Rifle and Bullet

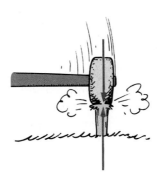

FIGURE 4.18

In the interaction between the hammer and the stake, each exerts the same amount of force on the other.

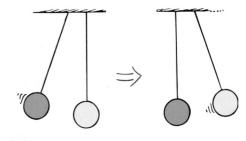

FIGURE 4.19

The impact forces between the blue and yellow ball move the yellow ball and stop the blue ball.

Simple Rule to Identify Action and Reaction

There is a simple rule for identifying action and reaction forces. First, identify the interaction—one thing (object A), interacts with another (object B). Then, action and reaction forces can be stated in the form:

Action: Object A exerts a force on object B.

Reaction: Object B exerts a force on object A.

The rule is easy to remember. If action is A acting on B, reaction is B acting on A. We see that A and B are simply switched around. Consider the case of your hand pushing on the wall. The interaction is between your hand and the wall. We'll say the action is your hand (object A) exerting a force on the wall (object B). Then the reaction is the wall exerting a force on your hand.

> Know that an action force and its reaction force always act on *different* objects. Two external forces acting on the same object, even if they are equal and opposite in direction, *cannot* be an action–reaction pair. That's the law!

Action: tire pushes on road   Reaction: road pushes on tire

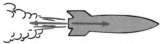

Action: rocket pushes on gas   Reaction: gas pushes on rocket

Action: man pulls on spring   Reaction: spring pulls on man

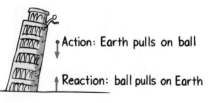

Action: Earth pulls on ball

Reaction: ball pulls on Earth

FIGURE 4.20

Action and reaction forces. Note that when action is "*A* exerts force on *B*," the reaction is then simply "*B* exerts force on *A*."

PRACTICING PHYSICS

Below we see two vectors on the sketch of the hand pushing the wall. The wall also pushes back on the hand. Note the others show only the action force. Draw appropriate vectors showing the reaction forces. Can you specify the action–reaction pairs in each case?

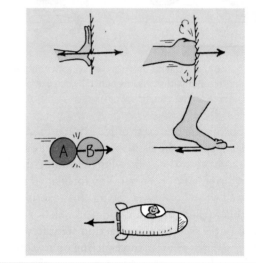

STOP AND CHECK YOURSELF

1. A car accelerates along a road. Identify the force that moves the car.

2. Identify the action and reaction forces for the case of an object in free fall (no air resistance).

CHECK YOUR ANSWERS

1. It is the road that pushes the car along. Really! Except for air resistance, only the road provides a horizontal force on the car. How does it do this? The rotating tires of the car push back on the road (action). The road simultaneously pushes forward on the tires (reaction). How about that!

2. To identify a pair of action–reaction forces in any situation, first identify the pair of interacting objects. In this case Earth interacts with the falling object via the force of gravity. So Earth pulls the falling object downward (call it *action*). Then *reaction* is the falling object pulling Earth upward. It is only because of Earth's enormous mass that you don't notice its upward acceleration.

When pushing my fingers together I see the same discoloration on each of them. Aha — evidence that each experiences the same amount of force!

Which falls toward the other, A or B? Do the accelerations of each relate to their relative masses?

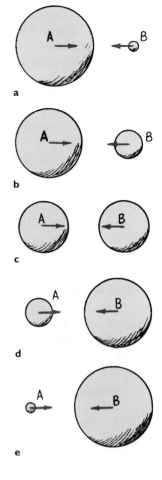

Action and Reaction on Different Masses

Quite interestingly, a falling object pulls upward on Earth with as much force as Earth pulls downward on it. The resulting acceleration of a falling object is evident, while the upward acceleration of Earth is too small to detect.

Consider the exaggerated examples of two planetary bodies, *a* through *e* in Figure 4.22. The forces between bodies A and B are equal in magnitude and oppositely directed in each case. If the acceleration of Planet A is unnoticeable in *a*, then it is more noticeable in *b*, where the difference between the masses is less extreme. In *c*, where both bodies have equal mass, acceleration of Planet A is as evident as it is for B. Continuing, we see the acceleration of A becomes even more evident in *d* and even more so in *e*. So, strictly speaking, when you step off the curb, the street rises ever so slightly to meet you.

When a cannon is fired, there is an interaction between the cannon and the cannonball. The sudden force that the cannon exerts on the cannonball is exactly equal and opposite to the force the cannonball exerts on the cannon. This is why the cannon recoils (kicks). But the effects of these equal forces are very different. This is because the forces act on different masses. Look at this in terms of Newton's second law,

$$a = \frac{F}{m}$$

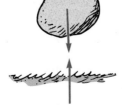

FIGURE **4.21**

Earth is pulled up by the boulder with just as much force as the boulder is pulled downward by Earth.

FIGURE **4.23**

The force exerted against the recoiling cannon is just as great as the force that drives the cannonball along the barrel. Why, then, does the cannonball undergo more acceleration than the cannon?

Let F represent both the action and reaction forces, m the mass of the cannon, and m the mass of the cannonball. Different sized symbols are used to indicate the relative masses and resulting accelerations. Then the acceleration of the cannonball and cannon can be represented in the following way.

$$\text{cannonball:} \frac{F}{m} = a$$

$$\text{cannon:} \frac{F}{\mathit{m}} = a$$

Thus we see why the change in velocity of the cannonball is so large compared with the change in velocity of the cannon. A given force exerted on a small mass produces a large acceleration, while the same force exerted on a large mass produces a small acceleration.

We can extend the idea of a cannon recoiling from the ball it fires to an understanding of rocket propulsion. Consider an inflated balloon recoiling when air is expelled (Figure 4.24). If the air is expelled downward, the balloon accelerates upward. The same principle applies to a rocket, which continually "recoils" from the ejected exhaust gas. Each molecule of exhaust gas is like a tiny cannonball shot from the rocket (Figure 4.25).

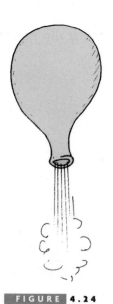

FIGURE 4.24

The balloon recoils from the escaping air and climbs upward.

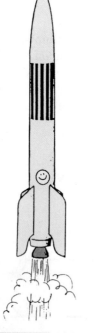

FIGURE 4.25

The rocket recoils from the "molecular cannonballs" it fires and rises.

A common misconception is that a rocket is propelled by the impact of exhaust gases against the atmosphere. In fact, before the advent of rockets, it was commonly thought that sending a rocket to the Moon was impossible. Why? Because there is no air above Earth's atmosphere for the rocket to push against. But this is like saying a cannon wouldn't recoil unless the cannonball had air to push against. Not true! Both the rocket and recoiling cannon accelerate because of the reaction forces by the material they fire—not because of any pushes on the air. In fact, a rocket operates better above the atmosphere where there is no air resistance.

fyi

Gases and fragments shoot out in all directions when a firecracker explodes. When fuel in a rocket burns, a slower explosion, exhaust gases shoot out in one direction.

STOP AND CHECK YOURSELF

1. Which pulls harder, the Moon on Earth, or Earth on the Moon?

2. A high-speed bus and an unfortunate bug have a head-on collision. The force of the bus on the bug splatters it all over the windshield. Is the corresponding force of the bug on the bus greater, less, or the same? Is the resulting deceleration of the bus greater than, less than, or the same as that of the bug?

CHECK YOUR ANSWERS

1. Each pull is the same in magnitude. This is like asking which distance is greater, from Reno to Miami or from Miami to Reno. So we see that Earth and the Moon simultaneously pull on each other, each with the *same* amount of force.

2. The magnitudes of the forces are the same, for they constitute an action–reaction force pair that makes up the interaction between the bus and the bug. The accelerations, however, are remarkably different because the masses are different! The bug undergoes an enormous and lethal deceleration, while the bus undergoes a very tiny deceleration—so tiny that the very slight slowing of the bus is unnoticed by its passengers. But if the bug were more massive, as massive as another bus, for example, the slowing down would be quite apparent.

■ **Tug-of-War**

Perform a tug-of-war between boys and girls on a polished floor that's somewhat slippery, with boys wearing socks and girls wearing rubber-soled shoes. Who will surely win, and why? (Hint: Who wins a tug-of-war, those who pull harder on the rope, or those who push harder against the floor?)

FIGURE 4.27
INTERACTIVE FIGURE

The force on the orange, provided by the apple, is not cancelled by the reaction force on the apple. The orange still accelerates.

Defining Your System

An interesting question often arises: since action and reaction forces are equal and opposite, why don't they cancel to zero? To answer this question we must consider the *system* involved. Consider, for example, a system made up of a single orange, Figure 4.26. The dashed line surrounding the orange encloses and defines the system. The vector that pokes outside the dashed line represents an external force on the system. The system accelerates

FIGURE 4.26
INTERACTIVE FIGURE

A force acts on the orange system and it accelerates to the right.

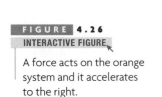

in accord with Newton's second law. In Figure 4.27 we see that this force is provided by an apple, which doesn't change our analysis. The apple is outside the system. The fact that the orange simultaneously exerts a force on the apple, which is external to the system, may affect the apple (another system), but not the orange. You can't cancel a force on the orange with a force on the apple. So, in this case, the action and reaction forces don't cancel.

A system may be as tiny as an atom or as large as the universe.

Now let's consider a larger system, enclosing both the orange and the apple. We see the system bounded by the dashed line in Figure 4.28. Notice that the force pair is *internal* to the orange–apple system. These forces *do* cancel each other. They play no role in accelerating the system. A force external to the

system is needed for acceleration. That's where friction with the floor plays a role (Figure 4.29). When the apple pushes against the floor, the floor simultaneously pushes on the apple—an external force on the system. The system accelerates to the right.

Inside a baseball are trillions and trillions of interatomic forces at play. They hold the ball together, but they play no role in accelerating the ball. Although every one of the interatomic forces is

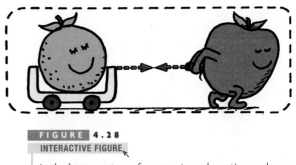

FIGURE 4.28
INTERACTIVE FIGURE

In the larger system of orange + apple, action and reaction forces are internal and do cancel. If these are the only horizontal forces, with no external force, no net acceleration of the system occurs.

FIGURE 4.29
INTERACTIVE FIGURE

An external horizontal force occurs when the floor pushes on the apple (reaction to the apple's push on the floor). The orange–apple system accelerates.

part of an action–interaction pair within the ball, they combine to zero, no matter how many of them there are. A force external to the ball, like batting it, is needed to accelerate it.

If this is confusing, it may be well to note that Newton had difficulties with the third law himself.

STOP AND
CHECK YOURSELF

1. On a cold, rainy day, your car battery is dead, and you must push the car to move it and get it started. Why can't you move the car by remaining comfortably inside and pushing against the dashboard?

2. Does a fast-moving baseball possess force?

CHECK YOUR ANSWERS

1. In this case, the system to be accelerated is the car. If you remain inside and push on the dashboard, the force pair you produce acts and reacts within the system. These forces cancel out, as far as any motion of the car is concerned. To accelerate the car, there must be an interaction between the car and something external—for example, you on the outside pushing against the road.

2. No, a force is not something an object *has*, like mass; it is part of an interaction between one object and another. A speeding baseball may possess the capability of exerting a force on another object when interaction occurs, but it does not possess force as a thing in itself. As we will see in the following chapter, moving things possess momentum and kinetic energy.

Using Newton's third law, we can understand how a helicopter gets its lifting force. The whirling blades are shaped to force air particles down (action), and the air forces the blades up (reaction). This upward reaction force is called *lift*. When lift equals the weight of the craft, the helicopter hovers in midair. When lift is greater, the helicopter climbs upward.

This is true for birds and airplanes. Birds fly by pushing air downward. The air, simultaneously, pushes the bird upward. When the bird is soaring, the wings must be shaped so that moving air particles are deflected downward. Slightly tilted wings that deflect oncoming air downward produce lift on an airplane. Air that is pushed downward continu-

FIGURE 4.30

Ducks fly in a V formation because air pushed downward at the tips of their wings swirls upward, creating an updraft that is strongest off to the side of the bird. A trailing bird gets added lift by positioning itself in this updraft, pushes air downward and creates another updraft for the next bird, and so on. The result is a flock flying in a V formation.

ously maintains lift. This supply of air is obtained by the forward motion of the aircraft, which results from propellers or jets that push air backward. When the propellers or jets push air backward, the air simultaneously pushes the propellers or jets forward. We will learn in Chapter 7 that the curved surface of a wing is an airfoil, which enhances the lifting force.

We see Newton's third law in action everywhere. A fish propels water backward with its fins, and the water propels the fish forward. The wind caresses the branches of a tree, and the branches caress back

FIGURE 4.31

You cannot touch without being touched—Newton's third law.

on the wind to produce whistling sounds. Forces are interactions between different things. Every contact requires at least a two-ness; there is no way that an object can exert a force on nothing. Forces, whether large shoves or slight nudges, always occur in pairs, each opposite to the other. Thus, we cannot touch without being touched.

4.5   Vectors

Recall that quantities such as velocity, force, and acceleration require both magnitude and direction for a complete description. Such a quantity is a *vector quantity*. By contrast, a quantity that can be described by magnitude only, not involving direction, is called a *scalar quantity*. Mass, volume, and speed are scalar quantities.

The Physics Place
Vectors

The Physics Place
Vector Representation:
How to Add and
Subtract Vectors
Geometric Addition of
Vectors

As briefly discussed in the previous chapter, a vector quantity is represented by an arrow. When the length of the arrow is scaled to represent the quantity's magnitude, and the direction of the arrow shows the direction of the quantity, we refer to the arrow as a **vector**.

> The valentine vector says, "I was only a scalar until you came along and gave me direction."

Adding vectors that act along parallel directions is simple enough: if they are in the same direction, they add; if they are in opposite directions, they subtract. The sum of two or more vectors is called their **resultant**. To find the resultant of two vectors that don't act in exactly the same or opposite direction, we use the *parallelogram rule.** Construct a parallelogram wherein the two vectors are adjacent sides—

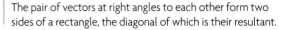

FIGURE 4.32

This vector, scaled so that 1 cm equals 20 N, represents a force of 60 N to the right.

---

* A parallelogram is a four-sided figure with opposite sides equal in length and parallel to each other. You can determine the length of the diagonal by measurement, but in the special case that the two vectors **V** and **H** are perpendicular, forming a square or rectangle, you can apply the Pythagorean theorem, $\mathbf{R}^2 = \mathbf{V}^2 + \mathbf{H}^2$, to give the resultant: $\mathbf{R} = \sqrt{\mathbf{V}^2 + \mathbf{H}^2}$. Note we express vector quantities in boldface.

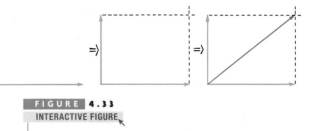

FIGURE 4.33
INTERACTIVE FIGURE

The pair of vectors at right angles to each other form two sides of a rectangle, the diagonal of which is their resultant.

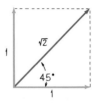

FIGURE 4.34

When a pair of equal-length vectors at right angles to each other are added, they form a square. The diagonal of the square is the resultant, $\sqrt{2}$ times the length of either side.

the diagonal of the parallelogram shows the resultant. In Figure 4.33 the parallelograms are rectangles.

In the special case of two perpendicular vectors that are equal in magnitude, the parallelogram is a square (Figure 4.34). Since for any square the length of a diagonal is $\sqrt{2}$, or 1.41, times one of the sides, the resultant is $\sqrt{2}$ times one of the vectors. For example, the resultant of two equal vectors of magnitude 100 acting at a right angle to each other is 141.

Force Vectors

Figure 4.35 shows the top view of a pair of horizontal forces acting on a box. One is 30 newtons and the other is 40 newtons. Simple measurement shows the resultant is 50 newtons.

Figure 4.36 shows Nellie Newton hanging at rest from a pair of ropes that form different angles with the vertical. Which rope has the greater tension?

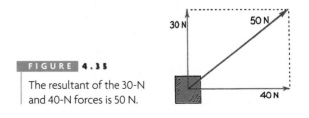

FIGURE 4.35

The resultant of the 30-N and 40-N forces is 50 N.

FIGURE 4.36

Nellie Newton hangs motionless by one hand from a clothesline. If the line is on the verge of breaking, which side is most likely to break?

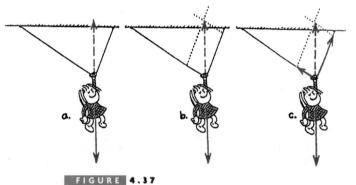

FIGURE 4.37

INTERACTIVE FIGURE

(a) Nellie's weight is shown by the downward vertical vector. An equal and opposite vector is needed for equilibrium, shown by the dashed vector. (b) This dashed vector is the diagonal of a parallelogram defined by the dotted lines. (c) Both rope tensions are shown by the constructed vectors. Tension is greater in the right rope, the one most likely to break.

Investigation will show there are three forces acting on Nellie: her weight, a tension in the left-hand rope, and a tension in the right-hand rope. Because the ropes hang at different angles, the rope tensions will be different from each other. Figure 4.37 shows a step-by-step solution. Because Nellie is suspended in equilibrium, her weight must be supported by the combination of rope tensions, which must add vectorially to equal her weight. Using the parallelogram rule, we find that the tension in the right-hand rope is greater than the tension in the left-hand rope. By measuring the vectors, you'll see that tension in the right rope is about twice the tension in the left rope. How does tension in the right rope compare with her weight? (Force vectors are treated

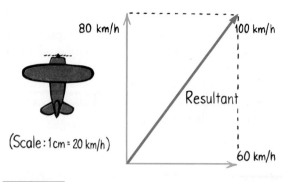

FIGURE 4.38

The 60-km/h crosswind blows the 80-km/h aircraft off course at 100 km/h.

further in Appendix C, and they are nicely developed in *Practice Book for Conceptual Physics Fundamentals.*)

Velocity Vectors

Recall that speed is a measure of "how fast" and velocity is a measure of both how fast *and* "which direction." If the speedometer in a car reads 100 kilometers per hour you know your *speed*. If there is also a compass on the dashboard, indicating that the car is moving due north, for example, you know your *velocity*—100 kilometers per hour north. To know your velocity is to know your speed *and* your direction.

> A pair of 6-unit and 8-unit vectors at right angles to each other say, "We may be a six and an eight, but together we're a perfect ten."

Consider an airplane flying due north at 80 kilometers per hour relative to the surrounding air. Suppose that the plane is caught in a 60-kilometer-per-hour crosswind (wind blowing at right angles to the direction of the airplane) that blows it off its intended course. This example is represented in Figure 4.38 with velocity vectors scaled so that 1 centimeter represents 20 kilometers per hour. Thus, the 80-kilometer-per-hour velocity of the airplane is shown by the 4-centimeter vector, and the 60-kilometer-per-hour tailwind is shown by the 3-centimeter vector. The diagonal of the constructed parallelogram (a rectangle, in this case) measures 5 cm, which represents 100 km/h. So the airplane flies at 100 km/h relative to the ground, in a direction between north and northeast.

PRACTICING PHYSICS

■ Hands-On Vectors

Here we see a top view of an airplane being blown off course by wind in various directions. With pencil and using the parallelogram rule, sketch the vectors that show the resulting velocities for each case. In which case does the airplane travel fastest across the ground? Slowest?

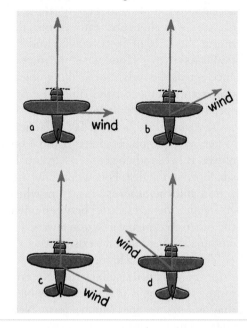

STOP AND CHECK YOURSELF

Consider a motorboat that normally travels 10 km/h in still water. If the boat heads directly across the river, which also flows at a rate of 10 km/h, what will be its velocity relative to the shore?

CHECK YOUR ANSWER

When the boat heads cross-stream (at right angles to the river flow) its velocity is 14.1 km/h, 45 degrees downstream (in accord with the diagram in Figure 4.34).

Vector Components

Just as two vectors at right angles can be combined into one resultant vector, in reverse any vector can be "resolved" into two *component* vectors perpendicular to each other. These two vectors are known as the components of the given vector they replace. The process of determining the components of a vector is

PRACTICING PHYSICS

Here we see top views of three motorboats crossing a river. All have the same speed relative to the water, and all experience the same water flow. Construct resultant vectors showing the speed and direction of the boats. Then answer the following:

(a) Which boat takes the shortest path to the opposite shore?

(b) Which boat reaches the opposite shore first?

(c) Which boat provides the fastest ride?

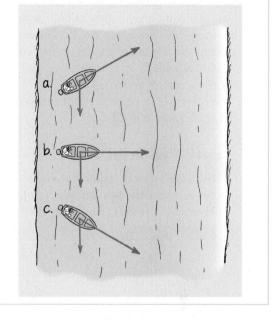

called *resolution*. Any vector drawn on a piece of paper can be resolved into a vertical and a horizontal component.

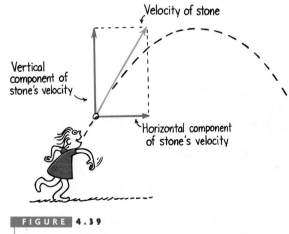

FIGURE 4.39

The horizontal and vertical components of a ball's velocity.

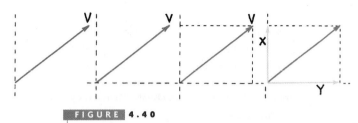

FIGURE 4.40

Construction of the vertical and horizontal components of a vector.

Vector resolution is illustrated in Figure 4.40. A vector **V** is drawn in the proper direction to represent a vector quantity. Then vertical and horizontal lines (axes) are drawn at the tail of the vector. Next, a rectangle is drawn that has **V** as its diagonal. The sides of this rectangle are the desired components, vectors **X** and **Y**. In reverse, note that the vector sum of vectors **X** and **Y** is **V**.

We'll return to vector components when we treat projectile motion in Chapter 8.

STOP AND
CHECK YOURSELF

With a ruler, draw the horizontal and vertical components of the two vectors shown. Measure the components and compare your findings with the answers below.

CHECK YOUR ANSWER

Left vector: the horizontal component is 1.8 cm; the vertical component is 2.3 cm. Right vector: the horizontal component is 3.5 cm; the vertical component is 2.3 cm.

4.6 Summary of Newton's Three Laws of Motion

The Physics Place
Newton's Third Law

Newton's first law, the law of inertia: An object at rest tends to remain at rest; an object in motion tends to remain in motion at constant speed along a straight-line path. This property of objects to resist changes in motion is called *inertia*. Mass is a measure of inertia. Objects will undergo changes in motion only in the presence of a net force.

Newton's second law, the law of acceleration: When a net force acts on an object, the object will accelerate. The acceleration is directly proportional to the net force and inversely proportional to the mass. Symbolically, $a \sim F/m$. Acceleration is always in the direction of the net force. When an object falls in a vacuum, the net force is simply the weight, and the acceleration is g (the symbol g denotes that acceleration is due to gravity alone.) When an object falls in air, the net force is equal to the weight minus the force of air resistance, and the acceleration is less than g. If and when the force of air resistance equals the weight of a falling object, acceleration terminates, and the object falls at constant speed (called the *terminal speed*).

Newton's third law, the law of action–reaction: Whenever one object exerts a force on a second object, the second object exerts an equal and opposite force on the first. Forces occur in pairs: one is an action and the other is a reaction, which together constitute the interaction between one object and the other. Action and reaction always act on different objects. Neither force exists without the other.

There has been a lot of new and exciting physics since the time of Isaac Newton. Nevertheless, and quite interestingly, it was primarily Newton's laws that got us to the Moon. Isaac Newton truly changed our way of viewing the world.

PRACTICING PHYSICS

If you drop a sheet of paper and a book side by side, the book will fall faster than the paper. Why? The book falls faster because of its greater weight compared to the air resistance it encounters. If you place the paper against the lower surface of the raised book and again drop them at the same time, it will be no surprise that they hit the surface below at the same time. The book simply pushes the paper with it as it falls. Now, repeat this, only with the paper on *top* of the book, not sticking over its edge. How will the accelerations of the book and paper compare? Will they separate and fall differently? Will they have the same acceleration? Try it and see! Then explain what happens.

■ ISAAC NEWTON (1642–1727)

On Christmas day 1642, the year that Galileo died, Isaac Newton was prematurely born and barely survived. Newton's birthplace was his mother's farmhouse in Woolsthorpe, England. His father died several months before his birth, and he grew up under the care of his mother and grandmother. As a child he showed no particular signs of brightness, and at the age of $14\frac{1}{2}$ he was taken out of school to work on his mother's farm. As a farmer he was a failure, preferring to read books that he borrowed from a neighboring pharmacist. An uncle sensed the scholarly potential in young Isaac and prompted him to study at the University of Cambridge, which he did for 5 years, graduating without particular distinction.

A plague swept through England, and Newton retreated to his mother's farm—this time to continue his studies. At the farm, at ages 23 and 24, he laid the foundations for the work that was to make him immortal. Seeing an apple fall to the ground led him to consider the force of gravity extending to the Moon and beyond. He formulated the law of universal gravitation. He invented the calculus, a very important mathematical tool in science. He extended Galileo's work and developed the three fundamental laws of motion. He also formulated a theory of the nature of light and showed with prisms that white light is composed of all the colors of the rainbow. It was his experiments with prisms that first made him famous.

When the plague subsided, Newton returned to Cambridge and soon established a reputation for himself as a first-rate mathematician. His mathematics teacher resigned in his favor and Newton was appointed the Lucasian Professor of Mathematics. He held this post for 28 years. In 1672 he was elected to the Royal Society, where he exhibited the world's first reflector telescope. It can still be seen, preserved at the library of the Royal Society in London with the inscription "The first reflecting telescope, invented by Sir Isaac Newton, and made with his own hands."

It wasn't until Newton was 42 that he began to write what is generally acknowledged as the greatest scientific book ever written, the *Principia Mathematica Philosophiae Naturalis.* He wrote the work in Latin and completed it in 18 months. It appeared in print in 1687, but wasn't printed in English until 1729, two years after his death. When asked how he was able to make so many discoveries, Newton replied that he solved his problems by continually thinking very long and hard about them—and not by sudden insight.

At the age of 46 he was elected a member of Parliament. He attended the sessions in Parliament for two years and never gave a speech. One day he rose and the House fell silent to hear the great man. Newton's "speech" was very brief; he simply requested that a window be closed because of a draft.

A further turn from his work in science was his appointment as warden and then as master of the mint. Newton resigned his professorship and directed his efforts toward greatly improving the workings of the mint, to the dismay of counterfeiters who flourished at that time. He maintained his membership in the Royal Society and was elected president, then was reelected each year for the rest of his life. At the age of 62, he wrote *Opticks,* which summarized his work on light. Nine years later he wrote a second edition of his *Principia.*

Although Newton's hair turned gray at 30, it remained full, long, and wavy all his life. Unlike others in his time, he did not wear a wig. He was a modest man, very sensitive to criticism, and never married. He remained healthy in body and mind into old age. At 80, he still had all his teeth, his eyesight and hearing were sharp, and his mind was alert. In his lifetime he was regarded by his countrymen as the greatest scientist who ever lived. In 1705 he was knighted by Queen Anne. Newton died at the age of 85 and was buried in Westminster Abbey along with England's kings and heroes.

Newton "opened up" the universe, showing that the same natural laws that apply to Earth govern the larger cosmos as well. For humankind this led to increased humility, but also to hope and inspiration because of the evidence of a rational order. Newton ushered in the Age of Reason. His ideas and insights truly changed the world and elevated the human condition.

SUMMARY OF TERMS

Newton's first law of motion Every object continues in a state of rest, or in a state of motion in a straight line at constant speed, unless acted upon by a net force.

Inertia The property of things to resist changes in motion.

Newton's second law of motion The acceleration produced by a net force on an object is directly proportional to the net force, is in the same direction as the net force, and is inversely proportional to the mass of the object.

Free fall Motion under the influence of the pull of gravity only.

Terminal speed The speed at which the acceleration of a falling object terminates when air resistance balances its weight.

Terminal velocity Terminal speed when direction is specified or implied.

Interaction Mutual action between objects where each object exerts an equal and opposite force on the other.

Force pair The action and reaction pair of forces that occur in an interaction.

Newton's third law of motion Whenever one object exerts a force on a second object, the second object exerts an equal and opposite force on the first.

Vector An arrow drawn to scale to represent a vector quantity.

Resultant The net result of a combination of two or more vectors.

Force vector An arrow drawn to scale so that its length represents the magnitude of a force and its direction represents the direction of the force.

Velocity vector An arrow drawn to scale so that its length represents the magnitude of a velocity and its direction represents the direction of motion.

Vector component Parts into which a vector can be separated and that act in different directions from the vector.

REVIEW QUESTIONS

4.1 Newton's First Law of Motion

1. State the law of inertia.
2. What concept was missing from people's minds in the sixteenth century when they couldn't believe Earth was moving?
3. When a bird lets go of a branch and drops to the ground below, why doesn't the moving Earth sweep away from the dropping bird?
4. What kind of path would the planets follow if suddenly their attraction to the Sun no longer existed?

4.2 Newton's Second Law of Motion

5. State Newton's second law.
6. Is acceleration directly proportional to force, or is it inversely proportional to force? Give an example.
7. Is acceleration directly proportional to mass, or is it inversely proportional to mass? Give an example.
8. What is the net force that acts on a 10-N freely falling object?
9. Why doesn't a heavy object accelerate more than a light object when both are freely falling?
10. What is the net force that acts on a 10-N falling object when it encounters 4 N of air resistance? 10 N of air resistance?
11. What two principal factors affect the force of air resistance on a falling object?
12. What is the acceleration of a falling object that has reached its terminal velocity?
13. If two objects of the same size fall through air at different speeds, which encounters the greater air resistance?
14. Why does a heavy parachutist fall faster than a lighter parachutist who wears the same size parachute?

4.3 Forces and Interactions

15. Previously, we said that a force was a push or pull; now we say it is an interaction. Which is it? A push or pull, or an interaction? And what does it mean to say *interaction*?
16. How many forces are required for a single interaction?
17. When you push against a wall with your fingers, they bend because they experience a force. Identify this force.
18. A boxer can hit a heavy bag with great force. Why can't he hit a sheet of newspaper in midair with the same amount of force?

4.4 Newton's Third Law of Motion

19. State Newton's third law.
20. Consider hitting a baseball with a bat. If we call the force on the bat against the ball the action force, identify the reaction force.
21. If the forces that act on a cannonball and the recoiling cannon from which it is fired are equal in magnitude, why do the cannonball and cannon have very different accelerations?
22. Do action and reaction forces always act on different bodies? Defend your answer.
23. Can you cancel a force on Body A with a force that acts on Body B? Defend your answer.
24. How does a helicopter get its lifting force?
25. What law of physics is inferred when we say you cannot touch without being touched?

4.5 Vectors

26. According to the parallelogram rule, what does the diagonal of a constructed parallelogram represent?

27. Consider Nellie in Figure 4.36. If the ropes were vertical, with no angle involved, what would be the tension in each rope?
28. Can it be said that, when a pair of vectors is at right angles to each other, the resultant is greater than either of the vectors separately? Defend your answer.
29. When a vector at an angle is resolved into horizontal and vertical components, can it be said that each component has less magnitude than the original vector? Defend your answer.

4.6 Summary of Newton's Three Laws of Motion

30. Briefly summarize Newton's three laws of motion.

ACTIVE EXPLORATIONS

1. Write a letter to grandpa, similar to the one of Active Exploration 1 in Chapter 3. Tell him that Galileo introduced the concepts of acceleration and inertia, and was familiar with forces, but didn't see the connection between these three concepts. Tell him how Isaac Newton did, and how the connection explains why heavy and light objects in free fall gain the same speed in the same time. In this letter, it's okay to use an equation or two, as long as you make it clear to grandpa that an equation is a shorthand notation of ideas you're explaining.
2. The net force acting on an object and the resulting acceleration are always in the same direction. You can demonstrate this with a spool. If the spool is pulled horizontally to the right, in which direction will it roll?

3. Hold your hand with the palm down like a flat wing outside the window of a moving automobile. Then slightly tilt the front edge of your hand upward and notice the lifting effect as air is deflected downward from the bottom of your hand. Can you see Newton's laws at work here?

ONE-STEP CALCULATIONS

Make these simple one-step calculations and familiarize yourself with the equations that link the concepts of force, mass, and acceleration.

Conversion factors: 1 kg weighs 10 N at Earth's surface; 1 N = 0.22 lb (You may express *g* as 10 N/kg or as 10 m/s^2, which are equivalent.)

Weight = *mg*

1. Calculate the weight of a person having a mass of 50 kg in newtons.
2. Calculate the weight of a 2000-kg elephant in newtons. What is its weight in pounds?
3. An apple weighs about 1 N. What is its mass in kilograms? What is its weight in pounds?
4. Susie Small finds she weighs 300 N. Calculate her mass.

Acceleration: $a = \dfrac{F_{net}}{m}$

5. Calculate the acceleration of a 2000-kg, single-engine airplane just before takeoff when the thrust of its engine is 500 N.

6. a. Calculate the acceleration of a 2-kg block on a horizontal friction-free air table when you exert a horizontal net force of 20 N.
   b. What acceleration occurs if the friction force is 4 N?

Force: *F = ma*

7. Calculate the horizontal force that must be applied to a 1-kg puck to make it accelerate on a horizontal friction-free air table with the same acceleration it would have if it were dropped and fell freely.
8. Calculate the horizontal force that must be applied to produce an acceleration of 1.8 *g* for a 1.2-kg puck on a horizontal friction-free air table.

Resultant of two vectors at right angles to each other: $R = \sqrt{V^2 + H^2}$

9. Calculate the resultant of a pair of 100-km/h velocity vectors that are at right angles to each other.
10. Calculate the resultant velocity of an airplane that normally flies at 200 km/h if it encounters a 50-km/h wind from the side (at a right angle to the airplane).

EXERCISES

Again, please do not be intimidated by the large number of exercises and problems in this and other meatier chapters. If your course work is to cover many chapters, your instructor will likely assign only a few exercises and/or problems from each.

1. In the orbiting space shuttle, you are handed two identical closed boxes, one filled with sand and the other filled with feathers. How can you tell which is which without opening the boxes?

2. Your empty hand is not hurt when it bangs lightly against a wall. Why does your hand hurt if it is carrying a heavy load? Which of Newton's laws is most applicable here?

3. Why is a massive cleaver more effective for chopping vegetables than an equally sharp knife?

4. Each of the vertebrae forming your spine is separated from its neighbors by disks of elastic tissue. What happens, then, when you jump heavily on your feet from an elevated position? Can you think of a reason why you are a little shorter in the evening than you are in the morning? (Hint: Think about the hammerhead in Figure 4.2.)

5. Before the time of Galileo and Newton, it was thought by many learned scholars that a stone dropped from the top of a tall mast on a moving ship would fall vertically and hit the deck behind the mast by a distance equal to how far the ship had moved forward while the stone was falling. In light of your understanding of Newton's laws, what do you think about this idea?

6. While standing at rest on a floor, does the floor exert an upward force against your feet? How much force does it exert? Why are you not moved upward by this force?

7. To pull a wagon across a lawn at a constant velocity, you must exert a steady force. Reconcile this fact with Newton's first law, which states that motion with a constant velocity indicates no force.

8. When your car moves along the highway at a constant velocity, the net force on it is zero. Why, then, do you continue running your engine?

9. A rocket becomes progressively easier to accelerate as it travels through space. Why is this so? (Hint: About 90 percent of the mass of a newly launched rocket is fuel.)

10. As you are leaping upward from the ground, how does the force that you exert on the ground compare with your weight?

11. A common saying goes, "It's not the fall that hurts you; it's the sudden stop." Translate this into Newton's laws of motion.

12. On which of these hills does the ball roll down with increasing speed and decreasing acceleration along the path? (Use this example if you wish to explain to someone the difference between speed and acceleration.)

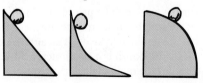

13. If you drop an object, its acceleration toward the ground is 10 m/s^2. If you throw it down instead, would its acceleration after throwing be greater than 10 m/s^2? Ignore air resistance. Why or why not?

14. Suppose the object in the preceding exercise were thrown downward in the presence of air resistance. Can you think of a reason why the acceleration of the object would actually be less than 10 m/s^2?

15. Two 100-N weights are attached to a spring scale as shown. Does the scale read 0 N, 100 N, or 200 N, or does it show some other reading? (Hint: Would it read any differently if one of the ropes were tied to the wall instead of to the hanging 100-N weight?)

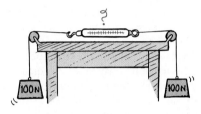

16. What is the net force on an apple that weighs 1 N when you hold it at rest above your head? What is the net force on it when you release it?

17. You hold an apple over your head.
    a. Identify all the forces acting on the apple and their reaction forces.
    b. When you drop the apple, identify all the forces acting on it as it falls and the corresponding reaction forces.

18. Aristotle claimed that the speed of a falling object depends on its weight. We now know that objects in free fall, whatever their weights, undergo the same gain in speed. Why does weight not affect acceleration?

19. Does a stick of dynamite contain force? Defend your answer.

20. Can a dog wag its tail without the tail in turn "wagging the dog"? (Consider a dog with a relatively massive tail.)

21. When the athlete holds the barbell overhead, the reaction force is the weight of the barbell on his hand. How does this force vary for the case in which the barbell is accelerated upward? Accelerated downward?

22. Why can you exert greater force on the pedals of a bicycle if you pull up on the handlebars?

23. If Earth exerts a gravitational force of 1000 N on an orbiting communications satellite, how much force does the satellite exert on Earth?

24. The strong man will push apart the two initially stationary freight cars of equal mass before he himself drops straight

to the ground. Is it possible for him to give either of the cars a greater speed than the other? Why or why not?

25. Suppose two carts, one twice as massive as the other, fly apart when the compressed spring that joins them is released. How fast does the heavier cart roll compared with the lighter cart?

26. If you exert a horizontal force of 200 N to slide a crate across a factory floor at a constant velocity, how much friction is exerted by the floor on the crate? Is the force of friction equal and oppositely directed to your 200-N push? Does the force of friction make up the reaction force to your push? Why not?

27. If a Mack truck and a motorcycle have a head-on collision, upon which vehicle is the impact force greater? Which vehicle undergoes the greater change in its motion? Explain your answers.

28. Two people of equal mass attempt a tug-of-war with a 12-m rope while standing on frictionless ice. When they pull on the rope, each person slides toward the other. How do their accelerations compare, and how far does each person slide before they meet?

29. Suppose that one person in the preceding exercise has twice the mass of the other. How far does each person slide before they meet?

30. Which team wins in a tug-of-war—the team that pulls harder on the rope, or the team that pushes harder against the ground? Explain.

31. The photo shows Steve Hewitt and his daughter Gretchen. Is Gretchen touching her dad, or is he touching her? Explain.

32. When your hand turns the handle of a faucet, water flows out. Does your push on the handle and the force of water flowing out comprise an action–reaction pair? Defend your answer.

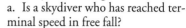

33. Why is it that a cat that falls from the top of a 50-story building will hit a safety net below no faster than if it fell from the twentieth story?

34. Free fall is motion in which gravity is the only force acting.
    a. Is a skydiver who has reached terminal speed in free fall?
    b. Is a satellite circling Earth above the atmosphere in free fall?

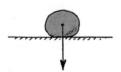

35. How does the weight of a falling body compare with the air resistance it encounters just before it reaches terminal velocity? Just after it reaches terminal velocity?

36. You tell your friend that the acceleration of a skydiver decreases as falling progresses. Your friend then asks if this means that the skydiver is slowing down. What is your response?

37. If and when Galileo dropped two balls from the top of the Leaning Tower of Pisa, air resistance was not really negligible. Assuming that both balls were the same size yet one was much heavier than the other, which ball struck the ground first? Why?

38. If you simultaneously drop a pair of tennis balls from the top of a building, they will strike the ground at the same time. If one of the tennis balls is filled with lead pellets, will it fall faster and hit the ground first? Which of the two will encounter more air resistance? Defend your answers.

39. Which is more likely to break, the ropes supporting a hammock stretched tightly between a pair of trees or one that sags more when you sit on it? Defend your answer.

40. When a bird alights upon a stretched power line wire, does the tension in the wire change? If so, is the increase more than, less than, or about equal to the bird's weight?

41. When you swim across a river, does the time to reach the opposite shore depend on the rate of flow of the water? Defend your answer.

42. Why does vertically falling rain make slanted streaks on the side windows of a moving automobile? If the streaks make an angle of 45°, what does this tell you about the relative speed of the car and the falling rain?

43. A stone is shown at rest on the ground.

    a. The vector shows the weight of the stone. Complete the vector diagram showing another vector that results in zero net force on the stone.
    b. What is the conventional name of the vector you have drawn?

44. Here is a stone at rest suspended by a string.

a. Draw force vectors for all the forces that act on the stone.

b. Should your vectors have a zero resultant?

c. Why, or why not?

45. Here the same stone is being accelerated vertically upward.

a. Draw force vectors to some suitable scale showing relative forces acting on the stone.

b. Which is the longer vector, and why?

46. Suppose that the string in the preceding exercise breaks and that the stone slows in its upward motion. Draw a force vector diagram of the stone when it reaches the top of its path.

47. What is the net force on the stone in the preceding exercise when it is at the top of its path? What is its instantaneous velocity? Its acceleration?

48. Here is the same stone sliding down a friction-free incline.

a. Identify the forces that act on it, and draw appropriate force vectors.

b. By the parallelogram rule, construct the resultant force on the stone (carefully showing that it has a direction parallel to the incline—the same direction as the stone's acceleration).

49. Here is the same stone at rest, interacting with both the surface of the incline and the block.

a. Identify all the forces that act on the stone and draw appropriate force vectors.

b. Show that the net force on the stone is zero. (Hint 1: There are two normal forces on the stone. Hint 2: Be sure the vectors you draw are for forces that act on the stone, not on the surfaces by the stone.)

50. Make up three multiple-choice questions, one for each of Newton's laws, that check a classmate's understanding of these laws.

PROGRAMS

PROBLEMS   ● **BEGINNER** ■ **INTERMEDIATE** ◆ **EXPERT**

1. ● When two horizontal forces are exerted on a cart, 600 N forward and 400 N backward, the cart undergoes acceleration. What additional force is needed to produce nonaccelerated motion?

2. ● You push with a 20-N horizontal force on a 2-kg box of cookies resting on a horizontal surface against a horizontal friction force of 12 N. Show that the acceleration of the box will be 4 m/s^2.

3. ● Suppose that you push with a 40-N horizontal force on a 4-kg ventriloquist's dummy resting on a horizontal tabletop. Further suppose you push against a horizontal friction force of 24 N. Show that the acceleration of the dummy will be 4 m/s^2.

4. ● An astronaut of mass 100 kg recedes from her spacecraft by activating a small propulsion unit attached to her back. The force generated by a spurt is 25 N. Show that her acceleration is 0.25 m/s^2.

5. ● A 747 jumbo jet of mass 330,000 kg experiences in takeoff a 250,000-N thrust for each of its four engines. Show that its acceleration is 3 m/s^2.

6. ● A 400-kg bear grasping a vertical tree slides down at a constant velocity. How much friction force acts on the bear?

7. ■ A firefighter of mass 80 kg slides down a vertical pole with an acceleration of 4 m/s^2. Show that the friction force that acts on the firefighter is 480 N.

8. ■ A boxer punches a sheet of paper in midair, and thereby brings it from rest up to a speed of 25 m/s in 0.05 second. The mass of the paper is 0.003 kg. Show that the force of the punch on the paper is only 1.5 N.

9. ● Suzie Skydiver with her parachute has a mass of 50 kg.

a. Before opening her chute, what force of air resistance will she encounter when she reaches terminal velocity?

b. What force of air resistance will she encounter when she reaches a lower terminal velocity after the chute is open?

c. Discuss why your answers are the same or different.

10. ● Suppose that you are standing on a skateboard near a wall and that you push on the wall with a force of 30 N. How hard does the wall push on you? If your mass is 60 kg, show that your acceleration while pushing on the wall will be 0.5 m/s^2.

11. ■ Consider raindrops that fall vertically at a speed of 3 m/s, while you are running horizontally at 4 m/s. Show that the raindrops hit your face at a speed of 5 m/s.

12. ■ Forces of 3 N and 4 N act at right angles on a block of mass 5 kg. Show that the resulting acceleration is 1 m/s^2.

13. ■ Consider an airplane that has an air speed of 120 km/h flying with its nose pointing north with a 90-km/h crosswind blowing from the west? Show that the ground speed of the airplane is 150 km/h.

14. ◆ A net force F acting on a mass m gives it an acceleration a.
a. Show that the same force acting on another mass M results in an acceleration equal to $a\left(\frac{m}{M}\right)$.
b. Suppose the net force F causes a 6.0-kg mass to accelerate at 2.5 m/s^2. Show that the same force acting on a 5.0-kg mass causes it to accelerate at 3.0 m/s^2.

15. ◆ Phil and his rocket-powered sled have a combined mass M and accelerate at a rate a. Then the sled runs into Zephram, mass m, who tumbles aboard. Ignore friction.
a. Show that the sled now accelerates at a rate equal to $\frac{M}{M+m}a$.

b. If Phil and his sled have a combined mass of 70 kg, Zephram's mass is 45 kg, and the initial acceleration of the sled was 3.6 m/s^2, show that when Zephram joins Phil the acceleration of the sled is 2.2 m/s^2.

16. ◆ A rock band's tour bus, mass M, is accelerating away from a STOP sign at rate a when a boulder, mass $M/6$, falls onto the top of the bus and remains there.
a. Show that the bus's acceleration is now $\frac{6}{7}a$.
b. If the initial acceleration of the bus was 1.2 m/s^2, show that when the bus carries the boulder with it, the acceleration will be 1.0 m/s^2.

CHAPTER 4 ONLINE RESOURCES

The Physics Place

Interactive Figures
4.1, 4.6, 4.16, 4.22, 4.26, 4.27, 4.28, 4.29, 4.33, 4.37

Tutorials
Parachuting and Newton's Second Law
Newton's Third Law
Vectors

Videos
Newton's Second Law
Free-Fall Acceleration Explained
Free Fall: How Fast?
Free Fall: How Far?
$V = gt$

Air Resistance and Falling Objects
Falling and Air Resistance
Forces and Interactions
Action and Reaction on Different Masses
Action and Reaction on Rifle and Bullet
Vector Representation: How to Add and Subtract Vectors
Geometric Addition of Vectors

Quiz

Flashcards

Links

Momentum and Energy

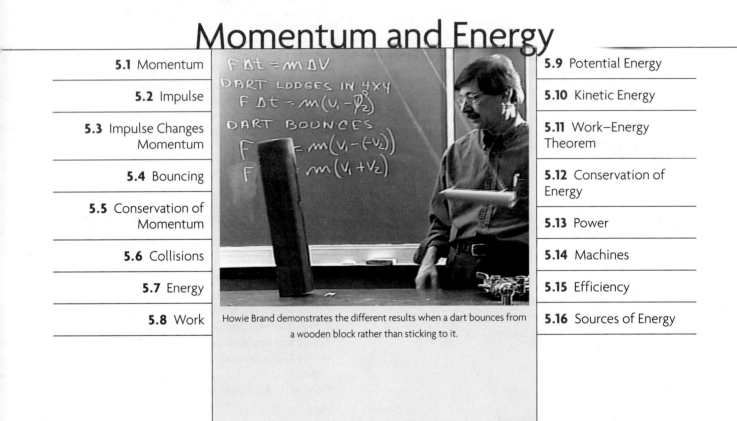

Howie Brand demonstrates the different results when a dart bounces from a wooden block rather than sticking to it.

I n Chapter 3, we introduced Galileo's concept of inertia and, in Chapter 4, we showed how it was incorporated into Newton's first law of motion. We discussed inertia in terms of objects at rest and objects in motion. In this chapter, we will concern ourselves only with the inertia of moving objects. When we combine the ideas of inertia and motion, we are dealing with momentum. *Momentum* is a property of moving things. Moving things also have energy of motion—*kinetic energy*. This chapter is about two of the most central concepts in mechanics—momentum and energy. We begin with the first of these concepts—momentum.

5.1 Momentum

W e know that it's more difficult to stop a large truck than a small car when both are moving at the same speed. We say the truck has more momentum than the car. By **momentum**, we mean *inertia in motion*, or, more specifically, the mass of an object multiplied by its velocity.

The Physics Place
Definition of Momentum

$$\text{Momentum} = \text{mass} \times \text{velocity}$$

Or, in shorthand notation,

$$\text{Momentum} = mv$$

When direction is not an important factor, we can say,

$$\text{Momentum} = \text{mass} \times \text{speed}$$

which we still abbreviate mv.*

* The symbol for momentum is p. In most physics textbooks, $p = mv$.

From the definition we can see that a moving object can have a large momentum if it has a large mass, a high speed, or both. A moving truck has more momentum than a car moving at the same speed because the truck has more mass. But a fast car can have more momentum than a slow truck can. And a truck at rest has no momentum at all.

The boulder, unfortunately, has more momentum than the runner.

5.2 Impulse

If the momentum of an object changes, then either the mass or the velocity or both change. If the mass remains unchanged while momentum changes, then the velocity changes and acceleration occurs. What produces acceleration? We know the answer is *force*. The greater the force acting on an object is, the greater its change in velocity and, hence, the greater its change in momentum will be.

When you push with the same force for twice the time, you impart twice the impulse and produce twice the change in momentum.

But something else is important in changing momentum: *time*—how long a time the force acts. If you apply a brief force to a stalled automobile, you produce a change in its momentum. Apply the same force over a longer period of time, and you produce a greater change in the automobile's momentum. A force sustained for a long time produces more change in momentum than does the same force applied briefly. So, both force and time interval are important in changing momentum.

The quantity *force × time interval* is called **impulse**. In shorthand notation,

$$\text{Impulse} = Ft$$

5.3 Impulse Changes Momentum

The greater the impulse exerted on something, the greater will be the change in momentum. The exact relationship is

$$\text{Impulse} = \text{change in momentum}$$

or*

$$Ft = \Delta(mv)$$

where Δ is the symbol for "change in."

The impulse–momentum relationship helps us to analyze a variety of situations where momentum changes. Here we will consider some ordinary examples in which impulse is related to increasing and decreasing momentum.

Case 1: Increasing Momentum

To increase the momentum of an object, it makes sense to apply the greatest force possible for as long as possible. A golfer teeing off and a baseball player trying for a home run do both of these things when they swing as hard as possible and follow through with their swings. Following through extends the time of contact.

The Physics Place

Changing Momentum:
Follow-Through
Decreasing Momentum
Over a Short Time

The forces involved in impulses usually vary from instant to instant. For example, a golf club that strikes a ball exerts zero force on the ball until it comes in contact; then the force increases rapidly as the ball is distorted (Figure 5.3). The force then diminishes as the ball comes up to speed and returns to its original shape. So, when we speak of such forces in this chapter, we mean the *average* force.

* This relationship is derived by rearranging Newton's second law to make the time factor more evident. If we equate the formula for acceleration, $a = F/m$, with what acceleration actually is, $a = \Delta v/\Delta t$, we get $F/m = \Delta v/\Delta t$. From this we derive $F\Delta t = \Delta(mv)$. Calling Δt simply t, the time interval, $Ft = \Delta(mv)$.

STOP AND
CHECK YOURSELF

1. Compare the momentum of a 1-ton car moving at 100 km/h with a 2-ton truck moving at 50 km/h.

2. Does a moving object have impulse?

3. Does a moving object have momentum?

4. For the same force, which cannon imparts a greater impulse to a cannonball—a long cannon or a short one?

CHECK YOUR ANSWERS

1. Both have the same momentum
(1 ton × 100 km/h = 2 ton × 50 km/h).

2. No, impulse is not something an object *has,* like momentum. Impulse is what an object can *provide* or what it can *experience* when it interacts with some other object. An object cannot possess impulse just as it cannot possess force.

3. Yes, but like velocity, in a relative sense—that is, with respect to a frame of reference, usually Earth's surface. The momentum possessed by a moving object with respect to a stationary point on Earth may be quite different from the momentum it possesses with respect to another moving object.

4. The long cannon will impart a greater impulse because the force acts over a longer time. (A greater impulse produces a greater change in momentum, so a long cannon will impart more speed to a cannonball than a short cannon.)

Case 2: Decreasing Momentum Over a Long Time

If you were in a car that was out of control and you had to choose between hitting a concrete wall or a haystack, you wouldn't have to call on your knowledge of physics to make a decision. Common sense tells you to choose the haystack. But, knowing the physics helps you to understand *why* hitting a soft object is entirely different from hitting a hard one. In the case of hitting either the wall or the haystack and coming to a stop, it takes the *same* impulse to decrease your momentum to zero. The same impulse does not mean the same amount of force or the same amount of time; rather, it means the same *product* of force and time. By hitting the haystack instead of the wall, you extend the *time during which your momentum is brought to zero.* A longer time interval reduces the force and decreases the resulting deceleration. For example, if the time interval is extended 100 times, the force is reduced to a hundredth. Whenever we wish the force to be small, we extend the time of contact. Hence the reason for padded dashboards and airbags in motor vehicles.

When jumping from an elevated position down to the ground, what happens if you keep your legs straight and stiff? Ouch! Instead, you bend your knees when your feet make contact with the

FIGURE 5.3

The force of impact on a golf ball varies throughout the duration of impact.

Timing is especially important when changing momentum.

FIGURE 5.4

If the change in momentum occurs over a long time, then the hitting force is small.

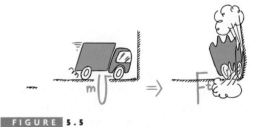

FIGURE 5.5

If the change in momentum occurs over a short time, then the hitting force is large.

ground. By doing so you extend the time during which your momentum decreases by 10 to 20 times that of a stiff-legged, abrupt landing. The resulting force on your bones is reduced by 10 to 20 times. A wrestler thrown to the floor tries to extend his time of impact with the mat by relaxing his muscles and spreading the impact into a series of smaller ones as his foot, knee, hip, ribs, and shoulder successively hit the mat. Of course, falling on a mat is preferable to falling on a solid floor because the mat also increases the time during which the force acts.

The safety net used by circus acrobats is a good example of how to achieve the impulse needed for a safe landing. The safety net reduces the force experienced by a fallen acrobat by substantially increasing the time interval during which the force acts.

If you're about to catch a fast baseball with your bare hand, you extend your hand forward so you'll have plenty of room to let your hand move backward after you make contact with the ball. You extend the time of impact and thereby reduce the force of impact. Similarly, a boxer rides or rolls with the punch to reduce the force of impact (Figure 5.6).

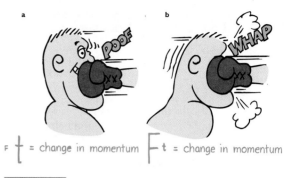

$F\ t$ = change in momentum $F\ t$ = change in momentum

FIGURE 5.6

In both cases, the impulse provided by the boxer's jaw reduces the momentum of the punch. (a) When the boxer moves away (rides with the punch), he extends the time and diminishes the force. (b) If the boxer moves into the glove, the time is reduced and he must withstand a greater force.

Case 3: Decreasing Momentum Over a Short Time

When boxing, if you move into a punch instead of away, you're in trouble. It's the same if you catch a high-speed baseball while your hand moves toward the ball instead of away upon contact. Or, when your car is out of control, if you drive it into a concrete wall instead of a haystack, you're really in trouble. In these cases of short impact times, the impact

forces are large. Remember that, for an object brought to rest, the impulse is the same, no matter how it is stopped. But, if the time is short, the force will be large.

The idea of short time of contact explains how a karate expert can split a stack of bricks with the blow of her bare hand (Figure 5.7). She brings her arm and hand swiftly against the bricks with considerable momentum. This momentum is quickly reduced when she delivers an impulse to the bricks. The impulse is the force of her hand against the bricks multiplied by the time during which her hand makes contact with the bricks. By swift execution, she makes the time of contact very brief and correspondingly makes the force of impact huge. If her hand is made to bounce upon impact, as we will soon see, the force is even greater.

STOP AND CHECK YOURSELF

1. If the boxer in Figure 5.6 is able to increase the duration of impact three times as long by riding with the punch, by how much will the force of impact be reduced?

2. If the boxer instead moves *into* the punch to decrease the duration of impact by half, by how much will the force of impact be increased?

3. A boxer being hit with a punch contrives to extend time for best results, whereas a karate expert delivers a force in a short time for best results. Isn't there a contradiction here?

CHECK YOUR ANSWERS

1. The force of impact will be only a third of what it would have been if he hadn't pulled back.

2. The force of impact will be two times greater than it would have been if he had held his head still. Impacts of this kind account for many knockouts.

3. There is no contradiction because the best results for each are quite different. The best result for the boxer is reduced force, accomplished by maximizing time, and the best result for the karate expert is increased force delivered in minimum time.

PROBLEM SOLVING

Problems

1. A chunk of rock of mass m breaks loose from the top of a rock climber's dome and falls for time t.

a. Neglecting air resistance, show that its momentum when it hits the ground below is **mgt**.

b. If it falls from a higher dome where the falling time is $2\,t$, show that its momentum when hitting the ground would be **$2\,mgt$**.

c. If the falling rock were more massive, how would this affect falling time?

d. Show that the height of a dome corresponding to a 3-second drop is **45 m**. (Use the acceleration of free fall, $g = 10$ m/s^2.)

2. An ostrich egg of mass m is thrown at speed v into a sagging bed sheet. The egg is brought to rest in time t.

a. Show that the average force of egg impact is $\dfrac{mv}{t}$.

b. If the mass of the egg is 1.0 kg, its speed when hitting the sheet is 2.0 m/s, and it is brought to rest in 0.2 s, show that the average force of impact is 10 N.

c. Why is breakage less likely with a sagging sheet than with a taut one?

Solutions

1. a. Its momentum hitting the ground is its mass $\times$ speed. As we learned in Chapter 3, falling speed

from rest is $v = gt$. So the momentum after time t is $mv = $ **mgt**.

b. Momentum $= ?$

Twice the time in the air $\Rightarrow$ twice the impulse on the chunk of rock $\Rightarrow$ twice the change in momentum. So the final momentum would be twice, **$2\,mgt$**.

c. Except for the effects of air resistance, falling time doesn't depend on mass (as we studied in Chapter 4). The rock would have greater momentum due to its greater mass, but not due to greater speed.

d. From Chapter 3, recall that when falling from rest, $d = 1/2\,gt^2 = 1/2\,(10\text{ m/s}^2)(3\text{ s})^2 = $ **45 m**.

2. a. From the impulse–momentum equation, $Ft = \Delta mv$, where in this case the egg ends up at rest, $\Delta mv = mv$, and simple algebraic rearrangement gives $F = \dfrac{mv}{t}$.

b. $F = \dfrac{mv}{t} = \dfrac{(1.0\text{ kg})\left(2.0\,\dfrac{\text{m}}{\text{s}}\right)}{(0.2\text{ s})} = 10\text{ kg} \cdot \dfrac{\text{m}}{\text{s}^2} = \textbf{10 N}.$

c. The time during which the tossed egg's momentum goes to zero is extended when it hits a sagging sheet. Extended time means less force in the impulse that brings the egg to a halt. Less force means less chance of breakage.

Cassy imparts a large impulse to the bricks in a short time and produces a considerable force.

5.4   Bouncing

If a flowerpot falls from a shelf onto your head, you may be in trouble. If it bounces from your head, you may be in more serious trouble. Why? Because impulses are greater when an object bounces. The impulse required to bring an object to a stop and then to "throw it back again" is greater than the impulse required merely to bring the object to a stop. Suppose, for example, that you catch the falling pot with your hands. You provide an impulse to reduce its momentum to zero. If you throw the pot upward again, you have to provide additional impulse. This increased amount of impulse is the same that your head supplies if the flowerpot bounces from it.

The photo opener to this chapter shows physics instructor Howie Brand swinging a dart against a

STOP AND
CHECK YOURSELF

1. In reference to Figure 5.7, how does the force that Cassy exerts on the bricks compare with the force exerted on her hand?

2. How will the impulse resulting from the impact differ if her hand bounces back upon striking the bricks?

CHECK YOUR ANSWERS

1. In accordance with Newton's third law, the forces will be equal. Only the resilience of the human hand and the training she has undergone to toughen her hand allow this feat to be performed without broken bones. Whapping bricks with bare hands is not recommended!

2. The impulse will be greater if her hand bounces from the bricks upon impact. If the time of impact is not correspondingly increased, a greater force is then exerted on the bricks (and her hand!).

wooden block. When the dart has a nail at its nose, the dart comes to a halt as it sticks to the block. The block remains upright. When the nail is removed and the nose of the dart is half of a solid rubber ball, the dart bounces upon contact with the block. The block topples over. Impulse against the block is greater when bouncing occurs.

The fact that impulses are greater when bouncing occurs was used with great success during the California Gold Rush. The waterwheels used in gold-mining operations were not very effective. A man named Lester A. Pelton recognized a problem with the flat paddles on the waterwheels. He designed a curved paddle that caused the incoming water to make a U-turn upon impact with the paddle. Because the water "bounced," the impulse exerted on the waterwheel was increased. Pelton patented his idea, and he probably made more money from his invention, the Pelton wheel, than any of the gold miners earned. Physics can indeed enrich your life in more ways than one.

5.5 Conservation of Momentum

Only an impulse external to a system can change the momentum of a system. Internal forces and impulses won't work. For example, the internal molecular forces within a baseball have no effect on the momentum of the baseball, just as a push against the dashboard of a car you're sitting in does not affect the momentum of the car. Molecular forces within the baseball and a push on the dashboard are internal forces. They come in balanced pairs that cancel to zero within the object. To change the momentum of the ball or the car, an external push or pull is required. If no external force is present, then no external impulse is present, and no change in momentum is possible.

As another example, consider the cannon being fired in Figure 5.9. The force on the cannonball inside the cannon barrel is equal and opposite to the force causing the cannon to recoil. Since these

FIGURE **5.8**

The Pelton wheel. The curved blades cause water to bounce and make a U-turn, which produces a greater impulse to turn the wheel.

FIGURE **5.9**
INTERACTIVE FIGURE

The net momentum before firing is zero. After firing, the net momentum is still zero, because the momentum of the cannon is equal and opposite to the momentum of the cannonball.

forces act for the same time, the impulses are also equal and opposite. Recall Newton's third law about action and reaction forces. It applies to impulses, too. These impulses are internal to the system comprising the cannon and the cannonball, so they don't change the momentum of the cannon–cannonball system. Before the firing, the system is at rest and the momentum is zero. After the firing, the net momentum, or total momentum, is *still* zero. Net momentum is neither gained nor lost.

Momentum, like the quantities velocity and force, has both direction and magnitude. It is a *vector quantity*. Like velocity and force, momentum can be cancelled. So, although the cannonball in the preceding example gains momentum when fired and the recoiling cannon gains momentum in the opposite direction, there is no gain in the cannon–cannonball *system*. The momenta (plural form of momentum) of the cannonball and the cannon are equal in magnitude and opposite in direction.* They cancel to zero for the system as a whole. *If no net force or net impulse acts on a system, the momentum of that system cannot change.*

When momentum, or any quantity in physics, does not change, we say it is *conserved*. The idea that momentum is conserved when no external force acts is elevated to a central law of mechanics, called the **law of conservation of momentum,** which states

In the absence of an external force, the momentum of a system remains unchanged.

In any system wherein all forces are internal—as, for example, cars colliding, atomic nuclei undergoing radioactive decay, or stars exploding—the net momentum of the system before and after the event is the same.

8-ball system   **cue-ball system**   **cue-ball + 8-ball system**

FIGURE 5.10

A cue ball hits an eight ball head-on. Consider this event in three systems: (a) An external force acts on the eight-ball system, and its momentum increases. (b) An external force acts on the cue-ball system, and its momentum decreases. (c) No external force acts on the cue-ball + eight-ball system, and momentum is conserved (simply transferred from one part of the system to the other).

---

* Here we neglect the momentum of ejected gases from the exploding gunpowder, which can be considerable. Firing a gun with blanks at close range is a definite no-no because of the considerable momentum of ejecting gases. More than one person has been killed by close-range firing of blanks. In 1998, a minister in Jacksonville, Florida, dramatizing his sermon before several hundred parishioners, including his family, shot himself in the head with a blank round from a .357-caliber Magnum. Although no slug emerged from the gun, exhaust gases did—enough to be lethal. So, strictly speaking, the momentum of the bullet (if any) + the momentum of the exhaust gases is equal to the opposite momentum of the recoiling gun.

5.6 Collisions

The collision of objects clearly illustrates the conservation of momentum. Whenever objects collide in the absence of external forces, the net momentum of both objects before the collision equals the net momentum of both objects after the collision.

net momentum $_{\text{before collision}}$ =

net momentum $_{\text{after collision}}$

The Physics Place
Momentum and Collisions

This is true no matter how the objects might be moving before they collide.

When a moving billiard ball makes a head-on collision with another billiard ball at rest, the moving ball comes to rest and the other ball moves with the speed of the colliding ball. We call this an **elastic collision**; ideally, the colliding objects rebound without lasting deformation or the generation of heat (Figure 5.11). But momentum is conserved even when the colliding objects become entangled during the collision. This is an **inelastic collision,** characterized by deformation, or the generation of heat, or both. In a perfectly inelastic collision, both objects stick together. Consider, for example, the case of a freight car moving along a track and colliding with another freight car at rest (Figure 5.12). If the freight cars are of equal mass and are coupled by

the collision, can we predict the velocity of the coupled cars after impact?

Suppose the single car is moving at 10 meters per second, and we consider the mass of each car to be m. Then, from the conservation of momentum,

$$(\text{net } mv)_{\text{before}} = (\text{net } mv)_{\text{after}}$$

$$(m \times 10 \text{ m/s})_{\text{before}} = (2m \times V)_{\text{after}}$$

By simple algebra, $V = 5$ m/s. This makes sense because, since twice as much mass is moving after the collision, the velocity must be half as much as the velocity before collision. Both sides of the equation are then equal.

Note the inelastic collisions shown in Figure 5.13. If A and B are moving with equal momenta in opposite directions (A and B colliding head-on), then one of these is considered to be negative, and

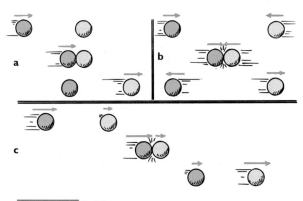

FIGURE 5.11
INTERACTIVE FIGURE

Elastic collisions of equally massive balls. (a) A green ball strikes a yellow ball at rest. (b) A head-on collision. (c) A collision of balls moving in the same direction. In each case, momentum is transferred from one ball to the other.

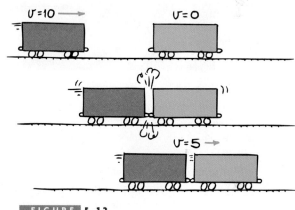

FIGURE 5.12
INTERACTIVE FIGURE

Inelastic collision. The momentum of the freight car on the left is shared with the same-mass freight car on the right after collision.

SCIENCE AND SOCIETY

■ CONSERVATION LAWS

A conservation law specifies that certain quantities in a system remain precisely constant, regardless of what changes may occur within the system. It is a law of constancy during change. In this chapter, we see that momentum is unchanged during collisions. We say that momentum is conserved. In the next chapter, we'll learn that energy is conserved as it transforms—the amount of energy in light, for example, transforms completely to thermal energy when the light is absorbed. In Appendix B we'll see that angular momentum is conserved—

whatever the rotational motion of a planetary system, its angular momentum remains unchanged so long as it is free of outside influences. In Chapter 10, we'll learn that electric charge is conserved, which means that it can neither be created nor destroyed. When we study nuclear physics, we'll see that these and other conservation laws rule in the submicroscopic world. Conservation laws are a source of deep insights into the simple regularity of nature and are often considered the most fundamental of physical laws. Can you think of things in your own life that remain constant as other things change?

Inelastic collisions. The net momentum of the trucks before and after collision is the same.

FIGURE 5.14

Will Maynez demonstrates his air track. Blasts of air from tiny holes provide a friction-free surface for the carts to glide upon.

the momenta add algebraically to zero. After collision, the coupled wreck remains at the point of impact, with zero momentum.

> Momentum is conserved for all collisions, elastic and inelastic (whenever external forces don't interfere).

If, on the other hand, A and B are moving in the same direction (A catching up with B), the net momentum is simply the addition of their individual momenta.

5.7 Energy

Perhaps the concept most central to all of science is energy. The combination of energy and matter makes up the universe: matter is substance, and energy is the mover of substance. The idea of matter is easy to grasp. Matter is stuff that we can see, smell, and feel. Matter has mass and it occupies space. Energy, on the other hand, is abstract. We cannot see, smell, or feel most forms of energy. Surprisingly, the idea of energy was unknown to Isaac Newton,

Consider the air track in Figure 5.14. Suppose a gliding cart with a mass of 0.5 kg bumps into, and sticks to, a stationary cart that has a mass of 1.5 kg. If the speed of the gliding cart before impact is v_{before}, how fast will the coupled carts glide after collision?

CHECK YOUR ANSWER

According to momentum conservation, the momentum of the 0.5-kg cart before the collision = momentum of both carts stuck together afterward.

$$0.5 \text{ kg } v_{before} = (0.5 \text{ kg} + 1.5 \text{ kg}) \, v_{after}$$

$$v_{after} = \frac{0.5 \text{ kg } v_{before}}{(0.5 \text{ kg} + 1.5 \text{ kg})} =$$

$$\frac{0.5 \, v_{before}}{2} = \frac{v_{before}}{4} \quad [\text{note kg in equation}]$$

This makes sense, because four times as much mass will be moving after the collision, so the coupled carts will glide more slowly. The same momentum means four times the mass glides 1/4 as fast.

So we see that changes in an object's motion depend both on force and on how long the force acts. When "how long" means time, we refer to the quantity "force × time" as impulse. But "how long" can mean distance also. When we consider the quantity "force × distance," we are talking about something entirely different—the concept of *energy*.

and its existence was still being debated in the 1850s. Although energy is familiar to us, it is difficult to define, because it is not only a "thing," but also both a thing and a process—similar to being both a noun and a verb. Persons, places, and things have energy, but we usually observe energy only when it is being transferred or being transformed. It appears in the form of electromagnetic waves from the Sun, and we feel it as thermal energy; it is captured by plants and binds molecules of matter together; it is in the foods we eat, and we transform it during metabolism. Even matter itself is condensed, bottled-up energy, as set forth in Einstein's famous formula, $E = mc^2$, which we'll return to in the last part of this book. In general, **energy** is the property of a system that enables it to do *work*.

The Physics Place
Energy

> An alternate definition of energy is anything that can be turned into heat.

5.8 Work

When you push a crate across a floor you're doing work. By definition, *force* × *distance* equals the concept we call **work**.

When we lift a load against Earth's gravity, work is done. The heavier the load or the higher we lift the load, the more work is being done. Two things enter the picture whenever work is done: (1) application of a force, and (2) the movement of something by that force. For the simplest case, where the force is constant and the motion is in a straight line in the direction of the force,* we define the work done on an object by an applied force as the product of the force and the distance through which the object is moved. In shorter form,

$$\text{Work} = \text{force} \times \text{distance}$$
$$W = Fd$$

If we lift two loads one story up, we do twice as much work as we do in lifting one load the same distance, because the *force* needed to lift twice the weight is twice as much. Similarly, if we lift a load two stories instead of one story, we do twice as much work because the *distance* is twice as great.

We see that the definition of work involves both a force and a distance. A weight lifter who holds a barbell weighing 1000 newtons overhead does no work on the barbell. He may get really tired holding the barbell, but, if it is not moved by the force he exerts, he does no work *on the barbell*. Work may be done on the muscles by stretching and contracting, which is force times distance on a biological scale, but this work is not done on the barbell. Lifting the barbell, however, is a different story. When the weight lifter raises the barbell from the floor, he does work on it.

The unit of measurement for work combines a unit of force (N) with a unit of distance (m); the unit of work is the newton-meter (N·m), also called the *joule* (J), which rhymes with *cool*. One joule of

* More generally, work is the product of only the component of force that acts in the direction of motion and the distance moved. For example, if a force acts at an angle to the motion, the component of force parallel to the motion is multiplied by the distance moved. When a force acts at right angles to the direction of motion, with no force component in the direction of motion, no work is done. A common example is a satellite in a circular orbit; the force of gravity is at right angles to its circular path and no work is done on the satellite. Hence, it orbits with no change in speed.

FIGURE 5.15
Work is done in lifting the barbell.

FIGURE 5.16
He may expend energy when he pushes on the wall, but, if the wall doesn't move, no work is done on the wall. Energy expended becomes *thermal energy*.

work is done when a force of 1 newton is exerted over a distance of 1 meter, as in lifting an apple over your head. For larger values, we speak of kilojoules (kJ, thousands of joules), or megajoules (MJ, millions of joules). The weight lifter in Figure 5.15 does work in kilojoules. To stop a loaded truck moving at 100 km/h takes megajoules of work.

The word *work*, in common usage, means physical or mental exertion. Don't confuse the physics definition of work with the everyday notion of work.

5.9 Potential Energy

An object may store energy by virtue of its position. The energy that is stored and held in readiness is called **potential energy** (PE) because in the stored state it has the potential for doing work. A stretched or compressed spring, for example, has the potential for doing work. When a bow is drawn, energy is stored in the bow. The bow can do work on the arrow. A stretched rubber band has potential energy because of the relative position of its parts. If the rubber band is part of a slingshot, it is capable of doing work.

The chemical energy in fuels is also potential energy. It is actually energy of position at the submicroscopic level. This energy is available when the positions of electric charges within and between molecules are altered—that is, when a chemical change occurs. Any substance that can do work through chemical action possesses potential energy.

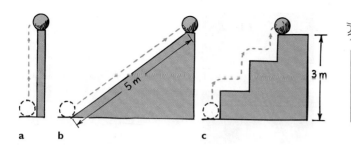

a   b   c

FIGURE 5.17

The potential energy of the 10-N ball is the same (30 J) in all three cases because the work done in elevating it 3 m is the same whether it is (a) lifted with 10 N of force, (b) pushed with 6 N of force up the 5-m incline, or (c) lifted with 10 N up each 1-m stair. No work is done in moving it horizontally (neglecting friction).

Potential energy is found in fossil fuels, electric batteries, and the foods we consume.

Work is required to elevate objects against Earth's gravity. The potential energy due to elevated positions is called *gravitational potential energy*. Water in an elevated reservoir and the raised ram of a pile driver both have gravitational potential energy. Whenever work is done, energy is exchanged.

The amount of gravitational potential energy possessed by an elevated object is equal to the work done against gravity in lifting it. The work done equals the force required to move it upward times the vertical distance it is moved (remember $W = Fd$). The upward force required while moving at constant velocity is equal to the weight, mg, of the object, so the work done in lifting it through a height h is the product mgh.

gravitational potential energy = weight × height

$$PE = mgh$$

Note that the height is the distance above some chosen reference level, such as the ground or the floor of a building. The gravitational potential energy, *mgh,* is relative to that level and depends only on mg and h. We can see, in Figure 5.17, that the potential energy of the elevated ball does not depend on the path taken to get it there.

FIGURE 5.18

Both do the same work in elevating the block.

Gravitational potential energy always involves *two* interacting objects—one relative to the other. The ram of a pile driver, for example, interacts via gravitational force with Earth.

5.10 Kinetic Energy

If you push on an object, you can set it in motion. If an object is moving, then it is capable of doing work. It has energy of motion. We say it has **kinetic energy** (KE). The kinetic energy of an object depends on the mass of the object as well as its speed. It is equal to the mass multiplied by the square of the speed, multiplied by the constant $\frac{1}{2}$.

$$\text{Kinetic energy} = \tfrac{1}{2}\,\text{mass} \times \text{speed}^2$$

$$KE = \tfrac{1}{2}mv^2$$

When you throw a ball, you do work on it to give it speed as it leaves your hand. The moving ball can then hit something and push it, doing work on what it hits. The kinetic energy of a moving object is equal to the work required to bring it from rest to that speed, or the work the object can do while being brought to rest:

$$\text{Net force} \times \text{distance} = \text{kinetic energy}$$

or, in equation notation,

$$Fd = \tfrac{1}{2}mv^2$$

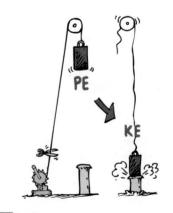

FIGURE 5.19

The potential energy of the elevated ram of the pile driver is converted to kinetic energy during its fall.

Note that the speed is squared, so, if the speed of an object is doubled, its kinetic energy is quadrupled ($2^2 = 4$). Consequently, it takes four times the work to double the speed. Whenever work is done, energy changes.

Potential energy *to* Potential + kinetic *to* Kinetic energy *to* Potential energy And so on

FIGURE 5.20

Energy transitions in a pendulum. PE is relative to the lowest point of the pendulum, when it is vertical.

FIGURE 5.21

The pendulum bob will swing to its original height whether or not the peg is present.

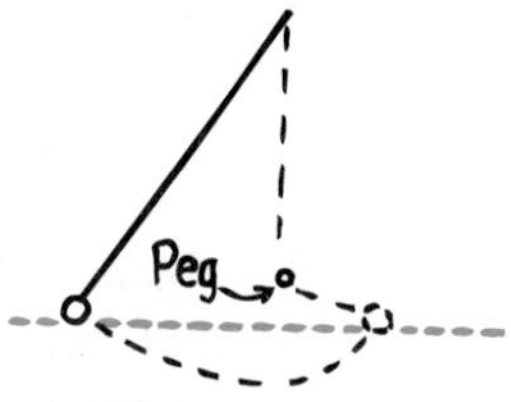

Peg

FIGURE 5.22

The potential energy of Tenny's drawn bow equals the work (average force × distance) that she did in drawing the arrow into position. When the arrow is released, most of the potential energy of the drawn bow will become the kinetic energy of the arrow.

FIGURE 5.23

The downhill "fall" of the roller coaster results in its roaring speed in the dip, and this kinetic energy sends it up the steep track to the next summit.

Work–Energy Theorem

When a car speeds up, its gain in kinetic energy comes from the work done on it. Or, when a moving car slows, work is done to reduce its kinetic energy. We can say*

$$\text{Work} = \Delta\text{KE}$$

Work equals *change* in kinetic energy. This is the **work–energy theorem**.

The work–energy theorem emphasizes the role of change. If there is no change in an object's energy, then we know no net work was done on it. This theorem applies to changes in potential energy also. Recall our previous example of the weight lifter raising the barbell. When work was being done on the barbell, its potential energy was being changed. But when it was held stationary, no further work was being done on the barbell, as evidenced by no further change in its energy.

Similarly, if you push against a box on a floor and it doesn't slide, then you are not doing work on the box. There is no change in kinetic energy. But if you push harder and it slides, then you're doing work on it. When the amount of work done to overcome friction is small, the amount of work done on the box is practically matched by its gain in kinetic energy.

The work–energy theorem applies to decreasing speed as well. Energy is required to reduce the speed of a moving object or to bring it to a halt. When we apply the brakes to slow a moving car, we do work on it. This work is the friction force supplied by the brakes multiplied by the distance over which the friction force acts. The more kinetic energy something has, the more work is required to stop it.

Interestingly, the friction supplied by the brakes is the same whether the car moves slowly or quickly. Friction between solid surfaces doesn't depend on speed. The variable that makes a difference is the braking distance. A car moving at twice the speed of another takes four times ($2^2 = 4$) as much work to stop. Therefore, it takes four times as much distance to stop. Accident investigators are well aware that an

---

* This can be derived as follows: If we multiply both sides of $F = ma$ (Newton's second law) by d, we get $Fd = mad$. Recall from Chapter 3 that, for constant acceleration, $d = \frac{1}{2}at^2$, so we can say $Fd = ma\left(\frac{1}{2}at^2\right) = \frac{1}{2}maat^2 = \frac{1}{2}m(at)^2$; and substituting $v = at$, we get $Fd = \frac{1}{2}mv^2$. That is, work = KE, or more specifically, $W = \Delta\text{KE}$.

FIGURE 5.24

Due to friction, energy is transferred both into the floor and into the tire when the bicycle skids to a stop. An infrared camera reveals the heated tire track (the red streak on the floor, *top*) and the warmth of the tire (*bottom*). (Courtesy of Michael Vollmer.)

automobile going 100 kilometers per hour has four times the kinetic energy as it would have at 50 kilometers per hour. So a car going 100 kilometers per hour will skid four times as far when its brakes are applied as when going 50 kilometers per hour. Kinetic energy depends on speed *squared*.

Automobile brakes convert kinetic energy to heat. Professional drivers are familiar with another way to slow a vehicle—shift to low gear to allow the engine to do the braking. Today's hybrid cars do the same and divert braking energy to electrical storage batteries, where it is used to complement the energy produced by gasoline combustion. (More about this in Chapter 11.) Hooray for hybrid cars!

Kinetic energy and potential energy are two among many forms of energy, and they underlie other forms of energy, such as chemical energy, nuclear energy, sound, and light. Kinetic energy of random molecular motion is related to temperature; potential energies of electric charges account for voltage; and kinetic and potential energies of vibrating air define sound intensity. Even light energy originates from the motion of electrons within atoms. Every form of energy can be transformed into every other form.

Kinetic Energy and Momentum Compared

Momentum and kinetic energy are properties of moving things, but they differ from each other. Like velocity, momentum is a vector quantity and is therefore directional and capable of being cancelled entirely. But kinetic energy is a nonvector (scalar) quantity, like mass, and can never be cancelled. The momenta of two firecrackers approaching each other may cancel, but, when they explode, there is no way their energies can cancel. Energies transform to other forms; momenta do not. Another difference is the velocity dependence of the two. Whereas momentum depends on velocity (mv), kinetic energy depends on the square of velocity ($1/2\ mv^2$). An object that moves with twice the velocity of another object of the same mass has twice the momentum but four times the kinetic energy. So when a car traveling twice as fast crashes, it crashes with four times the energy.

If the distinction between momentum and kinetic energy isn't really clear to you, you're in good company. Failure to make this distinction resulted in disagreements and arguments between the best British and French physicists for two centuries.

5.12 Conservation of Energy

Whenever energy is transformed or transferred, none is lost and none is gained. In the absence of work input or output or other energy exchanges,

---

CHECK YOURSELF

1. When you are driving at 90 km/h, how much more distance do you need to stop than if you were driving at 30 km/h?

2. For the same force, why does a longer cannon impart more speed to a cannonball?

CHECK YOUR ANSWERS

1. Nine times farther. The car has nine times as much kinetic energy when it travels three times as fast:

$$\tfrac{1}{2}m(3v)^2 = \tfrac{1}{2}m9v^2 = 9\left(\tfrac{1}{2}mv^2\right).$$

The friction force will ordinarily be the same in either case; therefore, nine times as much work requires nine times as much distance.

2. As learned earlier, a longer barrel imparts more impulse because of the longer *time* the force acts. The work–energy theorem similarly tells us that the longer the *distance* that the force acts, the greater the change will be in kinetic energy. So we see two reasons for cannons with long barrels producing greater cannonball speeds.

The Physics Place

Bowling Ball and
   Conservation of Energy
Conservation of Energy:
   Numerical Example

the total energy of a system before some process or event is equal to the total energy after.

Consider the changes in energy in the operation of the pile driver back in Figure 5.18. Work done to raise the ram, giving it potential energy, becomes kinetic energy when the ram is released. This energy transfers to the piling below. The distance the piling penetrates into the ground multiplied by the average force of impact is almost equal to the initial potential energy of the ram. We say *almost* because some energy goes into heating the ground and ram during penetration. Taking heat energy into account, we find energy transforms without net loss or net gain. Quite remarkable!

> Energy is nature's way of keeping score.

The study of various forms of energy and their transformations has led to one of the greatest generalizations in physics—the law of **conservation of energy:**

Energy cannot be created or destroyed; it may be transformed from one form into another, but the total amount of energy never changes.

When we consider any system in its entirety, whether it be as simple as a swinging pendulum or as complex as an exploding supernova, there is one quantity that isn't created or destroyed: energy. It may change form or it may simply be transferred from one place to another, but, conventional wisdom tells us, the total energy score stays the same. This energy score takes into account the fact that the atoms that make up matter are themselves concentrated bundles of energy. When the nuclei (cores) of atoms rearrange themselves, enormous amounts of energy can be released. The Sun shines because some of this nuclear energy is transformed into radiant energy.

Enormous compression due to gravity and extremely high temperatures in the deep interior of the Sun fuse the nuclei of hydrogen atoms together to form helium nuclei. This is *thermonuclear fusion,* a process that releases radiant energy, a small part of which reaches Earth. Part of the energy reaching Earth falls on plants (and on other photosynthetic organisms), and part of this, in turn, is later stored in the form of coal. Another part supports life in the food chain that begins with plants (and other photosynthesizers) and part of this energy later is stored in oil. Part of the energy from the Sun goes into the evaporation of water from the ocean, and part of this returns to Earth in rain that may be trapped

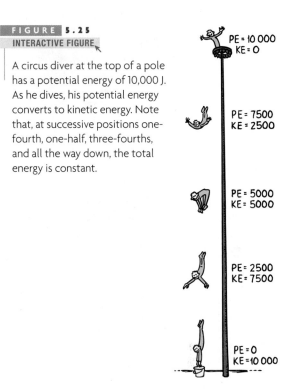

FIGURE 5.25
INTERACTIVE FIGURE

A circus diver at the top of a pole has a potential energy of 10,000 J. As he dives, his potential energy converts to kinetic energy. Note that, at successive positions one-fourth, one-half, three-fourths, and all the way down, the total energy is constant.

PE = 10 000
KE = 0

PE = 7500
KE = 2500

PE = 5000
KE = 5000

PE = 2500
KE = 7500

PE = 0
KE = 10 000

behind a dam. By virtue of its elevated position, the water behind a dam has energy that may be used to power a generating plant below, where it will be transformed to electric energy. The energy travels through wires to homes, where it is used for lighting, heating, cooking, and operating electrical gadgets. How wonderful that energy transforms from one form to another!

5.13 Power

The definition of work says nothing about how long it takes to do the work. The same amount of work is done when carrying a load up a flight of stairs, whether we walk up or run up. So why are we more out of breath after running upstairs in a few seconds than after walking upstairs in a few minutes? To understand this difference, we need to talk about a measure of how fast the work is done—*power.* **Power** is equal to the amount of work done per time it takes to do it:

$$\text{Power} = \frac{\text{work done}}{\text{time interval}}$$

The work done in climbing stairs requires more power when the worker is running up rapidly than it does when the worker is climbing slowly. A high-power automobile engine does work rapidly. An engine that delivers twice the power of another,

FIGURE 5.26

The three main engines of a space shuttle can develop 33,000 MW of power when fuel is burned at the enormous rate of 3400 kg/s. This is like emptying an average-size swimming pool in 20 s.

however, does not necessarily move a car twice as fast or twice as far. Twice the power means that the engine can do twice the work in the same amount of time—or it can do the same amount of

work in half the time. A powerful engine can produce greater acceleration.

Power is also the rate at which energy is changed from one form to another. The unit of power is the joule per second, called the *watt*. This unit was named in honor of James Watt, the eighteenth-century developer of the steam engine. One watt (W) of power is used when one joule of work is done in one second. One kilowatt (kW) equals one thousand watts. One megawatt (MW) equals one million watts.

PROBLEM SOLVING

Problems

1. Acrobat Art of mass m stands on the left end of a seesaw. Acrobat Bart of mass M jumps from a height h onto the right end of the seesaw, thus propelling Art into the air.

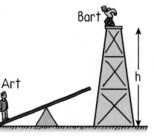

a. Neglecting inefficiencies, how will the PE of Art at the top of his trajectory compare with the PE of Bart just before Bart jumps?

b. Neglecting inefficiencies, show that Art reaches a height $\dfrac{M}{m}h$.

c. If Art's mass is 40 kg, Bart's mass is 70 kg, and the height of the initial jump was 4 m, show that Art rises a vertical distance of **7 m**.

2. A loaded elevator is lifted a distance h in a time t by a motor delivering power P.

a. Show that the force exerted by the motor is $\dfrac{Pt}{h}$.

b. If the elevator is raised 20 m in a time of 30 s, and the power of the motor is 60 kW, show that the force exerted by the motor is **90 kN**.

Solutions

1. a. Neglecting inefficiencies, the entire initial PE of acrobat Bart before he drops goes into the PE of acrobat Art rising to his peak—that is, at Art's moment of zero KE.

b. From $\text{PE}_{\text{Bart}} = \text{PE}_{\text{Art}} \Rightarrow Mgh_{\text{Bart}} = mgh_{\text{Art}} \Rightarrow h_{\text{Art}} = \dfrac{M}{m}h$.

c. $h_{\text{Art}} = \dfrac{M}{m}h = \left(\dfrac{70\ \text{kg}}{40\ \text{kg}}\right)4\ \text{m} = \mathbf{7\ m}$.

2. a. From $\text{Power} = \dfrac{\text{Work}}{\text{time}} = \dfrac{\text{force} \times \text{distance}}{\text{time}}$, we see that $P = \dfrac{Fh}{t}$. Rearranging, $F = \dfrac{Pt}{h}$.

b. $F = \dfrac{Pt}{h} = \dfrac{(60 \times 10^3\ \text{W})(30\ \text{s})}{20\ \text{m}} = 9.0 \times 10^4 \dfrac{\left(\dfrac{\text{N} \cdot \text{m}}{\text{s}}\right)\text{s}}{\text{m}} = \mathbf{9.0 \times 10^4\ N = 90\ kN}$.

5.14 Machines

A **machine** is a device for multiplying forces or simply changing the direction of forces. The principle underlying every machine is the *conservation of energy* concept. Consider one of the simplest machines, the **lever** (Figure 5.27). At the same time that we do work on one end of the lever, the other end does work on the load. We see that the direction of force is changed: if we push down, the load is lifted up. If the little work done by friction forces is small enough to neglect, the work input will be equal to the work output.

The Physics Place
Machines: Pulleys

Work input = work output

Since work equals force times distance, input force × input distance = output force output distance.

$$(\text{Force} \times \text{distance})_{\text{input}} = (\text{force} \times \text{distance})_{\text{output}}$$

The point of support on which a lever rotates is called a *fulcrum*. When the fulcrum of a lever is relatively close to the load, then a small input force will produce a large output force. This is because the input force is exerted through a large distance and the load is moved through a correspondingly short distance. So a lever can be a force multiplier. But no machine can multiply work or multiply energy. That's a conservation of energy no-no!

Today, a child can use the principle of the lever to jack up the front end of an automobile. By exerting a small force through a large distance, she can provide a large force that acts through a small distance. Consider the ideal example illustrated in Figure 5.28. Every time she pushes the jack handle down 25 centimeters, the car rises only a hundredth as far but with 100 times the force.

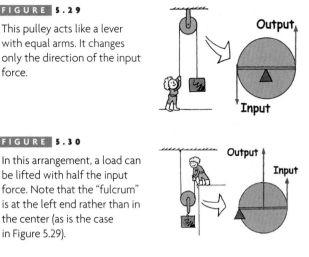

FIGURE 5.29

This pulley acts like a lever with equal arms. It changes only the direction of the input force.

FIGURE 5.30

In this arrangement, a load can be lifted with half the input force. Note that the "fulcrum" is at the left end rather than in the center (as is the case in Figure 5.29).

Another simple machine is a pulley. Can you see that it is a lever "in disguise"? When used as in Figure 5.29, it changes only the direction of the force; but, when used as in Figure 5.30, the output force is doubled. Force is increased and distance is decreased. As with any machine, forces can change while work input and work output are unchanged.

A block and tackle is a system of pulleys that multiplies force more than a single pulley can do. With the ideal pulley system shown in Figure 5.31, the man pulls 7 meters of rope with a force of 50 newtons and lifts a load of 500 newtons through a vertical distance of 0.7 meter. The energy the man expends in pulling the rope is numerically equal to the increased potential energy of the 500-newton block. Energy is transferred from the man to the load.

> A machine can multiply force, but never *energy*—no way!

Any machine that multiplies force does so at the expense of distance. Likewise, any machine that multiplies distance, such as your forearm and elbow, does so at the expense of force. No machine or device can put out more energy than is put into it.

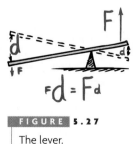

FIGURE 5.27

The lever.

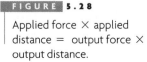

FIGURE 5.28

Applied force × applied distance = output force × output distance.

FIGURE 5.31

Applied force × applied distance = output force × output distance.

No machine can create energy; it can only transfer energy or transform it from one form to another.

5.15 Efficiency

The three previous examples were of *ideal machines*; 100% of the work input appeared as work output. An ideal machine would operate at 100% efficiency. In practice, this doesn't happen, and we can never expect it to happen. In any transformation, some energy is dissipated to molecular kinetic energy—thermal energy. This makes the machine and its surroundings warmer.

Efficiency can be expressed by the ratio

$$\text{Efficiency} = \frac{\text{useful energy output}}{\text{total energy input}}$$

Even a lever converts a small fraction of input energy into heat when it rotates about its fulcrum. We may do 100 joules of work but get out only 98 joules. The lever is then 98% efficient, and we waste 2 joules of work input as heat. In a pulley system, a larger fraction of input energy goes into heat. If we do 100 joules of work, the forces of friction acting through the distances through which the pulleys turn and rub about their axles may dissipate 60 joules of energy as heat. So the work output is only 40 joules, and the pulley system has an efficiency of 40%. The lower that the efficiency of a machine is, the greater will be the amount of energy wasted as heat.*

STOP AND CHECK YOURSELF

Consider an imaginary miracle car that has a 100% efficient engine and burns fuel that has an energy content of 40 megajoules per liter. If the air resistance and overall frictional forces on the car traveling at highway speed are 500 N, show that the distance the car could travel per liter at this speed is 80 kilometers per liter.

CHECK YOUR ANSWER

From the definition that work = force × distance, simple rearrangement gives distance = work/force. If all 40 million J of energy in 1 L were used to do the work of overcoming the air resistance and frictional forces, the distance would be:

$$\text{distance} = \frac{\text{work}}{\text{force}} = \frac{40,000,000 \text{ J/L}}{500 \text{ N}}$$
$$= 80,000 \text{ m/L} = 80 \text{ km/L}$$

(This is about 190 mpg.) The important point here is that, even with a hypothetically perfect engine, there is an upper limit of fuel economy dictated by the conservation of energy.

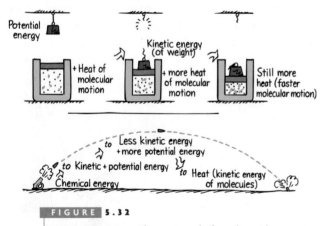

FIGURE 5.32

Energy transitions. The graveyard of mechanical energy is thermal energy.

5.16 Sources of Energy

Except for nuclear power, the source of practically all our energy is the Sun. Even the energy we obtain from petroleum, coal, natural gas, and wood originally came from the Sun. That's because these fuels are created by photosynthesis—the process by which plants trap solar energy and store it as plant tissue.

* When you study thermodynamics in Chapter 8, you'll learn that an internal combustion engine *must* transform some of its fuel energy into thermal energy. A fuel cell, on the other hand, which could power future automobiles and trucks, doesn't have this limitation. Watch for vehicles powered by fuel cells on the horizon!

Sunlight can be directly transformed into electricity by photovoltaic cells, like those found in solar-powered calculators and flexible solar-powered shingles for rooftops of buildings. They can collect energy to operate railroad trains (Figure 5.33). Sunlight evaporates water, which later falls as rain; rainwater flows into rivers and into dams where it is directed to generator turbines. Then it returns to the sea, where the cycle continues. Even the wind, caused by unequal warming of Earth's surface, is a form of solar power. The energy of wind can be used to turn generator turbines within specially equipped windmills. Because wind power can't be turned on and off at will, it is now only a supplement to fossil and nuclear fuels for large-scale power production. Harnessing the wind is most practical when the energy it produces is stored for future use, such as in the form of hydrogen.

Hydrogen holds much promise for the future, the least polluting of all fuels. Because it takes energy to make hydrogen (to extract it from water and carbon compounds), it is not a *source* of energy. Rather, like electricity, it needs an energy source and is a way of storing and transporting that energy. Wind or solar power to make hydrogen is an attractive, environmentally friendly option for the future. Again, for emphasis, hydrogen is *not* an energy source.

Most hydrogen in America is produced from natural gas, where high temperatures and pressures separate hydrogen from hydrocarbon molecules. The same is done with fossil fuels. A downside to hydrogen separation from carbon compounds is the unavoidable production of carbon dioxide, a greenhouse gas. A simpler and cleaner method that doesn't produce greenhouse gases is *electrolysis*—electrically splitting water into its constituent parts. Figure 5.34 shows how you can perform this in lab, or at home: Place two wires that are connected to the terminals of an ordinary battery into a glass of salted water. Be sure the wires don't touch each other. Bubbles of hydrogen form on one wire, and bubbles of oxygen form on the other. A fuel cell is similar, but runs backward. Hydrogen and oxygen gas are compressed at electrodes and electric current is formed, along with water. The space shuttle uses fuel cells to meet its electrical needs while producing drinking water for the astronauts. Here on Earth, fuel-cell researchers are developing fuel cells for buses, automobiles, and trains. Watch for the growth of fuel-cell technology. The major hurdle for this technology is not the device itself, but with economically acquiring hydrogen fuel.

FIGURE 5.33

The power harvested by photovoltaic cells, visible here on railroad ties, can be used to separate hydrogen for fuel-cell transportation. Plans for trains that run on solar power collected on ties, station roofs and trains, are now on the drawing-board (www.SuntrainUSA.com).

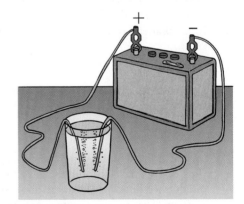

FIGURE 5.34

When electric current passes through salted water, ideally, bubbles of hydrogen form at one wire and bubbles of oxygen form at the other. This is *electrolysis*. A fuel cell does the opposite—hydrogen and oxygen enter the fuel cell and are combined to produce electricity and water.

The most concentrated form of usable energy is stored in uranium and plutonium, which are nuclear fuels. For the same weight of fuel, nuclear reactions release about 1 million times more energy than do chemical or food reactions. Watch for renewed interest in this form of power that doesn't pollute the atmosphere. Interestingly, Earth's interior is kept hot because of nuclear power, which has been with us since time zero.

A byproduct of nuclear power in Earth's interior is geothermal energy. Geothermal energy is held in underground reservoirs of hot water. Geothermal energy is predominantly limited to areas of volcanic activity, such as Iceland, New Zealand, Japan, and Hawaii. In these locations, heated water near Earth's surface is tapped to provide steam for driving turbogenerators.

In locations where heat from volcanic activity is near the ground surface and ground water is absent, another method holds promise for producing electricity. That's dry-rock geothermal power (Figure 5.35). With this method, water is put into cavities in deep, dry, hot rock. When the water turns to steam, it is piped to a turbine at the surface. After turning the turbine, it is returned to the cavity for reuse. In this way, electricity is produced inexpensively and cleanly.

> Inventors take heed: When introducing a new idea, first be sure it is consistent with the conservation of energy.

As the world population increases, so does our need for energy, especially since per capita demand is also growing. With the rules of physics to guide them, technologists are right now researching newer and cleaner ways to develop energy sources. But they race to keep ahead of a growing world population

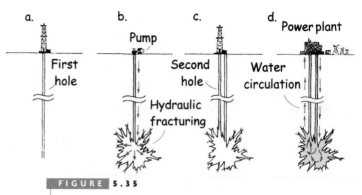

FIGURE 5.35

Dry-rock geothermal power. (a) A hole is sunk several kilometers into dry granite. (b) Water is pumped into the hole at high pressure and fractures surrounding rock to form a cavity with increased surface area. (c) A second hole is sunk to intercept the cavity. (d) Water is circulated down one hole and through the cavity, where it is superheated, before rising through the second hole. After driving a turbine, it is recirculated into the hot cavity again, making a closed cycle.

and greater demand in the developing world. Unfortunately, so long as controlling population is politically and religiously incorrect, human misery becomes the check to unrestrained population growth. H. G. Wells once wrote (in *The Outline of History*), "Human history becomes more and more a race between education and catastrophe."

> Another source of energy is tidal power, where the surging of tides turns turbines to produce power. Interestingly, this form of energy is neither nuclear nor from the Sun. Its source is the rotational energy of our planet.

SCIENCE AND SOCIETY

■ JUNK SCIENCE

Scientists have to be open to new ideas. That's how science develops. But there is a body of established knowledge that can't be easily overthrown. That includes energy conservation, which is woven into every branch of science and supported by countless experiments from the atomic to the cosmic scale. Yet no concept has inspired more "junk science" than energy. Wouldn't it be wonderful if we could get energy

for nothing, to possess a machine that puts out more energy than is put into it? That's what many practitioners of junk science offer. Gullible investors put their money into some of these schemes. But none of them pass the test of being real science. Perhaps some day a flaw in the law of energy conservation will be discovered. If it ever is, scientists will rejoice at the breakthrough. But so far, energy conservation is as solid as any knowledge we have. Don't bet against it.

SUMMARY OF TERMS

Momentum The product of the mass of an object and its velocity.

Impulse The product of the force acting on an object and the time during which it acts.

Relationship of impulse and momentum Impulse is equal to the change in the momentum of the object that the impulse acts upon. In symbol notation,

$$Ft = \Delta mv$$

Law of conservation of momentum In the absence of an external force, the momentum of a system remains unchanged. Hence, the momentum before an event involving only internal forces is equal to the momentum after the event:

$$mv_{\text{(before event)}} = mv_{\text{(after event)}}$$

Elastic collision A collision in which colliding objects rebound without lasting deformation or the generation of heat.

Inelastic collision A collision in which the colliding objects become distorted, generate heat, and possibly stick together.

Energy The property of a system that enables it to do work.

Work The product of the force and the distance moved by the force:

$$W = Fd$$

(More generally, work is the component of force in the direction of motion times the distance moved.)

Potential energy The energy that matter possesses because of its position:

$$\text{Gravitational PE} = mgh$$

Kinetic energy Energy of motion, quantified by the relationship

$$\text{Kinetic energy} = \tfrac{1}{2}mv^2$$

Work–energy theorem The work done on an object equals the change in kinetic energy of the object:

$$\text{Work} = \Delta KE$$

(Work can also transfer other forms of energy to a system.)

Conservation of energy Energy cannot be created or destroyed; it may be transformed from one form into another, but the total amount of energy never changes.

Power The rate of doing work:

$$\text{Power} = \frac{\text{work done}}{\text{time interval}}$$

(More generally, power is the rate at which energy is expended.)

Machine A device, such as a lever or pulley, that increases (or decreases) a force or simply changes the direction of a force.

Lever Simple machine consisting of a rigid rod pivoted at a fixed point called the fulcrum.

Efficiency The percentage of the work put into a machine that is converted into useful work output:

$$\text{Efficiency} = \frac{\text{useful energy output}}{\text{total energy input}}$$

SUGGESTED READING

Bodanis, David. *E = mc²: A Biography of the World's Most Famous Equation*. New York: Berkley Publishing Group, 2002. A delightful and engaging history of our understanding of energy, and insight into the personalities behind these discoveries.

REVIEW QUESTIONS

5.1 Momentum

1. Which has a greater momentum, a heavy truck at rest or a moving skateboard?

5.2 Impulse

2. For the same force, why does a long cannon impart more speed to a cannonball than a small cannon would impart input?

5.3 Impulse Changes Momentum

3. Why is it a good idea to have your hand extended forward when you are getting ready to catch a fast-moving baseball with your bare hand?
4. Why would it be a poor idea to have the back of your hand up against the outfield wall when you catch a long fly ball?
5. In karate, why is a force that is applied for a short time more advantageous?
6. In boxing, why is it advantageous to roll with the punch?

5.4 Bouncing

7. Which undergoes the greatest change in momentum: (a) a baseball that is caught, (b) a baseball that is thrown, or (c) a baseball that is caught and then thrown back, if all of the baseballs have the same speed just before being caught and just after being thrown?
8. In the preceding question, in which case is the greatest impulse required?

5.5 Conservation of Momentum

9. What does it mean to say that momentum (or any quantity) is *conserved*?
10. When a cannonball is fired, momentum is conserved for the *system* of cannon + cannonball. Would momentum be conserved for the system if momentum were not a vector quantity? Explain.

5.6 Collisions

11. Railroad car A rolls at a certain speed and makes a perfectly elastic collision with car B of the same mass. After the collision, car A is observed to be at rest. How does the speed of car B compare with the initial speed of car A?
12. If the equally massive cars of the previous question stick together after colliding inelastically, how does their speed after the collision compare with the initial speed of car A?

5.7 Energy

13. When is energy most evident?

5.8 Work

14. Cite an example in which a force is exerted on an object without doing work on the object.
15. Which, if either, requires more work—lifting a 50-kg sack a vertical distance of 2 m or lifting a 25-kg sack a vertical distance of 4 m?

5.9 Potential Energy

16. A car is raised a certain distance in a service-station lift and therefore has potential energy relative to the floor. If it were raised twice as high, how much potential energy would it have?

17. Two cars are raised to the same elevation on service-station lifts. If one car is twice as massive as the other, how do their potential energies compare?

5.10 Kinetic Energy

18. A moving car has kinetic energy. If it speeds up until it is going four times as fast, how much kinetic energy does it have in comparison?

5.11 Work–Energy Theorem

19. Compared with some original speed, how much work must the brakes of a car supply to stop a car that is moving four times as fast? How will the stopping distance compare?
20. If you push a crate horizontally with 100 N across a 10-m factory floor, and friction between the crate and the floor is a steady 70 N, how much kinetic energy is gained by the crate?

5.12 Conservation of Energy

21. What will be the kinetic energy of a pile driver ram when it undergoes a 10-kJ decrease in potential energy?
22. An apple hanging from a limb has potential energy because of its height. If the apple falls, what becomes of this energy just before it hits the ground? When it hits the ground?

5.13 Power

23. If two equal-mass sacks are lifted equal distances in the same time, how does the power required for each compare? How much power is required for the case in which the lighter sack is moved its distance in half the time?

5.14 Machines

24. Can a machine multiply input force? Input distance? Input energy? (If your three answers are the same, seek help, for the last question is especially important.)
25. If a machine multiplies force by a factor of four, what other quantity is diminished, and by how much?

5.15 Efficiency

26. What is the efficiency of a machine that miraculously converts all the input energy to useful output energy?
27. What happens to the percentage of useful energy as it is transformed from one form to another?

5.16 Sources of Energy

28. What is the ultimate source of energies for the burning of fossil fuels, dams, and windmills?
29. What is the ultimate source of geothermal energy?
30. Can we correctly say that a new source of energy is hydrogen? Why or why not?

ACTIVE EXPLORATION

When you get a bit ahead in your studies, cut classes some afternoon and visit your local pool or billiards parlor and bone up on momentum conservation. Note that, no matter how complicated the collision of balls, the momentum along the line of action of the cue ball before impact is the same as the combined momentum of all the balls along this direction after impact and that the components of momenta perpendicular to this line of action cancel to

zero after impact, the same value as before impact in this direction. You'll see both the vector nature of momentum and its conservation more clearly when rotational skidding, "English," is not imparted to the cue ball. When English is imparted by striking the cue ball off center, rotational momentum, which is also conserved, somewhat complicates analysis. But, regardless of how the cue ball is struck, in the absence of external forces, both linear and rotational momentum are always conserved. Both pool and billiards offer a first-rate exhibition of momentum conservation in action.

ONE-STEP CALCULATIONS

Momentum = *mv*

1. What is the momentum of an 8-kg bowling ball rolling at 2 m/s?
2. What is the momentum of a 50-kg carton that slides at 4 m/s across an icy surface?

Impulse = *Ft*

3. What impulse occurs when an average force of 10 N is exerted on a cart for 2.5 s?
4. What impulse occurs when the same force of 10 N acts on the cart for twice the time?

Impulse = change in momentum: *Ft* = Δ*mv*

5. What is the impulse on a 4-kg ball rolling at 3 m/s when it bumps into a pile of hay and stops?
6. How much impulse stops a 50-kg carton sliding at 4 m/s when it meets a rough surface?

Conservation of momentum: $mv_{before} = mv_{after}$

7. A 2-kg blob of putty moving at 3 m/s slams into a 2-kg blob of putty at rest. Calculate the speed of the two stuck-together blobs of putty immediately after colliding.
8. Calculate the speed of the two blobs if the one at rest is 4 kg.

Work = force × distance: *W = Fd*

9. Calculate the work done when a 20-N force pushes a cart 3.5 m.
10. Calculate the work done in lifting a 500-N barbell 2.2 m above the floor. (What is the potential energy of the barbell when it is lifted to this height?)

Gravitational potential energy = weight × height: PE = *mgh*

11. Calculate the increase in potential energy when a 20-kg block of ice is lifted a vertical distance of 2 m.
12. Calculate the change in potential energy of 8 million kg of water falling 50 m over Niagara Falls.

Kinetic energy = $\frac{1}{2}$mass × speed²: KE $\frac{1}{2}mv^2$

13. Calculate the kinetic energy of a 3-kg toy cart that moves at 4 m/s.
14. Calculate the kinetic energy of the same cart moving at twice the speed.

Work–energy theorem: Work = ΔKE

15. How much work is required to increase the kinetic energy of a car by 5000 J?
16. What change in kinetic energy does an airplane experience on takeoff if it is moved a distance of 500 m by a sustained net force of 5000 N?

Power = $\dfrac{\text{work done}}{\text{time interval}}$: $P = \dfrac{W}{t}$

17. Calculate the power expended when a 20-N force pushes a cart 3.5 m in a time of 0.5 s.
18. Calculate the power expended when a 500-N barbell is lifted 2.2 m in 2 s.

EXERCISES

1. To bring a supertanker to a stop, its engines are typically cut off about 25 km from port. Why is it so difficult to stop or turn a supertanker?
2. In terms of impulse and momentum, why do airbags in cars reduce the chances of injury in accidents?
3. Why do gymnasts use floor mats that are very thick?
4. In terms of impulse and momentum, why are nylon ropes, which stretch considerably under tension, favored by mountain climbers?
5. Automobiles were previously manufactured to be as rigid as possible, whereas today's autos are designed to crumple upon impact. Why?
6. In terms of impulse and momentum, why is it important that helicopter blades deflect air downward?
7. A lunar vehicle is tested on Earth at a speed of 10 km/h. When it travels as fast on the Moon, is its momentum more, less, or the same?
8. If you throw a raw egg against a wall, you'll break it; but, if you throw it with the same speed into a sagging sheet, it won't break. Explain, using concepts from this chapter.
9. If a ball is projected upward from the ground with 10 kg · m/s of momentum, what is the momentum of recoil of Earth? Why do we not feel this?
10. Why do 6-ounce boxing gloves hit harder than 16-ounce gloves?
11. A boxer can punch a heavy bag for more than an hour without tiring, but will tire quickly when boxing with an opponent for a few minutes. Why? (Hint: When the boxer's fist is aimed at the bag, what supplies the impulse to stop the punches? When the boxer's fist is aimed at the opponent, what or who supplies the impulse to stop the punches that don't connect?)
12. Railroad cars are loosely coupled so that there is a noticeable time delay from the time the first car is moved until the last cars are moved from rest by the locomotive. Discuss the advisability of this loose coupling and slack between cars from the point of view of impulse and momentum.

13. You are at the front of a floating canoe near a dock. You jump, expecting to land on the dock easily. Instead, you land in the water. Explain.
14. A fully dressed person is at rest in the middle of a pond on perfectly frictionless ice and must get to shore. How can this be accomplished? Explain in terms of momentum conservation.
15. If you throw a ball horizontally while standing on roller skates, you roll backward with a momentum that matches that of the ball. Will you roll backward if you go through the motions of throwing the ball, but instead hold on to it? Explain in terms of momentum conservation.

16. The examples of the two previous exercises can be explained both in terms of momentum conservation and in terms of Newton's third law. Explain your answers to Exercises 14 and 15 in terms of Newton's third law.
17. In the previous chapter, rocket propulsion was explained in terms of Newton's third law. That is, the force that propels a rocket is from the exhaust gases pushing against the rocket, the reaction to the force the rocket exerts on the exhaust gases. Explain rocket propulsion in terms of momentum conservation.
18. Your friend says that the law of momentum conservation is violated when a ball rolls down a hill and gains momentum. What do you say?
19. The momentum of an apple falling to the ground is not conserved because the external force of gravity acts on it. But momentum is conserved in a larger system. Explain.
20. Drop a stone from the top of a high cliff. Identify the system wherein the net momentum is zero as the stone falls.
21. Bronco dives from a hovering helicopter and finds his momentum increasing. Does this violate the conservation of momentum? Explain.
22. An ice sail craft is stalled on a frozen lake on a windless day. The skipper sets up a fan as shown. If all the wind bounces backward from the sail, will the craft be set in motion? If so, in what direction?

23. Will your answer to the preceding exercise be different if the air is brought to a halt by the sail without bouncing?
24. Discuss the advisability of simply removing the sail in the preceding exercises.
25. When vertically falling sand lands in a horizontally moving cart, the cart slows. Ignore any friction between the cart and the tracks. Give two reasons for this, one in terms of a horizontal force acting on the cart and one in terms of momentum conservation.

26. In a movie, the hero jumps straight down from a bridge onto a small boat that continues to move with no change in velocity. What physics is being violated here?
27. Suppose that there are three astronauts outside a spaceship and that they decide to play catch. All the astronauts weigh the same on Earth and are equally strong. The first

astronaut throws the second one toward the third one and the game begins. Describe the motion of the astronauts as the game proceeds. How long is the duration of the game?

28. To throw a ball, do you exert an impulse on it? Do you exert an impulse to catch it at the same speed? About how much impulse do you exert, in comparison, if you catch it and immediately throw it back again? (Imagine yourself on a skateboard.)

29. If your friend pushes a lawnmower four times as far as you do while exerting only half the force, which one of you does more work? How much more?

30. Which requires more work: stretching a strong spring a certain distance or stretching a weak spring the same distance? Defend your answer.

31. Two people who weigh the same climb a flight of stairs. The first person climbs the stairs in 30 s, while the second person climbs them in 40 s. Which person does more work? Which uses more power?

32. When a rifle with a longer barrel is fired, the force of expanding gases acts on the bullet for a longer distance. What effect does this have on the velocity of the emerging bullet? (Do you see why long-range cannons have such long barrels?)

33. Your friend says that the kinetic energy of an object depends on the reference frame of the observer. Explain why you agree or disagree.

34. You and a flight attendant toss a ball back and forth in an airplane in flight. Does the KE of the ball depend on the speed of the airplane? Carefully explain.

35. A baseball and a golf ball have the same momentum. Which has the greater kinetic energy?

36. At what point in its motion is the KE of a pendulum bob at a maximum? At what point is its PE at a maximum? When its KE is at half its maximum value, how much PE does it possess?

37. A physics instructor demonstrates energy conservation by releasing a heavy pendulum bob, as shown in the sketch, allowing it to swing to and fro. What would happen if, in his exuberance, he gave the bob a slight shove as it left his nose? Explain.

38. Why does the force of gravity do work on a car that rolls down a hill, but no work when it rolls along a level part of the road?

39. On a playground slide, a child has potential energy that decreases by 1000 J while her kinetic energy increases by 900 J. What other form of energy is involved, and how much?

40. Discuss the design of the roller coaster shown in the sketch in terms of the conservation of energy.

41. Suppose that you and two classmates are discussing the design of a roller coaster. One classmate says that each summit must be lower than the previous one. Your other classmate says this is nonsense, for as long as the first one is the highest, it doesn't matter what height the others are. What do you say?

42. Consider the identical balls released from rest on tracks A and B, as shown. When they reach the right ends of the tracks, which will have the greater speed? Why is this question easier to answer than the similar one (Exercise 31) in Chapter 3?

43. If a golf ball and a Ping-Pong ball both move with the same kinetic energy, can you say which has the greater speed? Explain in terms of the definition of KE. Similarly, in a gaseous mixture of massive molecules and light molecules with the same average KE, can you say which have the greater speed?

44. Does a car burn more gasoline when its lights are turned on? Does the overall consumption of gasoline depend on whether or not the engine is running while the lights are on? Defend your answer.

45. This may seem like an easy question for a physics type to answer: With what force does a rock that weighs 10 N strike the ground if dropped from a rest position 10 m high? In fact, the question cannot be answered unless you have more information. What information?

46. In the absence of air resistance, a ball thrown vertically upward with a certain initial KE will return to its original level with the same KE. When air resistance is a factor affecting the ball, will it return to its original level with the same, less, or more KE? Does your answer contradict the law of energy conservation?

47. You're on a rooftop and you throw one ball downward to the ground below and another upward. The second ball, after rising, falls and also strikes the ground below. If air resistance can be neglected, and if your downward and upward initial speeds are the same, how will the speeds of the balls compare upon striking the ground? (Use the idea of energy conservation to arrive at your answer.)

48. When a driver applies brakes to keep a car going downhill at constant speed and constant kinetic energy, the potential energy of the car decreases. Where does this energy go? Where does most of it appear in a hybrid vehicle?

49. Does the KE of a car change more when it goes from 10 to 20 km/h or when it goes from 20 to 30 km/h?

50. Can something have energy without having momentum? Explain. Can something have momentum without having energy? Explain.

51. When the mass of a moving object is doubled with no change in speed, by what factor is its momentum changed? By what factor is its kinetic energy changed?

52. When the velocity of an object is doubled, by what factor is its momentum changed? By what factor is its kinetic energy changed?

53. Which, if either, has greater momentum: a 1-kg ball moving at 2 m/s or a 2-kg ball moving at 1 m/s? Which has greater kinetic energy?

54. If two objects have equal kinetic energies, do they necessarily have the same momentum? Defend your answer.

55. Two lumps of clay with equal and opposite momenta have a head-on collision and come to rest. Is momentum conserved? Is kinetic energy conserved? Why are your answers the same or different?

56. Consider the swinging-balls apparatus. If two balls are lifted and released, momentum is conserved as two balls pop out the other side with the same speed as the released balls at impact. But momentum would also be conserved if one ball popped out at twice the speed. Explain why this never happens.

57. If an automobile were to have a 100% efficient engine, transferring all of the fuel's energy to work, would the engine be warm to your touch? Would its exhaust heat the surrounding air? Would it make any noise? Would it vibrate? Would any of its fuel go unused?

58. To combat wasteful habits, we often speak of "conserving energy," by which we mean turning off lights and hot water when they are not being used, and keeping thermostats at a moderate level. In this chapter, we also speak of "energy conservation." Distinguish between these two usages.

59. Your friend says that one way to improve air quality in a city is to have traffic lights synchronized so that motorists can travel long distances at constant speed. What physics principle supports this claim?

60. The energy we require to live comes from the chemically stored potential energy in food, which is transformed into other energy forms during the metabolism process. What happens to a person whose combined work and heat output is less than the energy consumed? What happens when the person's work and heat output is greater than the energy consumed? Can an undernourished person perform extra work without extra food? Defend your answers.

PROBLEMS

● BEGINNER ■ INTERMEDIATE ◆ EXPERT

1. ● In Chapter 3 we learned that acceleration is defined by $a = \frac{\Delta v}{\Delta t}$, and in Chapter 4 we learned the physics of acceleration: it is force divided by mass, $a = \frac{F}{m}$. Equate these two equations for acceleration and show that, for constant mass, $F\Delta t = \Delta(mv)$.

2. ● A 5-kg bag of groceries is tossed across the surface of a table at 4 m/s and slides to a stop in 3 s. Show that the average force of friction is 6.7 N.

3. ● An 8-kg ball rolling at 2 m/s bumps into a pillow and stops in 0.5 s.
a. Show that the average force exerted by the pillow is 32 N.
b. How much force does the ball exert on the pillow?

4. ● A car crashes into a wall at 25 m/s and its contents are brought to rest in 0.1 s. Show that the average force exerted on a 75-kg test dummy by the seat belt is more than 18,000 N.

5. ● At a ball game, consider a baseball of mass $m = 0.15$ kg that is moving at a speed $v = 40$ m/s as it is grabbed by a fan.
a. Show that the impulse supplied to bring the ball to rest is 6.0 N · s.
b. If the ball is stopped in 0.03 s, show that the average force of the ball on the catcher's hand is 200 N.

6. ■ Judy (mass 40.0 kg), standing on slippery ice, catches her leaping dog, Atti (mass 15 kg), moving horizontally at 3.0 m/s. Show that the speed of Judy and her dog after the catch is 0.8 m/s.

7. ■ A railroad diesel engine weighs four times as much as a freight car. The diesel engine coasts at 5 km/h into the freight car that is initially at rest. After they couple together, show that the two coast at 4 km/h.

8. ■ A 5-kg fish swimming 1 m/s swallows an absent-minded 1-kg fish swimming toward it at a velocity that brings both fish to a halt immediately after lunch. Show that the speed of the smaller fish before lunch was 5 m/s.

9. ■ Comic-strip hero Superman meets an asteroid in outer space and hurls it at 800 m/s, as fast as a bullet. The asteroid is a thousand times more massive than Superman. In the strip, Superman is seen at rest after the throw. Taking physics into account, show that his recoil speed would be 800,000 m/s.

10. ● The second floor of a house is 4 m above street level. Show that the work required to lift a 300-kg refrigerator to the second-floor level is 12,000 J.

11. ● Belly-flop Bernie dives from atop a tall flagpole into a swimming pool below. His potential energy at the top is 10,000 J and at the water surface it is zero.
    a. Show that when his potential energy reduces to 1000 J, his kinetic energy will be 9000 J.
    b. Compared with the height of the flagpole, how far above water level will Bernie be when his kinetic energy is 9000 J?

12. ● This question is similar to some on driver's-license exams: A car moving at 50 km/h skids 15 m with locked brakes. Show that with locked brakes at 150 km/h the car will skid 135 m.

13. ● A lever is used to lift a heavy load. When a 50-N force pushes one end of the lever down 1.2 m, the load rises 0.2 m. Show that the weight of the load is 300 N.

14. ● In raising a 5000-N piano with a pulley system, the workers note that, for every 2 m of rope pulled down, the piano rises 0.2 m. Ideally, show that the force required to lift the piano is 500 N.

15. ■ If we multiply both sides of the equation for Newton's second law, $F_{net} = ma$, by d, we get $F_{net}d = mad$. In this chapter we learn that Fd equals work, and in Chapter 3 we learned that the distance traveled by an object starting from rest and undergoing constant acceleration a is $d = \frac{1}{2}at^2$. Show that when this equation for distance is substituted for d on the right side of the equation above, the result is $F_{net}d$ (work) $= \frac{1}{2}mv^2$ (kinetic energy).

16. ■ A braking force is needed to bring a car of mass m moving at speed v to rest in time t.
    a. Show that the braking force is mv/t.
    b. The mass of the car is 1200 kg and its initial speed is 25 m/s. Show that the braking force needed to stop it in 12 s is 2500 N.

17. ■ A block of mass m moving at a speed v is stopped by a constant force F.
    a. Show that the time required to stop the block is mv/F.
    b. If the mass of the block is 20.0 kg, its initial speed is 3.0 m/s, and the stopping force is 15.0 N, show that the time to stop the block is 4.0 s.

18. ■ A goofy parrot of mass m drops vertically onto a skateboard of mass M that rolls horizontally at a speed v. The parrot grabs it tightly and moves with the skateboard.
    a. Show that the speed of the skateboard with the parrot is $\dfrac{M}{M + m}v$.
    b. If the parrot's mass is 2.0 kg, the mass of the skateboard is 8.0 kg, and the initial speed of the skateboard was 4.0 m/s, show that the final speed is 3.2 m/s.

19. ◆ An astronaut of mass M floating next to his spacecraft in deep space becomes untethered. To return to the craft, he throws a hammer of mass m at speed v away from the craft.
    a. Show that the astronaut will recoil back toward the spacecraft at a speed $\dfrac{mv}{M}$.
    b. If the astronaut's mass is 110 kg and the mass of the hammer is 15 kg, and the hammer is tossed at 4.5 m/s, show that the recoil speed of the astronaut is 0.6 m/s.

20. ◆ A lump of putty of mass m_1 and velocity v_1 catches up with and bumps into a slower lump of putty of mass m_2 and velocity v_2 heading in the same direction. They share a common velocity after they stick together.
    a. Show that this common velocity is $\dfrac{m_1v_1 + m_2v_2}{m_1 + m_2}$.
    b. If the mass of the first lump of putty is 2.2 kg with an initial speed of 3.2 m/s, and the second lump of putty is 2.8 kg with an initial speed of 1.2 m/s, show that the final speed of the combined lumps is 2.1 m/s.

21. ◆ An oil tanker of mass M travels a distance x at constant speed in time t.
    a. Show that the momentum of the tanker is Mx/t.
    b. Show that the KE of the oil tanker is $\dfrac{Mx^2}{2t^2}$.
    c. If the mass of the oil tanker is 9.0×10^7 kg and it sails 250 km in 8.0 hours at a constant speed, show that its KE is 3.4×10^9 J. (Useful information: 1 km = 1000 m and 1 hour = 3600 s.)

22. ◆ Hank bats a baseball that weighs w and leaves the bat with a speed v.
    a. Show that the baseball's momentum is wv/g.
    b. Show that the baseball's KE is $\dfrac{wv^2}{2g}$.
    c. If the weight of the baseball is 1.5 N and it leaves the bat with a speed of 38 m/s, show that its KE is 110 J.

23. ◆ When an average force F is exerted over a certain distance on a shopping cart of mass m, its kinetic energy increases by $\frac{1}{2}mv^2$.
    a. Show that the distance over which the force acts is $\dfrac{mv^2}{2F}$.
    b. If twice the force is exerted over twice the distance, how does the resulting increase in kinetic energy compare with the original increase in kinetic energy?

24. ◆ Manuel drops a water balloon of mass m from rest atop the roof of a building of unknown height. The balloon takes time t to hit the ground below.
    a. Show that if air resistance is negligible, its kinetic energy just before it hits is $\frac{1}{2}mg^2t^2$.
    b. If the water balloon has a mass of 1.2 kg and the dropping time from rest is 2.0 s, show that its kinetic energy when it hits the ground is 230 J.
    c. Why can't the force of impact be found with the information given?

25. ◆ A block of ice of mass m is at rest atop an inclined plane of vertical height h. It then slides down the incline, reaching the floor below at speed v.
    a. If friction can be ignored, show that the speed of the ice when it reaches the floor is $\sqrt{2gh}$.
    b. If the block's mass is 27 kg and it slides down an incline of vertical height 1.5 m, show that its speed at the bottom is 5.4 m/s.

26. ◆ A motor of peak power P raises an elevator of mass m.
    a. Show that the maximum speed at which the motor can raise the elevator is P/mg.
    b. The elevator has a mass 900 kg and is powered by a 100-kW motor. Show that the maximum speed at which the elevator can be raised is 11 m/s.

CHAPTER 5 ONLINE RESOURCES

The Physics Place

Interactive Figures
5.9, 5.11, 5.12, 5.13, 5.22, 5.25

Tutorials
Momentum and Collisions
Energy

Videos
Definition of Momentum
Changing Momentum: Follow-Through

Decreasing Momentum Over a Short Time
Bowling Ball and Conservation of Energy
Conservation of Energy: Numerical Example
Machines: Pulleys

Quiz

Flashcards

Links

CHAPTER 6

Gravity, Projectiles, and Satellites

Neil deGrasse Tyson emphasizes the universal nature of gravity.

Newton did not discover gravity. That discovery dates back to earlier times, when Earth dwellers experienced the consequences of tripping and falling. What Newton found was that gravity is universal—that it is not unique to Earth, as others of his time assumed.

From the time of Aristotle, the circular motion of heavenly bodies was regarded as natural. The ancients believed that the stars, the planets, and the Moon move in divine circles, free from any impelling forces. As far as the ancients were concerned, this circular motion required no explanation. Isaac Newton, however, recognized that a force of some kind must act on the planets (whose orbits followed elliptical paths); otherwise, their paths would be straight lines.

Others of his time, influenced by Aristotle, supposed that any force on a planet would be directed along its path. Newton, however, reasoned that the force on each planet would be directed toward a fixed central point—toward the Sun. This force of gravity was the same force that pulls an apple off a tree. Newton's stroke of intuition, that the force between Earth and an apple is the same as the force that acts between moons and planets and everything else in our universe, was a revolutionary break with the prevailing notion that there were two sets of natural laws: one for earthly events, and another for motion in the heavens. This union of terrestrial laws and cosmic laws is called the Newtonian synthesis.

6.1 # The Universal Law of Gravity

According to popular legend, Newton was sitting under an apple tree when the idea struck him that gravity extends beyond Earth. Perhaps he looked up through tree branches toward the origin of the falling apple and noticed the Moon. Perhaps the apple hit him on the head, as popular stories tell us. In any event, Newton had the insight to see that the force between Earth and a falling apple is the same force that pulls the Moon in an orbital path around Earth, a path similar to a planet's path around the Sun.

The Physics Place
Motion and Gravity

To test this hypothesis, Newton compared the fall of an apple with the "fall" of the Moon. He realized that the Moon falls in the sense that *it falls away from the straight line it would follow if there were no forces acting on it.* Because of its tangential velocity, it "falls around" the round Earth (as we shall investigate later in this chapter). By simple geometry, the Moon's distance of fall per second could be compared with the distance that an apple or anything that far away would fall in one second. Newton's calculations didn't agree. Disappointed, but recognizing that brute fact must always win over a beautiful hypothesis, he placed his papers in a drawer, where they remained for nearly 20 years. During this period, he founded and developed the field of geometric optics, for which he first became famous.

Newton's interest in mechanics was rekindled with the advent of a spectacular comet in 1680 and another comet two years later. He returned to the Moon problem at the prodding of his astronomer friend, Edmund Halley, for whom the second comet was later named. He made corrections in the experimental data used in his earlier method and obtained excellent results. Only then did he publish what is one of the most far-reaching generalizations of the human mind: the **law of universal gravitation.**[*]

Everything pulls on everything else in a beautifully simple way that involves only mass and distance. According to Newton, any body attracts any other body with a force that is directly proportional to the product of their masses and inversely proportional to the square of the distance separating them.

This statement can be expressed as

$$\text{Force} \sim \frac{\text{mass}_1 \times \text{mass}_2}{\text{distance}^2}$$

or symbolically as

$$F \sim \frac{m_1 m_2}{d^2}$$

where m_1 and m_2 are the masses of the bodies and d is the distance between their centers. Thus, the greater the masses m_1 and m_2 are, the greater the force of attraction between them will be, in direct proportion to the masses.[**] The greater that the distance of separation d is, the weaker the force of attraction will be, in inverse proportion to the square of the distance between their centers of mass.

> Just as sheet music guides a musician playing music, equations guide a physics student to understand how concepts are connected.

[*] This is a dramatic example of the painstaking effort and cross-checking that go into the formulation of a scientific theory. Contrast Newton's approach with the failure to "do one's homework," the hasty judgments, and the absence of cross-checking that so often characterize the pronouncements of people advocating less-than-scientific theories.

[**] Note the different role of mass here. Thus far, we have treated mass as a measure of inertia, which is called *inertial mass.* Now we see mass as a measure of gravitational force, which in this context is called *gravitational mass.* It is experimentally established that the two are equal, and, as a matter of principle, the equivalence of inertial and gravitational mass is the foundation of Einstein's general theory of relativity.

FIGURE 6.1

Could the gravitational pull on the apple reach to the Moon?

FIGURE 6.2

The tangential velocity of the Moon about Earth allows it to fall around Earth rather than directly into it.

As the rocket gets farther from Earth, gravitational strength between the rocket and Earth decreases.

STOP AND CHECK YOURSELF

1. In Figure 6.2, we see that the Moon falls around Earth rather than straight into it. If the Moon's tangential velocity were suddenly reduced to zero, how would it move?

2. According to the equation for gravitational force, what happens to the force between two bodies if the mass of one of the bodies is doubled? If both masses are doubled?

3. Gravitational force acts on all bodies in proportion to their masses. Why, then, doesn't a heavy body fall faster than a light body?

CHECK YOUR ANSWERS

1. If the Moon's tangential velocity became zero, it would fall straight down and crash into Earth!

2. When one mass is doubled, the force between it and the other one doubles. If both masses double, the force is four times as much.

3. The answer goes back to Chapter 4. Recall Figure 4.10, in which heavy and light bricks fall with the same acceleration because both have the same ratio of weight to mass. Newton's second law ($a = F/m$) reminds us that a greater force acting on a greater mass does not result in a greater acceleration.

6.2 The Universal Gravitational Constant, *G*

The proportionality form of the universal law of gravitation can be expressed as an exact equation when the constant of proportionality *G* is introduced. *G* is called the *universal gravitational constant*. Then the equation is

$$F = G\frac{m_1 m_2}{d^2}$$

Just as π relates circumference and diameter for circles, *G* relates gravitational force with mass and distance.

The Physics Place
von Jolly's Method of Measuring the Attraction Between Two Masses

In words, the force of gravity between two objects is found by multiplying their masses, dividing by the square of the distance between their centers, and then multiplying this result by the constant *G*. The magnitude of *G* is identical to the magnitude of the force between a pair of 1-kg masses that are 1 meter apart: 0.0000000000667 newton. This small magnitude indicates an extremely weak force. In standard units and in scientific notation,*

$$G = 6.67 \times 10^{-11}\,\text{N} \cdot \text{m}^2/\text{kg}^2$$

Interestingly, Newton could calculate the product of *G* and Earth's mass, but not either one alone. Calculating *G* alone was first done by the English physicist Henry Cavendish in the eighteenth century, a century after Newton's time.

Cavendish measured *G* by measuring the tiny force between lead masses with an extremely sensitive torsion balance. A simpler method was later developed by Philipp von Jolly, who attached a spherical flask of mercury to one arm of a sensitive balance (Figure 6.4). After the balance was put in equilibrium, a 6-ton lead sphere was rolled beneath the mercury flask. The gravitational force between the two masses was measured by the weight needed on the opposite end of the balance

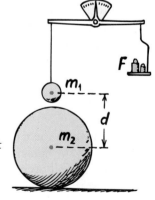

Jolly's method of measuring G. Balls of mass m_1 and m_2 attract each other with a force *F* equal to the weights needed to restore balance.

---

* The numerical value of *G* depends entirely on the units of measurement we choose for mass, distance, and time. The international system of choice is for mass, the kilogram; for distance, the meter; and for time, the second. Scientific notation is discussed in Appendix A at the end of the book.

to restore equilibrium. All the quantities, m_1, m_2, F, and d were known, from which the constant G was calculated:

$$G = \frac{F}{\left(\dfrac{m_1 m_2}{d^2}\right)} = 6.67 \times 10^{-11}\ \text{N/kg}^2/\text{m}^2$$

$$= 6.67 \times 10^{-11}\ \text{N} \cdot \text{m}^2/\text{kg}^2$$

The value of G indicates that the force of gravity is a very weak force. It is the weakest of the known four fundamental forces. (The other three are the electromagnetic force and two kinds of nuclear forces.) We sense gravitation only when masses similar to that of Earth are involved. If you stand on a large ship, the force of attraction between you and the ship is too weak for ordinary measurement. The force of attraction between you and Earth, however, can be measured. It is your weight.

> You can never change only one thing! Every equation reminds us of this—you can't change a term on one side without affecting the other side.

Your weight depends not only on your mass but also on your distance from the center of Earth. At the top of a mountain, your mass is the same as it is anywhere else, but your weight is slightly less than it is at ground level. That's because your distance from Earth's center is greater.

Once the value of G was known, the mass of Earth was easily calculated. The force that Earth exerts on a mass of 1 kilogram at its surface is 9.8 newtons. The distance between the 1-kilogram mass and the center of Earth is Earth's radius, 6.4×10^6 meters. Therefore, from $F = G(m_1 m_2/d^2)$, where m_1 is the mass of Earth,

$$9.8\,\text{N} = 6.67 \times 10^{-11}\,\text{N} \cdot \text{m}^2/\text{kg}^2 \frac{1\,\text{kg} \times m_1}{(6.4 \times 10^6\text{m})^2}$$

from which the mass of Earth, m_1, is calculated to be 6×10^{24} kilograms.

People all over the world in 1798 were excited about the discovery of G. Newspapers everywhere announced the discovery as one that measured the mass of the planet Earth. How exciting that Newton's formula gives the mass of the entire planet, with all its oceans, mountains, and inner parts yet to be discovered. G and the mass of Earth were measured when a great portion of Earth's surface was still undiscovered.

6.3 Gravity and Distance: The Inverse-Square Law

The space surrounding all objects with mass is energized with a *gravitational field*.* The gravitational field weakens with increased distance from the object. How a gravitational field becomes less intense with distance is similar to how paint from a paint gun spreads with increasing distance (Figure 6.5). Suppose we position a paint gun at the center of a sphere with a radius of 1 meter, and a burst of paint spray travels 1 meter to produce a square patch of paint that is 1 millimeter thick. How thick would the patch be if the experiment were done in a sphere with twice the radius? If the same amount of paint travels in straight lines for 2 meters, it will spread to a patch twice as tall and twice as wide. The paint would then be spread over an area four times as big, and its thickness would be only $\frac{1}{4}$ millimeter.

The Physics Place
Inverse-Square Law

Can you see from the figure that, for a sphere of radius 3 meters, the thickness of the paint patch would be only $\frac{1}{9}$ millimeter? Can you see the thickness

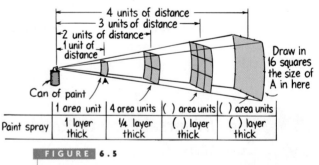

FIGURE 6.5

The inverse-square law. Paint spray travels radially away from the nozzle of the can in straight lines. Like gravity, the "strength" of the spray obeys the inverse-square law.

* The gravitational field about a massive object is defined to be the gravitational force per mass on an object in the vicinity of the massive object. The symbol for the gravitational field is a boldfaced *g* (with units N/kg, having the same magnitude as gravitational acceleration at that point, *g*).

of the paint decreases as the square of the distance increases? This is known as the **inverse-square law**. The inverse-square law holds for gravity and for all phenomena wherein the effect from a localized source spreads uniformly throughout the surrounding space: the electric field about an isolated electron, light from a match, radiation from a piece of uranium, and sound from a cricket.

Newton's law of gravity as written applies to particle and spherical bodies, as well as to nonspherical bodies sufficiently far apart. The distance term d in Newton's equation is the distance between the centers of masses of the objects. Note in Figure 6.6 that the apple that normally weighs 1 newton at Earth's surface weighs only $\frac{1}{4}$ as much when it is twice the distance from Earth's center. The greater the distance from Earth's center that an object is, the less the weight of that object will be. A child who weighs 300 newtons at sea level will weigh only 299 newtons atop Mt. Everest. For greater distances, force is less. For very great distances, Earth's gravitational force approaches zero. The force *approaches* zero, but it never reaches zero. Even if you were transported to the far reaches of the universe, the gravitational field of home would still be with you. It may be overwhelmed by the gravitational fields of nearer and/or more massive bodies, but it is there.

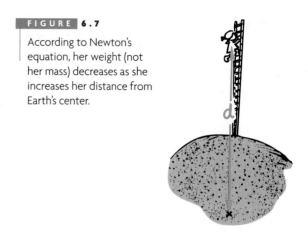

FIGURE 6.7
According to Newton's equation, her weight (not her mass) decreases as she increases her distance from Earth's center.

The gravitational field of every material object, however small or however far, extends through all of space.

> Saying that *F* is inversely proportional to the square of *d* means, for example, that if *d* gets bigger by 3, *F* gets *smaller* by 9.

STOP AND
CHECK YOURSELF

1. By how much does the gravitational field between two objects decrease when the distance between their centers is doubled? Tripled? Increased tenfold?

2. Consider an apple at the top of a tree that is pulled by Earth's gravity with a force of 1 N. If the tree were twice as tall, would the gravitational field be only $\frac{1}{4}$ as strong? Defend your answer.

CHECK YOUR ANSWERS

1. It decreases to one-fourth, one-ninth, and one-hundredth the original value.

2. No, because an apple at the top of the twice-as-tall apple tree is not twice as far from Earth's center. The taller tree would need a height equal to the radius of Earth (6,370 km) for the apple's weight at its top to reduce to $\frac{1}{4}$ N. Before its weight decreases by 1%, an apple or any object must be raised 32 km—nearly four times the height of Mt. Everest. So, as a practical matter, we disregard the effects of everyday changes in elevation.

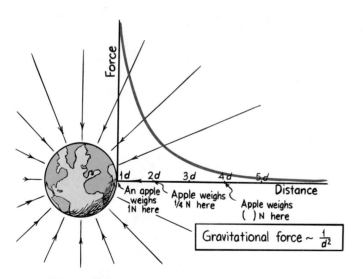

$$\text{Gravitational force} \sim \frac{1}{d^2}$$

FIGURE 6.6
INTERACTIVE FIGURE

If an apple weighs 1 N at Earth's surface, it would weigh only $\frac{1}{4}$ N twice as far from the center of Earth. At three times the distance, it would weigh only $\frac{1}{9}$ N. Gravitational force versus distance is plotted in color. What would the apple weigh at four times the distance? Five times?

6.4 Weight and Weightlessness

When you step on a bathroom scale, you effectively compress a spring inside. When the pointer stops, the elastic force of the deformed spring balances the gravitational attraction between you and Earth—nothing moves as you and the scale are in static equilibrium. The pointer is calibrated to show your **weight**. If you stand on a bathroom scale in a moving elevator, you'll find variations in your weight. If the elevator accelerates upward, the springs inside the bathroom scale are more compressed and your weight reading is greater. If the elevator accelerates downward, the springs inside the scale are less compressed and your weight reading is less. If the elevator cable breaks and the elevator falls freely, the reading on the scale goes to zero. According to the scale's reading, you would be

The Physics Place
Weight and Weightlessness
Apparent Weightlessness

weightless. Would you really be weightless? We can answer this question only if we agree on what we mean by *weight*.

In Chapters 3 and 4, we treated the weight of an object as the force due to gravity upon it. When in equilibrium on a firm surface, weight is evidenced by a support force, or, when in suspension, by a supporting rope tension. In either case, with no acceleration, weight equals *mg*. In future rotating habitats in space, where rotating environments act as giant centrifuges, support force can occur without regard to gravity. So a broader definition of the weight of something is the force it exerts against a supporting floor or a weighing scale. According to

FIGURE 6.9

Your weight equals the force with which you press against the supporting floor. If the floor accelerates up or down, your weight varies (even though the gravitational force *mg* that acts on you remains the same).

this definition, you are as heavy as you feel; so, in an elevator that accelerates downward, the supporting force of the floor is less and you weigh less. If the elevator is in free fall, your weight is zero (Figure 6.9). Even in this weightless condition, however, there is still a gravitational force acting on you, causing your downward acceleration. But gravity now is not felt as weight because there is no support force.

Astronauts in orbit are without a support force and are in a continual state of weightlessness. They sometimes experience "space sickness" until they become accustomed to a state of sustained weightlessness. Astronauts in orbit are in a state of continual free fall.

The International Space Station in Figure 6.11 provides a weightless environment. The station facility and astronauts all accelerate equally toward Earth, at somewhat less than 1 *g* because of their

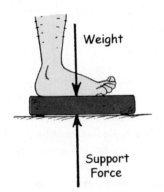

FIGURE 6.8

When you step on a weighing scale, two forces act on it; a downward force of gravity (your ordinary weight, *mg*, if there is no acceleration) and an upward support force. These equal and opposite forces squeeze a springlike device inside the scale that is calibrated to show weight.

FIGURE 6.10

Both are weightless.

FIGURE 6.11

FIGURE 6.11

The inhabitants in this laboratory and docking facility continually experience weightlessness. They are in free fall around Earth. Does a force of gravity act on them?

altitude. This acceleration is not sensed at all. With respect to the station, the astronauts experience zero *g*. Over extended periods of time, this causes loss of muscle strength and other detrimental changes in the body. Future space travelers, however, need not be subjected to weightlessness. Habitats that lazily rotate as giant wheels or pods at the end of a tether will likely take the place of today's nonrotating space habitats. Rotation effectively supplies a support force and nicely provides weight.

> Astronauts inside an orbiting space vehicle have no weight, even though the force of gravity between them and Earth is only slightly less than it is at ground level.

STOP AND
CHECK YOURSELF

In what sense is drifting in space far away from all celestial bodies like stepping off the edge of a table?

CHECK YOUR ANSWER

In both cases, you'd experience weightlessness. Drifting in deep space, you would remain weightless because no discernable force acts on you. Stepping off a table, you would be only momentarily weightless because of a momentary lapse of support force.

6.5 Universal Gravitation

We all know that Earth is round. But why is Earth round? It is round because of gravitation. Everything attracts everything else, and so all parts of Earth have attracted together as far as they can! Any "corners" of our planet have been pulled in; as a result, every part of the surface is equidistant from the center. This makes it a sphere. Therefore, we see, from the law of gravity, that the Sun, the Moon, and Earth are spherical because they have to be (although rotational effects make them slightly ellipsoidal).

The Physics Place

Gravitational Field Inside a Hollow Planet
The Weight of an Object Inside a Hollow Planet but Not at its Center
Discovery of Neptune

If everything pulls on everything else, then the planets must pull on each other. The force that controls Jupiter, for example, is not just the force from the Sun; there are also the pulls from the other planets. Their effect is small in comparison with the pull of the much more massive Sun, but it still shows. When Saturn is near Jupiter, its pull disturbs the otherwise smooth path traced by Jupiter. Both planets "wobble" about their expected orbits. The interplanetary forces causing this wobbling are called *perturbations*. By the 1840s, studies of the most recently discovered planet at the time, Uranus, showed that the deviations of its orbit could not be explained by perturbations from all other known planets. Either the law of gravitation was failing at this great distance from the Sun or an unknown eighth planet was perturbing the orbit of Uranus. An Englishman and a Frenchman, J. C. Adams and Urbain Leverrier, respectively, each assumed Newton's law to be valid, and they independently calculated where an eighth planet should be. At about the same time, Adams sent a letter to the Greenwich Observatory in England and Leverrier sent a letter to the Berlin Observatory in Germany, both suggesting that a certain area of the sky be searched for a new planet. The request by Adams was delayed by misunderstandings at Greenwich, but Leverrier's request was heeded immediately. The planet Neptune was discovered that very night!

Subsequent tracking of the orbits of both Uranus and Neptune led to the discovery of Pluto in 1930 at the Lowell Observatory in Arizona. Whatever you may have learned in your early schooling, Pluto is no longer a planet. Since 2006 its planetary status has been declassified. Other objects of Pluto's size

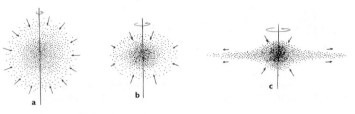

FIGURE 6.12

Formation of the solar system. A slightly rotating ball of interstellar gas (a) contracts due to mutual gravitation and (b) conserves "angular momentum" by speeding up. The increased momentum of individual particles and clusters of particles causes them (c) to sweep in wider paths about the rotational axis, producing an overall disk shape. The greater surface area of the disk promotes cooling and condensation of matter in swirling eddies—the birthplace of the planets.

continue to be discovered beyond Neptune.* Astronomers were faced with whether to classify the increasing list of Pluto's neighbors as planets, or reclassify Pluto. Pluto is now classified as a *dwarf* planet. Pluto takes 248 years to make a single revolution about the Sun, so no one will see it in its discovered position again until the year 2178.

Recent evidence suggests that the universe is expanding and accelerating outward, pushed by an antigravity *dark energy* that makes up some 73% of the universe. Another 23% is composed of the yet-to-be-discovered particles of exotic *dark matter*. Ordinary matter, the stuff of stars, cabbages, and kings, makes up only about 4%. The concepts of dark energy and dark matter are late-twentieth- and twenty-first-century confirmations. The present view of the universe has progressed appreciably beyond what Newton and other scientists of his time perceived.

Yet few theories have affected science and civilization as much as Newton's theory of gravity. The successes of Newton's ideas ushered in the Enlightenment. Newton had demonstrated that, by observation and reason, people could uncover the workings of the physical universe. How profound that all the moons, planets, stars, and galaxies have such a beautifully simple rule to govern them, namely,

$$F = G\frac{m_1 m_2}{d^2}$$

The formulation of this simple rule is one of the major reasons for the success in science that followed,

* Quaoar has a moon, Eris, which is 30% wider than Pluto and also has a moon. Object 2003 EL61 has two moons. Objects nicknamed Sedna and Buffy, discovered in 2005, are nearly the size of Pluto.

for it provided hope that other phenomena of the world might also be described by equally simple and universal laws.

This hope nurtured the thinking of many scientists, artists, writers, and philosophers of the 1700s. One of these was the English philosopher John Locke, who argued that observation and reason, as demonstrated by Newton, should be our best judge and guide in all things. Locke urged that all of nature and even society should be searched to discover any "natural laws" that might exist. Using Newtonian physics as a model of reason, Locke and his followers modeled a system of government that found adherents in the thirteen British colonies across the Atlantic. These ideas culminated in the Declaration of Independence and the Constitution of the United States of America.

6.6 Projectile Motion

Without gravity, you could toss a rock at an angle skyward and it would follow a straight-line path. Because of gravity, however, the path curves. A tossed rock, a cannonball, or any object that is projected by some means and continues in motion by its own inertia is called a **projectile**. To the cannoneers of earlier centuries, the curved paths of projectiles seemed very complex. Today these paths are surprisingly simple when we look at the horizontal and vertical components of velocity separately.

The Physics Place
Projectile Motion

The Physics Place
Projectile Motion Demo
More Projectile Motion

The horizontal component of velocity for a projectile is no more complicated than the horizontal velocity of a bowling ball rolling freely on the lane of a bowling alley. If the retarding effect of friction can be ignored, there is no horizontal force on the ball and its velocity is constant. It rolls of its own

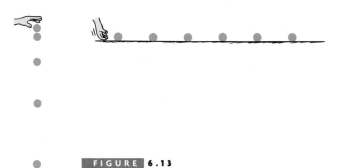

FIGURE 6.13

(Top) Roll a ball along a level surface, and its velocity is constant because no component of gravitational force acts horizontally. *(Left)* Drop it, and it accelerates downward and covers a greater vertical distance each second.

inertia and covers equal distances in equal intervals of time (Figure 6.13, top). The horizontal component of a projectile's motion is just like the bowling ball's motion along the lane.

The vertical component of motion for a projectile following a curved path is just like the motion described in Chapter 4 for a freely falling object. The vertical component is exactly the same as for an object falling freely straight down, as shown at the left in Figure 6.13. The faster the object falls, the greater the distance covered in each successive second. Or, if the object is projected upward, the vertical distances of travel decrease with time on the way up.

The curved path of a projectile is a combination of horizontal and vertical motion. When air resistance is small enough to ignore, the horizontal and vertical components of a projectile's velocity are completely independent of one another. Their combined effect produces the trajectories of projectiles.

Projectiles Launched Horizontally

Projectile motion is nicely analyzed in Figure 6.15, which shows a simulated multiple flash exposure of a ball rolling off the edge of a table. Investigate it carefully, for there's a lot of good physics there. On the left we notice equally timed sequential positions of the ball without the effect of gravity. Only the effect of the ball's horizontal component of motion is shown. Next we see vertical motion without a horizontal component. The curved path in the third view is best analyzed by considering the horizontal and vertical components of motion separately. There are two important things to notice. The first is that the ball's horizontal component of velocity doesn't change as the falling ball moves forward. The ball travels the same horizontal distance in equal times between each flash. That's because there is no component of gravitational force acting horizontally. Gravity acts only *downward,* so the only acceleration of the ball is *downward.* The second thing to notice is that the vertical positions become farther apart with time. The vertical distances traveled are the same as if the ball were simply dropped. Note the curvature of the ball's path is the combination of horizontal motion, which remains constant, and vertical motion, which undergoes acceleration due to gravity.

The trajectory of a projectile that accelerates only in the vertical direction while moving at a constant horizontal velocity is a **parabola.** When air resistance is small enough to neglect, as it is for a heavy object without great speed, the trajectory is parabolic.

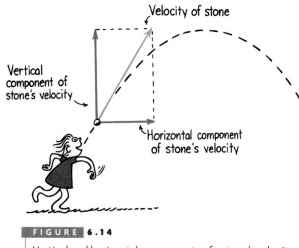

FIGURE 6.14

Vertical and horizontal components of a stone's velocity.

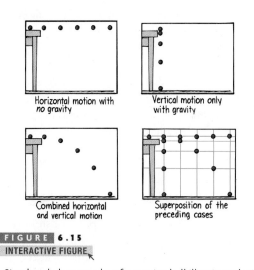

Horizontal motion with *no* gravity

Vertical motion only with gravity

Combined horizontal and vertical motion

Superposition of the preceding cases

FIGURE 6.15

INTERACTIVE FIGURE

Simulated photographs of a moving ball illuminated with a strobe light.

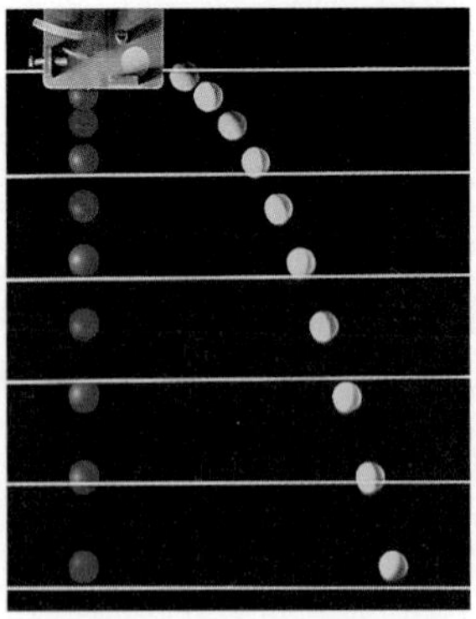

FIGURE 6.16
INTERACTIVE FIGURE

A strobe-light photograph of two golf balls released simultaneously from a mechanism that allows one ball to drop freely while the other is projected horizontally.

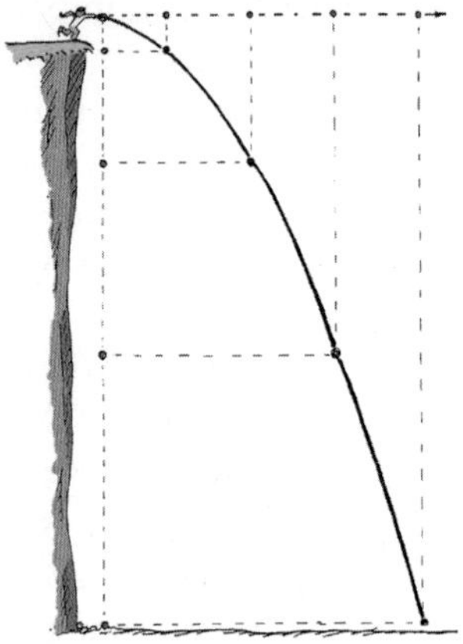

FIGURE 6.17

The vertical dashed line is the path of a stone dropped from rest. The horizontal dashed line would be its path if there were no gravity. The curved solid line shows the resulting trajectory that combines horizontal and vertical motion.

vertical distance that the stone falls beneath the idealized straight-line paths is the same for equal times. This vertical distance is independent of what's happening horizontally.

Figure 6.19 shows specific vertical distances for a cannonball shot at an upward angle. If there were

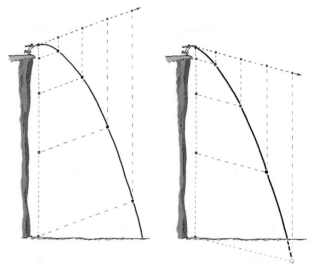

FIGURE 6.18

Whether launched at an angle upward or downward, the vertical distance of fall beneath the idealized straight line path is the same for equal times.

STOP AND CHECK YOURSELF

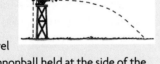

At the instant a cannon fires a cannonball horizontally over a level range, another cannonball held at the side of the cannon is released and drops to the ground. Which ball, the one fired downrange or the one dropped from rest, strikes the ground first?

CHECK YOUR ANSWER

Both cannonballs hit the ground at the same time, for both fall *the same vertical distance*. Can you see that the physics is the same as the physics of Figures 6.15 through 6.17?

Projectiles Launched at an Angle

In Figure 6.18, we see the paths of stones thrown at an angle upward (left) and downward (right). The dashed straight lines show the ideal trajectories of the stones in the absence of gravity. Notice that the

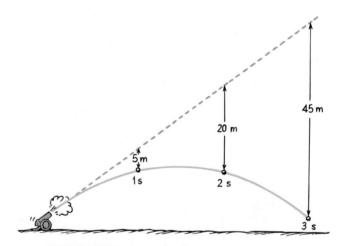

FIGURE 6.19

With no gravity, the projectile would follow a straight-line path (dashed line). But, because of gravity, the projectile falls beneath this line the same vertical distance it would fall if it were released from rest. Compare the distances fallen with those given in Table 3.2 in Chapter 3. (With $g = 9.8$ m/s^2, these distances are more precisely 4.9 m, 19.6 m, and 44.1 m.)

PROBLEM SOLVING

Problems

1. A ball of mass m rolls off of a y-meter-high lab table and hits the floor a distance x from the base of the table.

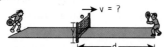

a. Show that the ball takes $\sqrt{2y/g}$ seconds to hit the floor.

b. Show that the ball leaves the table at a speed $x/\sqrt{2y/g}$ meters per second.

c. The ball has a mass of 0.010 kg, the height of the table is 1.25 m, and the ball hits the floor 3.0 m from the base of the table. Using $g = 10$ m/s^2, show that the speed of the ball leaving the table is 6.0 m/s.

2. A horizontally moving tennis ball barely clears the net, a distance y above the surface of the court. To land within the tennis court the ball must not be moving too fast.

a. To remain within the court's border a horizontal distance d from the bottom of the net, show that the ball's maximum speed over the net is

$$v = \frac{d}{\sqrt{\frac{2y}{g}}}$$

b. Suppose the height of the net is 1.00 m, and the court's border is 12.0 m from the bottom of the net. Use $g = 10$ m/s^2 and show that the maximum speed of the horizontally moving ball clearing the net is about 27 m/s (about 60 mi/h).

c. Does the mass of the ball make a difference? Defend your answer.

Solutions

1. a. We want the time of the ball in the air. First, some physics. The time t it takes for the ball to hit the floor would be the same as if it were dropped from rest a vertical distance y. We say from rest because the initial vertical component of velocity is zero.

$$\text{From } y = \frac{1}{2}gt^2 \Rightarrow t^2 = \frac{2y}{g} \Rightarrow t = \sqrt{\frac{2y}{g}}.$$

b. The horizontal speed of the ball as it leaves the table, using the time found in (a), is

$$v = \frac{d}{t} = \frac{x}{t} = \frac{x}{\sqrt{\frac{2y}{g}}}.$$

c. $v = \dfrac{x}{\sqrt{\dfrac{2y}{g}}} = \dfrac{3.0 \text{ m}}{\sqrt{\dfrac{2(1.25 \text{ m})}{10\frac{\text{m}}{\text{s}^2}}}} = \mathbf{6.0 \frac{m}{s}}.$

Notice how the terms of the equations guide the solution. Notice also that the mass of the ball, not showing up in the equations, is extraneous information (as would be the color of the ball).

2. a. As with Problem 1, the physics concept here involves projectile motion in the absence of air resistance, where horizontal and vertical components of velocity are independent. We're asked for horizontal speed, so we write,

$$v_{\text{x}} = \frac{d}{t}$$

where d is horizontal distance traveled in time t. As with Problem 1, the time t of the ball in flight will be the same as if we had just dropped it from rest a vertical distance y from the top of the net. As the ball clears the net, its highest point in its path, its vertical component of velocity is zero.

$$\text{From } y = \frac{1}{2}gt^2 \Rightarrow t^2 = \frac{2y}{g} \Rightarrow t = \sqrt{\frac{2y}{g}}.$$

$$\text{So } v = \frac{d}{t} = \frac{d}{\sqrt{\frac{2y}{g}}}.$$

Can you see that solving in terms of symbols better shows that these two problems are one in the same? All the physics occurs in parts (a) and (b) in Problem 1. These steps are combined in part (a) of Problem 2.

b. $v = \dfrac{d}{\sqrt{\dfrac{2y}{g}}} = \dfrac{12.0 \text{ m}}{\sqrt{\dfrac{2(1.00 \text{ m})}{10\frac{\text{m}}{\text{s}^2}}}} = 26.8 \dfrac{\text{m}}{\text{s}} \approx \mathbf{27\frac{m}{s}}.$

c. We can see that the mass of the ball (in both problems) doesn't show up in the equations for motion, which tells us that mass is irrelevant. Recall from Chapter 4 that mass has no effect on a freely falling object—and the tennis ball is a freely falling object (as is every projectile when air resistance can be neglected).

no gravity the cannonball would follow the straight-line path shown by the dashed line. But there is gravity, so this doesn't occur. What happens is that the cannonball continuously falls beneath the imaginary line until it finally strikes the ground. Note that the vertical distance it falls beneath any point on the dashed line is the same vertical distance it would have fallen if it had been dropped from rest and had been falling for the same amount of time. This distance, as introduced in Chapter 3, is given by $d = \frac{1}{2}gt^2$, where t is the elapsed time. For $g = 10 \text{ m/s}^2$, this becomes $d = 5t^2$.

We can put it another way: Shoot a projectile skyward at some angle and pretend there is no gravity. After so many seconds t, it should be at a certain point along a straight-line path. But, because of gravity, it isn't. Where is it? The answer is that it's directly below this point. How far below? The answer in meters is $5t^2$ (or, more precisely, $4.9t^2$). How about that!

PRACTICING PHYSICS

■ Hands-On Dangling Beads

Make your own model of projectile paths. Divide a ruler or a stick into five equal spaces. At position 1, hang a bead from a string that is 1 cm long, as shown. At position 2, hang a bead from a string that is 4 cm long. At position 3, do the same with a 9-cm length of string. At position 4, use 16 cm of string, and for position 5, use 25 cm of string. If you hold the stick horizontally, you will have a version of Figure 6.17. Hold it at a slight upward angle to show a version of Figure 6.18, left. Hold it at a downward angle to show a version of Figure 6.18, right.

In Figure 6.20, we see vectors representing both the horizontal and vertical components of velocity for a projectile following a parabolic trajectory. Notice that the horizontal component everywhere along the trajectory is the same, and only the vertical component changes. Note also that the actual velocity is represented by the vector that forms the diagonal of the rectangle formed by the vector components. At the top of the trajectory, the vertical component is zero, so the velocity at the zenith is

only the horizontal component of velocity. Everywhere else along the trajectory, the magnitude of velocity is greater (just as the diagonal of a rectangle is greater than either of its sides). Figure 6.21 shows the trajectory traced by a projectile launched with the same speed at a steeper angle.

Figure 6.22 shows the paths of several projectiles, all with the same initial speed but different launching

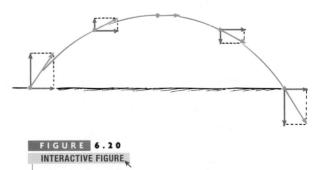

FIGURE 6.20

INTERACTIVE FIGURE

The velocity of a projectile at various points along its trajectory. Note that the vertical component changes and that the horizontal component is the same everywhere.

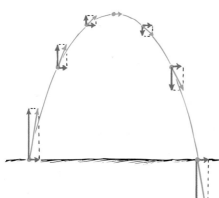

FIGURE 6.21

Trajectory for a steeper projection angle.

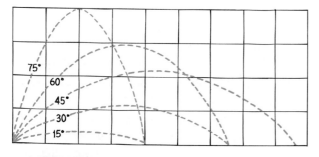

FIGURE 6.22
INTERACTIVE FIGURE

Ranges of a projectile shot at the same speed at different projection angles.

angles. The figure neglects the effects of air resistance, so the trajectories are all parabolas. Notice that these projectiles reach different *altitudes,* or heights above the ground. They also have different *horizontal ranges,* or distances traveled horizontally. The remarkable thing to note from Figure 6.22 is that the same range is obtained from two different launching angles when the angles add up to 90°! An object thrown into the air at an angle of 60°, for example, will have the same range as if it were thrown at the same speed at an angle of 30°. For the smaller angle, of course, the object remains in the air for a shorter time. The greatest range occurs when the launching angle is 45°—and when air resistance is negligible.

Without the effects of air, the maximum range for a baseball would occur when it is batted 45° above the horizontal. Because of air resistance and lift due to spinning of the ball (next chapter), the best range occurs at batting angles noticeably less than 45°. Air resistance and spin are more significant for golf balls, where angles less than 38° or so result in maximum range. For heavy projectiles like

javelins and the shot, air has less effect on range. A javelin, being heavy and presenting a very small cross section to the air, follows an almost perfect parabola when thrown. So does a shot. For such projectiles, maximum range for equal launch speeds would occur for a launch angle of about 45° (slightly less because the launching height is above ground level). Aha, but launching speeds are *not* equal for such a projectile thrown at different angles. In throwing a javelin or putting a shot, a significant part of the launching *force* goes into combating gravity—the steeper the angle, the less speed it has when leaving the thrower's hand. So gravity plays a role before and after launching. You can test this yourself: Throw a heavy boulder horizontally, then vertically—you'll find the horizontal throw to be considerably faster than the vertical throw. So maximum range for heavy projectiles thrown by humans is attained for angles of less than 45°—and not because of air resistance.

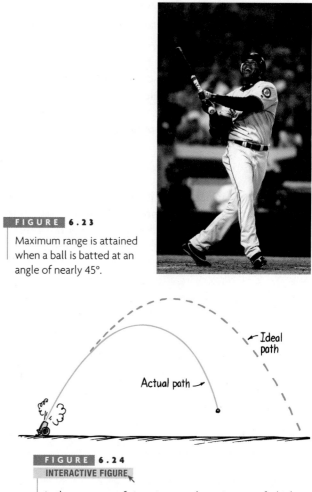

FIGURE 6.23

Maximum range is attained when a ball is batted at an angle of nearly 45°.

Ideal path

Actual path

FIGURE 6.24
INTERACTIVE FIGURE

In the presence of air resistance, the trajectory of a high-speed projectile falls short of the idealized parabolic path.

EVERDAY APPLICATION

■ HANG TIME REVISITED

In Chapter 3, we stated that airborne time during a jump is independent of horizontal speed. Now we see why this is so—horizontal and vertical components of motion are independent of each other. The rules of projectile motion apply to jumping. Once one's feet are off the ground, only the force of gravity acts on the jumper (neglecting air resistance). Hang time depends only on the vertical component of lift-off velocity. It so happens, however, that the action of running can make a difference. When running, the lift-off force during jumping can be somewhat increased by pounding of the feet against the ground (and the ground pounding against the feet in action–reaction fashion), so hang time for a running jump can often exceed hang time for a standing jump. But, once the runner's feet are off the ground, only the vertical component of lift-off velocity determines hang time.

STOP AND CHECK YOURSELF

1. A baseball is batted at an angle into the air. Once the ball is airborne, and neglecting air resistance, what is the ball's acceleration vertically? Horizontally?

2. At what part of its trajectory does the baseball have minimum speed?

3. Consider a batted baseball following a parabolic path on a day when the Sun is directly overhead. How does the speed of the ball's shadow across the field compare with the ball's horizontal component of velocity?

CHECK YOUR ANSWERS

1. Vertical acceleration is g because the force of gravity is vertical. Horizontal acceleration is zero because no horizontal force acts on the ball.

2. A ball's minimum speed occurs at the top of its trajectory. If it is launched vertically, its speed at the top is zero. If launched at an angle, the vertical component of velocity is zero at the top, leaving only the horizontal component. So the speed at the top is equal to the horizontal component of the ball's velocity at any point. Doesn't this make sense?

3. They are the same!

When air resistance is small enough to be negligible, the time that a projectile takes to rise to its maximum height will be the same as the time it takes to fall back to its initial level (Figure 6.25).

This is because its deceleration by gravity while going up is the same as its acceleration by gravity while coming down. The speed it loses while going up is therefore the same as the speed gained while coming down. So the projectile arrives at its initial level with the same speed it had when it was initially projected.

FIGURE 6.25

Without air resistance, speed lost while going up equals speed gained while coming down: Time going up equals time coming down.

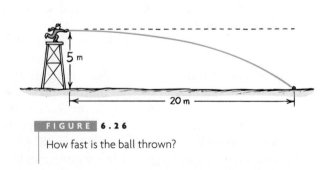

FIGURE 6.26

How fast is the ball thrown?

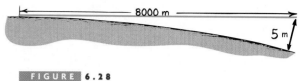

FIGURE 6.28

Earth's curvature—not to scale!

STOP AND CHECK YOURSELF

The boy on the tower throws a ball 20 m down-range, as shown in Figure 6.26. What is his pitching speed?

CHECK YOUR ANSWER

The ball is thrown horizontally, so the pitching speed is horizontal distance divided by time. A horizontal distance of 20 m is given, but the time is not stated. However, knowing the vertical drop is 5 m, you remember that a 5-m drop takes 1 s! From the equation for constant speed (which applies to horizontal motion), $v = d/t = (20 \text{ m})/(1 \text{ s}) = 20 \text{ m/s}$. It is interesting to note that the equation for constant speed, $v = d/t$, guides our thinking about the crucial factor in this problem—the *time*.

Baseball games normally take place on level ground. For the short-range projectile motion on the playing field, Earth can be considered to be flat because the flight of the baseball is not affected by Earth's curvature. For very long-range projectiles, however, the curvature of Earth's surface must be taken into account. We'll now see that, if an object is projected fast enough, it will fall all the way around Earth and become an Earth satellite.

6.7 Fast-Moving Projectiles— Satellites

Consider the baseball pitcher on the tower in Figure 6.26. If gravity did not act on the ball, the ball would follow a straight-line path shown by the dashed line. But gravity does act, so the ball falls below this straight-line path. In fact, as just discussed, 1 second after the ball leaves the pitcher's hand it will have fallen a vertical distance of

5 meters below the dashed line—whatever the pitching speed. It is important to understand this, for it is the crux of satellite motion.

An Earth **satellite** is simply a projectile that falls *around* Earth rather than *into* it. The speed of the satellite must be great enough to ensure that its falling distance matches Earth's curvature. A geometrical fact about the curvature of Earth is that its surface drops a vertical distance of 5 meters for every 8000 meters tangent to the surface (Figure 6.28). If a baseball could be thrown fast enough to travel a horizontal distance of 8 kilometers during the one second it takes to fall 5 meters, then it would follow the curvature of Earth. This is a speed of 8 kilometers per second. If this doesn't seem fast, convert it to kilometers per hour and you get an impressive 29,000 kilometers per hour (or 18,000 miles per hour)!

> Earth's curvature, dropping 5 m for each 8-km tangent, means that, if you were floating in a calm ocean, you'd be able to see only the top of a 5-m mast on a ship 8 km away.

At this speed, atmospheric friction would burn the baseball—or even a piece of iron—to a crisp. This is the fate of bits of rock and other meteorites that enter Earth's atmosphere and burn up, appearing as "falling stars." That is why satellites, such as the space shuttles, are launched to altitudes of 150 kilometers or more—to be above almost all of the atmosphere and to be nearly free of air resistance. A common misconception is that satellites orbiting at high altitudes are free from gravity. Nothing could be further from the truth. The force of gravity on a

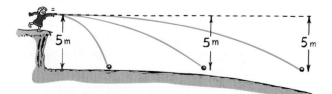

FIGURE 6.27

If you throw a stone at any speed, one second later it will have fallen 5 m below where it would have been without gravity.

FIGURE 6.29

If the speed of the stone and the curvature of its trajectory are great enough, the stone may become a satellite.

FIGURE 6.30

If Superman threw a rock fast enough, it would orbit Earth, if there were no air resistance.

satellite 200 kilometers above Earth's surface is nearly as strong as it is at the surface. Otherwise, the satellite would go in a straight line and leave Earth. The high altitude is not to position the satellite beyond Earth's gravity, but beyond Earth's atmosphere, where air resistance is almost totally absent.

A space shuttle is a projectile in a constant state of free fall. Because of its tangential velocity, it falls around Earth rather than vertically into it.

Satellite motion was understood by Isaac Newton, who reasoned that the Moon was simply a projectile circling Earth under the attraction of gravity. This concept is illustrated in a drawing by Newton (Figure 6.31). He compared the motion of the Moon to that of a cannonball fired from the top of a high mountain. He imagined that the mountaintop was above Earth's atmosphere, so that air resistance would not impede the motion of the

cannonball. If fired with a low horizontal speed, a cannonball would follow a curved path and soon hit Earth below. If it were fired faster, its path would be less curved and it would hit Earth farther away. If the cannonball were fired fast enough, Newton reasoned, the curved path would become a circle and the cannonball would circle Earth indefinitely. It would be in orbit.

Both cannonball and Moon have tangential velocity (parallel to Earth's surface) sufficient to ensure motion *around* Earth rather than *into* it. If there is no resistance to reduce its speed, the Moon or any Earth satellite "falls" around and around Earth indefinitely. Similarly, the planets continuously fall around the Sun in closed paths. Why don't the planets crash into the Sun? They don't because of sufficient tangential velocities. What would happen if their tangential velocities were reduced to zero? The answer is simple enough: Their falls would be straight toward the Sun, and they would indeed crash into it. Any objects in the solar system without sufficient tangential velocities have long ago crashed into the Sun. What remains is the harmony we observe.

fyi

When a spacecraft comes into the atmosphere at too steep an angle, more than about 6 degrees, it can burn up. If it comes in at too shallow an angle, it stands a chance of bouncing back into space like a pebble skipped across water.

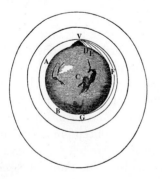

FIGURE 6.31

"The greater the velocity . . . with which [a stone] is projected, the farther it goes before it falls to Earth. We may therefore suppose the velocity to be so increased, that it would describe an arc of 1, 2, 5, 10, 100, 1000 miles before it arrived at Earth, till at last, exceeding the limits of Earth, it should pass into space without touching."
—Isaac Newton, *System of the World*.

STOP AND
CHECK YOURSELF

One of the beauties of physics is that there are usually different ways to view and explain a given phenomenon. Is the following explanation valid? "Satellites remain in orbit instead of falling to Earth because they are beyond the main pull of Earth's gravity."

CHECK YOUR ANSWER

No, no, a thousand times no! If any moving object were beyond the pull of gravity, it would move in a straight line and would not curve around Earth. Satellites remain in orbit because they *are* being pulled by gravity, not because they are beyond it. For the altitudes of most Earth satellites, Earth's gravitational field is only a few percent weaker than it is at Earth's surface.

6.8 Circular Satellite Orbits

An 8-kilometers-per-second cannonball fired horizontally from Newton's mountain would follow Earth's curvature and glide in a circular path around Earth again and again (provided the cannoneer and the cannon got out of the way). Fired at a slower speed, the cannonball would strike Earth's surface; fired at a faster speed, it would overshoot a circular orbit, as we will discuss shortly. Newton calculated the speed for circular orbit, and because such a cannon-muzzle velocity was clearly impossible, he did not foresee the possibility of humans launching satellites (and likely didn't consider multistage rockets).

The Physics Place
Orbits and Kepler's Laws

The Physics Place
Circular Orbits

Note that, in circular orbit, the speed of a satellite is not changed by gravity: only the direction changes. We can understand this by comparing a satellite in circular orbit with a bowling ball rolling along a bowling lane. Why doesn't the gravity that acts on the bowling ball change its speed? The answer is that gravity pulls straight downward with no component of force acting forward or backward.

Consider a bowling lane that completely surrounds Earth, elevated high enough to be above the atmosphere and air resistance. The bowling ball will roll at constant speed along the lane. If a part of the lane were cut away, the ball would roll off its edge and would hit the ground below. A faster ball encountering the gap would hit the ground farther along the gap. Is there a speed at which the ball will clear the gap (like a motorcyclist who drives off a ramp and clears a gap to meet a ramp on the other side)? The answer is yes: 8 kilometers per second will be enough to clear that gap—and any gap—even a 360° gap. The ball would be in circular orbit.

Note that a satellite in circular orbit is always moving in a direction perpendicular to the force of gravity that acts upon it. There is no component of

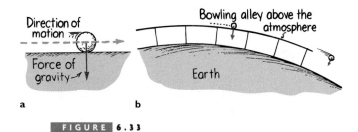
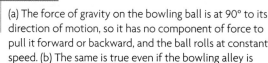

FIGURE 6.33

(a) The force of gravity on the bowling ball is at 90° to its direction of motion, so it has no component of force to pull it forward or backward, and the ball rolls at constant speed. (b) The same is true even if the bowling alley is larger and remains "level" with the curvature of Earth.

force acting in the direction of satellite motion to change its speed. Only a change in direction occurs. So we see why a satellite in circular orbit moves parallel to the surface of Earth at constant speed—a very special form of free fall.

For a satellite close to Earth, the period (the time for a complete orbit about Earth) is about 90 minutes. For higher altitudes, the orbital speed is less, distance is more, and the period is longer. For example, communication satellites located in orbit 5.5 Earth radii above the surface of Earth have a period of 24 hours. This period matches the period of daily Earth rotation. For an orbit around the equator, these satellites remain always above the same point on the ground. The Moon is even farther away and has a period of 27.3 days. So the higher the orbit of a satellite is, the less its speed, the longer its path, and the longer its period will be.*

Putting a payload into Earth orbit requires control over the speed and direction of the rocket that

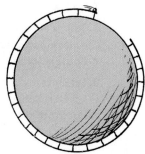

FIGURE 6.34

What speed will allow the ball to clear the gap?

FIGURE 6.32
INTERACTIVE FIGURE

Fired fast enough, the cannonball will go into orbit.

---

* The speed of a satellite in circular orbit is given by $v = \sqrt{GM/d}$ and the period of satellite motion is given by $T = 2\pi\sqrt{d^3/GM}$, where G is the universal gravitational constant, M is the mass of Earth (or whatever body the satellite orbits), and d is the distance of the satellite from the center of Earth or other parent body.

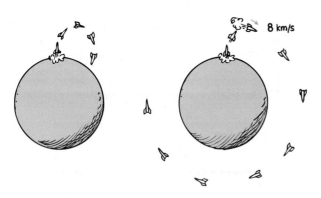

FIGURE 6.35

The initial thrust of the rocket pushes it up above the atmosphere. Another thrust to a tangential speed of at least 8 km/s is required if the rocket is to fall around rather than into Earth.

carries it above the atmosphere. A rocket initially fired vertically is intentionally tipped from the vertical course. Then, once above the drag of the atmosphere, it is aimed horizontally, whereupon the payload is given a final thrust to orbital speed. We see this in Figure 6.35, where, for the sake of simplicity, the payload is the entire single-stage rocket. With the proper tangential velocity, it falls around Earth, rather than into it, and becomes an Earth satellite.

The initial vertical climb gets a rocket quickly through the denser part of the atmosphere. Eventually, the rocket must acquire enough tangential speed to remain in orbit without thrust, so it must tilt until its path is parallel to Earth's surface.

6.9 Elliptical Orbits

If a projectile just above the drag of the atmosphere is given a horizontal speed somewhat greater than 8 kilometers per second, it will overshoot a circular path and trace an oval path called an **ellipse**.

An ellipse is a specific curve: the closed path taken by a point that moves in such a way that the sum of its distances from two fixed points (called *foci*) is constant. For a satellite orbiting a planet, one focus is at the center of the planet; the other focus could be internal or external to the planet. An ellipse can be easily constructed by using a pair of tacks (one at each focus), a loop of string, and a pencil (Figure 6.36). The closer the foci are to each other, the closer the ellipse is to a circle. When both foci are together, the ellipse *is* a circle. So we can see that a circle is a special case of an ellipse.

Whereas the speed of a satellite is constant in a circular orbit, its speed varies in an elliptical orbit. For an initial speed greater than 8 kilometers per second, the satellite overshoots a circular path and moves away from Earth, against the force of gravity. It therefore loses speed. The speed it loses in receding is regained as it falls back toward Earth, and it

FIGURE 6.36
INTERACTIVE FIGURE

A simple method for constructing an ellipse.

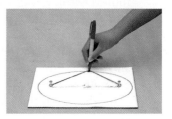

STOP AND CHECK YOURSELF

1. True or false: The space shuttle orbits at altitudes in excess of 150 kilometers to be above both gravity and Earth's atmosphere.

2. Satellites in close circular orbit fall about 5 meters during each second of orbit. Why doesn't this distance accumulate and send satellites crashing into Earth's surface?

CHECK YOUR ANSWER

1. False. What satellites are above is the atmosphere and air resistance—*not* gravity! It's important to note that Earth's gravity extends throughout the universe in accord with the inverse-square law.

2. In each second, the satellite falls about 5 m below the straight-line tangent it would have followed if there were no gravity. Earth's surface also curves 5 m beneath a straight-line 8-km tangent. The process of falling with the curvature of Earth continues from tangent line to tangent line, so the curved path of the satellite and the curve of Earth's surface "match" all the way around Earth. Satellites do, in fact, crash to Earth's surface from time to time when they encounter air resistance in the upper atmosphere that decreases their orbital speed.

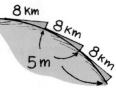

finally rejoins its original path with the same speed it had initially (Figure 6.37). The procedure repeats over and over, and an ellipse is traced each cycle.

Interestingly enough, the parabolic path of a projectile, such as a tossed baseball or a cannonball, is actually a tiny segment of a skinny ellipse that extends within and just beyond the center of Earth (Figure 6.38a). In Figure 6.38b, we see several paths of cannonballs fired from Newton's mountain. All these ellipses have the center of Earth as one focus. As the cannon's muzzle velocity is increased, the ellipses are less eccentric (more nearly circular); and, when muzzle velocity reaches 8 kilometers per second, the ellipse rounds into a circle and does not intercept Earth's surface. The cannonball coasts in circular orbit. At greater muzzle velocities, orbiting cannonballs trace the familiar external ellipses.

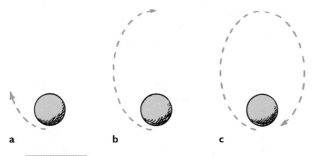

FIGURE 6.37

Elliptical orbit. An Earth satellite that has a speed somewhat greater than 8 km/s overshoots a circular orbit (a) and travels away from Earth. Gravitation slows it to a point where it no longer moves farther from Earth (b). It falls toward Earth, gaining the speed it lost in receding (c), and follows the same path as before in a repetitious cycle.

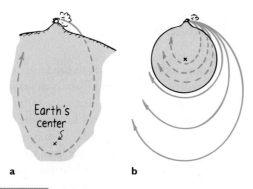

FIGURE 6.38

(a) The parabolic path of the cannonball is part of an ellipse that extends within Earth. Earth's center is the far focus. (b) All paths of the cannonball are ellipses. For less than orbital speeds, the center of Earth is the far focus; for a circular orbit, both foci are Earth's center; for greater speeds, the near focus is Earth's center.

STOP AND CHECK YOURSELF

The orbital path of a satellite is shown in the sketch. In which of the positions A through D does the satellite have the greatest speed? The lowest speed?

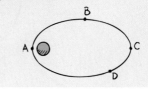

CHECK YOUR ANSWER

The satellite has its greatest speed as it whips around A and has its lowest speed at position C. After passing C, it gains speed as it falls back to A to repeat its cycle.

6.10 Energy Conservation and Satellite Motion

Recall, from Chapter 5, that an object in motion possesses kinetic energy (KE) due to its motion. An object above Earth's surface possesses potential energy (PE) by virtue of its position. Everywhere in its orbit, a satellite has both KE and PE. The sum of the KE and PE is a constant all through the orbit. The simplest case occurs for a satellite in circular orbit.

In a circular orbit, the distance between the satellite and the center of the attracting body does not change, which means that the PE of the satellite is the same everywhere in its orbit. From conservation of energy, the KE must also be constant. So a satellite in circular orbit moves with an unchanging PE, KE, and speed (Figure 6.39).

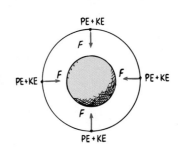

FIGURE 6.39

The force of gravity on the satellite is always toward the center of the body it orbits. For a satellite in circular orbit, no component of force acts along the direction of motion. The speed and thus the KE do not change.

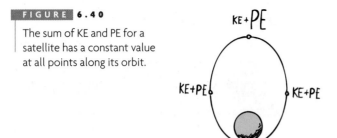

FIGURE 6.40

The sum of KE and PE for a satellite has a constant value at all points along its orbit.

In an elliptical orbit, the situation is different. Both speed and distance vary. PE is greatest when the satellite is farthest away (at the *apogee*) and least when the satellite is closest (at the *perigee*). Note that the KE will be least when the PE is most, and the KE will be most when the PE is least. At every point in the orbit, the sum of KE and PE is the same (Figure 6.40).

At all points along the elliptical orbit, except at the apogee and the perigee, there is a component of gravitational force parallel to the direction of motion of the satellite. This component of force changes the speed of the satellite. Or we can say that (this component of force) × (distance moved) = ΔKE. Either way, when the satellite gains altitude and moves against this component, its speed and KE decrease. The decrease continues to the apogee. Once past the apogee, the satellite moves in the same direction as the component, and the speed and KE increase. The increase continues until the satellite whips past the perigee and repeats the cycle.

Wouldn't Newton have relished seeing satellite motion in terms of *energy*—a concept that came much later?

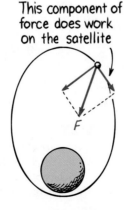

This component of force does work on the satellite

FIGURE 6.41

In elliptical orbit, a component of force exists along the direction of the satellite's motion. This component changes the speed and, thus, the KE. (The perpendicular component changes only the direction.)

CHECK YOURSELF

1. The orbital path of another satellite is shown in the sketch. In which positions A through D does the satellite have the greatest KE? The greatest PE? The greatest total energy?

2. Why does the force of gravity change the speed of a satellite when it is in an elliptical orbit but not when it is in a circular orbit?

CHECK YOUR ANSWERS

1. KE is maximum at the perigee A; PE is maximum at the apogee C; the total energy is the same everywhere in the orbit.

2. In circular orbit, the gravitational force is always perpendicular to the orbital path. With no component of gravitational force along the path, only the direction of motion changes—not the speed. In elliptical orbit, however, the satellite moves in directions that are not perpendicular to the force of gravity. Then components of force do exist along the path, which change the speed of the satellite. A component of force along (parallel to) the direction the satellite moves does work to change its KE.

6.11 Escape Speed

We know that a cannonball fired horizontally at 8 kilometers per second from Newton's mountain would find itself in orbit. But what would happen if the cannonball were instead fired at the same speed *vertically*? It would rise to some maximum height, reverse direction, and then fall back to Earth. Then the old saying "What goes up must come down" would hold true, just as surely as a stone tossed skyward will be returned by gravity (unless, as we shall see, its speed is great enough).

In today's space-faring age, it is more accurate to say, "What goes up *may* come down," for there is a critical starting speed that permits a projectile to escape Earth. This critical speed is called the **escape**

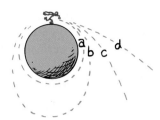

If Superman tosses a ball 8 km/s horizontally from the top of a mountain high enough to be just above air resistance (a), then about 90 minutes later he can turn around and catch it (neglecting Earth's rotation). Tossed slightly faster (b), it will take an elliptical orbit and return in a slightly longer time. Tossed at more than 11.2 km/s (c), it will escape Earth. Tossed at more than 42.5 km/s (d), it will escape the solar system.

speed or, if direction is involved, the *escape velocity*. From the surface of Earth, escape speed is 11.2 kilometers per second. If you launch a projectile at any speed greater than that, it will *leave Earth, traveling slower and slower, never stopping due to Earth's gravity.** *We can understand the magnitude of this speed from an energy point of view.*

How much work would be required to lift a payload against the force of Earth's gravity to a distance very, very far ("infinitely far") away? We might think that the change of PE would be infinite because the distance is infinite. But gravity diminishes with distance by the inverse-square law. The force of gravity on the payload would be strong only near Earth. Most of the work done in launching a rocket occurs within 10,000 km or so of Earth. It turns out that the change of PE of a 1-kilogram body moved from the surface of Earth to an infinite distance is 62 million joules (62 MJ). So, to put a payload infinitely far from Earth's surface requires at least 62 million joules of energy per kilogram of load. We won't go through the calculation here, but 62 million joules per kilogram corresponds to a speed of 11.2 kilometers per second, whatever the total mass

involved. This is the escape speed from the surface of Earth.**

If we give a payload any more energy than 62 million joules per kilogram at the surface of Earth or, equivalently, any more speed than 11.2 kilometers per second, then, neglecting air resistance, the payload will escape from Earth, never to return. As the payload continues outward, its PE increases and its KE decreases. Earth's gravitational pull continuously slows it down but never reduces its speed to zero. The payload escapes.

The escape speeds from the surfaces of various bodies in the solar system are shown in Table 6.1. Note that the escape speed from the surface of the Sun is 620 kilometers per second. Even at a 150,000,000-kilometer distance from the Sun (Earth's distance), the escape speed to break free of the Sun's influence is 42.2 kilometers per second—considerably more than the escape speed from Earth. An object projected from Earth at a speed greater than 11.2 kilometers per second but less than 42.5 kilometers per second will escape Earth but not the Sun. Rather than recede forever, it will take up an orbit around the Sun.

The first probe to escape the solar system, *Pioneer 10,* was launched from Earth in 1972 with a speed of only 15 kilometers per second. The escape was accomplished by directing the probe into the

TABLE 6.1

Escape Speeds at the Surfaces of Bodies in the Solar System

Astronomical Body	Mass (Earth masses)	Radius (Earth radii)	Escape Speed (km/s)
Sun	333,000	109	620
Sun (at a distance of Earth's orbit)		23,500	42.2
Jupiter	318	11	60.2
Saturn	95.2	9.2	36.0
Neptune	17.3	3.47	24.9
Uranus	14.5	3.7	22.3
Earth	1.00	1.00	11.2
Venus	0.82	0.95	10.4
Mars	0.11	0.53	5.0
Mercury	0.055	0.38	4.3
Moon	0.0123	0.27	2.4

* Escape speed from any planet or any body is given by $v = \sqrt{2GM/d}$, where G is the universal gravitational constant, M is the mass of the attracting body, and d is the distance from its center. (At the surface of the body, d would simply be the radius of the body.) For a bit more mathematical insight, compare this formula with the one for orbital speed in the footnote on page 121.

** Interestingly enough, this might well be called the *maximum falling speed.* Any object, however far from Earth, released from rest and allowed to fall to Earth only under the influence of Earth's gravity, would not exceed 11.2 km/s. (With air friction, it would be less.)

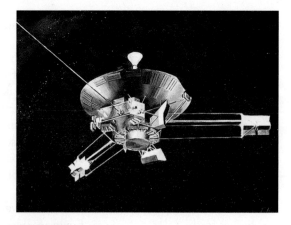

FIGURE 6.43

Pioneer 10, launched from Earth in 1972, was the first spacecraft to travel to an outer planet, providing data and images of Jupiter. Eleven years later, it became the first human-made object to leave the solar system. Its last, very weak signal was received on 23 January 2003. *Pioneer 10,* out of contact with its makers, is now wandering in our galaxy.

FIGURE 6.44

The European–U.S. spacecraft *Cassini* beams close-up images of Saturn and its giant moon Titan to Earth. It also measures surface temperatures, magnetic fields, and the size, speed, and trajectories of tiny surrounding space particles.

path of oncoming Jupiter. It was whipped about by Jupiter's great gravitational field, picking up speed in the process—similar to the increase in the speed of a baseball encountering an oncoming bat. Its speed of departure from Jupiter was increased enough to exceed the escape speed from the Sun at the distance of Jupiter. *Pioneer 10* passed the orbit of Pluto in 1984. Unless it collides with another body, it will wander indefinitely through interstellar space. Like a note inside a bottle cast into the sea,

Pioneer 10 contains information about Earth that might be of interest to extraterrestrials, in hopes that it will one day "wash up" and be found on some distant "seashore."

It is important to point out that the escape speed of a body is the initial speed given by a brief thrust, after which there is no force to assist motion. One could escape Earth at *any* sustained speed more than zero, given enough time. For example, suppose a rocket is launched to a destination such as the Moon. If the rocket engines burn out when still close to Earth, the rocket needs a minimum speed of 11.2 kilometers per second. But if the rocket engines can be sustained for long periods of time, the rocket could reach the Moon without ever attaining 11.2 kilometers per second.

It is interesting to note that the accuracy with which an unmanned rocket reaches its destination is not accomplished by staying on a preplanned path or by getting back on that path if the rocket strays off course. No attempt is made to return the rocket to its original path. Instead, the control center in effect asks, "Where is it now and what is its velocity? What is the best way to reach its destination, given its present situation?" With the aid of high-speed computers, the answers to these questions are used in finding a new path. Corrective thrusters direct the rocket to this new path. This process is repeated over and over again all the way to the goal.*

> The mind that encompasses the universe is as marvelous as the universe that encompasses the mind.

> **fyi**
> Just as planets fall around the Sun, stars fall around the centers of galaxies. Those with insufficient tangential speeds are pulled into, and are gobbled up by, the galactic nucleus— usually a black hole.

---

* Is there a lesson to be learned here? Suppose you find that you are off course. You may, like the rocket, find it more fruitful to follow a course that leads to your goal as best plotted from your present position and circumstances, rather than try to get back on the course you plotted from a previous position and under, perhaps, different circumstances.

SUMMARY OF TERMS

Law of universal gravitation Every body in the universe attracts every other body with a force that, for two bodies, is directly proportional to the product of their masses and inversely proportional to the square of the distance separating them:

$$F = G\frac{m_1 m_2}{d^2}$$

Inverse-square law A law relating the intensity of an effect to the inverse square of the distance from the cause:

$$\text{Intensity} = \frac{1}{\text{distance}^2}$$

Gravity follows an inverse-square law, as do the effects of electric, magnetic, light, sound, and radiation phenomena.

Weight The force that an object exerts on a supporting surface (or, if suspended, on a supporting string), which is often, but not always, due to the force of gravity.

Weightless Being without a support force, as in free fall.

Projectile Any object that moves through the air or through space under the influence of gravity.

Parabola The curved path followed by a projectile under the influence of constant gravity only.

Satellite A projectile or small celestial body that orbits a larger celestial body.

Ellipse The oval path followed by a satellite. The sum of the distances from any point on the path to two points called foci is a constant. When the foci are together at one point, the ellipse is a circle. As the foci get farther apart, the ellipse becomes more "eccentric."

Escape speed The speed that a projectile, space probe, or similar object must reach to escape the gravitational influence of Earth or of another celestial body to which it is attracted.

SUGGESTED READING

Cole, K. C. *The Hole in the Universe: How Scientists Peered over the Edge of Emptiness and Found Everything.* New York: Harcourt, 2001.

Einstein, A., and L. Infeld. *The Evolution of Physics.* New York: Simon & Schuster, 1938.

Gamow, G. *Gravity.* Science Study Series. Garden City, NY: Doubleday (Anchor), 1962.

For information on space-faring projections, visit the Web site of the National Space Society (NSS) at www.nss.org

REVIEW QUESTIONS

1. What did Newton discover about gravity?
2. What is the Newtonian synthesis?

6.1 The Universal Law of Gravity

3. In what sense does the Moon "fall"?
4. State Newton's law of universal gravitation in words. Then do the same with one equation.

6.2 The Universal Gravitational Constant, G

5. What is the magnitude of gravitational force between two 1-kilogram bodies that are 1 meter apart?
6. What is the magnitude of the gravitational force between Earth and a 1-kilogram body?

6.3 Gravity and Distance: The Inverse-Square Law

7. How does the gravitational field about a planet vary with distance from the planet?
8. Where do you weigh more—at sea level or atop one of the peaks of the Sierra Nevada? Defend your answer.

6.4 Weight and Weightlessness

9. Would the springs inside a bathroom scale be more compressed or less compressed if you weighed yourself in an elevator that accelerated upward? Downward?
10. Would the springs inside a bathroom scale be more compressed or less compressed if you weighed yourself in an elevator that moved upward at *constant velocity*? Downward at *constant velocity*?
11. When is your weight equal to *mg*?

6.5 Universal Gravitation

12. What was the cause of perturbations discovered in the orbit of the planet Uranus? What greater discovery did this lead to?
13. Why was the status of Pluto recently demoted to that of a dwarf planet?
14. What percentage of the universe is speculated to be composed of dark matter and dark energy?

6.6 Projectile Motion

15. What exactly is a projectile?
16. Why does the vertical component of velocity for a projectile change with time, whereas the horizontal component of velocity doesn't?
17. A stone is thrown upward at an angle. What happens to the horizontal component of its velocity as it rises? As it falls?
18. A stone is thrown upward at an angle. What happens to the vertical component of its velocity as it rises? As it falls?
19. A projectile is launched upward at an angle of 75° from the horizontal and strikes the ground a certain distance downrange. For what other angle of launch at the same speed would this projectile land just as far away?
20. A projectile is launched vertically at 100 m/s. If air resistance can be neglected, at what speed will it return to its initial level?

6.7 Fast-Moving Projectiles—Satellites

21. Why will a projectile that moves horizontally at 8 km/s follow a curve that matches the curvature of Earth?
22. Why is it important that the projectile in the previous question be above Earth's atmosphere?

6.8 Circular Satellite Orbits

23. Why doesn't the force of gravity change the speed of a satellite in circular orbit?
24. For orbits of greater altitude, is the period longer or shorter?

6.9 Elliptical Orbits

25. Why does the force of gravity change the speed of a satellite in an elliptical orbit?
26. At what part of an elliptical orbit does a satellite have the greatest speed? The slowest speed?

6.10 Energy Conservation and Satellite Motion

27. Why is kinetic energy a constant for a satellite in a circular orbit but not for a satellite in an elliptical orbit?
28. With respect to the apogee and perigee of an elliptical orbit, where is the gravitational potential greatest? Where is it the least?
29. Is the sum of kinetic and potential energies a constant for satellites in circular orbits, in elliptical orbits, or in both?

6.11 Escape Speed

30. What happens to a satellite close to Earth's surface if it is given a speed exceeding 11.2 km/s?

ACTIVE EXPLORATIONS

1. Hold your hands outstretched in front of you, one twice as far from your eyes as the other, and make a casual judgment as to which hand looks bigger. Most people see them to be about the same size, while many see the nearer hand as slightly bigger. Almost no one, upon casual inspection, sees the nearer hand as four times as big; but, by the inverse-square law, the nearer hand should appear to be twice as tall and twice as wide and therefore seem to occupy four times as much of your visual field as the farther hand. Your belief that your hands are the same size is so strong that you likely overrule this information. Now, if you overlap your hands slightly and view them with one eye closed, you'll see the nearer hand as clearly bigger. This raises an interesting question: What other illusions do you have that are not so easily checked?

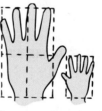

2. Repeat the eyeballing experiment, only this time use two dollar bills—one regular, and the other folded along its middle lengthwise, and again width-wise, so it has 1/4 the area. Now hold the two in front of your eyes. Where do you hold the folded one so that it looks the same size as the unfolded one? Nice?
3. With stick and strings, make a "trajectory stick" as shown on page 116.

ONE-STEP CALCULATIONS

$$F = G\frac{m_1 m_2}{d^2}$$

1. Calculate the force of gravity on a 1-kg mass at Earth's surface. Earth's mass is 6×10^{24} kg, and its radius is 6.4×10^6 m.

2. Calculate the force of gravity on the same 1-kg mass if it were 6.4×10^6 m above Earth's surface (that is, if it were two Earth radii from Earth's center).
3. Calculate the force of gravity between Earth (mass = 6.0×10^{24} kg) and the Moon (mass = 7.4×10^{22} kg). The average Earth–Moon distance is 3.8×10^8 m.

4. Calculate the force of gravity between Earth and the Sun (the Sun's mass = 2.0×10^{30} kg; average Earth–Sun distance = 1.5×10^{11} m).

5. Calculate the force of gravity between a newborn baby (mass = 3 kg) and the planet Mars (mass = 6.4×10^{23} kg), when Mars is at its closest to Earth (distance = 5.6×10^{10} m).

6. Calculate the force of gravity between a newborn baby of mass 3 kg and the obstetrician of mass 100 kg, who is 0.5 m from the baby. Which exerts more gravitational force on the baby, Mars or the obstetrician? By how much?

EXERCISES

1. Comment on whether or not the following label on a consumer product should be cause for concern. *CAUTION: The mass of this product pulls on every other mass in the universe, with an attracting force that is proportional to the product of the masses and inversely proportional to the square of the distance between them.*

2. Gravitational force acts on all bodies in proportion to their masses. Why, then, doesn't a heavy body fall faster than a light body?

3. Is the force of gravity stronger on a piece of iron than it is on a piece of wood if both have the same mass? Defend your answer.

4. Is the force of gravity stronger on a crumpled piece of paper than on an identical piece of paper that has not been crumpled? Defend your answer.

5. A friend says that astronauts in orbit are weightless because they're beyond the pull of Earth's gravity. Correct your friend's ignorance.

6. Somewhere between Earth and the Moon, gravity from these two bodies on a space pod would cancel. Is this location nearer to Earth or to the Moon?

7. Is the acceleration due to gravity more or less atop Mt. Everest than it is at sea level? Defend your answer.

8. An astronaut lands on a planet that has the same mass as Earth but twice the diameter. How does the astronaut's weight differ from that on Earth?

9. An astronaut lands on a planet that has twice Earth's and twice Earth's diameter. How does the astronaut's weight differ from that on Earth?

10. If Earth somehow expanded to a larger radius, with no change in mass, how would your weight be affected? How would it be affected if instead Earth shrunk? (Hint: Let the equation for gravitational force guide your thinking.)

11. A small light source located 1 m in front of a $1 - m^2$ opening illuminates a wall behind. If the wall is 1 m behind the opening (2 m from the light source), the illuminated area covers $4\ m^2$. How many square meters will be illuminated if the wall is 3 m from the light source? 5 m? 10 m?

12. The intensity of light from a central source varies inversely as the square of the distance. If you lived on a planet only half as far from the Sun as our Earth, how would the light intensity compare with that on Earth? How about a planet ten times farther away than Earth?

13. The planet Jupiter is more than 300 times as massive as Earth, so it might seem that a body on the surface of Jupiter would weigh 300 times as much as it would weigh on Earth. But it so happens that a body would scarcely weigh three times as much on the surface of Jupiter as it would on the surface of Earth. Can you think of an explanation for why this is so? (Hint: Let the terms in the equation for gravitational force guide your thinking.)

14. Why do the passengers in high-altitude jet planes feel the sensation of weight while passengers in an orbiting space vehicle, such as a space shuttle, do not?

15. If you were in a car that drove off the edge of a cliff, why would you be momentarily weightless? Would gravity still be acting on you?

16. What two forces act on you while you are in a moving elevator? When are these forces of equal magnitude and when are they not?

17. If you were in a freely falling elevator and you dropped a pencil, it would hover in front of you. Is there a force of gravity acting on the pencil? Defend your answer.

18. Your friend says that the primary reason astronauts in orbit feel weightless is that they are beyond the main pull of Earth's gravity. Why do you agree or disagree?

19. Explain why the following reasoning is wrong. "The Sun attracts all bodies on Earth. At midnight, when the Sun is directly below, it pulls on you in the same direction as Earth pulls on you; at noon, when the Sun is directly overhead, it pulls on you in a direction opposite to Earth's pull on you. Therefore, you should be somewhat heavier at midnight and somewhat lighter at noon."

20. Which requires more fuel—a rocket going from Earth to the Moon or a rocket coming from the Moon to Earth? Why?

21. Some people dismiss the validity of scientific theories by saying they are "only" theories. The law of universal gravitation is a theory. Does this mean that scientists still doubt its validity? Explain.

22. Suppose you roll a ball off a tabletop. Will the time to hit the floor depend on the speed of the ball? (Will a faster ball take a longer time to hit the floor?) Defend your answer.

23. A heavy crate accidentally falls from a high-flying airplane just as it flies directly above a shiny red Porsche smartly

parked in a car lot. Relative to the Porsche, where will the crate crash?

24. In the absence of air resistance, why does the horizontal component of a projectile's motion not change, while the vertical component does?

25. At what point in its trajectory does a batted baseball have its minimum speed? If air resistance can be neglected, how does this compare with the horizontal component of its velocity at other points?

26. A friend claims that bullets fired by some high-powered rifles travel for many meters in a straight-line path before they start to fall. Another friend disputes this claim and states that all bullets from any rifle drop beneath a straight-line path a vertical distance given by $\frac{1}{2}gt^2$ and that the curved path is apparent for low velocities and less apparent for high velocities. Now it's your turn: Will all bullets drop the same vertical distance in equal times? Explain.

27. Two golfers each hit a ball at the same speed, but one at 60° with the horizontal and the other at 30°. Which ball goes farther? Which hits the ground first? (Ignore air resistance.)

28. A park ranger shoots a monkey hanging from a branch of a tree with a tranquilizing dart. The ranger aims directly at the monkey, not realizing that the dart will follow a parabolic path and thus will fall below the monkey. The monkey, however, sees the dart leave the gun and lets go of the branch to avoid being hit. Will the monkey be hit anyway? Does the velocity of the dart affect your answer, assuming that it is great enough to travel the horizontal distance to the tree before hitting the ground? Defend your answer.

29. A projectile is fired straight upward at 141 m/s. How fast is it moving at the instant it reaches the top of its trajectory? Suppose instead that it were fired upward at 45°. What would be its speed at the top of its trajectory?

30. When you jump upward, your hang time is the time your feet are off the ground. Does hang time depend on your vertical component of velocity when you jump, your horizontal component of velocity, or both? Defend your answer.

31. The hang time of a basketball player who jumps a vertical distance of 2 feet (0.6 m) is about 2/3 second. What will be the hang time if the player reaches the same height while jumping 4 feet (1.2 m) horizontally?

32. Since the Moon is gravitationally attracted to Earth, why doesn't it simply crash into Earth?

33. Does the speed of a falling object depend on its mass? Does the speed of a satellite in orbit depend on its mass? Defend your answers.

34. If you have ever watched the launching of an Earth satellite, you may have noticed that the rocket starts vertically upward, then departs from a vertical course and continues its climb at an angle. Why does it start vertically? Why does it not continue vertically?

35. If a cannonball is fired from a tall mountain, gravity changes its speed all along its trajectory. But if it is fired fast enough to go into circular orbit, gravity does not change its speed at all. Explain.

36. A satellite can orbit at 5 km above the Moon, but not at 5 km above Earth. Why?

37. Would the speed of a satellite in close circular orbit about Jupiter be greater than, equal to, or less than 8 km/s?

38. Why are satellites normally sent into orbit by firing them in an easterly direction, the direction in which Earth spins?

39. Of all the United States, why is Hawaii the most efficient launching site for nonpolar satellites? (Hint: Look at the spinning Earth from above either pole and compare it to a spinning turntable.)

40. Earth is closer to the Sun in December than it is in June. In which of these two months is Earth moving faster around the Sun?

41. What is the shape of the orbit when the velocity of the satellite is everywhere perpendicular to the force of gravity?

42. A communications satellite with a 24-hour period hovers over a fixed point on Earth. Why is it placed in orbit only in the plane of Earth's equator? (Hint: Think of the satellite's orbit as a ring around Earth.)

43. If a flight mechanic drops a wrench from a high-flying jumbo jet, it crashes to Earth. If an astronaut on the orbiting space shuttle drops a wrench, does it crash to Earth also? Defend your answer.

44. How could an astronaut in a space shuttle "drop" an object vertically to Earth?

45. If you stopped an Earth satellite dead in its tracks, it would simply crash into Earth. Why, then, don't the communications satellites that "hover motionless" above the same spot on Earth crash into Earth?

46. The orbital velocity of Earth about the Sun is 30 km/s. If Earth were suddenly stopped in its tracks, it would simply fall radially into the Sun. Devise a plan whereby a rocket loaded with radioactive wastes could be fired into the Sun

for permanent disposal. How fast and in what direction with respect to Earth's orbit should the rocket be fired?

47. In an accidental explosion, a satellite breaks in half while in circular orbit about Earth. One half is brought momentarily to rest. What is the fate of the half brought to rest? What happens to the other half?

48. If Pluto were somehow stopped short in its orbit, it would fall into, rather than around, the Sun. How fast would it be moving when it hit the Sun?

49. At which of the indicated positions does the satellite in elliptical orbit experience the greatest gravitational force? Have the greatest speed? The greatest velocity? The greatest momentum? The greatest kinetic energy? The greatest gravitational potential energy?

The greatest total energy? The greatest acceleration?

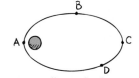

50. A rocket coasts in an elliptical orbit around Earth. To attain the greatest amount of KE for escape using a given amount of fuel, should it fire its engines at the apogee or at the perigee? (Hint: Let the formula $Fd = \Delta KE$ be your guide to thinking. Suppose the thrust F is brief and of the same duration in either case. Then consider the distance d the rocket would travel during this brief burst at the apogee and at the perigee.)

PROBLEMS

● BEGINNER ■ INTERMEDIATE ◆ EXPERT

1. ● Consider a pair of planets for which the distance between them is decreased by a factor of 5. Show that the force between them becomes 25 times greater.

2. ● Many people mistakenly believe that the astronauts who orbit Earth are "above gravity." Earth's mass is 6×10^{24} kg, and its radius is 6.38×10^{6} m (6380 km). Use the inverse-square law to show that in space-shuttle territory, 200 kilometers above Earth's surface, the force of gravity on a shuttle is about 94% of that at Earth's surface.

3. ■ The mass of a certain neutron star is 3.0×10^{30} kg (1.5 solar masses) and its radius is 8000 m (8 km). Show that the force of gravity at the surface of this condensed, burned-out star is about 300 billion times that of Earth.

4. ■ A ball is thrown horizontally from a cliff at a speed of 10 m/s. Show that its speed one second later is 14.1 m/s.

5. ■ An airplane is flying horizontally with speed 1000 km/h (280 m/s) when an engine falls off. Neglecting air resistance, assume it takes 30 s for the engine to hit the ground.
   a. Show that the altitude of the airplane is 4500 m.
   b. Show that the horizontal distance that the aircraft engine falls is 8400 m.
   c. If the airplane somehow continues to fly as if nothing had happened, where is the engine relative to the airplane at the moment the engine hits the ground?

6. ■ A cannonball shot with an initial velocity of 141 m/s at an angle of 45° follows a parabolic path and hits a balloon at the top of its trajectory. Neglecting air resistance, show that the cannonball hits the balloon at a speed of 100 m/s.

7. ■ A certain satellite has a kinetic energy of 8 billion joules at perigee (the point at which it is closest to Earth) and 5 billion joules at apogee (the point at which it is farthest from Earth). As the satellite travels from apogee to perigee, how much work does the gravitational force do on it? Does its potential energy increase or decrease during this time, and by how much?

8. ■ The force that pulls moving objects into circular paths is called *centripetal force* and is given by the equation $F_c = \dfrac{mv^2}{d}$, where m is the mass of an object moving in a circular path at speed v and distance d from the center of the circular path. For the Moon circling Earth, gravity supplies the centripetal force. Equate centripetal force to gravitational force and show that the speed of the Moon in its orbit about Earth is $v = \sqrt{\dfrac{GM}{d}}$, where M is the mass of Earth and d is the distance between the centers of the Moon and Earth.

9. ■ Calculate the speed in m/s at which Earth revolves about the Sun. You may assume the orbit is nearly circular.

10. ■ The Moon is about 3.8×10^{5} km from Earth. Show that its average orbital speed about Earth is 1026 m/s.

11. ◆ The force of gravity on you by Earth is GmM/d^2, where G is the universal gravitational constant, m is your mass, M is the mass of Earth, and d is your distance from Earth's center.
    a. Use Newton's second law to show that your gravitational acceleration toward Earth at distance d from its center is $a = GM/d^2$.
    b. How does this equation support the finding that the acceleration due to gravity doesn't depend on the mass of an object in free fall?

12. ◆ The gravitational field about a massive object is defined to be the gravitational force per mass on an object in the vicinity of the massive object. The symbol for the gravitational field is boldfaced $\mathbf{g}$ (with magnitude the same as the magnitude of gravitational acceleration at that point, g).
    a. Show that the gravitational field a distance d from Earth's center is GM/d^2, where G is the universal gravitational constant and M is the mass of Earth.
    b. The value of $\mathbf{g}$ at Earth's surface is about 9.8 N/kg. Show that the value of $\mathbf{g}$ at a distance from Earth's center that is four times Earth's radius would be 0.6 m/s².

13. ◆ A rock thrown horizontally from a bridge hits the water below at a horizontal distance x directly below the throwing point. The rock travels a smooth parabolic path in time t.
    a. Show that the vertical distance of the bridge above water is $\frac{1}{2}gt^2$.
    b. What is the height of the bridge if the time the rock is airborne is 2 seconds?
    c. What information is provided in Chapter 6 that wasn't provided in Chapter 3 for the solution of this problem?

14. ◆ A baseball is tossed at a steep angle into the air and makes a smooth parabolic path. Its time in the air is t and it reaches a maximum height h. Assume that air resistance is negligible.
    a. Show that the height reached by the ball is $gt^2/8$.
    b. If the ball is in the air for 4 seconds, show that the ball reaches a height of 19.6 m.
    c. If the ball reached the same height as when tossed at some other angle, would the time of flight be the same?

15. ◆ A penny moving at speed v slides off the horizontal surface of a coffee table a vertical distance y from the floor.
    a. Show that the penny lands a distance $v\sqrt{\dfrac{2y}{g}}$ from the base of the coffee table.
    b. If the speed is 3.5 m/s and the coffee table is 0.4 m tall, show that the distance the coin lands from the base of the table is 1.0 m.

16. ◆ Students in a lab measure the speed of a steel ball launched horizontally from a tabletop to be v. The tabletop is distance y above the floor. They place a tall tin coffee can of height $0.1y$ on the floor to catch the ball.
    a. Show that the can should be placed a horizontal distance from the base of the table of $v\sqrt{\dfrac{2(0.9)y}{g}}$.
    b. If the ball leaves the tabletop at a speed of 4.0 m/s, the tabletop is 1.5 m above the floor, and the can is 0.15 m tall, show that the center of the can should be placed a horizontal distance of 0.52 m from the base of the table.

CHAPTER 6 ONLINE RESOURCES

The Physics Place

Interactive Figures
6.6, 6.15, 6.16, 6.20, 6.22, 6.24, 6.32, 6.36, 6.42

Tutorials
Motion and Gravity
Projectile Motion
Orbits and Kepler's Laws

Videos
von Jolly's Method of Measuring the Attraction
  Between Two Masses
Inverse-Square Law
Weight and Weightlessness

Apparent Weightlessness
Gravitational Field Inside a Hollow Planet
The Weight of an Object Inside a Hollow Planet but
  Not at its Center
Discovery of Neptune
Projectile Motion Demo
More Projectile Motion
Circular Orbits

Quiz

Flashcards

Links

Fluid Mechanics

The forces due to atmospheric pressure are nicely shown by Swedish father-and-son physics professors, P.O. and Johan Zetterberg, who pull on a classroom model of the Magdeburg hemispheres.

Liquids and gases have the ability to flow; hence, they are called *fluids*. Because they are both fluids we find that they obey similar mechanical laws. How is it that iron boats don't sink in water or that helium balloons don't sink from the sky? Why is it impossible to breathe through a snorkel when you're under more than a meter of water? Why do your ears pop when riding an elevator? How do hydrofoils and airplanes attain lift? To discuss fluids, it is important to introduce two concepts—*density* and *pressure*.

7.1 Density

An important property of materials, whether in the solid, liquid, or gaseous phases is the measure of compactness: **density**. We think of density as the "lightness" or "heaviness" of materials of the same size. It is a measure of how much mass occupies a given space; it is the amount of matter per unit volume:

$$\text{Density} = \frac{\text{mass}}{\text{volume}}$$

Or in shorthand notation,

$$\rho = \frac{m}{V}$$

where ρ (rho) is the symbol for density, m is mass, and V is volume.

The densities of a few materials are listed in Table 7.1. Mass is measured in grams or kilograms, and volume in cubic centimeters (cm^3) or

TABLE 7.1

Densities of Some Materials

Material	Grams per Cubic Centimeter (g/cm³)	Kilograms per Cubic Meter (kg/m³)
Liquids		
Mercury	13.6	13,600
Glycerin	1.26	1,260
Seawater	1.03	1,025
Water at 4°C	1.00	1,000
Benzene	0.90	899
Ethyl alcohol	0.81	806
Solids		
Iridium	22.6	22,650
Osmium	22.6	22,610
Platinum	21.1	21,090
Gold	19.3	19,300
Uranium	19.0	19,050
Lead	11.3	11,340
Silver	10.5	10,490
Copper	8.9	8,920
Brass	8.6	8,600
Iron	7.8	7,874
Tin	7.3	7,310
Aluminum	2.7	2,700
Ice	0.92	919
Gases (atmospheric pressure at sea level)		
Dry air		
0°C	0.00129	1.29
10°C	0.00125	1.25
20°C	0.00121	1.21
30°C	0.00116	1.16
Helium	0.000178	0.178
Hydrogen	0.000090	0.090
Oxygen	0.00143	1.43

cubic meters (m³).* A gram of any material has the same mass as 1 cubic centimeter of water at a temperature of 4°C. So water has a density of 1 gram per cubic centimeter. Mercury's density is 13.6 grams per cubic centimeter, which means that it has 13.6 times as much mass as an equal volume of water. Iridium, a hard, brittle, silvery-white metal in the platinum family, is the densest substance on Earth.

A quantity known as weight density, commonly used when discussing liquid pressure, is

---

* A cubic meter is a sizable volume and contains a million cubic centimeters, so there are a million grams of water in a cubic meter (or, equivalently, a thousand kilograms of water in a cubic meter). Hence, 1 g/cm³ = 1000 kg/m³.

FIGURE 7.1

When the volume of a loaf of bread is reduced, its density increases.

expressed by the amount of weight of a body per unit volume:**

$$\text{Weight density} = \frac{\text{weight}}{\text{volume}}$$

> The metals lithium, sodium, and potassium (not listed in Table 7.1) are all less dense than water and will float in water.

STOP AND CHECK YOURSELF

1. Which has the greater density—1 kg of water or 10 kg of water?

2. Which has the greater density—5 kg of lead or 10 kg of aluminum?

3. Which has the greater density—an entire candy bar or half a candy bar?

CHECK YOUR ANSWERS

1. The density of any amount of water is the same: 1 g/cm³ or, equivalently, 1000 kg/m³, which means that the mass of water that would exactly fill a thimble of volume 1 cubic centimeter would be 1 gram; or the mass of water that would fill a 1-cubic-meter tank would be 1000 kg. One kg of water would fill a tank only a thousandth as large, 1 liter, whereas 10 kg would fill a 10-liter tank. Nevertheless, the important concept is that the ratio of mass/volume is the same for *any* amount of water.

2. Density is a *ratio* of weight or mass per volume, and this ratio is greater for any amount of lead than for any amount of aluminum—see Table 7.1.

3. Both the half and the entire candy bar have the same density.

** Weight density is common to United States Customary System (USCS) units in which one cubic foot of freshwater (nearly 7.5 gallons) weighs 62.4 pounds. So freshwater has a weight density of 62.4 lb/ft³. Saltwater is slightly denser, 64 lb/ft³.

Pressure

Place a book on a bathroom scale and, whether you place it on its back, on its side, or balanced on a corner, it still exerts the same force. The weight reading is the same. Now balance the book on the palm of your hand and you sense a difference—the *pressure* of the book depends on the area over which the force is distributed (Figure 7.2). You'll see a distinction between force and pressure. **Pressure** is defined as the force exerted over a unit of area, such as a square meter or square foot:*

$$\text{Pressure} = \frac{\text{force}}{\text{area}}$$

A dramatic illustration of pressure is shown in Figure 7.3 The author applies appreciable force when he breaks the cement block with a sledgehammer. Yet his teaching buddy, who is sandwiched between two beds of sharp nails, is unharmed. This is because the force is distributed over more than 200 nails making contact with his body. The combined surface area of the nails results in a tolerable pressure that does not puncture the skin. Again, force and pressure are different from each other.

STOP AND
CHECK YOURSELF

Does a bathroom scale measure weight, pressure, or both?

CHECK YOUR ANSWER

A bathroom scale measures weight, the force that compresses an internal spring or equivalent. The weight reading is the same whether you stand on one or both feet (although the pressure on the scale is twice as much when standing on one foot).

FIGURE **7.2**

Although the weight of both books is the same, the upright book exerts greater pressure against the table.

---

* Pressure may be measured in any unit of force divided by any unit of area. The standard international (SI) unit of pressure, the newton per square meter, is called the *pascal* (Pa), after the seventeenth-century theologian and scientist Blaise Pascal. A pressure of 1 Pa is very small and approximately equals the pressure exerted by a dollar bill resting flat on a table. Physics types prefer kilopascals (1 kPa = 1000 Pa).

FIGURE **7.3**

The author applies a force to physics teacher Pablo Robinson, who is bravely sandwiched between beds of sharp nails. The driving force per nail is not enough to puncture the skin. From an inertia point of view, is Pablo safer if the block is massive? From the point of view of energy, is he in danger if the block doesn't break?

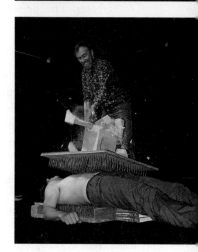

Pressure in a Liquid

When you swim under water, you can feel the water pressure acting against your eardrums: the deeper you swim, the greater the pressure. What causes this pressure? It is simply the weight of the fluids directly above you—water plus air—pushing against you. As you swim deeper, there is more water above you. Therefore, there's more pressure. If you swim twice as deep, there is twice the weight of water above you, so the water's contribution to the pressure you feel is doubled. Added to the water pressure is the pressure of the atmosphere, which is equivalent to an extra 10.3-meter depth of water. Because atmospheric pressure at Earth's surface is nearly constant, pressure differences you feel under water depend only on changes in depth.

If you were submerged in a liquid denser than water, the pressure would be correspondingly

 The Physics Place
Dam Keeps Water in Place
Water Keeps Dam in Place

FIGURE 7.4

This water tower does more than store water. The depth of water above ground level insures substantial and reliable water pressure to the many homes it serves.

FIGURE 7.5

Tsing Bardin shows her class that liquid pressure is the same for any given depth below the surface, regardless of the shape of the containing vessel.

greater. The pressure due to a liquid is precisely equal to the product of weight density and depth:*

Liquid pressure = weight density × depth

It is important to note that pressure does not depend on the volume of liquid. You feel the same pressure a meter deep in a small pool as you do a meter deep in the middle of the ocean. This is illustrated by the connecting vases shown in Figure 7.5. If the pressure at the bottom of a large vase were greater than the pressure at the bottom of a neighboring narrower vase, the greater pressure would force water sideways and then up the narrower vase to a higher level. We find, however, that this doesn't happen. Pressure depends on depth, not volume.

Water seeks its own level. This can be demonstrated by filling a garden hose with water and holding the two ends upright. The water levels will be equal whether the ends are held close together or far apart. Pressure is depth dependent, not volume

> When measuring blood pressure, notice that you measure it in your upper arm—level with your heart.

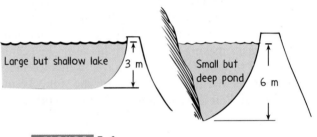

Large but shallow lake | 3 m

Small but deep pond | 6 m

FIGURE 7.6

The average water pressure acting against the dam depends on the average depth of the water and not on the volume of water held back. The large shallow lake exerts only one-half the average pressure that the small deep pond exerts.

* This is derived from the definitions of pressure and density. Consider an area at the bottom of a vessel that contains liquid. The weight of the column of liquid directly above this area produces pressure. From the definition *weight density = weight/volume*, we can express this weight of liquid as *weight = weight density × volume*, where the volume of the column is simply the area multiplied by the depth. Then we get

$$\text{Pressure} = \frac{\text{force}}{\text{area}} = \frac{\text{weight}}{\text{area}} = \frac{\text{weight density} \times \text{density}}{\text{area}}$$

$$= \frac{\text{weight density} \times (\text{area} \times \text{depth})}{\text{area}}$$

$$= \text{weight density} \times \text{depth}.$$

For the total pressure we should add to this equation the pressure due to the atmosphere on the surface of the liquid.

dependent. So we see there is an explanation for why water seeks its own level.

In addition to being depth dependent, liquid pressure is exerted equally in all directions. For example, if we are submerged in water, it makes no difference which way we tilt our heads—our ears feel the same amount of water pressure. Because a liquid can flow, the pressure isn't only downward. We know pressure acts upward when we try to push a beach ball beneath the water's surface. The bottom of a boat is certainly pushed upward by water pressure. And we know water pressure acts sideways when we see water spurting sideways from a leak in an upright can. Pressure in a liquid at any point is exerted in equal amounts in all directions.

When liquid presses against a surface, there is a net force directed perpendicular to the surface

FIGURE 7.7

The forces due to liquid pressure against a surface combine to produce a net force that is perpendicular to the surface.

FIGURE 7.8

The force vectors act in a direction perpendicular to the inner container surface and increase with increasing depth.

(Figure 7.7). If there is a hole in the surface, the liquid spurts at right angles to the surface before curving downward due to gravity (Figure 7.8). At greater depths the pressure is greater and the speed of the exiting liquid is greater.*

7.4   Buoyancy in a Liquid

Anyone who has ever lifted a submerged object out of water is familiar with buoyancy, the apparent loss of weight of submerged objects. For example, lifting a large boulder off the bottom of a riverbed is a relatively easy task as long as the boulder is below the surface. When it is lifted above the surface, however, the force required to lift it is considerably more. This is because, when the boulder is submerged, the water exerts an upward force on it that is exactly opposite in direction to gravity. This upward force is called the **buoyant force**, and is a consequence of greater pressure increasing with greater depth. Figure 7.9 shows why the buoyant force acts upward. Pressure is exerted everywhere against the object in a direction perpendicular to its surface. The arrows represent the magnitude and direction of forces at different places. Forces that produce pressures against the sides due to equal depths cancel one another. Pressure is greatest against the bottom of the boulder simply because the bottom of the boulder is deeper.

Since the upward forces against the bottom are greater than the downward forces against the top, the forces do not cancel, and there is a net force upward. This net force is the buoyant force.

If the weight of the submerged object is greater than the buoyant force, the object will sink. If the weight is equal to the buoyant force acting up on the submerged object, it will remain at any level, like a fish. If the buoyant force is greater than the weight of the completely submerged object, it will rise to the surface and float.

Understanding buoyancy requires understanding the meaning of the expression "volume of water displaced." If a stone is placed in a container that is already up to its brim with water, some water will overflow (Figure 7.10). Water is *displaced* by the stone. A little thought will tell us that the *volume of the stone*—that is, the amount of space it occupies or its number of cubic centimeters—is equal to the *volume of water displaced*. Place any object in a container partially filled with water, and the level of the surface rises (Figure 7.11). By how much? That would be to exactly the level that would be reached by pouring in a volume of water equal to the volume of the submerged object. This is a good method for determining the volume of irregularly shaped objects: *A completely submerged object always displaces a volume of liquid equal to its own volume.*

> Stick your foot in a swimming pool and your foot is immersed. Jump in and sink below the surface and immersion is total—you're submerged.

The Physics Place
Buoyancy

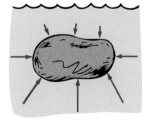

FIGURE 7.9

The greater pressure against the bottom of a submerged object produces an upward buoyant force.

* The speed of liquid exiting the hole is $\sqrt{2gh}$, where h is the depth below the free surface. Interestingly, this is the same speed that water or anything else would have if freely falling the same distance h.

Water displaced

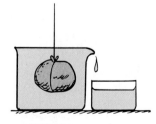

FIGURE 7.10

When a stone is submerged, it displaces a volume of water equal to the volume of the stone.

FIGURE 7.11

The raised level due to placing a stone in the container is the same as if a volume of water equal to the volume of the stone were poured in.

7.5 Archimedes' Principle

The relationship between buoyancy and displaced liquid was first discovered in the third century BC by the Greek scientist Archimedes. It is stated as follows:

An immersed body is buoyed up by a force equal to the weight of the fluid it displaces.

This relationship is called **Archimedes' principle**. It applies to liquids and gases, which are both fluids. If an immersed body displaces 1 kilogram of fluid, the buoyant force acting on it is equal to the weight of 1 kilogram.* By *immersed*, we mean either *completely* or *partially submerged*. If we immerse a sealed 1-liter container halfway into the water, it will displace one half-liter of water and be buoyed up by the weight of one half-liter of water. If we immerse it completely (submerge it), it will be buoyed up by the weight of a full liter (or 1 kilogram) of water. Unless the completely submerged container is compressed, the buoyant force will equal the weight of 1 kilogram at *any* depth. This is because, at any depth, it can displace no greater volume of water than its own volume. And the weight of this volume of water (not the weight of the submerged object!) is equal to the buoyant force.

If a 25-kilogram object displaces 20 kilograms of fluid upon immersion, its apparent weight will equal the weight of 5 kilograms. Notice in Figure 7.13 that the 3-kilogram block has an apparent weight equal to the weight of 1 kilogram when submerged. The apparent weight of a submerged object is its weight out of water minus the buoyant force.

The Physics Place
Archimedes' Principle
Flotation

FIGURE 7.12

A liter of water occupies a volume of 1000 cm³, has a mass of 1 kg, and weighs 9.8 N. Its density may therefore be expressed as 1 kg/L and its weight density as 9.8 N/L. (Seawater is slightly denser, about 10 N/L.)

* A kilogram is not a unit of force but a unit of mass. So, strictly speaking, the buoyant force is not 1 kg, but the *weight* of 1 kg, which is 9.8 N. We could as well say that the buoyant force is 1 *kilogram weight*, not simply 1 kg.

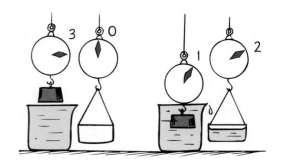

FIGURE 7.13

A 3-kg block weighs more in air than it does in water. When submerged in water, the block's loss in weight is the buoyant force, which equals the weight of water displaced.

STOP AND CHECK YOURSELF

1. Does Archimedes' principle tell us that if an immersed block displaces 10 N of fluid, the buoyant force on the block is 10 N?

2. A 1-liter container completely filled with lead has a mass of 11.3 kg and is submerged in water. What is the buoyant force acting on it?

3. A boulder is thrown into a deep lake. As it sinks deeper and deeper into the water, does the buoyant force on it increase? Decrease?

CHECK YOUR ANSWERS

1. Yes. Looking at it in a Newton's-third-law way, when the immersed block pushes 10 N of fluid aside, the fluid reacts by pushing back on the block with 10 N.

2. The buoyant force is 9.8 N (the weight of 1 kg of water). That's because the volume of water displaced is 1 L, which has a mass of 1 kg and a weight of 9.8 N. The 11.3 kg of the lead is irrelevant; 1 L of anything submerged in water will displace 1 L and be buoyed upward with a force 9.8 N, the weight of 1 kg. (Get this straight before going further!)

3. Buoyant force does not change as the boulder sinks because the boulder displaces the same volume of water at any depth. Since water is practically incompressible, its density is very nearly the same at all depths; hence, the weight of water displaced, or the buoyant force, is practically the same at all depths.

Perhaps your instructor will summarize Archimedes' principle by way of a numerical example to show the difference between the upward-acting and the downward-acting forces on a submerged

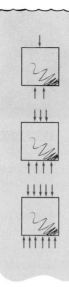

FIGURE 7.14

The difference in the upward and downward forces acting on the submerged block is the same at any depth.

CHECK YOURSELF

1. Drop a boulder in a deep well. As it descends beneath the surface, pressure on it increases. Does this imply that buoyant force likewise increases?

2. Since buoyant force is the upward force that a fluid exerts on a body, and we learned in Chapter 4 that forces produce accelerations, why doesn't a submerged body accelerate?

CHECK YOUR ANSWERS

1. No! Once the boulder is beneath the water surface, it has displaced all the water it can. The water level in the well stays the same as the boulder further descends, showing that water displacement, and therefore buoyant force on the boulder, remains the same—even though water pressure on the boulder increases with depth. Buoyancy and pressure are different concepts.

2. It does accelerate if the buoyant force is not balanced by other forces that act on it—the force of gravity and fluid resistance. The net force on a submerged body is the result of the force the fluid exerts (buoyant force), the weight of the body, and, if the body is moving, the force of fluid friction. When the net force is zero, the body is in equilibrium.

cube (due to differences of pressure). You'll see that the force difference is numerically identical to the weight of fluid displaced. It makes no difference how deep the cube is placed, because, although the pressures are greater with increasing depths, the *difference* between the pressure exerted upward against the bottom of the cube and the pressure exerted downward against the top of the cube is the same at any depth (Figure 7.14). Whatever the shape of the submerged body, the buoyant force is equal to the weight of fluid displaced.

Flotation

Iron is much denser than water and therefore sinks, but an iron ship floats. Why is this so? Consider a solid 1-ton block of iron. Iron is nearly eight times

as dense as water, so when it is submerged it will displace only 1/8 ton of water, which is certainly not enough to prevent it from sinking. Suppose we reshape the same iron block into a bowl, as shown

HISTORY OF SCIENCE

■ ARCHIMEDES AND THE GOLD CROWN

According to legend, Archimedes (287–212 BC) had been given the task of determining whether a crown made for King Hiero II of Syracuse was of pure gold or whether it contained some less expensive metals such as silver. Archimedes' problem was to determine the density of the crown without destroying it. He could weigh the gold, but determining its volume was a problem. The story has it that Archimedes realized the solution when he noted the rise in water level while immersing his body in the

public baths of Syracuse. Legend reports that he excitedly rushed naked through the streets shouting, "Eureka! Eureka!" ("I have found it! I have found it!").

What Archimedes had discovered was a simple and accurate way of finding the volume of an irregular object—the displacement method of determining volume. Once he knew both the weight and volume, he could calculate the density. Then the density of the crown could be compared with the density of gold. Archimedes' insight preceded Newton's law of motion, from which Archimedes' principle can be derived, by almost 2000 years.

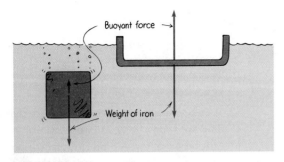

FIGURE 7.15

An iron block sinks, while the same quantity of iron shaped like a bowl floats.

FIGURE 7.18

The same ship empty and loaded. How does the weight of its load compare to the weight of additional water displaced?

> Only in the special case of floating does the buoyant force acting on an object equal the object's weight.

in Figure 7.15. It still weighs 1 ton. When we place it in the water, it settles into the water, displacing a greater volume of water than before. The deeper it is immersed, the more water it displaces and the greater the buoyant force acting on it. When the buoyant force equals 1 ton, it will sink no further.

When the iron boat displaces a weight of water equal to its own weight, it floats. This is called the **principle of flotation**, which states:

A floating object displaces a weight of fluid equal to its own weight.

Every ship, submarine, and dirigible must be designed to displace a weight of fluid equal to its own weight. Thus, a 10,000-ton ship must be built wide enough to displace 10,000 tons of water before it immerses too deep in the water. The same

applies to vessels in air. A dirigible or huge balloon that weighs 100 tons displaces at least 100 tons of air. If it displaces more, it rises; if it displaces less, it descends. If it displaces exactly its weight, it hovers at constant altitude.

Since the buoyant force upon a body equals the weight of the fluid it displaces, denser fluids will exert a greater buoyant force upon a body than less-dense fluids of the same volume. A ship therefore floats higher in saltwater than in freshwater because saltwater is slightly denser than freshwater. In the same way, a solid chunk of iron will float in mercury even though it sinks in water.

> People who can't float are, nine times out of ten, males. Most males are more muscular and slightly denser than females. Also, cans of diet soda float whereas cans of regular soda sink in water. What does this tell you about their relative densities?

FIGURE 7.16

The weight of a floating object equals the weight of the water displaced by the submerged part.

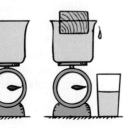

FIGURE 7.17

A floating object displaces a weight of fluid equal to its own weight.

STOP AND CHECK YOURSELF

Fill in the blanks for these statements:

1. The volume of a submerged body is equal to the _____ of the fluid displaced.
2. The weight of a floating body is equal to the _____ of the fluid displaced.
3. Why is it easier to float in saltwater than in freshwater?

CHECK YOUR ANSWERS

1. Volume.
2. Weight.
3. When you're floating, the weight of water you displace equals your weight. Saltwater is denser, so you don't "sink" as far to displace your weight. You'd float even higher in mercury (density 13.6 g/cm^3), and you'd sink completely in alcohol (density 0.8 g/cm^3).

▪ FLOATING MOUNTAINS

Mountains float on Earth's semiliquid mantle just as icebergs float in water. Both the mountains and icebergs are less dense than the material they float upon. Just as most of an iceberg is below the water surface (90%), most of a mountain (about 85%) extends into the dense semiliquid mantle. If you could shave off the top of an iceberg, the iceberg would be lighter and be buoyed up to nearly its original height before its top was shaved. Similarly, when mountains erode they are lighter, and are pushed up from below to float to nearly their original heights. So when a kilometer of mountain erodes away, some 85% of a kilometer of mountain returns. That's why it takes so long for mountains to weather away. Mountains, like icebergs, are bigger than they appear to be. The concept of floating mountains is *isostacy*—Archimedes' principle for rocks.

Notice in our discussion of liquids that Archimedes' principle and the law of flotation were stated in terms of *fluids*, not liquids. That's because, although liquids and gases are different phases of matter, they are both fluids, with much the same mechanical principles. Let's turn our attention to the mechanics of gases in particular.

7.6 Pressure in a Gas

The primary difference between a gas and a liquid is the distance between molecules. In a gas, the molecules are far apart and free from the cohesive forces that dominate their motions when in the liquid and solid phases. Molecular motions in a gas are less restricted. A gas expands, fills all space available to it, and exerts a pressure against its container. Only when the quantity of gas is very large, such as Earth's atmosphere or a star, do the gravitational forces limit the size or determine the shape of the mass of gas.

> Liquids and gases are both fluids. A gas takes the shape of its container. A liquid does so only below its surface.

The Physics Place
Air Has Pressure

Boyle's Law

The air pressure inside the inflated tires of an automobile is considerably greater than the atmospheric pressure outside. The density of air inside is also more than that of the air outside. To understand the relation between pressure and density, think of the molecules of air (primarily nitrogen and oxygen) inside the tire. The air molecules behave like tiny billiard balls, randomly moving and bumping against the inner walls producing a jittery force that appears to our coarse senses as a steady push. This pushing force, averaged over the wall area, provides the pressure of the enclosed air.

Suppose there are twice as many molecules in the same volume (Figure 7.19). Then the air density is doubled. If the molecules move at the same average speed—or, equivalently, if they have the same temperature—then the number of collisions will be doubled. This means that the pressure is doubled. So pressure is proportional to density.

We double the density of air in the tire by doubling the amount of air. We can also double the density of a *fixed* amount of air by compressing it to half its volume. Consider the cylinder with the movable piston in Figure 7.20. If the piston is pushed downward so that the volume is half the original volume, the density of molecules is doubled, and the pressure is correspondingly doubled. Decrease the volume to a third of its original value,

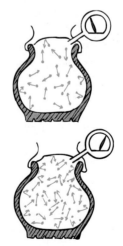

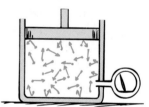

FIGURE 7.19
When the density of gas in the tire is increased, pressure is increased.

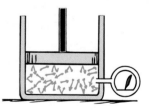

FIGURE 7.20
When the volume of gas is decreased, density and therefore pressure are increased.

and the pressure is increased by three, and so forth (provided the temperature remains the same).

Notice in these examples with the piston that the product of pressure and volume remains the same. For example, a doubled pressure multiplied by a halved volume gives the same value as a tripled pressure multiplied by a one-third volume. In general, we can state that the product of pressure and volume for a given mass of gas is a constant as long as the temperature does not change. "Pressure × volume" for a quantity of gas at some initial time is equal to any "different pressure × different volume" at some later time. In shorthand notation,

$$P_1 V_1 = P_2 V_2$$

where P_1 and V_1 represent the original pressure and volume, respectively, and P_2 and V_2 the second pressure and volume. This relationship is called **Boyle's law**, after Robert Boyle, the seventeenth-century physicist who is credited with its discovery.*

Boyle's law applies to ideal gases. An ideal gas is one in which the disturbing effects of the forces between molecules and the finite size of the individual molecules can be neglected. Air and other gases under normal pressures and temperatures approach ideal gas conditions.

STOP AND CHECK YOURSELF

1. A piston in an airtight pump is withdrawn so that the volume of the air chamber is increased three times. What is the change in pressure?

2. A scuba diver breathes compressed air beneath the surface of water. If for some reason she abandons her scuba gear and returns to the surface while holding her breath, what happens to the volume of her lungs?

CHECK YOUR ANSWERS

1. The pressure in the piston chamber is reduced to one-third. This is the principle that underlies a mechanical vacuum pump.

2. As she rises the surrounding water pressure on her body decreases, which allows the volume of air in her lungs to increase—ouch! A first lesson in scuba diving is to not hold your breath when ascending. To do so can be fatal.

---

* A general law that takes temperature changes into account is $P_1 V_1 / T_1 = P_2 V_2 / T_2$, where T_1 and T_2 represent the initial and final *absolute* temperatures, measured in SI units called kelvins (Chapter 8).

7.7 Atmospheric Pressure

We live at the bottom of an ocean of air. The atmosphere, much like the water in a lake, exerts a pressure. One of the most celebrated experiments demonstrating the pressure of the atmosphere was conducted in 1654 by Otto von Guericke, burgermeister of Magdeburg and inventor of the vacuum pump. Von Guericke placed together two copper hemispheres about 1/2 meter in diameter to form a sphere, as shown in Figure 7.21. He set a gasket made of a ring of leather soaked in oil and wax between them to make an airtight joint. When he evacuated the sphere with his vacuum pump, two teams of eight horses each were unable to pull the hemispheres apart. (The photo that opens this chapter on page 133 shows a much smaller pair of similar hemispheres that the Zetterberg team can't pull apart.)

The Physics Place
Air Has Weight
Air Is Matter

Interestingly, von Guericke's demonstration preceded knowledge of Newton's third law. The forces on the hemispheres would have been the same if he used only one team of horses and tied the other end of the rope to a tree!

FIGURE 7.21

The famous "Magdeburg hemispheres" experiment of 1654, demonstrating atmosphere pressure. The two teams of horses couldn't pull the evacuated hemisphere apart. Were the hemispheres sucked together or pushed together? By what?

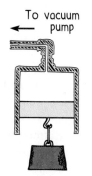

To vacuum
pump ←

FIGURE 7.22

Is the piston pulled up
or pushed up?

FIGURE 7.23

You don't notice the weight of
a bag of water while you're
submerged in water. Similarly,
you don't notice that the air
around you has weight.

When the air pressure inside a cylinder similar to that shown in Figure 7.22 is reduced, there is an upward force on the piston. This force is large enough to lift a heavy weight. If the inside diameter of the cylinder is 12 centimeters or greater, a person can be lifted by this force.

What do the experiments of Figures 7.21 and 7.22 demonstrate? Do they show that air exerts pressure or that there is a "force of suction"? If we say there is a force of suction, then we assume that a vacuum can exert a force. But what is a vacuum? It is an absence of matter; it is a condition of nothingness. How can nothing exert a force? The hemispheres are not sucked together, nor is the piston holding the weight sucked upward. The hemispheres and the piston are being pushed against by the pressure of the atmosphere.

Just as water pressure is caused by the weight of water, **atmospheric pressure** is caused by the weight of air. We have adapted so completely to the invisible air that we sometimes forget it has weight. Perhaps a fish "forgets" about the weight of water in the same way. The reason we don't feel this weight crushing against our bodies is that the pressure inside our bodies equals that of the surrounding air. There is no net force for us to sense.

At sea level, 1 cubic meter of air at 20°C has a mass of about 1.2 kilograms. To estimate the mass of air in your room, estimate the number of cubic meters there, multiply by 1.2 kg/m^3, and you'll have the mass. Don't be surprised if it's heavier than your kid sister. If your kid sister doesn't believe air has weight, maybe it's because she's always surrounded by air. Hand her a plastic bag of water and she'll tell you it has weight. But hand her the same bag of water while she's submerged in a swimming pool, and she won't feel the weight. We don't notice that air has weight because we're submerged in air.

Unlike the constant density of water in a lake, the density of air in the atmosphere decreases with altitude. At 10 kilometers, 1 cubic meter of air has a mass of about 0.4 kilogram. To compensate for this, airplanes are pressurized; the additional air needed to fully pressurize a 747 jumbo jet, for example, is more than 1000 kilograms. Air is heavy, if you have enough of it.

Consider the mass of air in an upright 30-kilometer-tall bamboo pole that has an inside cross-sectional area of 1 square centimeter. If the density of air inside the pole matches the density of air outside, the enclosed mass of air would be about one kilogram. The weight of this much air is about 10 newtons. So air pressure at the bottom of the bamboo pole would be about 10 newtons per square centimeter (10 N/cm^2). Of course, the same is true without the bamboo pole. There are 10,000 square centimeters in 1 square meter, so a column of air 1-square meter in cross section that extends up through the atmosphere has a mass of about 10,000 kilograms. The weight of this air is about 100,000 newtons (10^5 N). This weight produces a pressure of 100,000 newtons per square meter—or equivalently, 100,000 pascals, or 100 kilopascals. To be more precise, the average atmospheric pressure at sea level is 101.3 kilopascals (101.3 kPa).*

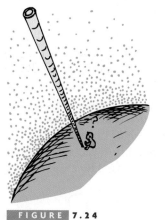

FIGURE 7.24

The mass of air that would
occupy a bamboo pole that
extends to the "top" of the
atmosphere is about 1 kg.
This air has a weight of
about 10 N.

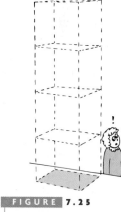

FIGURE 7.25

The weight of air that presses
down on a one-square-meter
surface at sea level is about
100,000 newtons. So atmospheric pressure is about
10^5 N/m^2, or about 100 kPa.

* As mentioned earlier, the pascal, named after Blaise Pascal, is the SI unit of measurement. The average pressure at sea level (101.3 kPa) is often called 1 atmosphere. In British units, the average atmospheric pressure at sea level is 14.7 lb/in^2 (psi).

The pressure of the atmosphere is not uniform. Besides altitude variations, there are variations in atmospheric pressure at any one locality due to moving fronts and storms. Measurement of changing air pressure is important to meteorologists in predicting weather.

STOP AND CHECK YOURSELF

1. Estimate the mass of air in kilograms in a classroom that has a 200-m² floor area and a 4-m-high ceiling. (Assume a chilly 10°C temperature.)

2. Why doesn't the pressure of the atmosphere break windows?

CHECK YOUR ANSWERS

1. The mass of air is 1000 kg. The volume of air is 200 m² × 4 m = 800 m³; each cubic meter of air has a mass of about 1.25 kg, so 800 m³ × 1.25 kg/m³ = 1000 kg.

2. Atmospheric pressure is exerted on both sides of a window, so no net force is exerted on the window. If for some reason the pressure is reduced or increased on one side only, as in a strong wind, then watch out!

Barometers

An instrument used for measuring the pressure of the atmosphere is called a **barometer**. A simple mercury barometer is illustrated in Figure 7.26. A glass tube, longer than 76 centimeters and closed at one end, is filled with mercury and tipped upside down in a dish of mercury. The mercury in the tube flows out of the submerged open bottom until the difference in the mercury levels in the tube and the dish is 76 centimeters. The empty space trapped above, except for some mercury vapor, is a pure vacuum.

The explanation for the operation of such a barometer is similar to that of children balancing on a seesaw. The barometer "balances" when the weight of liquid in the tube exerts the same pressure as the atmosphere outside. Whatever the width of the tube, a 76-centimeter column of mercury weighs the same as the air that would fill a vertical 30-kilometer tube of the same width. If the atmospheric pressure increases, then the atmosphere pushes down harder on the mercury in the dish and pushes mercury higher in the tube. Then the increased height of the mercury column exerts an equal balancing pressure.

Water could instead be used to make a barometer, but the glass tube would have to be much longer—13.6 times as long, to be exact. The density of mercury is 13.6 the density of water. That's why a tube of water 13.6 times longer than one of mercury (of the same cross section) is needed to provide the same weight as mercury in the tube. A water barometer would have to be 13.6 × 0.76 meter, or 10.3 meters high—too tall to be practical.

What happens in a barometer is similar to what happens when you are drinking through a straw. By sucking, you reduce the air pressure in the straw when it is placed in a drink. Atmospheric pressure on the drink then pushes the liquid up into the reduced-pressure region. Strictly speaking, the liquid is not sucked up; it is pushed up by the pressure of the atmosphere. If the atmosphere is prevented from pushing on the surface of the drink, as in the party-trick bottle with the straw through an airtight cork stopper, one can suck and suck and get no drink.

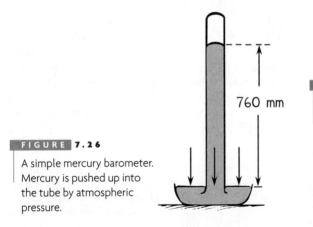

FIGURE 7.26

A simple mercury barometer. Mercury is pushed up into the tube by atmospheric pressure.

760 mm

FIGURE 7.27

Strictly speaking, they do not suck the soda up the straws. They instead reduce pressure in the straws, which allows the weight of the atmosphere to press the liquid up into the straws. Could they drink a soda this way on the Moon?

FIGURE 7.28

The atmosphere pushes water from below up into a pipe that is evacuated of air by the pumping action.

If you understand these ideas, you can understand why there is a 10.3-meter limit on the height water can be lifted with vacuum pumps. The old fashioned farm-type pump, shown in Figure 7.28, operates by producing a partial vacuum in a pipe that extends down into the water below. Atmospheric pressure on the surface of the water simply pushes the water up into the region of reduced pressure inside the pipe. Can you see that, even with a perfect vacuum, the maximum height to which water can be lifted in this way is 10.3 meters?

A small portable instrument that measures atmospheric pressure is the *aneroid barometer* (Figure 7.29). A metal box partially exhausted of air with a slightly flexible lid bends in or out with changes in atmospheric pressure. Motion of the lid is indicated on a scale by a mechanical spring-and-lever system. Atmospheric pressure decreases with increasing altitude, so a barometer can be used to determine elevation. An aneroid barometer

FIGURE 7.29

The aneroid barometer.

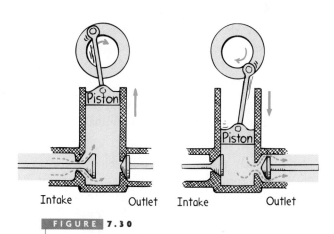

Intake   Outlet   Intake   Outlet

FIGURE 7.30

A mechanical vacuum pump. When the piston is lifted, the intake valve opens and air moves in to fill the empty space. When the piston is moved downward, the outlet valve opens and the air is pushed out. What changes would you make to convert this pump into an air compressor?

calibrated for altitude is called an *altimeter* (altitude meter). Some of these instruments are sensitive enough to indicate a change in elevation as you walk up a flight of stairs.*

Reduced air pressures are produced by pumps, which work by virtue of a gas tending to fill its container. If a space with less pressure is provided, gas will flow from the region of higher pressure to the one of lower pressure. A vacuum pump simply provides a region of lower pressure into which the normally fast-moving gas molecules randomly move. The air pressure is repeatedly lowered by piston and valve action (Figure 7.30).

> When the pump handle is raised, air in the pipe is "thinned" as it expands to fill a larger volume. Atmospheric pressure on the well surface pushes water up into the pipe, causing water to overflow at the spout.

7.8 Pascal's Principle

One of the most important facts about fluid pressure is that a change in pressure at one part of the fluid will be transmitted undiminished to other parts. For example, if the pressure of city water is increased at the pumping station by 10 units of

* Evidence of a noticeable pressure difference over a 1-m or less difference in elevation is any small helium-filled balloon that rises in air. The atmosphere really does push with more force against the lower bottom than against the higher top!

pressure, the pressure everywhere in the pipes of the connected system will be increased by 10 units of pressure (providing the water is at rest). This rule is called **Pascal's principle:**

A change in pressure at any point in an enclosed fluid at rest is transmitted undiminished to all points in the fluid.

Pascal's principle was discovered in the seventeenth century by theologian and scientist Blaise Pascal (who was an invalid at the age of 18 and remained so until his death at the age of 39). Recall that the SI unit of pressure, the pascal ($1\,\text{Pa} = 1\,\text{N/m}^2$), is named after him.

Fill a U-tube with water and place pistons at each end, as shown in Figure 7.31. Pressure exerted against the left piston will be transmitted throughout the liquid and against the bottom of the right piston. (The pistons are simply "plugs" that can slide freely but snugly inside the tube.) The pressure that the left piston exerts against the water will be exactly equal to the pressure the water exerts against the right piston. This is nothing to write home about. But suppose you make the tube on the right side wider and use a piston of larger area; then the result is impressive. In Figure 7.32 the piston on the right has 50 times the area of the piston on the left (say, the left has 100 square centimeters and the right 5000 square centimeters). Suppose a 10-kg load is placed on the left piston. Then an additional pressure (nearly $1\,\text{N/cm}^2$) due to the weight of the load is transmitted throughout the liquid and up against the larger piston. Here is where the difference between force and pressure comes in. The additional pressure is exerted against every square centimeter of the larger piston. Since there is 50 times the area, 50 times as much force is exerted on the larger piston. Thus, the larger piston will support a 500-kg load—50 times the load on the smaller piston!

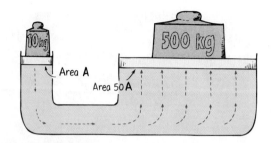

FIGURE 7.32

A 10-kg load on the left piston will support 500 kg on the right piston.

This *is* something to write home about, for we can multiply forces using such a device. One newton input produces 50 newtons output. By further increasing the area of the larger piston (or reducing the area of the smaller piston), we can multiply force, in principle, by any amount. Pascal's principle underlies the operation of the hydraulic press.

The hydraulic press does not violate energy conservation, because a decrease in the distance moved compensates for the increase in force. When the small piston in Figure 7.32 is moved downward 10 centimeters, the large piston will be raised only one-fiftieth of this, or 0.2 centimeter. The input force multiplied by the distance moved by the smaller piston is equal to the output force multiplied by the distance moved by the larger piston; this is one more example of a simple machine operating on the same principle as a mechanical lever.

Pascal's principle applies to all fluids, whether gases or liquids. A typical application of Pascal's principle for gases and liquids is the automobile lift seen in many service stations (Figure 7.33). Increased air pressure produced by an air compressor is transmitted through the air to the surface of oil in an underground reservoir. The oil in turn transmits the pressure to a piston, which lifts the automobile. The relatively low pressure that exerts the lifting force against the piston is about the same as the air pressure in automobile tires.

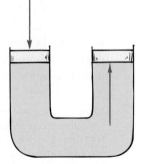

FIGURE 7.31

The force exerted on the left piston increases the pressure in the liquid and is transmitted to the right piston.

Pascal is remembered scientifically for hydraulics, which changed the technological landscape more than he imagined. He is remembered theologically for his many assertions, one of which relates to centuries of human landscape: "Men never do evil so cheerfully and completely as when they do so from religious conviction."

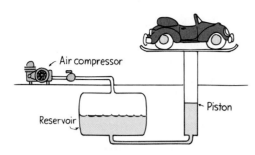

FIGURE **7.33**

Pascal's principle in a service station.

Hydraulics is employed by modern devices ranging from very small to enormous. Note the hydraulic pistons in almost all construction machines where heavy loads are involved (Figure 7.34).

FIGURE **7.34**

Pascal's principle at work in the hydraulic devices on this common but incredible machine. We can only wonder whether Pascal envisioned the extent to which his principle would allow huge loads to be so easily lifted.

STOP AND CHECK YOURSELF

1. As the automobile in Figure 7.33 is being lifted, how does the change in oil level in the reservoir compare to the distance the automobile moves?

2. If a friend commented that a hydraulic device is a common way of multiplying energy, what would you say?

CHECK YOUR ANSWERS

1. The car moves up a greater distance than the oil level drops, since the area of the piston is smaller than the surface area of the oil in the reservoir.

2. No, no, no! Although a hydraulic device, like a mechanical lever, can multiply *force*, it always does so at the expense of distance. Energy is the product of force and distance. Increase one, decrease the other. *No device has ever been found that can multiply energy!*

7.9 Buoyancy in a Gas

A crab lives at the bottom of its ocean floor and looks upward at jellyfish and other lighter-than-water marine life drifting above it. Similarly, we live at the bottom of our ocean of air and look upward at balloons and other lighter-than-air objects drifting above us. A balloon is suspended in air and a jellyfish is suspended in water for the same reason: each is buoyed upward by a displaced weight of fluid equal to its own weight. We've learned that objects in water are buoyed upward because the pressure acting up against the bottom of the object exceeds the pressure acting down against the top. Likewise, air pressure acting up against an object immersed in air is greater than the pressure above pushing down. The buoyancy in both cases is numerically equal to the weight of fluid displaced. Archimedes' principle applies to air just as it does for water:

The Physics Place
Buoyancy of Air

An object surrounded by air is buoyed up by a force equal to the weight of the air displaced.

We know that a cubic meter of air at ordinary atmospheric pressure and room temperature has a mass of about 1.2 kilograms, so its weight is about 12 newtons. Therefore, any 1-cubic-meter object in air is buoyed up with a force of 12 newtons. If the mass of the 1-cubic-meter object is greater than 1.2 kilograms (so that its weight is greater than 12 newtons), it falls to the ground when released. If an object of this size has a mass of less than 1.2 kilograms, buoyant force is greater than weight and it rises in the air. Any object that has a mass that is less than the mass of an equal volume of air rises in the air. Stated another way, any object less dense than air will rise in air. Gas-filled balloons that rise in air are less dense than air.

No gas at all in a balloon would mean no weight (except for the weight of the balloon's material), but such a balloon would collapse. The gas used in balloons prevents the atmosphere from collapsing them. Hydrogen is the lightest gas, but it is seldom used because it is highly flammable. In sport balloons, the gas is simply heated air. In balloons intended to reach very high altitudes or to remain aloft for a long time, helium is usually used. Its density is small enough that the combined weight of the helium, the balloon, and the cargo is less than the weight of air they displace. Low-density gas is used in a balloon for the same reason that cork is used in life preservers. The cork possesses no strange tendency to be drawn toward the water's surface, and the gas possesses no strange tendency to rise. Cork and gases are buoyed upward like anything else. They are simply light enough for the buoyancy to be significant.

Unlike water, there is no sharp surface at the "top" of the atmosphere. Furthermore, unlike water, the atmosphere becomes less dense with altitude. Whereas cork will float to the surface of the water, a released helium-filled balloon does not rise to any atmospheric surface. Will a lighter-than-air balloon rise indefinitely? How high will a balloon rise? We can state the answer in several ways. A gas-filled balloon will rise only so long as it displaces a weight of air greater than its own weight. Because air becomes less dense with altitude, a lesser weight of air is displaced per given volume as the balloon rises. When the weight of displaced air equals the total weight of the balloon, upward motion of the balloon will cease. We can also say that, when the buoyant force on the balloon equals its weight, the balloon will cease rising. Equivalently, when the density of the balloon (including its load) equals the density of the surrounding air, the balloon will cease rising. Helium-filled toy balloons usually break some time after being released into the air when the expansion of the helium they contain stretches the rubber until it ruptures.

Large dirigible airships are designed so that, when they are loaded, they will slowly rise in air; that is, their total weight is a little less than the weight of air displaced. When in motion, the ship may be raised or lowered by means of horizontal "elevators."

Thus far we have treated pressure only as it applies to stationary fluids. Motion produces an additional influence.

7.10 Bernoulli's Principle

Consider a continuous flow of liquid or gas through a pipe: the volume of fluid that flows past any cross section of the pipe in a given time is the same as that flowing past any other section of the pipe—even if the pipe widens or narrows. For continuous flow, a fluid speeds up when it goes from a wide to a narrow part of the pipe. This is evident for a broad, slow-moving river that flows more swiftly as it enters a narrow gorge. It is also evident when water flowing from a garden hose speeds up when you squeeze the end of the hose to make the stream narrower.

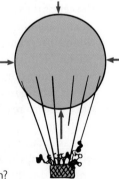

FIGURE 7.35

All bodies are buoyed up by a force equal to the weight of air they displace. Why, then, don't all objects float like this balloon?

FIGURE 7.36

Because the flow is continuous, water speeds up when it flows through the narrow and/or shallow part of the brook.

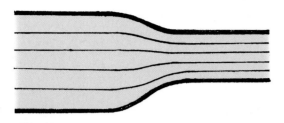

FIGURE 7.37

Water speeds up when it flows into the narrower pipe. The close-together streamlines indicate increased speed and decreased internal pressure.

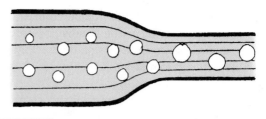

FIGURE 7.38

Internal pressure is greater in slower moving water in the wide part of the pipe, as evidenced by the more-squeezed air bubbles. The bubbles are bigger in the narrow part because internal pressure there is less.

The motion of a fluid in steady flow follows imaginary *streamlines*, represented by thin lines in Figure 7.36 and in other figures that follow. Streamlines are the smooth paths of bits of fluid. The lines are closer together in narrower regions, where the flow speed is greater. (Streamlines are visible when smoke or other visible fluids are passed through evenly spaced openings, as in a wind tunnel.)

> **fyi**
>
> Because the volume of water flowing through a pipe of different cross-sectional areas A remains constant, speed of flow v is high where the area is small, and the speed is low where the area is large. This is stated in the equation of continuity:
>
> $A_1v_1 = A_2v_2$
>
> The product A_1v_1 at point 1 equals the product A_2v_2 at point 2.
>
>

Daniel Bernoulli, an eighteenth-century Swiss scientist, studied fluid flow in pipes. His discovery, now called **Bernoulli's principle**, can be stated as follows:

Where the speed of a fluid increases, internal pressure in the fluid decreases.

Where streamlines of a fluid are closer together, flow speed is greater and pressure within the fluid is less. Changes in internal pressure are evident for water containing air bubbles. The volume of an air bubble depends on the surrounding water pressure. Where water gains speed, pressure is lowered and bubbles become bigger. In water that slows, pressure is greater and bubbles are squeezed to a smaller size.

Bernoulli's principle is a consequence of the conservation of energy, although, surprisingly, he developed it long before the concept of energy was formalized.* The full energy picture for a fluid in motion is quite complicated. Simply stated, more speed and kinetic energy mean less pressure, and more pressure means less speed and kinetic energy.

Bernoulli's principle applies to a smooth, steady flow (called *laminar* flow) of constant-density fluid.

> **fyi**
>
> The friction of both liquids and gases sliding over one another is called *viscosity* and is a property of all fluids.

At speeds above some critical point, however, the flow may become chaotic (called *turbulent* flow) and follow changing, curling paths called *eddies*. This exerts friction on the fluid and dissipates some of its energy. Then Bernoulli's equation doesn't apply well.

The decrease of fluid pressure with increasing speed may at first seem surprising, particularly if you fail to distinguish between the pressure *within* the fluid, internal pressure, and the pressure *by* the fluid on something that interferes with its flow. Internal pressure within flowing water and the external pressure it can exert on whatever it encounters are two different pressures. When the momentum of moving water or anything else is suddenly reduced, the impulse it exerts is relatively huge. A dramatic example is the use of high-speed

* In mathematical form: $1/2\ mv^2 + mgy + pV =$ constant (along a streamline); where m is the mass of some small volume V, v its speed, g the acceleration due to gravity, y its elevation, and p its internal pressure. If mass m is expressed in terms of density ρ, where $r = m/V$, and each term is divided by V, Bernoulli's equation reads $1/2\ \rho v^2 + \rho gy + p =$ constant. Then all three terms have units of pressure. If y does not change, an increase in v means a decrease in p, and vice versa. Note when v is zero Bernoulli's equation reduces to $\Delta p = -\rho g\Delta y$ (weight density × depth).

jets of water to cut steel in modern machine shops. The water has very little internal pressure, but the pressure the stream exerts on the steel interrupting its flow is enormous.

> Recall from Chapter 5 that a large change in momentum is associated with a large impulse. So when water from a firefighter's hose hits you, the impulse can knock you off your feet. Interestingly, the pressure *within* that water is relatively small!

Applications of Bernoulli's Principle

Hold a sheet of paper in front of your mouth, as shown in Figure 7.39. When you blow across the top surface, the paper rises. That's because the internal pressure of moving air against the top of the paper is less than the atmospheric pressure beneath it.

Anyone who has ridden in a convertible car with the canvas top up has noticed that the roof puffs upward as the car moves. This is Bernoulli's principle again. Pressure outside against the top of the fabric, where air is moving, is less than the static atmospheric pressure inside the car. The result is an upward net force on the fabric.

Consider wind blowing across a peaked roof. The wind gains speed as it flows over the roof, as the crowding of streamlines in the sketch indicates. Pressure along the streamlines is reduced where they are closer together. The greater pressure inside the roof can lift it off the house. During a severe storm, the difference in outside and inside pressure doesn't need to be very much. A small pressure difference over a large area produces a force that can be formidable.

If we think of the blown-off roof as an airplane wing, we can better understand the lifting force that supports a heavy aircraft. In both cases, a greater pressure below pushes the roof or the wing into a region of lesser pressure above. Wings come in a variety of designs. What they all have in common is that air is made to flow faster over the wing's top surface than

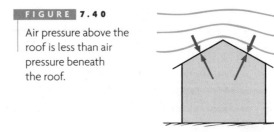

FIGURE 7.40

Air pressure above the roof is less than air pressure beneath the roof.

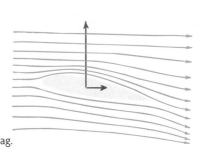

FIGURE 7.41

The vertical vector represents the net upward force (lift) that results from more air pressure below the wing than above the wing. The horizontal vector represents air drag.

under its lower surface. This is mainly accomplished by a tilt in the wing, called its *angle of attack*. Then air flows faster over the top surface for much the same reason that air flows faster in a narrowed pipe or in any other constricted region. Most often, but not always, different speeds of airflow over and beneath a wing are enhanced by a difference in the curvature (*camber*) of the upper and lower surfaces of the wing. The result is more crowded streamlines along the top wing surface than along the bottom. When the average pressure difference over the wing is multiplied by the surface area of the wing, we have a net upward force—lift. Lift is greater when there is a large wing area and when the plane is traveling fast. A glider has a very large wing area relative to its weight, so it does not have to be going very fast for sufficient lift. At the other extreme, a fighter plane designed for high-speed flight has a small wing area relative to its weight. Consequently, it must take off and land at high speeds.

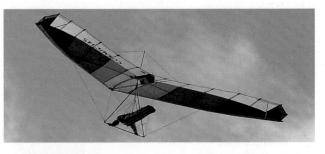

FIGURE 7.42

Where is air pressure greater—on the top or bottom surface of the hang glider?

FIGURE 7.39

The paper rises when Tim blows air across its top surface.

FIGURE 7.43

FIGURE 7.43

(a) The streamlines are the same on either side of a nonspinning baseball. (b) A spinning ball produces a crowding of streamlines. The resulting "lift" (red arrow) causes the ball to curve as shown by the blue arrow.

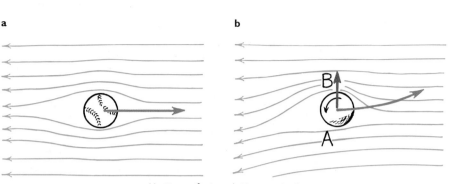

a   b

Motion of air relative to ball

We all know that a baseball pitcher can impart a spin on a ball to make it curve off to one side as it approaches home plate. Similarly, a tennis player can hit a ball so it curves. A thin layer of air is dragged around the spinning ball by friction, which is enhanced by the baseball's threads or the tennis ball's fuzz. The moving layer of air produces a crowding of streamlines on one side. Note, in Figure 7.43b, that the streamlines are more crowded at B than at A for the direction of spin shown. Air pressure is greater at A, and the ball curves as shown.

Recent findings show that many insects increase lift by employing motions similar to those of a curving baseball. Interestingly, most insects do not flap their wings up and down. They flap them forward and backward, with a tilt that provides an angle of attack. Between flaps, their wings make semicircular motions to create lift.

A familiar sprayer, such as a perfume atomizer, utilizes Bernoulli's principle. When you squeeze the bulb, air rushes across the open end of a tube inserted into the perfume. This reduces the pressure in the tube, whereupon atmospheric pressure on the liquid below pushes it up into the tube, where it is carried away by the stream of air.

Bernoulli's principle explains why trucks passing closely on the highway are drawn to each other, and why passing ships run the risk of a sideways collision. Water flowing between the ships travels faster than water flowing past the outer sides. Streamlines are closer together between the ships than outside, so water pressure acting against the hulls is reduced

between the ships. Unless the ships are steered to compensate for this, the greater pressure against the outer sides of the ships forces them together. Figure 7.45 shows how to demonstrate this in your kitchen sink or bathtub.

Bernoulli's principle plays a small role when your bathroom shower curtain swings toward you in the shower when the water is on full blast. The pressure in the shower stall is reduced with fluid in motion, and the relatively greater pressure outside the curtain pushes it inward. Like so much in the complex real world, this is but one physics principle that applies. More important is the convection of air in the shower. In any case, the next time you're taking a shower and the curtain swings in against your legs, think of Daniel Bernoulli.

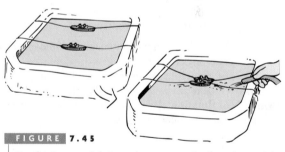

FIGURE 7.45

Try this in your sink. Loosely moor a pair of toy boats side by side. Then direct a stream of water between them. The boats will draw together and collide. Why?

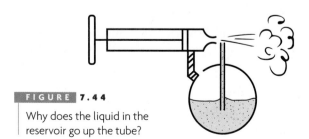

FIGURE 7.44

Why does the liquid in the reservoir go up the tube?

FIGURE 7.46

Why can't you blow the bent filing card off the table when you blow through the arch?

Rats to you too, Daniel Bernoulli!

FIGURE 7.47
The curved shape of an umbrella can be disadvantageous on a windy day.

CHECK YOURSELF

Blimps, airplanes, and rockets operate under three very different principles. Which operates by way of buoyancy? Bernoulli's principle? Newton's third law?

CHECK YOUR ANSWERS

Blimps function by way of buoyancy, airplanes by the Bernoulli effect, and rockets by way of Newton's third law. Interesting, Newton's third law also plays a significant role in airplane flight—wing pushes air downward; air pushes wing upward.

SUMMARY OF TERMS

Density The amount of matter per unit volume.

$$\text{Density} = \frac{\text{mass}}{\text{volume}}$$

Weight density is expressed as weight per unit volume.

Pressure The ratio of force to the area over which that force is distributed:

$$\text{Pressure} = \frac{\text{force}}{\text{area}}$$

$$\text{Liquid pressure} = \text{weight density} \times \text{depth}$$

Buoyant force The net upward force that a fluid exerts on an immersed object.

Archimedes' principle An immersed body is buoyed up by a force equal to the weight of the fluid it displaces.

Principle of flotation A floating object displaces a weight of fluid equal to its own weight.

Boyle's law The product of pressure and volume is a constant for a given mass of confined gas regardless of changes in either pressure or volume individually, so long as temperature remains unchanged. Pressure times volume in region 1 will equal pressure times volume in region 2:

$$P_1 V_1 = P_2 V_2$$

Atmospheric pressure The pressure exerted against bodies immersed in the atmosphere resulting from the weight of air pressing down from above. At sea level, atmospheric pressure is about 101 kPa.

Barometer Any device that measures atmospheric pressure.

Pascal's principle A change in pressure at any point in an enclosed fluid at rest is transmitted undiminished to all points in the fluid.

Bernoulli's principle The pressure in a fluid moving steadily without friction or outside energy input decreases when the fluid velocity increases.

REVIEW QUESTIONS

1. Give two examples of a fluid.

7.1 Density

2. What happens to the volume of a loaf of bread that is squeezed? What happens to the mass? What happens to the density?

3. Distinguish between mass density and weight density. What are the mass density and the weight density of water?

7.2 Pressure

4. Distinguish between force and pressure.

7.3 Pressure in a Liquid

5. How does the pressure exerted by a liquid change with depth in the liquid? How does the pressure exerted by a liquid change as the density of the liquid changes?
6. Discounting the pressure of the atmosphere, if you swim twice as deep in water, how much more water pressure is exerted on your ears? If you swim in saltwater, will the pressure be greater than in freshwater at the same depth? Why or why not?
7. How does water pressure one meter below the surface of a small pond compare to water pressure one meter below the surface of a huge lake?
8. If you punch a hole in the side of a container filled with water, in what direction does the water initially flow outward from the container?

7.4 Buoyancy in a Liquid

9. Why does buoyant force act upward on an object submerged in water?
10. How does the volume of a completely submerged object compare with the volume of water displaced?

7.5 Archimedes' Principle

11. Cite Archimedes' principle.
12. What is the difference between being immersed and being submerged?
13. How does the buoyant force on a fully submerged object compare with the weight of water displaced?
14. What is the mass in kilograms of 1 L of water? What is its weight in newtons?
15. If a 1-L container is immersed halfway in water, what is the volume of water displaced? What is the buoyant force on the container?
16. Does the buoyant force on a fully submerged object depend on the weight of the object or on the weight of the fluid displaced by the object? Does the force depend on the weight of the object or on its volume? Defend your answers.
17. There is a condition in which the buoyant force on an object does equal the weight of the object. What is this condition?
18. Does the buoyant force on a submerged object depend on the volume of the object?
19. Does the buoyant force on a floating object depend on the weight of the object or on the weight of the fluid displaced by the object? Or are these two weights the same for the special case of floating? Defend your answer.
20. What weight of water is displaced by a 100-ton ship? What is the buoyant force that acts on this ship?

7.6 Pressure in a Gas

21. Describe the primary differences between liquids and gases.

22. By how much does the density of air increase when it is compressed to half its volume?
23. What happens to the air pressure inside a balloon when the balloon is squeezed to half its volume at constant temperature?
24. Define Boyle's law, and give an application.

7.7 Atmospheric Pressure

25. What is the mass in kilograms of a cubic meter of air at room temperature (20°C)?
26. What is the approximate mass in kilograms of a column of air that has a cross-sectional area of 1 cm^2 and extends from sea level to the upper atmosphere? What is the weight in newtons of this amount of air?
27. How does the downward pressure of the 76-cm column of mercury in a barometer compare with the air pressure at the bottom of the atmosphere?
28. How does the weight of mercury in a barometer tube compare with the weight of an equal cross section of air from sea level to the top of the atmosphere?
29. Why would a water barometer have to be 13.6 times taller than a mercury barometer?
30. When you drink liquid through a straw, is it more accurate to say that the liquid is pushed up the straw rather than sucked up? What exactly does the pushing? Defend your answer.

7.8 Pascal's Principle

31. What happens to the pressure in all parts of a confined fluid when the pressure in one part is increased?

7.9 Buoyancy in a Gas

32. A balloon that weighs 1 N is suspended in air, drifting neither up nor down. How much buoyant force acts upon it? What happens if the buoyant force decreases? Increases?

7.10 Bernoulli's Principle

33. Cite Bernoulli's principle.
34. What are streamlines? Is pressure greater or less in regions of crowded streamlines?
35. Does Bernoulli's principle refer to internal pressure changes in a fluid or to pressures that a fluid can exert on objects it encounters?
36. What do peaked roofs, convertible tops, and airplane wings have in common when air moves faster across their top surfaces?

ACTIVE EXPLORATIONS

1. Try to float an egg in water. Then dissolve salt in the water until the egg floats. How does the density of an egg compare to that of tap water? To saltwater?

2. Punch a couple of holes in the bottom of a water-filled container, and water will spurt out because of water pressure. Now drop the container, and, as it freely falls, note that the water no longer spurts out. If your friends don't understand this, could you figure it out and then explain it to them?

3. Place a wet Ping-Pong ball in a can of water held high above your head. Then drop the can on a rigid floor. Because of surface tension, the ball will be pulled beneath the surface as the can falls. What happens when the can comes to an abrupt stop is worth watching!

4. Try this when you're washing dishes: lower a drinking glass, mouth downward, over a small floating object. What do you observe? How deep will the glass have to be pushed in order to compress the enclosed air to half its volume? (You won't be able to do this in your sink unless it's 10.3 m deep!)

5. You can find the pressure exerted by the tires of your car on the road and compare it with the air pressure in the tires. For this project, you need to get the weight of your car from the manual or a dealer, and then divide by four to get the approximate weight held up by one tire. You can closely approximate the area of contact of a tire with the road by tracing the edges of tire contact on a sheet of paper marked with 1-inch^2 squares beneath the tire. Compare the pressure of the tire on the road with the air pressure in the tire. Are they nearly equal, or is one greater? Defend your answer.

6. You ordinarily pour water from a full glass into an empty glass simply by placing the full glass above the empty glass and tipping. Have you ever poured air from one glass to another? The procedure is similar. Lower two glasses in water, mouths downward. Let one fill with water by tilting its mouth upward. Then hold the water-filled glass mouth downward above the air-filled glass. Slowly tilt the lower glass and let the air escape, filling the upper glass. You will be pouring air from one glass into another!

7. Raise a filled glass of water above the waterline, but with its mouth beneath the surface. Why does the water not flow out? How tall would a glass have to be before water began to flow out? (You won't be able to do this indoors unless you have a ceiling that is at least 10.3 m higher than the waterline.)

8. Place a card over the open top of a glass filled to the brim with water, and then invert it. Why does the card stay in place? Try it sideways and explain what happens.

9. Invert a water-filled pop bottle or small-necked jar. Notice that the water doesn't simply fall out, but it gurgles out of the container instead. Air pressure won't allow the water out until some air has pushed its way up inside the bottle to occupy the space above the liquid. How would an inverted, water-filled bottle empty if you tried this on the Moon?

10. Heat a small amount of water to boiling in an aluminum soda-pop can, and invert it quickly into a dish of cold water. What happens is surprisingly dramatic!

11. Make a small hole near the bottom of an open tin can. Fill the can with water, which then proceeds to spurt from the hole. If you cover the top of the can firmly with the palm of your hand, the flow stops. Explain.

12. Lower a narrow glass tube or drinking straw into water and place your finger over the top of the tube. Lift the tube from the water and then lift your finger from the top of the tube. What happens? (You'll do this often in chemistry experiments.)

13. Blow across the top of a sheet of paper as Tim does in Figure 7.39. Try this with those of your friends who are not enrolled in a physics course. Then explain it to them!

14. Push a pin through a small card and place it over the hole of a thread spool. Try to blow the card from the spool by blowing through the hole. Try it in all directions.

15. Hold a spoon in a stream of water as shown and feel the effect of the differences in pressure.

EXERCISES

1. Stand on a bathroom scale and read your weight. When you lift one foot up so you're standing on one foot, does the reading change? Does a scale read force or pressure?

2. The photo shows physics teacher Marshall Ellenstein walking barefoot on broken glass bottles in his class. What physics concept is Marshall demonstrating, and why is he careful that the broken pieces are small and numerous?

3. In a deep dive, a whale is appreciably compressed by the pressure of the surrounding water. What happens to the whale's density?

4. The density of a rock doesn't change when it is submerged in water. Does your density change when you are submerged in water? Defend your answer.

5. Why are persons who are confined to bed less likely to develop bedsores on their bodies if they lie on a waterbed rather than on an ordinary mattress?

6. If water faucets upstairs and downstairs are turned fully on, will more water per second flow out the downstairs faucet? Or will the flows from the faucets be the same?

7. Which do you suppose exerts more pressure on the ground—an elephant or a woman standing on spike heels? (Which will be more likely to make dents in a linoleum floor?) Can you approximate a rough calculation for each?

8. Suppose you wish to lay a level foundation for a home on hilly and bushy terrain. How can you use a garden hose filled with water to determine equal elevations for distant points?

9. When you are bathing on a stony beach, why do the stones hurt your feet less when you get in deep water?

10. If liquid pressure were the same at all depths, would there be a buoyant force on an object submerged in the liquid? Explain.

11. The Himalayas are slightly less dense than the mantle material upon which they "float." Do you suppose that, like floating icebergs, these mountains are deeper than they are high?

12. How much force is needed to push a nearly weightless but rigid 1-L carton beneath a surface of water?

13. Why is it inaccurate to say that heavy objects sink and that light objects float? Give exaggerated examples to support your answer.

14. Compared to an empty ship, would a ship loaded with a cargo of Styrofoam sink deeper into water or rise in water? Defend your answer.

15. A barge filled with scrap iron is in a canal lock. If the iron is thrown overboard, does the water level at the side of the lock rise, fall, or remain unchanged? Explain.

16. Would the water level in a canal lock go up or down if a battleship in the lock were to sink?

17. A balloon is weighted so that it is barely able to float in water. If it is pushed beneath the surface, will it return to the surface, stay at the depth to which it is pushed, or sink? Explain. (Hint: Does the balloon's density change?)

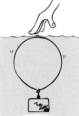

18. A ship sailing from the ocean into a freshwater harbor sinks slightly deeper into the water. Does the buoyant force on it change? If so, does it increase or decrease?

19. Suppose you are given the choice between two life preservers that are identical in size, the first a light one filled with Styrofoam and the second a very heavy one filled with lead pellets. If you submerge these life preservers in the water, upon which will the buoyant force be greater? Upon which will the buoyant force be ineffective? Why are your answers different?

20. The relative densities of water, ice, and alcohol are 1.0, 0.9, and 0.8, respectively. Do ice cubes float higher or lower in a mixed alcoholic drink? What can you say about a cocktail in which the ice cubes lie submerged at the bottom of the glass?

21. When an ice cube in a glass of water melts, does the water level in the glass rise, fall, or remain unchanged? Does your answer change if the ice cube contains many air bubbles? Does your answer change if the ice cube contains many grains of heavy sand?

22. A half-filled bucket of water is on a spring scale. Will the reading of the scale increase or remain the same if a fish is placed in the bucket? (Will your answer be different if the bucket is initially filled to the brim?)

23. We say that the shape of a liquid is that of its container. But with no container and no gravity, what is the natural shape of a blob of water? Why?

24. If you release a Ping-Pong ball beneath the surface of water, it will rise to the surface. Would it do the same if it were submerged in a big blob of water floating weightless in an orbiting spacecraft?

25. It is said that a gas fills all the space available to it. Why, then, doesn't the atmosphere go off into space?

26. Count the tires on a large tractor trailer that is unloading food at your local supermarket, and you may be surprised to count 18 tires. Why so many tires? (Hint: See Active Exploration 5.)

27. How does the density of air in a deep mine compare with the air density at Earth's surface?

28. Two teams of eight horses each were unable to pull the Magdeburg hemispheres apart (Figure 7.21). Why? Suppose two teams of nine horses each could pull them apart. Then would one team of nine horses succeed if the other team were replaced with a strong tree? Defend your answer.

29. Before boarding an airplane, you buy a bag of chips (or any item packaged in an airtight bag) and, while in flight, you notice that the bag is puffed up. Explain why this occurs.

30. Why do you suppose that airplane windows are smaller than bus windows?

31. A half cup or so of water is poured into a 5-L can, which is placed on a source of heat until most of the water has boiled away. Then the top of the can is screwed on tightly and the can is removed from the source of heat and allowed to cool. What happens to the can and why?

32. We can understand how pressure in water depends on depth by considering a stack of bricks. The pressure below the bottom brick is determined by the weight of the entire stack. Halfway up the stack, the pressure is half because the weight of the bricks above is half. To explain atmospheric pressure, we should consider compressible bricks, like foam rubber. Why is this so?

33. The "pump" in a vacuum cleaner is merely a high-speed fan. Would a vacuum cleaner pick up dust from a rug on the Moon? Explain.

34. If you could somehow replace the mercury in a mercury barometer with a denser liquid, would the height of the liquid column be greater or less than with mercury? Why?

35. Would it be slightly more difficult to draw soda through a straw at sea level or on top of a very high mountain? Explain.

36. Your friend says that the buoyant force of the atmosphere on an elephant is significantly greater than the buoyant force of the atmosphere on a small helium-filled balloon. What do you say?

37. Why is it so difficult to breathe when snorkeling at a depth of 1 m, and practically impossible at a 2-m depth? Why can't a diver simply breathe through a hose that extends to the surface?

38. When you replace helium in a balloon with hydrogen, which is less dense, does the buoyant force on the balloon change if the balloon remains the same size? Explain.

39. A steel tank filled with helium gas doesn't rise in air, but a balloon containing the same helium easily does? Why?

40. Two identical balloons of the same volume are pumped up with air to more than atmospheric pressure and suspended on the ends of a stick that is horizontally balanced. One of the balloons is then punctured. Is there a change in the stick's balance? If so, which way does it tip?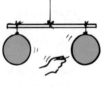

41. The force of the atmosphere at sea level against the outside of a 10-square-meter store window is about a million N. Why does this not shatter the window? Why might the window shatter in a strong wind blowing past the window?

42. In the hydraulic arrangement shown, the larger piston has an area that is 50 times that of the smaller piston. The strong man hopes to exert enough force on the large piston to raise the 10 kg that rest on the small piston. Do you think he will be successful? Defend your answer.

43. When a steadily flowing gas flows from a larger-diameter pipe to a smaller-diameter pipe, what happens to (a) its speed, (b) its pressure, and (c) the spacing between its streamlines?

44. How is an airplane able to fly upside down?

45. When a jet plane is cruising at a high altitude, the flight attendants have more of a "hill" to climb as they walk forward along the aisle than when the plane is cruising at a lower altitude. Why does the pilot have to fly with a greater "angle of attack" at a high altitude than at a lower one?

46. What physics principle underlies the following three observations? When passing an oncoming truck on the highway, your car tends to sway toward the truck. The canvas roof of a convertible automobile bulges upward when the car is traveling at high speeds. The windows of older passenger trains sometimes break when a high-speed train passes by on the next track.

47. On a windy day, waves in a lake or the ocean are higher than their average height. How does Bernoulli's principle contribute to the increased height?

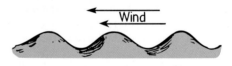

48. Wharves are made with pilings that permit the free passage of water. Why would a solid-walled wharf be disadvantageous to ships attempting to pull alongside?

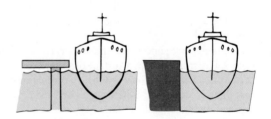

PROBLEMS

1. ● Suppose that you balance a 5-kg ball on the tip of your finger, which has an area of 1 cm^2? Show that the pressure on your finger is 49 N/cm^2, which is 490 kPa.

2. ● A 6-kg piece of metal displaces 1 liter of water when submerged. Show that its density is 6000 kg/m^3. How does this compare with the density of water?

3. ● In the deepest part of Lake Superior the water depth is 406 m. Show that the water pressure at this depth is 3978.8 kPa, and the total pressure there is 4080.1 kPa.

4. ● A rectangular barge, 5 m long and 2 m wide, floats in freshwater. Suppose that its load is a 400-kg granite block. Show that the barge floats 4 cm deeper.

5. ■ Suppose that the barge in the preceding problem can only be pushed 15 cm deeper into the water before the water overflows to sink it. Show that it could carry three, but not four, 400-kg blocks.

6. ● A merchant in Katmandu sells you a solid gold 1-kg statue for a very reasonable price. When you arrive home, you wonder whether or not you got a bargain, so you lower the statue into a container of water and measure the volume of displaced water. Show that, for pure gold, the volume of water displaced will be 51.8 cm^3.

7. ■ An ice cube measures 10 cm on a side, and it floats in water. One cm extends above water level. Show that if you shaved off the 1-cm part, 0.9 cm of the remaining ice would extend above water level.

8. ■ A vacationer floats lazily in the ocean with 90 percent of his body below the surface. The density of the ocean water is 1025 kg/m^3. Show that the vacationer's average density is 923 kg/m^3.

9. ● Air in a cylinder is compressed to one-tenth its original volume with no change in temperature. What is the change in its pressure?

10. ● On a perfect fall day, you are hovering at low altitude in a hot-air balloon, accelerated neither upward nor downward. The total weight of the balloon, including its load and the hot air in it, is 20,000 N. What is the weight of the displaced air?

11. ■ For the previous problem, show that the volume of the displaced air is 1700 m^3.

12. ■ In the hydraulic pistons shown in the sketch, the small piston has a diameter of 2 cm and the large piston has a diameter of 6 cm. How much more force can the larger piston exert compared with the force applied to the smaller piston?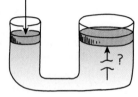

13. ■ The wings of a certain airplane have a total surface area of 100 m^2. At a particular speed, the difference in air pressure below and above the wings is 4% of atmospheric pressure. Show that the lift on the airplane is 4 × 10^5 N.

CHAPTER 7 ONLINE RESOURCES

The Physics Place

Videos
Dam Keeps Water in Place
Water Keeps Dam in Place
Buoyancy
Archimedes' Principle
Flotation
Air Has Pressure

Air Has Weight
Air Is Matter
Buoyancy of Air

Quiz

Flashcards

Links

▪ WATER DOWSING

The practice of water dowsing, which goes back to ancient times in Europe and Africa, was carried across the Atlantic to America by some of the earliest settlers. Dowsing refers to the practice of using a forked stick, or a rod, or some similar device to locate underground water, minerals, or hidden treasure. In the classic method of dowsing, each hand grasps a fork, with palms upward. The pointed end of the stick points skyward at an angle of about 45°. The dowser walks back and forth over the area to be tested, and, when passing over a source of water (or whatever else is being sought), the stick is supposed to rotate downward. Some dowsers have reported an attraction so great that blisters formed on their hands. Some claim special powers that enable them to "see" through soil and rock, and some are mediums who go into trances when conditions are especially favorable. Although most dowsing is done at the actual site, some dowsers claim to be able to locate water simply by passing the stick over a map.

Since drilling a well is an expensive process, the dowser's fee is usually seen as reasonable.

The practice is widespread, with thousands of dowsers active in the United States. This is because dowsing works. The dowser can hardly miss—not because of special powers, but because groundwater is within 100 meters of the surface at almost every spot on Earth.

If you drill a hole into the ground, you'll find that the wetness of the soil varies with depth. Near the surface, pores and open spaces are filled mostly with air. Deeper, the pores are saturated with water. The upper boundary of this water-saturated zone is called the *water table*. It usually rises and falls with the contour of the surface topography. Wherever you see a natural lake or pond, you're seeing a place where the water table extends above the surface of the land.

The depth, quantity, and quality of water below the water table are studied by hydrologists, who rely on a variety of technological techniques—water dowsing *not* being one of them. Findings of the U.S. Geological Survey conclude that water dowsing falls into the category of pseudoscience. As mentioned in Chapter 1, the real test of a water dowser would be finding a location in which water can't be found.

PART TWO

Heat

Although the temperature of these sparks exceeds 2000°C, the heat they impart when striking my skin is very small—which illustrates that *temperature* and *heat* are different concepts. Learning to distinguish between closely related concepts is the challenge and essence of *Conceptual Physics Fundamentals*.

CHAPTER 8

Temperature, Heat, and Thermodynamics

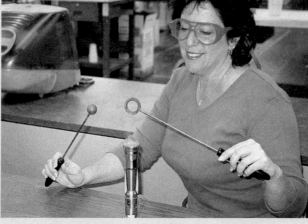

Ellyn Daugherty asks her class to predict whether the hole in the ring expands or contracts when heated.

Matter in all forms is made up of constantly jiggling atoms or molecules. When jiggling slowly, the particles form solids. When jiggling faster, they slide over one another and we have a liquid. When the same particles move so fast that they disconnect and fly loose, we have a gas. When moved still faster, atoms dissociate and we have a plasma. Although we are more familiar with solids, liquids, and gases, plasma is the dominant phase of matter in the universe. The phase of a substance, whether a solid, liquid, gas, or a plasma, depends on the motion of its particles.

8.1 Temperature

The quantity that indicates how warm or cold an object is relative to some standard is called **temperature.** We express the temperature of matter by a number that corresponds to the degree of hotness on some chosen scale. A common thermometer measures temperature by means of the expansion and contraction of a liquid, usually mercury or colored alcohol.

The Physics Place
Low Temperature with
Liquid Nitrogen

The most common temperature scale in the world is the Celsius scale, named in honor of the Swedish astronomer Anders Celsius (1701–1744), who first suggested the scale of 100 equal parts (*degrees*) between the freezing point and boiling point of water. The number 0 is assigned to the temperature at which water freezes, and the number 100 to the temperature at which water boils (at standard atmospheric pressure).

The most common temperature scale used in the United States is the Fahrenheit scale, named after its

FIGURE 8.1

Can we trust our sense of hot and cold? Will both fingers feel the same temperature when they are dipped in the warm water? Try this and see (feel) for yourself.

FIGURE 8.2

A testament to Fahrenheit outside his home (now in Gdansk, Poland).

FIGURE 8.3

Fahrenheit and Celsius scales on a thermometer.

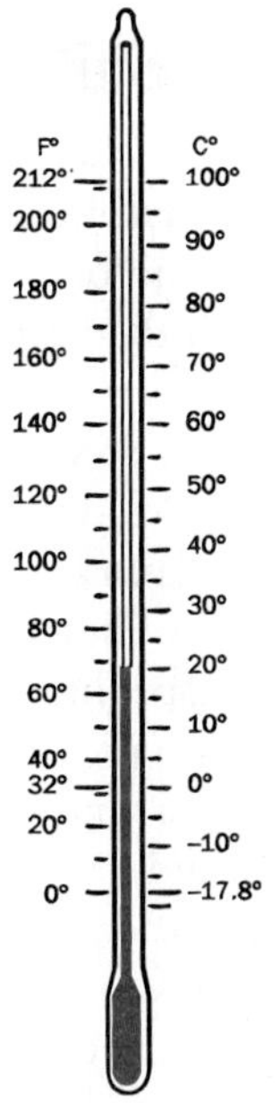

originator, the German physicist G. D. Fahrenheit (1686–1736). On this scale the number 32 is assigned to the temperature at which water freezes, and the number 212 is assigned to the temperature at which water boils. The Fahrenheit scale will become obsolete if and when the United States changes to the metric system.*

Arithmetic formulas used for converting from one temperature scale to the other are common in classroom exams. Because such arithmetic exercises are not really physics, we won't be concerned with these conversions (perhaps important in a math class, but not here). Besides, the conversion between Celsius and Fahrenheit temperatures is closely approximated in the side-by-side scales of Figure 8.3.**

Temperature is proportional to the average translational kinetic energy per particle that makes up a substance. By translational we mean to-and-fro linear motion. For a gas, we are referring to how fast the gas particles are bouncing back and forth; for a liquid, we are referring to how fast they slide and jiggle past each other; and for a solid we are referring to how fast the particles move as they vibrate and jiggle in place. It is important to note that temperature does *not* depend on how much of the substance you have. If you have a cup of hot water and then pour half of the water onto the floor, the water remaining in the cup hasn't changed its temperature. The water remaining in the cup contains half the *thermal energy* that the full cup of water contained, because there are only half as many water molecules in the cup as before. Temperature is a *per-particle property; thermal energy* is related to the sum total kinetic energy of all of the particles in your sample. Twice as much hot water will have twice the thermal energy, even though its temperature (the average KE per particle) will be the same.

When we measure the temperature of something with a conventional thermometer, thermal energy flows between the thermometer and the object whose temperature we are measuring. When both the object and the thermometer have the same average kinetic energy per particle, we say that they are in *thermal equilibrium.* When we measure something's temperature, we are really reading the temperature of the thermometer when it and the object have reached thermal equilibrium.

fyi Thermal contact is not required with infrared thermometers that show digital temperature readings by measuring the infrared radiation emitted by all bodies.

* Changing any long-established custom is difficult, and the Fahrenheit scale does have some advantages in everyday use. For example, its degrees are smaller (1°F = 5/9°C), which gives greater accuracy when reporting the weather in whole-number temperature readings. Then, too, people somehow attribute a special significance to numbers increasing by an extra digit, so that, when the air temperature on a hot day is reported as having reached 100°F, the idea of heat is conveyed more dramatically than it would be by announcing that it is 38°C. Like so much of the British system of measurement, the Fahrenheit scale is geared to human beings.

** Okay, if you really want to know, the formulas for temperature conversion are: C = 5/9F − 32; F = 9/5 C + 32, where C is the Celsius temperature and F is the Fahrenheit temperature.

8.2 Absolute Zero

As thermal motion increases, a solid object first melts and becomes a liquid. With more thermal motion it then vaporizes. As the temperature is further increased, molecules dissociate into atoms, and atoms lose some or all of their electrons, thereby forming a cloud of electrically charged particles—a *plasma*. Plasmas exist in stars, where the temperature is many millions of degrees Celsius. Temperature has no upper limit.

In contrast, there is a definite limit at the lower end of the temperature scale. Gases expand when heated, and they contract when cooled. Nineteenth-century experiments found something quite amazing. They found that if one starts out with a gas, any gas, at 0°C and then changes its temperature, while pressure is held constant, the volume changes by 1/273 for each degree Celsius change in temperature. When a gas was cooled from 0°C to −10°C, its volume decreased by 10/273 and contracted to 263/273 of its original volume. If a gas at 0°C could be cooled down by 273°C, it would contract 273/273 volumes and be reduced to zero volume. Clearly, we cannot have a substance with zero volume.

Experimenters got similar results for pressure. Starting at 0°C, the pressure of a gas held in a container of fixed volume decreased by 1/273 for each Celsius degree its temperature was lowered. If it were cooled to 273°C below zero, it would have no pressure at all. In practice, every gas converts to a liquid before becoming this cold. Nevertheless, these decreases by 1/273 increments suggested the idea of a lowest temperature: −273°C. That's the lower limit of temperature, **absolute zero**. At this temperature, molecules have lost all available kinetic energy.* No more energy can be removed from a substance at absolute zero. It can't get any colder.

The absolute temperature scale is called the *Kelvin scale*, named after the famous British mathematician and physicist William Thomson, First Baron Kelvin. Absolute zero is 0 K (short for "0 kelvin"; note that the word "degrees" is not used with Kelvin temperatures).** There are no negative numbers on the Kelvin scale. Its temperature divisions are identical to the divisions on the Celsius scale. Thus, the melting point of ice is 273 K, and the boiling point of water is 373 K.

> Absolute zero isn't the coldest you can get. It's the coldest you can hope to approach.

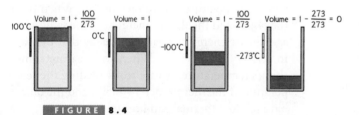

FIGURE **8.4**

When pressure is held constant, the volume of a gas changes by 1/273 of its volume at 0°C with each 1°C change in temperature. At 100°C, the volume is 100/273 greater than it is at 0°C. When the temperature is reduced to −100°C, the volume is reduced by 100/273. At −273°C, the volume of the gas would be reduced by 273/273 and therefore would be zero.

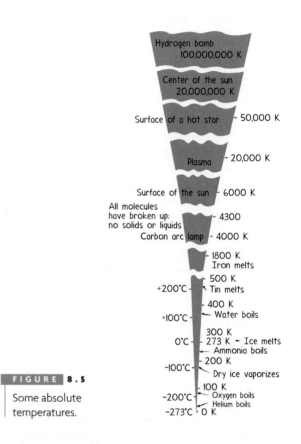

FIGURE **8.5**

Some absolute temperatures.

* Even at absolute zero, molecules still possess a small amount of kinetic energy, called the *zero-point energy*. Helium, for example, has enough motion at absolute zero to prevent it from freezing. The explanation for this involves quantum theory.

** When Thomson became a baron he took his title from the Kelvin River that ran through his estate. In 1968 the term *degrees Kelvin* (°K) was officially changed to simply *kelvin* (lowercase k), which is abbreviated K (capital K). The precise value of absolute zero (0 K) is −273.15°C.

8.3   Internal Energy

When you strike a penny with a hammer, it becomes warm. Why? Because the hammer's blow causes the atoms in the coin to jiggle faster. Some of the kinetic energy of the hammer goes into increasing the kinetic energy of the individual atoms making up the penny. If an object becomes warmer, that's because each of the particles making it up possesses, on average, more kinetic energy than before. We say that the warmer an object becomes, the more thermal energy it contains. So when you warm up by a fire on a cold winter night, you are increasing the motions of atoms and molecules in your body. Thermal energy includes both the kinetic energy and the potential energy of the particles in a substance as they wiggle and jiggle, twist and turn, vibrate, or race back and forth. Up to this point and in previous chapters we called the energy resulting from heat flow *thermal energy*, to make clear its link to heat and temperature. Henceforth we will use the term that physicists prefer, *internal energy*. **Internal energy** is the grand total of all energies in a substance.

8.4   Heat

If you touch your finger to a hot stove, energy enters your finger because the stove is warmer than your finger. When you touch a piece of ice, however, energy passes out of your finger and into the colder ice. The direction of energy flow is always from a warmer object to a neighboring cooler

FIGURE 8.6

The temperature of the sparks is very high, about 2000°C. That's a lot of energy per molecule of spark. But because there are only a few molecules per spark, the total amount of internal energy in the sparks is safely small. Temperature is one thing; transfer of internal energy is another.

object. A physicist defines **heat** as the internal energy transferred from one thing to another due to a temperature difference.

According to this definition, matter contains *internal energy*—not heat. Once internal energy has been transferred to an object or substance, it ceases to be heat. Stating again, for emphasis—a substance does not contain heat; it contains internal energy. Heat is internal energy in transit.

For substances in thermal contact, internal energy flows from the higher-temperature substance into the lower-temperature one until thermal equilibrium is reached. This does not mean that internal energy necessarily flows from a substance with more internal energy into one with less internal energy. For example, there is more internal energy in a bowl of warm water than there is in a red-hot thumbtack. If the tack is placed into the water, internal energy doesn't flow from the warm water to the tack.

> Just as dark is the absence of light, cold is the absence of internal energy.

Instead, it flows from the hot tack to the cooler water. Internal energy never flows unassisted from a low-temperature substance into a higher-temperature one.

If heat is internal energy that transfers in a direction from hot to cold, what is cold? Does a cold substance contain something opposite to internal energy? Not so. An object is cold because each of its constituent particles on

FIGURE 8.7

The left pot contains 1 liter of water. The right one contains 3 liters. Although both pots absorb the same quantity of heat, the temperature increases three times as much in the pot with the smaller amount of water.

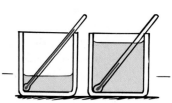

Hot stove

average has less kinetic energy than the particles in a warmer object. When you find yourself outdoors on a near-zero winter day, you feel cold not because something called cold gets to you. You feel cold because you lose heat. Your molecules are transferring energy to less energetic molecules around them. The purpose of your coat is to slow down the heat flow between your body and your surroundings. Cold is not a thing in itself, but the result of reduced molecular kinetic energy.

fyi
> Temperature is measured in degrees. Heat is measured in joules (or calories). In the United States we speak of low-calorie foods and drinks. Most of the world speaks of low-joule foods and drinks.

STOP AND
CHECK YOURSELF

1. Suppose you apply a flame to 1 liter of water for a certain time and its temperature rises by 2°C. If you apply the same flame for the same time to 2 liters of water, by how much will its temperature rise?

2. If a fast marble hits a random scatter of slow marbles, does the fast marble usually speed up or slow down? Which lose(s) kinetic energy and which gain(s) kinetic energy, the initially fast-moving marble or the initially slow ones? How do these questions relate to the direction of heat flow?

CHECK YOUR ANSWERS

1. Its temperature will rise by only 1°C, because there are twice as many molecules in 2 liters of water, and each molecule receives only half as much energy on the average. So the average kinetic energy, and thus the temperature, increases by half as much.

2. A fast-moving marble slows when it hits slower-moving marbles. It gives up some of its kinetic energy to the slower ones. Likewise with heat. Molecules with more kinetic energy that make contact with slower molecules give some of their excess kinetic energy to the slower ones. The direction of energy transfer is from hot to cold. For both the marbles and the molecules, however, the total energy of the system before and after contact is the same.

8.5 Quantity of Heat

Heat is measured in joules. It takes 4.18 joules (or, equivalently, 1 calorie) of heat to change the temperature of 1 gram of water by 1 Celsius degree.*

The energy ratings of foods and fuels are determined from the energy released when they are burned. (Metabolism is really "burning" at a slow rate.) The heat unit for labeling food is the kilocalorie (which is 1000 calories, the heat needed to change the temperature of 1 kilogram of water by 1°C). To differentiate this unit and the smaller calorie, the food unit is usually called a *Calorie*, with a capital C. So 1 Calorie is really 1000 calories.

What we've learned thus far about heat and internal energy is summed up in the *laws of thermodynamics*. The word **thermodynamics** stems from Greek words meaning "movement of heat."

STOP AND
CHECK YOURSELF

Which will raise the temperature of water more, adding 4.18 joules or 1 calorie?

CHECK YOUR ANSWER

Both are the same. This is like asking which is longer, a 1-mile-long track or a 1.6-kilometer-long track. They're the same length, just expressed in different units.

FIGURE 8.8

To the weight watcher, the peanut contains 10 Calories; to the physicist, it releases 10,000 calories (41,800 joules) of energy when burned or digested.

* Another common unit of heat is the British thermal unit (Btu). The Btu is defined as the amount of heat required to change the temperature of 1 lb of water by 1 degree Fahrenheit. One Btu is equal to 1054 J.

8.6 The Laws of Thermodynamics

When internal energy transfers as heat, the energy lost in one place is gained in another in accord with conservation of energy. When the law of energy conservation is applied to thermal systems we call it the **first law of thermodynamics**. We state it generally in the following form:

Whenever heat is added to a system, it transforms to an equal amount of some other form of energy.

When we add heat energy to a system, whether it be a steam engine, Earth's atmosphere, or the body of a living creature, this added energy increases the internal energy of the system if it remains in the system and/or does external work if it leaves the system. More specifically, the first law of thermodynamics states:

Heat added = increase in internal energy + external work done by the system

Suppose that you put an air-filled, rigid, airtight can on a hot stove and apply heat to the can. **Warning**: *Do not actually do this.* Since the can has a fixed volume, the walls of the can don't move, so no work is done. All of the heat going into the can increases the internal energy of the enclosed air, so its temperature rises. Now suppose instead that the can is able to expand. The heated air does work as the sides of the can expand, exerting a force for some distance on the surrounding atmosphere. Since some of the added heat goes into doing work, less of the added heat goes into increasing the internal energy of the enclosed air. Can you see that the temperature of the enclosed air will be lower when it does work than when it doesn't do work? The first law of thermodynamics makes good sense.

The **second law of thermodynamics** restates what we've learned about the direction of heat flow:

Heat never spontaneously flows from a cold substance to a hot substance.

FIGURE 8.9

When you push down on the piston, you do work on the air inside. What happens to its temperature?

When heat flow is spontaneous—that is, without the assistance of external work—the direction of flow is always from hot to cold. In winter, heat flows from inside a warm home to the cold air outside. In summer, heat flows from the hot air outside into the cooler interior. Heat can be made to flow the other way *only* when work is done on the system or by adding energy from another source. This occurs with heat pumps and air conditioners. In these devices, internal energy is pumped from a cooler to a warmer region. But without external effort, the direction of heat flow is always from hot to cold. The second law, like the first, makes logical sense.*

The **third law of thermodynamics** restates what we've learned about the lowest limit of temperature:

No system can reach absolute zero.

As investigators attempt to reach this lowest temperature, it becomes more difficult to get closer to

---

* The laws of thermodynamics were the rage back in the 1800s. At that time, horses and buggies were yielding to steam-driven locomotives. There is the story of the engineer who explained the operation of a steam engine to a peasant. The engineer cited in detail the operation of the steam cycle, how expanding steam drives a piston that in turn rotates the wheels. After some thought, the peasant asked, "Yes, I understand all that. But where's the horse?" This story illustrates how difficult it is to abandon our way of thinking about the world when a newer method comes along to replace established ways. Are we different today?

it. Physicists have been able to record temperatures that are less than a millionth of 1 kelvin—but never as low as 0 K.

Order Tends to Disorder

The first law of thermodynamics states that energy can neither be created nor destroyed. It speaks of the *quantity* of energy. The second law qualifies this by adding that the form that energy takes when transforming "deteriorates" to less useful forms. It speaks of *quality* of energy, as energy becomes more diffuse and ultimately degenerates into waste.

With this broader perspective, the second law can be stated another way:

In natural processes, high-quality energy tends to transform into lower-quality energy—order tends to disorder.

Processes in which disorder returns to order without external help don't occur in nature. Interestingly, time is given a direction via this thermodynamic rule. Time's arrow always points from order to disorder.*

> The laws of thermodynamics can be stated this way: You can't win (because you can't get any more energy out of a system than you put into it); you can't break even (because you can't get as much useful energy out as you put in); and you can't get out of the game (entropy in the universe is always increasing).

8.7 Entropy

The idea of ordered energy tending to disordered energy is embodied in the concept of *entropy*.** **Entropy** is the measure of the amount of disorder in a system. When disorder increases, entropy increases.

More disorder means more entropy. The molecules of an automobile's exhaust, for example, cannot spontaneously recombine to form more highly organized gasoline molecules. Warm air that escapes to a room when the oven door is open cannot spontaneously return to the oven. Whenever a physical system is allowed to distribute its energy freely, it always does so in a manner such that entropy increases, while the energy of the system available for doing work decreases.†

In a system left to itself, entropy increases. Entropy can decrease only if work is put into the system. This occurs in living organisms, where input energy allows a decrease in the entropy of the system. That is, it can become more ordered. All living things, from bacteria to trees to human beings, extract energy from their surroundings and use this energy to increase their own organization. The process of extracting energy (for instance,

FIGURE 8.10

Entropy.

---

* In the previous century when movies were new, audiences were amazed to see a train come to a stop inches away from a heroine tied to the tracks. This was filmed by starting with the train at rest, inches away from the heroine, and then moving *backward*, gaining speed. When the film was reversed, the train was seen to move *toward* the heroine. (Next time, watch closely for the telltale smoke that *enters* the smokestack.)

** Entropy can be expressed mathematically. The increase in entropy ΔS of a thermodynamic system is equal to the amount of heat added to the system ΔQ divided by the temperature T at which the heat is added: $\Delta S = \Delta Q/T$.

† Interestingly enough, the American writer Ralph Waldo Emerson, who lived during the time the second law of thermodynamics was the new science topic of the day, philosophically speculated that not everything becomes more disordered with time and cited the example of human thought. Ideas about the nature of things grow increasingly refined and better organized as they pass through the minds of succeeding generations. Human thought is evolving toward more order.

breaking down a highly organized food molecule into smaller molecules) increases entropy elsewhere, so life forms plus their waste products have a net increase in entropy. Energy must be transformed within the living system to support life. When it is not, the organism soon dies and tends toward disorder.

8.8 Specific Heat Capacity

While eating, you've likely noticed that some foods remain hotter much longer than others. Whereas the filling of hot apple pie can burn your tongue, the crust does not, even when the pie has just been removed from the oven. Or a piece of toast may be comfortably eaten a few seconds after coming from the hot toaster, whereas you must wait several minutes before eating soup that has the same high temperature.

Different substances have different thermal capacities for storing energy. If we heat a pot of water on a stove, we might find that it requires 15 minutes to raise it from room temperature to its boiling temperature. But an equal mass of iron on the same stove would rise through the same temperature range in only about 2 minutes. For silver, the time would be less than a minute. Equal masses of different materials require different quantities of heat to change their temperatures by a specified number of degrees.*

As mentioned earlier, a gram of water requires 1 calorie of energy to raise the temperature 1 degree Celsius. It takes only about one-eighth as much

energy to raise the temperature of a gram of iron by the same amount. Water absorbs more heat than iron for the same change in temperature. We say water has a higher **specific heat capacity** (sometimes simply called *specific heat*).

The specific heat capacity of any substance is defined as the quantity of heat required to change the temperature of a unit mass of the substance by 1 degree Celsius.

We can think of specific heat capacity as thermal inertia. Recall that inertia is a term used in mechanics to signify the resistance of an object to a change in its state of motion. Specific heat capacity is like thermal inertia since it signifies the resistance of a substance to a change in temperature.

> Water is very useful in the cooling systems of automobiles and other engines because it absorbs a great quantity of heat for small increases in temperature. Water also takes longer to cool.

FIGURE 8.11

The filling of hot apple pie may be too hot to eat, even though the crust is not.

* In the case of silver and iron, silver atoms are about twice as massive as iron atoms. A given mass of silver contains only about half as many atoms as an equal mass of iron, so only about half the heat is needed to raise the temperature of the silver. Hence, the specific heat of silver is about half that of iron's.

STOP AND CHECK YOURSELF

1. Which has a higher specific heat capacity, water or sand? In other words, which takes longer to warm in sunlight (or longer to cool at night)?

2. Why does a piece of watermelon stay cool for a longer time than sandwiches do when both are removed from a picnic cooler on a hot day?

CHECK YOUR ANSWERS

1. Water has the higher specific heat capacity. In the same sunlight, the temperature of water increases more slowly than does the temperature of sand. And water will cool more slowly at night. (Walking or running barefoot across scorching sand in daytime is a different experience than doing the same across sand in the evening!) The low specific heat capacity of sand and soil, as evidenced by how quickly they warm in the morning Sun and how quickly they cool at night, affects local climates.

2. Water in the melon has more "thermal inertia" than do sandwich ingredients, and it resists changes in temperature much more. This thermal inertia is specific heat capacity.

If the specific heat capacity c is known for a substance, then the heat transferred = specific heat capacity × mass × change in temperature. This can be expressed by the formula

$$Q = cm\Delta T$$

where Q is the quantity of heat, c is the specific heat of the substance, m is the mass, and ΔT is the corresponding change in temperature of the substance. When mass m is in grams, using the specific heat capacity of water as 1.0 cal/gram°C gives Q in calories.

Problems

1. What would be the final temperature of a mixture of 50 grams of 20°C water and 50 grams of 40°C water?

2. Consider mixing 100 grams of 25°C water with 75 grams of 40°C water. Show that the final temperature of the mixture is 31.4°C.

3. Radioactive decay in Earth's interior provides enough energy to keep the interior hot, generate magma, and provide warmth to natural hot springs. This is due to the average release of about 0.03 J per kilogram each year. Show that the time it takes for a chunk of thermally insulated rock to increase 500°C in temperature (assuming the specific heat of the rock sample is 800 J/kg · °C) will be 13.3 million years.

Solutions

1. The heat gained by the cooler water = heat lost by the warmer water. Since the masses of water are the same, the final temperature is midway, 30°C. So we'll end up with 100 grams of **30°C** water.

2. Here we have different masses of water that are mixed together. We equate the heat gained by the cool water to the heat lost by the warm water. We can express this equation formerly, then let the expressed terms lead to a solution:

Heat gained by cool water = heat lost by warm water

$$cm_1\Delta T_1 = cm_2\Delta T_2$$

ΔT_1 doesn't equal ΔT_2 as in Problem 1 because of different masses of water. Some thinking will show that ΔT_1 will be the final temperature T minus 25°, since T will be greater than 25°. ΔT_2 is 40° minus T, because T will be less than 40°. Then,

$$c(100\,\text{g})(T - 25) = c(75\,\text{g})(40 - T)$$
$$100T - 2500 = 3000 - 75T$$
$$T = \textbf{31.4°C}$$

3. Here we switch to rock, but the same concept applies. And we switch to specific heat expressed in joules per kilogram °C. No particular mass is specified, so we'll work with quantity of heat/mass (for our answer should be the same for a small chunk of rock or a huge chunk). From $Q = cm\Delta T \Rightarrow Q/m = c\Delta T = (800\,\text{J/kg} \cdot °\text{C})(500°\text{C}) = 400,000$ J/kg. The time required is $(400,000\,\text{J/kg}) \div (0.03\,\text{J/kg} \cdot \text{yr}) = \textbf{13.3 million years}$. Small wonder it remains hot down there!

The High Specific Heat Capacity of Water

Water has a much higher capacity for storing energy than almost any other substance. The reason for water's high specific heat involves the various ways that energy can be absorbed. Energy absorbed by any substance increases the jiggling motion of molecules, which raises the temperature. Or absorbed energy may increase the amount of internal vibration or rotation within the molecules, which becomes potential energy, which does not raise the temperature. Usually absorption of energy involves a combination of both. When we compare water molecules with atoms in a metal, we find there are many more ways for water molecules to absorb energy without increasing translational kinetic energy. So, water has a much higher specific heat capacity than metals—and most other common materials.

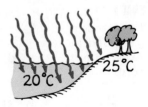

FIGURE **8.12**

Because water has a high specific heat capacity and is transparent, it takes more energy to warm the water than to warm the land. Solar energy striking the land is concentrated at the surface, but that striking the water extends beneath the surface and so is "diluted."

Water's high specific heat capacity affects the world's climate. Look at a world globe and notice the high latitude of Europe. Water's high specific heat helps to keep Europe's climate appreciably milder than regions of the same latitude in northeastern regions of Canada. Both Europe and Canada receive about the same amount of sunlight per square kilometer. Fortunately for Europeans, the Atlantic Ocean current known as the Gulf Stream carries warm water northeast from the Caribbean Sea, retaining much of its internal energy long enough to reach the North Atlantic Ocean off the coast of Europe. There the water releases 4.18 joules of energy for each gram of water that cools by 1°C. The released energy is carried by westerly winds over the European continent.*

A similar effect occurs in the United States. The winds in North America are mostly westerly. On the West Coast, air moves from the Pacific Ocean to the land. In winter months, the ocean water is warmer than the air. Air blows over the warm water and then moves over the coastal regions. This produces a warm climate. In summer, the opposite occurs. Air blowing over the water carries cooler air to the coastal regions. The East Coast does not benefit from the moderating effects of water because the direction of air is from the land to the Atlantic Ocean. Land, with a lower specific heat capacity, gets hot in the summer but cools rapidly in the winter.

Islands and peninsulas do not have the temperature extremes that are common in interior regions of a continent. The high summer and low winter temperatures common in Manitoba and the Dakotas,

for example, are largely due to the absence of large bodies of water. Europeans, islanders, and people living near ocean air currents should be glad that water has such a high specific heat capacity. San Franciscans certainly are!

STOP AND CHECK YOURSELF

Bermuda is close to North Carolina, but, unlike North Carolina, it has a tropical climate year-round. Why?

CHECK YOUR ANSWER

Bermuda is an island. The surrounding water warms it when it might otherwise be too cold, and cools it when it might otherwise be too warm.

8.9 Thermal Expansion

As the temperature of a substance increases, its molecules jiggle faster and move farther apart. The result is thermal expansion. Most substances expand when heated and contract when cooled. Sometimes the changes aren't noticeable, and sometimes they are. Telephone wires are longer and sag more on a hot summer day than they do in winter. Railroad tracks that were laid on cold winter days expand and buckle in the hot summer (Figure 8.14). Metal lids on glass fruit jars can often be loosened by heating them under hot water. If one part of a piece of glass is heated or cooled more rapidly than adjacent parts, the resulting expansion or

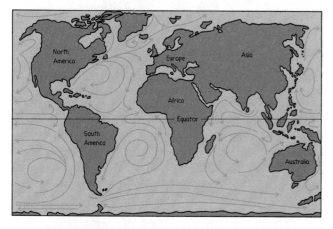

FIGURE 8.13

Many ocean currents, shown in blue, distribute heat from the warmer equatorial regions to the colder polar regions.

* Additionally, jet streams high in the atmosphere are a major contributor to the warming of Europe.

FIGURE 8.14

Thermal expansion. Extreme heat on a July day caused the buckling of these railroad tracks.

contraction may break the glass. This is especially true of thick glass. Pyrex glass is an exception because it is specially formulated to expand very little with increasing temperature.

Thermal expansion must be taken into account in structures and devices of all kinds. A civil engineer uses reinforcing steel with the same expansion rate as concrete. A long steel bridge usually has one end anchored while the other rests on rockers (Figure 8.15). Notice also that many bridges have tongue-and-groove gaps called *expansion joints* (Figure 8.16). Similarly, concrete roadways and sidewalks are intersected by gaps, which are sometimes filled with tar, so that the concrete can expand freely in summer and contract in winter.

> *fyi*
> Thermal expansion accounts for the creaky noises often heard in the attics of old houses on cold nights.

The fact that different substances expand at different rates is nicely illustrated with a bimetallic strip (Figure 8.17). This device is made of two strips of different metals welded together, one of brass and the other of iron. When heated, the greater expansion of the brass bends the strip. This bending

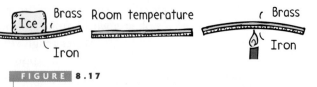

FIGURE 8.17

A bimetallic strip. Brass expands more when heated than iron does, and it contracts more when cooled. Because of this behavior, the strip bends as shown.

may be used to turn a pointer, to regulate a valve, or to close a switch.

A practical application of a bimetallic strip wrapped into a coil is the thermostat (Figure 8.18). When a room becomes too cold, the coil bends toward the brass side and activates an electrical switch that turns on the heater. When the room gets too warm, the coil bends toward the iron side, which breaks the electrical circuit and turns off the heater. Bimetallic strips are used in oven thermometers, refrigerators, electric toasters, and various other devices.

With increases in temperature, liquids expand more than solids. We notice this when gasoline overflows from a car's tank on a hot day. If the tank and its contents expanded at the same rate, no overflow would occur. This is why a gas tank being filled shouldn't be "topped off," especially on a hot day.

FIGURE 8.15

One end of the bridge rides on rockers to allow for thermal expansion. The other end (not shown) is anchored.

FIGURE 8.16

This gap in the roadway of a bridge is called an expansion joint; it allows the bridge to expand and contract. (Was this photo taken on a warm or a cold day?)

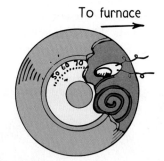

FIGURE 8.18

A thermostat. When the bimetallic coil expands, the drop of liquid mercury rolls away from the electrical contacts and breaks the electrical circuit. When the coil contracts, the mercury rolls against the contacts and completes the circuit.

Thermal Expansion of Water

Water, like most other substances, expands when heated. But interestingly, it *doesn't* expand in the temperature range between 0°C and 4°C. Something quite fascinating happens in this range.

Ice has a crystalline structure, with open-structured crystals. Water molecules in this open structure have more space between them than they do in the liquid phase (Figure 8.19). This means that ice is less dense than water. When ice melts, not all the open-structured crystals collapse. Some remain in the ice–water mixture, making up a microscopic slush that slightly "bloats" the water—increases its volume slightly (Figure 8.21). This results in ice water being less dense than slightly warmer water. As the temperature of water at 0°C is increased, more of the remaining ice crystals collapse. The melting of these ice crystals further decreases the volume of the water. Two opposite processes occur for the water at the same time—contraction and expansion. Volume decreases as ice crystals collapse, while volume increases due to greater molecular motion. The collapsing effect dominates until the temperature reaches 4°C. After that, expansion overrides contraction because most of the ice crystals have melted (Figure 8.22).

When ice water freezes to become solid ice, its volume increases tremendously. As solid ice cools further, like most substances, it contracts. The density of ice at any temperature is much lower than the density of water, which is why ice floats on water. This behavior of water is very important in nature. If water were most dense at 0°C, it would settle to the bottom of a pond or lake instead of forming at the surface.

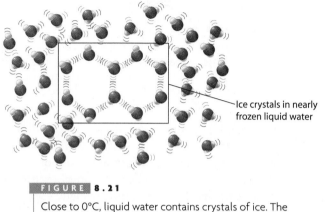

FIGURE 8.21

Close to 0°C, liquid water contains crystals of ice. The open structure of these crystals increases the volume of the water slightly.

Ice crystals in nearly frozen liquid water

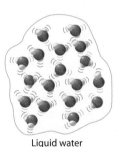

Liquid water (dense)

Ice (less dense)

FIGURE 8.19

Water molecules in a liquid are denser than water molecules frozen in ice, where they have an open crystalline structure.

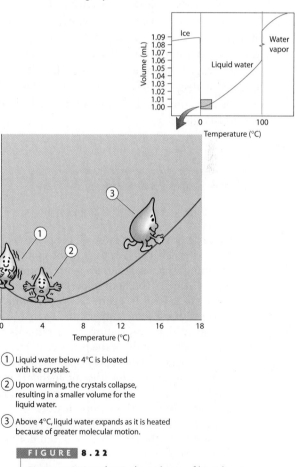

① Liquid water below 4°C is bloated with ice crystals.

② Upon warming, the crystals collapse, resulting in a smaller volume for the liquid water.

③ Above 4°C, liquid water expands as it is heated because of greater molecular motion.

FIGURE 8.22

Between 0°C and 4°C, the volume of liquid water decreases as temperature increases. Above 4°C, thermal expansion exceeds contraction and volume increases as temperature increases.

FIGURE 8.20

The six-sided structure of a snowflake is a result of the six-sided ice crystals that compose it. The crystals are made mostly from water vapor, not liquid water. (Most snowflakes are not as symmetrical as this one.)

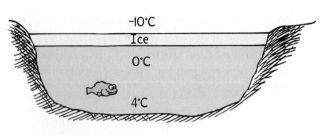

Because water is most dense at 4°C, colder water rises and freezes on the surface. This means that fish remain in relative warmth!

FIGURE 8.23

As water cools, it sinks until the entire pond is at 4°C. Then, as water at the surface is cooled further, it floats on top and can freeze. Once ice is formed, temperatures lower than 4°C can extend down into the pond.

A pond freezes from the surface downward. In a cold winter the ice will be thicker than it would be in a milder winter. Water at the bottom of an ice-covered pond is 4°C, which is relatively warm for organisms that live there. Interestingly, very deep bodies of water are not ice covered even in the coldest of winters. This is because all the water must be cooled to 4°C before lower temperatures can be reached. For deep water, the winter is not long enough to reduce an entire pond to 4°C. Any 4°C water lies at the bottom. Because of water's high specific heat and poor ability to conduct heat, the bottom of deep bodies of water in cold regions remains at a constant 4°C year-round. Fish should be glad that this is so.

STOP AND CHECK YOURSELF

1. What was the precise temperature at the bottom of Lake Michigan on New Year's Eve in 1901?

2. What's inside the open spaces of the ice crystals shown in Figure 8.19? Is it air, water vapor, or nothing?

CHECK YOUR ANSWERS

1. The temperature at the bottom of any body of water that has 4°C water in it is 4°C at the bottom, for the same reason that rocks are at the bottom. Both 4°C water and rocks are more dense than water at any other temperature. Water is a poor heat conductor, so, if the body of water is deep and in a region of long winters and short summers, the water at the bottom is likely to remain a constant 4°C year-round.

2. There's nothing at all in the open spaces. It's empty space—a void. If there were air or vapor in the open spaces, the illustration should show molecules there—oxygen and nitrogen for air and H_2O for water vapor.

SCIENCE AND SOCIETY

■LIFE AT THE EXTREMES

Some deserts, such as those on the plains of Spain, the Sahara in Africa, and the Gobi Desert in central Asia, reach surface temperatures of 60°C (140°F). Too hot for life? Not for certain species of ants of the genus *Cataglyphis*, which thrive at this searing temperature. At this extremely high temperature, the desert ants can forage for food without the presence of lizards, which would otherwise prey upon them. Resilient to heat, these ants can withstand higher temperatures than any other creatures in the desert. How they are able to do this is currently being researched. They scavenge the desert surface for the corpses of those creatures that did not find cover in time, touching the hot sand as little as possible while often sprinting on four legs with two held high in the air. Although their foraging paths zigzag over the desert floor, their return paths are almost straight lines to their nest holes.

They attain speeds of 100 body lengths per second. During an average six-day lifespan, most of these ants retrieve 15 to 20 times their weight in food.

From deserts to glaciers, a variety of creatures have invented ways to survive the harshest corners of the world. A species of worm thrives in the glacial ice in the Arctic. There are insects in the Antarctic ice that pump their bodies full of antifreeze to ward off becoming frozen solid. Some fish that live beneath the ice are able to do the same. Then there are bacteria that thrive in boiling hot springs as a result of having heat-resistant proteins.

An understanding of how creatures survive at the extremes of temperature can provide clues for practical solutions to the physical challenges faced by humans. Astronauts who venture from Earth, for example, will need all the techniques available for coping with unfamiliar environments.

SUMMARY OF TERMS

Temperature A measure of the hotness or coldness of substances, related to the average kinetic energy per molecule in a substance, measured in degrees Celsius, or in degrees Fahrenheit, or in kelvins.

Absolute zero The theoretical temperature at which a substance possesses no kinetic energy.

Internal energy The total energy (kinetic plus potential) of the submicroscopic particles that make up a substance.

Heat The internal energy that flows from a substance of higher temperature to a substance of lower temperature, commonly measured in calories or joules.

Thermodynamics The study of heat and its transformation to different forms of energy.

First law of thermodynamics A restatement of the law of energy conservation, usually as it applies to systems involving changes in temperature: Whenever heat flows into or out of a system, the gain or loss of internal energy equals the amount of heat transferred.

Second law of thermodynamics Heat never spontaneously flows from a cold substance to a hot substance. Also, in natural processes, high-quality energy tends to transform into lower-quality energy—order tends to disorder.

Third law of thermodynamics No system can reach absolute zero.

Entropy The measure of energy dispersal of a system. Whenever energy freely transforms from one form to another, the direction of transformation is toward a state of greater disorder and, therefore, toward one of greater entropy.

Specific heat capacity The quantity of heat required to change the temperature of a unit mass of the substance by 1 degree Celsius.

REVIEW QUESTIONS

8.1 Temperature

1. What are the temperatures for freezing water on the Celsius and Fahrenheit scales? For boiling water?
2. Is the temperature of an object a measure of the total kinetic energy of molecules in the object or a measure of the average kinetic energy per molecule in the object?
3. What is meant by the statement "a thermometer measures its own temperature"?

8.2 Absolute Zero

4. By how much does the pressure of a gas in a rigid vessel decrease when the temperature is decreased by 1°C?
5. What pressure would you expect in a rigid container of 0°C gas if you cooled it by 273 Celsius degrees?
6. What are the temperatures for freezing water and boiling water on the Kelvin temperature scale?
7. How much energy can be taken from a system at 0 K?

8.3 Internal Energy

8. Why does a penny become warmer when it is struck by a hammer?
9. How does internal energy differ from kinetic energy?

8.4 Heat

10. When you touch a cold surface, does cold travel from the surface to your finger or does energy travel from your finger to the cold surface? Explain.
11. Distinguish between temperature and heat.
12. Distinguish between heat and internal energy.
13. What determines the direction of heat flow?

8.5 Quantity of Heat

14. How is the energy value of foods determined?
15. Distinguish between a calorie and a Calorie.
16. Distinguish between a calorie and a joule.

8.6 The Laws of Thermodynamics

17. Cite the first law of thermodynamics.
18. How does the law of the conservation of energy relate to the first law of thermodynamics?
19. Cite the second law of thermodynamics.
20. How does the second law of thermodynamics relate to the direction of heat flow?
21. Cite the third law of thermodynamics.
22. In what sense is the direction of time related to the second law of thermodynamics?

8.7 Entropy

23. When the order of a system decreases, does entropy increase or decrease?
24. What must occur in cases where entropy decreases?

8.8 Specific Heat Capacity

25. Which warms up faster when heat is applied—iron or silver?
26. Does a substance that heats up quickly have a high or a low specific heat capacity?
27. How does the specific heat capacity of water compare with the specific heat capacities of other common materials?

8.9 Thermal Expansion

28. Why does a bimetallic strip bend with changes in temperature?
29. Which generally expands more for an equal increase in temperature—solids or liquids?
30. When the temperature of ice-cold water is increased slightly, does it undergo a net expansion or a net contraction?
31. What is the reason for ice being less dense than water?
32. Does "microscopic slush" in water tend to make it more dense or less dense? What happens to the slush when temperature increases?
33. At what temperature do the combined effects of contraction and expansion produce the smallest volume for water?
34. Why does ice form at the surface of a pond instead of at the bottom?

ACTIVE EXPLORATIONS

How much energy is in a nut? Burn it and find out. The heat of the flame is energy released upon the formation of chemical bonds (carbon dioxide, CO_2, and water, H_2O). Pierce a nut (pecan or walnut halves work best) with a bent paper clip that holds the nut above the table surface. Above this, secure a can of water so that you can measure its temperature change when the nut burns. Use about 10 cubic centimeters (10 milliliters) of water and a Celsius thermometer. As soon as you ignite the nut with a match, place the can of water above it and record the increase in water temperature as soon as the flame burns out. The number of calories released by the burning nut can be calculated by the formula $Q = cm\Delta T$, where c is its specific heat (1 cal/g°C), m is the mass of water, and ΔT is the change in temperature. The energy in food is expressed in terms of the dietetic Calorie, which is one thousand of the calories you'll measure. So to find the number of dietetic Calories, divide your result by 1000.

EXERCISES

1. Why wouldn't you expect all the molecules in a gas to have the same speed?
2. In your room, there are things such as tables, chairs, other people, and so forth. Which of these things has a temperature (a) lower than, (b) greater than, and (c) equal to the temperature of the air?
3. Why can't you establish whether you are running a high temperature by touching your own forehead?
4. Which is greater, an increase in temperature of 1°C or an increase of 1°F?
5. Which has the greater amount of internal energy, an iceberg or a cup of hot coffee? Explain.
6. On which temperature scale does the average kinetic energy of molecules double when the temperature doubles?
7. The temperature of the Sun's interior is about 10^7 degrees. Does it matter whether this is degrees Celsius or kelvins? Defend your answer.
8. Use the laws of thermodynamics to defend the statement that 100% of the electrical energy that goes into lighting a lamp is converted to internal energy.
9. When air is rapidly compressed, why does its temperature increase?
10. Which of the laws of thermodynamics has exceptions?
11. If you vigorously shake a can of liquid back and forth for more than a minute, will there be a noticeable temperature increase? (Try it and see.)
12. What happens to the gas pressure within a sealed gallon can when it is heated? Cooled? Why?
13. After driving a car for some distance, why does the air pressure in the tires increase?
14. If you drop a hot rock into a pail of water, the temperature of the rock and the water will change until both are equal. The rock will cool and the water will warm. Does this hold true if the hot rock is dropped into the Atlantic Ocean? Explain.
15. In the old days, on a cold winter night, it was common to bring a hot object to bed with you. Which would be better to keep you warm through the cold night—a 10-kilogram iron brick or a 10-kilogram jug of hot water at the same temperature? Explain.
16. Desert sand is very hot in the day and very cool at night. What does this tell you about its specific heat?
17. Why does adding the same amount of heat to two different objects not necessarily produce the same increase in temperature?
18. What role does specific heat capacity play in a watermelon staying cool after removal from a cooler on a hot day?
19. When a 1-kg metal pan containing 1 kg of cold water is removed from the refrigerator and set on a table, which absorbs more heat from the room—the pan or the water? Defend your answer.
20. Iceland, so named to discourage conquest by expanding empires, is not at all ice covered like Greenland and parts of Siberia, even though it is nearly on the Arctic Circle. The average winter temperature of Iceland is considerably higher than regions at the same latitude in eastern Greenland and central Siberia. Why is this so?
21. Why does the presence of large bodies of water tend to moderate the climate of nearby land—making it warmer in cold weather, and cooler in hot weather?

22. If the winds at the latitude of San Francisco and Washington, D.C., were from the east rather than from the west, why might San Francisco be able to grow only cherry trees and Washington, D.C., only palm trees?

23. Cite an exception to the claim that all substances expand when heated.

24. Would a bimetallic strip function if the two different metals happened to have the same rates of expansion? Is it important that they expand at different rates? Defend your answer.

25. Steel plates are commonly attached to each other with rivets, which are slipped into holes in the plates and rounded over with hammers. The hotness of the rivets makes them easier to round over, but their hotness has another important advantage in providing a tight fit. What is it?

26. A method for breaking boulders used to be putting them in a hot fire, then dousing them with cold water. Why would this fracture the boulders?

27. An old remedy for a pair of nested drinking glasses that stick together is to run water at different temperatures into the inner glass and over the surface of the outer glass. Which water should be hot, and which cold?

28. Would you or the gas company gain by having gas warmed before it passed through your gas meter?

29. A metal ball is just able to pass through a metal ring. But if the ball is first held in a hot flame, it expands and won't be able to pass through the ring. Suppose instead that the ring is heated, and not the ball. Will the ball then be able to pass through the hot ring? (See the chapter-opener photo on page 160.)

30. After a machinist very quickly slips a hot, snugly fitting iron ring over a very cold brass cylinder, there is no way that the two can be separated intact. Can you explain why this is so?

31. Suppose you cut a small gap in a metal ring. If you heat the ring, will the gap become wider or narrower?

32. Why do long steam pipes often have one or more relatively large U-shaped sections of pipe?

33. Suppose that water is used in a thermometer instead of mercury. If the temperature is at 4°C and then changes, why can't the thermometer indicate whether the temperature is rising or falling?

34. How does the combined volume of the billions and billions of hexagonal open spaces in the structures of ice crystals in a piece of ice compare with the portion of ice that floats above the water line?

35. State whether water at the following temperatures will expand or contract when warmed a little: 0°C; 4°C; 6°C.

36. Why is it important to protect water pipes so they don't freeze?

37. If cooling occurred at the bottom of a pond instead of at the surface, would a lake freeze from the bottom up? Explain.

38. Make up a multiple-choice question that distinguishes between heat and temperature.

PROBLEMS

● **BEGINNER** ■ **INTERMEDIATE** ◆ **EXPERT**

The quantity of heat Q *released or absorbed from a substance of specific heat* c *and mass* m, *undergoing a change in temperature* ΔT *is* $Q = cm\Delta T$

1. ● Will Maynez burns a 0.6-gram peanut beneath 50 grams of water, which increases in temperature from 22°C to 50°C.

   a. Assuming 40% efficiency, show that the peanut's food value is 3500 calories.

   b. Then show how the food value in calories per gram is 5.8 kilocalories per gram (or 5.8 Calories per gram).

2. ■ Pounding a nail into wood makes the nail warmer. Consider a 5-gram steel nail 6 cm long and a hammer that exerts an average force of 500 N on the nail when it is being driven into a piece of wood. The nail becomes hotter. Show that the increase in the nail's temperature is 13.3°C. (Assume that the specific heat capacity of steel is 450 J/kg °C.)

3. ■ If you wish to warm 100 kg of water by 20°C for your bath, show that the amount of heat is 2000 kilocalories (2000 Calories). Then show that this is equivalent to 8370 kilojoules.

4. ■ The specific heat capacity of copper is 0.092 calories per gram per degree Celsius. Show that the amount of heat needed to raise the temperature of a 10-gram piece of copper from 0°C to 100°C is 92 calories. How does this compare with the heat needed to raise the temperature of the same mass of water through the same temperature difference?

5. ● In lab you submerge 100 grams of 40°C nails in 100 grams of 20°C water. (The specific heat of iron is 0.12 cal/g C.) Equate the heat gained by the water to the heat lost by the nails and show that the final temperature of the water becomes 31.4°C.

To solve the problems below, you will need to know about the average coefficient of linear expansion, α, which differs for different materials. We define α to be the change in length per unit length—or the fractional change in length—for a temperature change of one degree Celsius. That is, $\Delta L/L$ per °C. For aluminum, $\alpha = 24 \times 10^{26}/°C$, and for steel, $\alpha = 11 \times 10^{26}/°C$. The change in length ΔL of a material is given by $\Delta L = L\alpha\Delta T$.

6. ● Consider a bar 1 m long that expands 0.5 cm when heated. Show that when similarly heated, a 100-m bar of the same material becomes 100.5 m long.

7. ● Suppose that the 1.3-km main span of steel for the Golden Gate Bridge had no expansion joints. Show that for an increase in temperature of 15°C the bridge would be 0.21 m longer.

8. ■ Imagine a 40,000-km steel pipe that forms a ring to fit snugly all around the circumference of Planet Earth. Suppose that people along its length breathe on it so as to raise its temperature by 1°C. The pipe gets longer. It is also no longer snug. How high does it stand above ground level? Show that the answer is an astounding 70 m higher! (To simplify, consider only the expansion of its radial distance from the center of Earth, and apply the geometry formula that relates circumference C and radius r, $C = 2\pi r$.)

CHAPTER 8 ONLINE RESOURCES

The Physics Place

Videos
Low Temperature with Liquid Nitrogen
How a Thermostat Works

Quiz

Flashcards

Links

CHAPTER 9

Heat Transfer and Change of Phase

John Suchocki demonstrates the low conductivity of wood by walking barefoot on red-hot wooden coals.

H eat transfers from warmer to cooler things. If several objects with different temperatures come into contact, those that are warm become cooler and those that are cool become warmer. They tend to reach a common temperature. This process occurs in three ways: by *conduction*, by *convection*, and by *radiation*.

9.1 Conduction

W hen you hold one end of an iron nail in a flame, the nail quickly becomes too hot to hold. If you hold one end of a short glass rod in a flame, the rod takes much longer before it becomes too hot to hold. In both cases, heat at the hot end travels along the entire length. This method of heat transfer is called **conduction**. Thermal conduction occurs by collisions between particles and their immediate neighbors. Because the heat travels quickly through the nail we say that it is a good *conductor* of heat. Materials that are poor conductors are called *insulators*.

The Physics Place

The Secret to Walking on Hot Coals
Air Is a Poor Conductor

Solids (such as metals) whose atoms or molecules have loosely held electrons are good conductors of heat. These mobile electrons move quickly and transfer energy to other electrons, which migrate quickly throughout the solid. Poor conductors

FIGURE 9.1

The tile floor feels colder than the wooden floor, even though both are at the same temperature. Tile is a better heat conductor than wood, and it more quickly conducts internal energy from your feet.

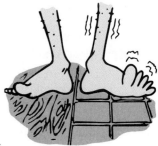

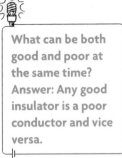

(such as glass, wool, wood, paper, cork, and plastic foam) are made up of molecules that hold tightly to their electrons. In these materials, molecules vibrate in place, and transfer energy only through interactions with their immediate neighbors. Since the electrons are not mobile, energy is transferred much more slowly in insulators.

Wood is a good insulator, and it is often used for cookware handles. Even when a pot is hot, you can briefly grasp the wooden handle with your bare hand without harm. An iron handle of the same temperature would surely burn your hand. Wood is a good insulator even when it's red hot. This explains how fire-walking Conceptual Chemistry author John Suchocki can walk barefoot on red-hot wood coals without burning his feet (as shown in the chapter-opener photo). (**Caution:** Don't try this on your own; even experienced firewalkers sometimes receive bad burns when conditions aren't just right.) The main factor here is the poor conductivity of wood—even red-hot wood. Although its temperature is high, very little internal energy is conducted to the feet. A firewalker must be careful that no iron nails or other good conductors are among the hot coals. Ouch!

Air is a very poor conductor. Hence, you can briefly put your hand in a hot pizza oven without harm. But don't touch the metal in the hot oven. Ouch again! The good insulating properties of such things as wool, fur, and feathers are largely due to the air spaces they contain. Be glad that air is a poor conductor; if it weren't, you'd feel quite chilly on a 20°C (68°F) day!

FIGURE 9.3

Conduction of heat from Lil's hand to the wine is minimized by the long stem of the wine glass.

Snow is a poor conductor because its flakes are formed of crystals that trap air and provide insulation. That's why a blanket of snow keeps the ground warm in winter. Animals in the forest find shelter from the cold in snow banks and in holes in the snow. The snow doesn't provide them with energy—it simply slows down the loss of body heat that the animals generate. The same principle explains why igloos, Arctic dwellings built from compacted snow, can shield inhabitants from the cold.

Interestingly, insulation doesn't prevent the flow of internal energy. Insulation simply slows down the *rate* at which internal energy flows. Even a warm, well-insulated house gradually cools. Insulation such as rock wool or fiberglass placed in the walls and ceiling of a house slows down the transfer of

FIGURE 9.2

When you stick a nail into ice, does cold flow from the ice to your hand, or does energy flow from your hand to the ice?

FIGURE 9.4

Snow patterns on the roof of a house show areas of conduction and insulation. Bare parts show where heat from inside has conducted through the roof and melted the snow.

internal energy from a warm house to a cooler outside in winter, and in summer, from the warm outside to the cool inside.

> **Thermal conduction is the process by which energy is transferred by heat through a material between two points at different temperatures.**

CHECK YOURSELF

1. In desert regions that are hot in the day and cold at night, the walls of houses are often made of mud. Why is it important that the mud walls be thick?

2. Wood is a better insulator than glass. Yet fiberglass is commonly used to insulate buildings. Why?

CHECK YOUR ANSWERS

1. A wall of appropriate thickness retains the warmth of the house at night by slowing the flow of internal energy from inside to outside, and it keeps the house cool in the daytime by slowing the flow of internal energy from outside to inside. Such a wall has "thermal inertia."

2. Fiberglass is a good insulator, many times better than glass, because of the air that is trapped among its fibers.

9.2 Convection

On a hot day you can see ripples in the air as hot air rises from an asphalt road. Likewise, if you put an ice cube into a clear glass of hot water you can see ripples as the cold water from the melting ice cube descends in the glass. Transfer of heat by the motion of fluid as it rises or sinks, is called **convection**. Unlike conduction, convection occurs only in fluids (liquids and gases). Convection involves bulk motion of a fluid (currents) rather than interactions at the molecular level.

> *fyi*
>
> Convection ovens are simply ovens with a fan inside. Cooking is speeded up by the circulation of heated air.

FIGURE **9.5**
Convection currents in a gas (air) (a) and a liquid (b).

a

b

FIGURE **9.6**
The tip of a heater element submerged in water produces convection currents, which are revealed as shadows (caused by deflections of light in water of different temperatures).

We can see why warm air rises. When warmed, it expands, becomes less dense, and is buoyed upward in the cooler surrounding air like a balloon buoyed upward. When the rising air reaches an altitude at which the air density is the same, it no longer rises. We see this occurring when smoke from a fire rises and then settles off as it cools and its density matches that of the surrounding air.

To see for yourself that expanding air cools, do the experiment shown in Figure 9.7. Expanding air really does cool.*

A dramatic example of cooling by expansion occurs with steam expanding through the nozzle of a pressure cooker (Figure 9.8). The combined cooling effects of expansion and rapid mixing with cooler air will allow you to hold your hand comfortably in the jet of condensed vapor. (**Caution:** If you try this, be sure to place your hand high above the

---

* Where does the energy go in this case? It goes into work done on the surrounding air as the expanding air pushes outward.

FIGURE **9.7**

Blow warm air onto your hand from your wide-open mouth. Now reduce the opening between your lips so the air expands as you blow. Try it now. Do you notice a difference in the temperature of exhaled air? Does air cool as it expands?

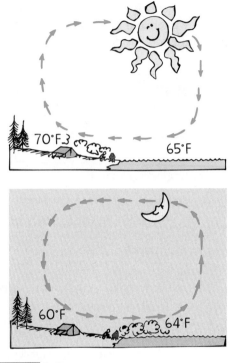

FIGURE **9.8**

The hot steam expands from the pressure cooker and is cool to Millie's touch.

FIGURE **9.9**

Convection currents produced by unequal heating of land and water. During the day, warm air above the land rises, and cooler air over the water moves in to replace it. At night, the direction of air flow is reversed, because now the water is warmer than the land.

nozzle at first and then lower it slowly to a comfortable distance above the nozzle. If you put your hand directly at the nozzle where no steam is visible, watch out! Steam is invisible, and is clear of the nozzle before it expands and cools. The cloud of "steam" you see is actually condensed water vapor, which is much cooler than live steam.)

Cooling by expansion is the opposite of what occurs when air is compressed. If you've ever compressed air with a tire pump, you probably noticed that both air and pump became quite hot. Compression of air warms it.

Convection currents stir the atmosphere and produce winds. Some parts of Earth's surface absorb energy from the Sun more readily than others. This results in uneven heating of the air near the ground. We see this effect at the seashore, as Figure 9.9 shows. In the daytime, the ground warms up more than the water. Then warmed air close to the ground rises and is replaced by cooler air that moves in from above the water. The result is a sea breeze. At night, the process reverses because the shore cools off more quickly than the water, and then the warmer air is

over the sea. If you build a fire on the beach, you'll see that the smoke sweeps inland during the day and then seaward at night.

fyi

Opening a refrigerator door lets warm air in, which then takes energy to cool. The more empty your fridge, the more cold air is swapped with warm air. So keep your fridge full for lower operating costs—especially if you're an excessive open-and-close-the-door type.

STOP AND
CHECK YOURSELF

Explain why you can hold your fingers beside the candle flame without harm, but not above the flame.

CHECK YOUR ANSWER

Hot air travels upward by air convection. Since air is a poor conductor, very little energy travels sideways to your fingers.

FIGURE **9.10**

Types of radiant energy (electromagnetic waves).

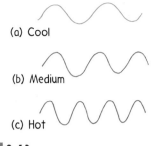

(a) Cool

(b) Medium

(c) Hot

FIGURE **9.12**

(a) A low-temperature (cool) source emits primarily low-frequency, long-wavelength waves. (b) A medium-temperature source emits primarily medium-frequency, medium-wavelength waves. (c) A high-temperature (hot) source emits primarily high-frequency, short-wavelength waves.

9.3 Radiation

Energy travels from the Sun through space and then through Earth's atmosphere and warms Earth's surface. This transfer of energy cannot involve conduction or convection, for there is no medium between the Sun and Earth. Energy must be transmitted some other way—by **radiation**.* The transferred energy is called *radiant energy*.

Radiant energy exists in the form of *electromagnetic waves*, ranging from the longest wavelengths to the shortest: radio waves, microwaves, infrared waves (invisible waves below red in the visible spectrum), visible waves, ultraviolet waves, X-rays, and gamma rays. We'll treat waves further in Chapters 11 and 12.

The wavelength of radiation is related to the frequency of vibration. Frequency is the rate of vibration of a wave source. Nellie Newton in Figure 9.11 shakes a rope at a low frequency (left), and at a higher frequency (right). Note that shaking at a low frequency produces a long lazy wave, and shaking at a higher frequency produces a wave of shorter wavelength. We shall see in later chapters that vibrating electrons emit electromagnetic waves. Low-frequency vibrations produce long wavelength waves, and high-frequency vibrations produce waves with shorter wavelengths.

FIGURE **9.11**

A wave of long wavelength is produced when the rope is shaken gently (at a low frequency). When shaken more vigorously (at a high frequency), a wave of shorter wavelength is produced.

* The radiation we are talking about here is electromagnetic radiation, including visible light. Don't confuse this with *radioactivity*, a process of the atomic nucleus that we'll discuss in Chapter 16.

Emission of Radiant Energy

Every object at any temperature above absolute zero emits radiant energy. The peak frequency $\bar{f}$ of radiant energy is directly proportional to the Kelvin temperature T of the emitter:

$$\bar{f} \sim T$$

If an object is hot enough, some of the radiant energy it emits is in the range of visible light. At a temperature of about 500°C an object begins to emit the longest waves we can see, red light. Higher temperatures produce a yellowish light. At about 1500°C all the different waves to which the eye is sensitive are emitted and we see an object as "white hot." A blue-hot star is hotter than a white-hot star, and a red-hot star is less hot. Since a blue-hot star has twice the light frequency of a red-hot star, it therefore has twice the surface temperature of a red-hot star.**

So whereas a blue-hot star with twice the radiation frequency of a red-hot star has twice the Kelvin temperature, a blue-hot star with twice the temperature emits 16 times as much energy as a same-size red-hot star.

The amount of radiation emitted also depends on surface characteristics, and is referred to as the *emissivity* of the object—ranging from close to 0 for very shiny surfaces and close to 1 for very black ones. A perfectly black surface emits what is called *black body radiation*, and has an emissivity of 1.

** The *amount* of radiant energy Q, emitted by an object is proportional to the fourth power of the Kelvin temperature T:

$$Q \sim T^4$$

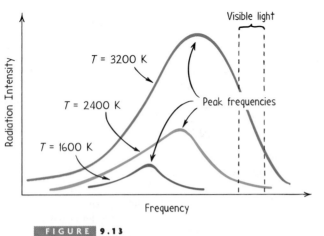

FIGURE 9.13
INTERACTIVE FIGURE

Radiation curves for different temperatures. The average frequency of radiant energy is directly proportional to the absolute temperature of the emitter.

Because the surface of the Sun has a high temperature (by earthly standards) it therefore emits radiant energy at a high frequency—much of it in the visible portion of the *electromagnetic spectrum*. The surface of Earth, by comparison, is relatively cool, and so the radiant energy it emits has a frequency lower than that of visible light. The radiation emitted by Earth is in the form of *infrared waves*—below our threshold of sight. Radiant energy emitted by Earth is called **terrestrial radiation**.

The Sun's radiant energy stems from nuclear reactions in its deep interior. Likewise, nuclear reactions in Earth's interior warm Earth (visit the depths of any mine and you'll find it's warm down there—year-round). Much of this internal energy conducts to the surface to become terrestrial radiation.

All objects—you, your instructor, and everything in your surroundings—continually emit radiant energy over a range of frequencies. Objects with everyday temperatures emit mostly low-frequency infrared waves. When the higher-frequency infrared waves are absorbed by your skin, as when standing

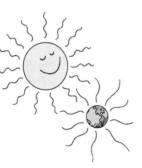

FIGURE 9.14

Both the Sun and Earth emit the same kind of radiant energy. The Sun's glow is visible to the eye; Earth's glow consists of longer waves and isn't visible to the eye.

beside a hot stove, you feel the sensation of heat. So it is common to refer to infrared radiation as *heat radiation*. Common infrared sources that give the sensation of heat are the Sun, a lamp filament, and burning embers in a fireplace.

Heat radiation underlies infrared thermometers. You simply point the thermometer at something whose temperature you want, press a button, and a digital temperature reading appears. The radiation emitted by the object in question provides the reading. Typical classroom infrared thermometers operate in the range of about −30°C to 200°C.

> Radiation by Earth is *terrestrial radiation*. Radiation by the Sun is *solar radiation*. Both are regions in the electromagnetic spectrum. (What do you call radiation from that special someone?)

STOP AND CHECK YOURSELF

Which of these do not emit radiant energy?
(a) the Sun; (b) lava from a volcano; (c) red-hot coals; (d) this textbook.

CHECK YOUR ANSWER

All the above emit radiant energy—even your textbook, which, like the other substances listed, has a temperature. According to the rule $\bar{f} \sim T$, the book therefore emits radiation whose peak frequency $\bar{f}$ is quite low compared with the radiation frequencies emitted by the other substances. Everything with any temperature above absolute zero emits radiant energy. That's right—*everything*!

Absorption of Radiant Energy

If everything is radiating energy, why doesn't everything finally run out of it? The answer is, everything is also *absorbing* energy. Good emitters of radiant energy are also good absorbers; poor emitters are

FIGURE 9.15

When the black rough-surfaced container and the shiny polished one are filled with hot (or cold) water, the blackened one cools (or warms) faster.

fyi
Everything around you both radiates and absorbs energy continuously!

poor absorbers. For example, a radio dish antenna constructed to be a good emitter of radio waves is also, by its very design, a good receiver (absorber) of them. A poorly designed transmitting antenna is also a poor receiver.

The surface of any material, hot or cold, both absorbs and emits radiant energy. If the surface absorbs more energy than it emits, it is a net absorber and its temperature rises. If it emits more than it absorbs, it is a net emitter and its temperature drops. Whether a surface plays the role of net emitter or net absorber depends on whether its temperature is above or below that of its surroundings. In short, if it's hotter than its surroundings, the surface will be a net emitter and will cool; if it's colder than its surroundings, it will be a net absorber and will become warmer.

A hot pizza put outside on a winter day is a net emitter. The same pizza placed in a hotter oven is a net absorber.

STOP AND CHECK YOURSELF

1. If a good absorber of radiant energy were a poor emitter, how would its temperature compare with the temperature of its surroundings?
2. A farmer turns on the propane burner in his barn on a cold morning and heats the air to 20°C (68°F). Why does he still feel cold?

CHECK YOUR ANSWERS

1. If a good absorber were not also a good emitter, there would be a net absorption of radiant energy and the temperature of the absorber would remain higher than the temperature of the surroundings. Things around us approach a common temperature only because good absorbers are, by their very nature, also good emitters.
2. The walls of the barn are still cold. He radiates more energy to the walls than the walls radiate back at him, and he feels chilly. (On a winter day, you are comfortable inside your home or classroom only if the walls are warm—not just the air.)

Reflection of Radiant Energy

Absorption and reflection are opposite processes. A good absorber of radiant energy reflects very little of it, including visible light. Hence, a surface that reflects very little or no radiant energy looks dark. So a good absorber appears dark, and a perfect absorber reflects no radiant energy and appears completely black. The pupil of the eye, for example, allows light to enter with no reflection, which is why it appears black. (An exception occurs in flash photography when pupils appear pink, which occurs when very bright light is reflected off the eye's pink inner surface and back through the pupil.)

Look at the open ends of pipes in a stack; the holes appear black. Look at open doorways or windows of distant houses in the daytime, and they, too, look black. Openings appear black because the light that enters them is reflected back and forth on the inside walls many times and is partly absorbed at each reflection. As a result, very little or none of the light remains to come back out of the opening and travel to your eyes (Figure 9.16).

Good reflectors, on the other hand, are poor absorbers. Clean snow is a good reflector and therefore does not melt rapidly in sunlight. If the snow is dirty, it absorbs radiant energy from the Sun and melts faster. Dropping black soot from an aircraft onto snow-covered mountains is a technique sometimes used in flood control. Controlled melting at favorable times, rather than a sudden runoff of melted snow, is thereby accomplished.

fyi
Emission and absorption in the visible part of the spectrum are affected by color. But not so in the infrared part of the spectrum where surface texture has more effect. In the infrared, a dull finish emits/absorbs better than a polished one, whatever the color.

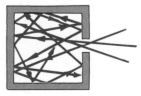

FIGURE 9.16
Radiation that enters the opening has little chance of leaving because most of it is absorbed. For this reason, the opening to any cavity appears black to us.

FIGURE 9.17

The hole looks perfectly black and indicates a black interior, when in fact the interior has been painted a bright white.

STOP AND CHECK YOURSELF

Which would be more effective in heating the air in a room, a heating radiator painted black, or silver?

CHECK YOUR ANSWER

Interestingly, the color of paint is a small factor, so either color can be used. That's because radiators do very little heating by radiation. Their hot surfaces warm surrounding air by conduction, the warmed air rises, and warmed convection currents heat the room. (A better name for this type of heater would be a *convector*.) Now if you're interested in *optimum* efficiency, a silver-painted radiator will radiate less, become and remain hotter, and do a better job of heating the air.

9.4 Newton's Law of Cooling

Left to themselves, objects hotter than their surroundings eventually cool to match the surrounding temperature. The rate of cooling depends on how much hotter the object is than its surroundings. A hot apple pie will cool more each minute if it is put in a cold freezer than if it is left on the kitchen table. That's because, in the freezer, the temperature difference between the pie and its surroundings is greater. Similarly, the rate at which a warm house leaks internal energy to the cold outdoors depends on the difference between the inside and outside temperatures.

The rate of cooling of an object—whether by conduction, by convection, or by radiation—is approximately proportional to the temperature difference ΔT between the object and its surroundings.

$$\text{Rate of cooling} \sim \Delta T$$

This is known as **Newton's law of cooling**. (Guess who is credited with discovering this?)

The law applies also to warming. If an object is cooler than its surroundings, then its rate of warming up is also proportional to ΔT.* Frozen food will warm up faster in a warm room than it would in a cold room.

STOP AND CHECK YOURSELF

Since a hot cup of tea loses internal energy more rapidly than a lukewarm cup of tea, would it be correct to say that a hot cup of tea will cool to room temperature before a lukewarm cup of tea will?

CHECK YOUR ANSWER

No! Although the *rate* of cooling is greater for the hotter cup, it has further to cool to reach thermal equilibrium. The extra time is equal to the time it takes to cool to the initial temperature of the lukewarm cup of tea. Cooling rate and cooling time are not the same thing.

9.5 Global Warming and the Greenhouse Effect

An automobile parked in the street in the bright Sun on a hot day with closed windows can get very hot inside—appreciably hotter than the outside air. This is an example of the greenhouse effect, so named for the same temperature-raising effect in florists' glass greenhouses. Understanding the greenhouse effect requires knowing about two concepts.

---

* A warm object that contains a source of energy may remain warmer than its surroundings indefinitely. The internal energy it emits doesn't necessarily cool it, and Newton's law of cooling doesn't apply. Thus, an automobile engine that is running remains warmer than the automobile's body and the surrounding air. But after the engine is turned off, it cools in accordance with Newton's law of cooling and gradually approaches the same temperature as its surroundings. Likewise, the Sun will remain hotter than its surroundings as long as its nuclear furnace is functioning—another 5 billion years or so.

■ THE THERMOS BOTTLE

A common thermos bottle, a double-walled glass container with a vacuum between its silvered walls, nicely summarizes heat transfer. When a hot or cold liquid is poured into such a bottle, it remains at very nearly the same temperature for many hours. This is because the transfer of internal energy by conduction, convection, and radiation is severely inhibited.

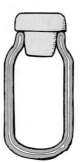

1. Heat transfer by conduction through the vacuum is impossible. Some internal energy escapes by conduction through the glass and stopper, but this is a slow process, because glass, plastic, and cork are poor conductors.
2. The vacuum also prevents heat loss through the walls by convection, since there is no air between the walls.
3. Heat loss by radiation is prevented by the silvered surfaces of the walls, which reflect radiant energy back into the bottle.

The first concept has been previously stated—that all things radiate, and the wavelength of radiation depends on the temperature of the object emitting the radiation. High-temperature objects radiate short waves; low-temperature objects radiate long waves. The second concept we need to know is that the transparency of things such as air and glass depends on the wavelength of radiation. Air is transparent to both infrared (long) waves and visible (short) waves, unless the air contains excess water vapor and carbon dioxide, in which case it is opaque to infrared. Glass is transparent to visible light waves, but is opaque to infrared waves. (The physics of transparency and opacity is discussed in Chapter 12.)

Now to why that car gets so hot in bright sunlight: Compared with the car, the Sun's temperature is very high. This means the waves that the Sun radiates are very short. These short waves easily pass through both Earth's atmosphere and the glass windows of the car. So energy from the Sun gets into the car interior, where, except for reflection, it is absorbed. The interior of the car warms up. The car interior radiates its own waves, but, since it is not as hot as the Sun, the waves are longer. The reradiated long waves encounter glass that isn't transparent to them. So the reradiated energy remains in the car, which makes the car's interior even warmer (which is why leaving your pet in a car on a hot sunny day is a no-no).

The same effect occurs in Earth's atmosphere, which is transparent to solar radiation. The surface of Earth absorbs this energy, and reradiates part of this as longer-wavelength terrestrial radiation. Atmospheric gases (mainly water vapor and carbon dioxide) absorb and re-emit much of this long-wavelength terrestrial radiation back to Earth. Terrestrial radiation that cannot escape Earth's atmosphere warms Earth. This global warming process is very nice, for Earth would be a frigid −18°C otherwise. Over the last 500,000 years the average temperature of Earth has fluctuated between 19°C and 27°C and is now at the high point—27°C—and climbing. Our present environmental concern is that increased levels of carbon dioxide and other

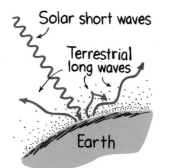

FIGURE 9.18

The hot Sun emits short waves, and the cool Earth emits long waves. Water vapor, carbon dioxide, and other "greenhouse gases" in the atmosphere retain heat that would otherwise be radiated from Earth to space.

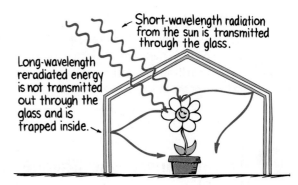

FIGURE 9.19

Glass is transparent to short-wavelength radiation but opaque to long-wavelength radiation. Reradiated energy from the plant is of long wavelength because the plant has a relatively low temperature.

atmospheric gases in the atmosphere may further increase the temperature and produce a new thermal balance unfavorable to the biosphere.

An important credo is, "You can never change only one thing." Change one thing, and you change another. A slightly higher Earth temperature means slightly warmer oceans, which means changes in weather and storm patterns. Warmer oceans also mean slightly increased evaporation, which means slightly increased snowfall in polar regions. The fraction of Earth at present beneath ice and snow is greater than the total area used for farmlands—and is shrinking at a historically unprecedented rate. These white areas reflect more solar radiation, which potentially could lead to a significant drop in global temperature. So overheating Earth today conceivably could cool it tomorrow and trigger the next ice age! Or it might not. We don't know.

What we do know is that energy consumption is related to population size. We are seriously questioning the idea of continued growth. (Please take the time to read Appendix E, "Exponential Growth and Doubling Time"—very important material.)

> A significant role of glass in a florist's greenhouse is preventing convection of cooler outside air with warmer inside air. So the greenhouse effect actually plays a bigger role in global warming than it does in the warming of florists' greenhouses.

STOP AND CHECK YOURSELF

What does it mean to say that the greenhouse effect is like a one-way valve?

CHECK YOUR ANSWER

Both the atmosphere of Earth and glass in a florist's greenhouse are transparent only to incoming short-wavelength light, and block outgoing long waves. Because of the blockage, radiation travels only in one direction.

9.6 Heat Transfer and Change of Phase

Matter exists in four common phases (states). Ice, for example, is the *solid* phase of water. When internal energy is added, the increased molecular motion breaks down the frozen structure and it becomes the *liquid* phase, water. When more energy is added, the liquid changes to the *gaseous* phase. Add still more energy, and the molecules break into ions and electrons, giving the *plasma* phase. Plasma (not to be confused with blood plasma) is the illuminating gas found in TV screens and fluorescent and other vapor lamps. The Sun, stars, and much of the space between them is in the plasma phase. Whenever matter changes phase, a transfer of internal energy is involved.

The Physics Place
Condensation Is a Warming Process

Evaporation

Water changes to the gaseous phase by the process of **evaporation**. In a liquid, molecules move randomly at a wide variety of speeds. Think of the water molecules as tiny billiard balls, moving helter-skelter, continually bumping into one another. During their bumping, some molecules gain kinetic energy, while others lose kinetic energy. Molecules at the surface that gain kinetic energy by being bumped from below are the ones to break free from the liquid. They leave the surface and escape into the space above the liquid. In this way, they become gas.

> *fyi*
> Water evaporating from your body takes energy with it, cooling you when emerging from water on a warm and windy day.

When fast-moving molecules leave the water, the molecules left behind are the slower-moving ones. What happens to the overall kinetic energy in a liquid when the high-energy molecules leave? The answer: The average kinetic energy of molecules left in the liquid decreases. The temperature (which measures the average kinetic energy of the molecules) decreases and the water is cooled.

When our bodies begin to overheat, our sweat glands produce perspiration, and evaporation cools us. This is part of nature's thermostat, for the

FIGURE 9.20

When wet, the cloth covering on the canteen promotes cooling. As the faster-moving water molecules evaporate from the wet cloth, the temperature of the cloth decreases and cools the metal. The metal, in turn, cools the water within. The water in the canteen can become a lot cooler than the outside air.

FIGURE 9.21

Like other dogs, Tammy's dogs have no sweat glands (except between their toes). They cool by panting. In this way, evaporation occurs in the mouth and within the bronchial tract.

FIGURE 9.22

Pigs have no sweat glands and therefore cannot cool by the evaporation of perspiration. Instead, they wallow in the mud to cool themselves.

evaporation of perspiration cools us and helps us maintain a stable body temperature. Many animals do not have sweat glands and must cool themselves by other means (Figures 9.21 and 9.22).

STOP AND
CHECK YOURSELF

Would evaporation be a cooling process if there were no transfer of molecular kinetic energy from water to the air above?

CHECK YOUR ANSWER

No. A liquid cools only when kinetic energy is carried away by evaporating molecules. This is similar to billiard balls that gain speed at the expense of others that lose speed. Those that leave (evaporate) are gainers, while losers remain behind and lower the temperature of the water.

In solid carbon dioxide (dry ice) molecules jump directly from the solid to the gaseous phase—that's why it's called dry ice. This form of evaporation is called **sublimation**. Moth balls are well known for their sublimation. Even frozen water undergoes sublimation. Because water molecules are so tightly held in a solid, frozen water sublimes much more slowly than liquid water evaporates. Sublimation accounts for the loss of much snow and ice, especially on high, sunny mountain tops. Sublimation also explains why ice cubes left in the freezer for a long time get smaller.

Condensation

The opposite of evaporation is **condensation**—the changing of a gas to a liquid. When gas molecules near the surface of a liquid are attracted to the liquid, they strike the surface with increased kinetic energy and become part of the liquid. This kinetic energy is absorbed by the liquid. The result is increased temperature. So whereas the liquid left behind is cooled with evaporation, with condensation, the object upon which the vapor condenses is warmed. Condensation is a warming process.

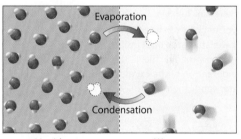

FIGURE 9.23

The exchange of molecules at the interface between liquid and gaseous water.

FIGURE 9.24

Internal energy is released by steam when it condenses inside the "radiator."

A dramatic example of warming by condensation is the energy released by steam when it condenses. The steam gives up a lot of energy when it condenses to a liquid and moistens the skin. That's why a burn from 100°C steam is much more damaging than a burn from 100°C boiling water. This energy release by condensation is utilized in steam-heating systems.

When taking a shower, you may have noticed that you feel warmer in the moist shower region than outside the shower. This difference is quickly sensed when you step outside. Away from the moisture, the rate of evaporation is much higher than the rate of condensation, and you feel chilly. When you remain in the moist shower stall, the rate of condensation is higher and you feel warmer. So now you know why you can dry yourself with a towel much more comfortably if you remain in the shower stall. If you're in a hurry and don't mind the chill, dry yourself off in the hallway.

On a July afternoon in dry Phoenix or Santa Fe, you'll feel a lot cooler than in New York City or New Orleans, even when the temperatures are the same. In the drier cities, the rate of evaporation from your skin is much greater than the rate of condensation of

water molecules from the air onto your skin. In humid locations, the rate of condensation is greater than the rate of evaporation. You feel the warming effect as vapor in the air condenses on your skin. You are literally being bombarded by the impact of H_2O molecules in the air slamming into you.

Although water molecules tend to stick to one another, in the air they move fast enough to avoid sticking. When they collide they bounce from one another and remain in the gaseous phase. At any moment some water molecules may move more slowly than on average and are more likely to stick upon collision (Figure 9.27). (This can be understood by thinking of a fly making a grazing contact with sticky flypaper. At a high speed it has enough momentum and energy to rebound from the flypaper without sticking, but, at a low speed, it is more likely to get stuck.) So slow-moving molecules are the most likely to condense and form droplets of water. That's why a cold soda pop can is wet when in warm air. Water molecules slow when making contact with the cold surface and condense.

FIGURE 9.25

If you're chilly outside the shower stall, step back inside and be warmed by the condensation of the excess water vapor there.

FIGURE 9.26

The toy drinking bird operates by the evaporation of ether inside its body and by the evaporation of water from the outer surface of its head. The lower body contains liquid ether, which evaporates rapidly at room temperature. As it (a) vaporizes, it (b) creates pressure (inside arrows), which pushes ether up the tube. Ether in the upper part does not vaporize because the head is cooled by the evaporation of water from the outer felt-covered beak and head. When the weight of ether in the head is sufficient, the bird (c) pivots forward, permitting the ether to run back to the body. Each pivot wets the felt surface of the beak and head, and the cycle is repeated.

STOP AND CHECK YOURSELF

Place a dish of water anywhere in your room. If the water level in the dish remains unchanged from one day to the next, can you conclude that no evaporation or condensation is occurring?

CHECK YOUR ANSWER

Not at all, for significant evaporation and condensation occur continuously at the molecular level. The fact that the water level remains constant indicates equal rates of evaporation and condensation.

Fast-moving H_2O molecules rebound upon collision

Slow-moving H_2O molecules coalesce upon collision

FIGURE 9.27

Condensation of water vapor.

■ CONDENSATION CRUNCH

Put a small amount of water in an aluminum soda pop can and heat it on a stove until steam issues from the opening. When this occurs, air has been driven out and replaced by steam. Then, with a pair of tongs, quickly invert the can into a pan of water. Crunch! The can is crushed by atmospheric pressure! Why? When the molecules of steam inside the can hit the inner wall they

bounce—the metal certainly doesn't absorb them. But when steam molecules encounter water in the pan, they stick to the water surface. Condensation occurs, leaving a very low pressure in the can, whereupon the surrounding atmospheric pressure crunches the can. Here we see, dramatically, how pressure is reduced by condensation. (This demonstration nicely underlies the condensation cycle of a steam engine—perhaps something for future study.)

9.7 Boiling

Evaporation occurs at the surface of a liquid. A change of phase from liquid to gas can also occur beneath the surface under proper conditions. The gas that forms beneath the surface of a liquid produces bubbles. The bubbles are buoyed upward to the surface, where they escape into the surrounding air. This change of phase is called **boiling**.

The Physics Place
Boiling Is a Cooling Process
Pressure Cooker: Boiling and Freezing at the Same Time

The pressure of the vapor within the bubbles in a boiling liquid must be great enough to resist the pressure of the surrounding liquid. Unless the vapor pressure is great enough, the surrounding pressures will collapse any bubbles that tend to form. At temperatures below the boiling point, vapor pressure is not great enough, so bubbles do not form until the boiling point is reached.

> When boiling water, it is common to say we are heating it. Actually, boiling cools the water.

Boiling, like evaporation, is a cooling process. At first thought, this may seem surprising—perhaps

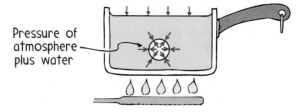

Pressure of atmosphere plus water

FIGURE 9.28

The motion of vapor molecules in the bubble of steam (much enlarged) creates a gas pressure (called the vapor pressure) that counteracts the atmospheric and water pressure against the bubble.

because we usually associate boiling with heating. However, heating water is one thing; boiling it is another. When 100°C water at atmospheric pressure is boiling, it is in thermal equilibrium. The water in the pot is being cooled by boiling as fast as it is being heated by energy from the heat source (Figure 9.29). If cooling did not occur, continued application of heat to a pot of boiling water would raise its temperature.

When pressure on the surface of a liquid increases, boiling is hampered. Then the temperature needed for boiling rises. The boiling point of a

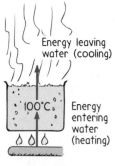

FIGURE **9.29**

Heating warms the water from below, and boiling cools it from above.

FIGURE **9.30**

The tight lid of a pressure cooker holds pressurized vapor above the water surface, and this inhibits boiling. In this way, the boiling temperature of the water is increased to above 100°C.

FIGURE **9.31**

Apparatus to demonstrate that water freezes and boils at the same time in a vacuum. A gram or two of water is placed in a dish that is insulated from the base by a polystyrene cup.

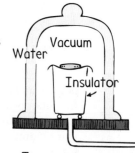

FIGURE **9.32**

Ron Hipschman at the Exploratorium removes a freshly frozen piece of ice from the "Water Freezer" exhibit, a vacuum chamber as depicted in Figure 9.31.

liquid depends on the pressure on the liquid—which is most evident in a pressure cooker (Figure 9.30). In such a device, vapor pressure builds up inside and prevents boiling, which results in a higher water temperature. It is important to note that it is the high temperature of the water that cooks the food, not the boiling process itself.

Lower atmospheric pressure (as at high altitudes) decreases the boiling temperature. For example, in Denver, Colorado, the "mile-high city," water boils at 95°C, instead of at 100°C. If you try to cook food in boiling water that is cooler than 100°C, you must wait a longer time for proper cooking. A three-minute boiled egg in Denver is yucky. If the temperature of the boiling water is very low, food does not cook at all.

A dramatic demonstration of the cooling effect of evaporation and boiling is shown in Figure 9.31. Here we see a shallow dish of room-temperature water in a vacuum jar. When the pressure in the jar is slowly reduced by a vacuum pump, the water begins to boil. As in all evaporation, the highest-energy molecules escape from the water, and the water left behind is cooled. As the pressure is further reduced, more and more of the faster-moving molecules boil away until the remaining liquid water reaches approximately 0°C. Continued cooling by boiling causes ice to form over the surface of the bubbling water. Boiling and freezing occur at the same time! Frozen bubbles of boiling water are a remarkable sight.

fyi

Mountaineering pioneers in the nineteenth century, without altimeters, used the boiling point of water to determine their altitudes.

STOP AND CHECK YOURSELF

1. Since boiling is a cooling process, would it be a good idea to cool your hot, sticky hands by dipping them into boiling water?

2. Rapidly boiling water has the same temperature as simmering water, both 100°C. Why, then, do the directions for cooking spaghetti often call for rapidly boiling water?

CHECK YOUR ANSWERS

1. No, no, no! When we say boiling is a cooling process, we mean that the water left behind in the pot (and not your hands!) is being cooled relative to the higher temperature it would attain otherwise. Because of the cooling effect of the boiling, the water remains at 100°C instead of getting hotter. A dip in 100°C water would be extremely uncomfortable for your hands!

2. Good cooks know that the reason for the rapidly boiling water is not higher temperature, but simply a way to keep the spaghetti strands from sticking together.

If you spray some drops of coffee into a vacuum chamber, they will boil until they freeze. Even after they are frozen, the water molecules continue to evaporate into the vacuum, until little crystals of coffee solids remain. This is how freeze-dried coffee is produced. The low temperature of this process tends to keep the chemical structure of the coffee solids from changing. When hot water is added, much of the original flavor of the coffee is retained.

9.8 Melting and Freezing

Melting occurs when a substance changes phase from a solid to a liquid. To visualize what happens, imagine a group of people holding hands and jumping around. The more violent the jumping, the more difficult it is to keep holding hands. If the jumping is violent enough, continuing to hold hands might become impossible. A similar thing happens to the molecules of a solid when it is heated. As heat is absorbed by the solid, its molecules vibrate more and more violently. If enough heat is absorbed, the attractive forces between the molecules no longer hold them together. The solid melts.

Freezing occurs when a liquid changes to a solid phase—the opposite of melting. As energy is removed from a liquid, molecular motion slows until molecules move so slowly that attractive forces between them bind them together. The liquid freezes when its molecules vibrate about fixed positions and form a solid.

At atmospheric pressure, ice forms at 0°C. With impurities in the water, the freezing point is lowered. "Foreign" molecules get in the way and interfere with crystal formation. In general, adding anything to water lowers its freezing temperature. Antifreeze is a practical application of this process.

fyi

> **Why is rock salt spread on icy roads in winter?** A short answer is that salt makes ice melt. Salt in water separates into sodium and chlorine ions. When these ions join water molecules, heat is given off, which melts microscopic parts of an icy surface. The melting process is enhanced by the pressure of automobiles rolling along the salt-covered icy surface, which forces the salt into the ice. The only difference between the rock salt applied to roads in winter and the substance you sprinkle on popcorn is the size of the crystals.

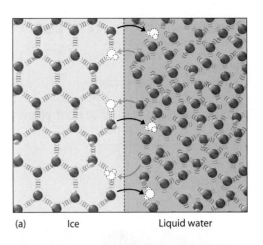

(a)   Ice   Liquid water

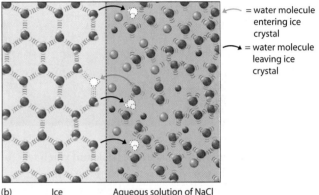

= water molecule entering ice crystal

= water molecule leaving ice crystal

(b)   Ice   Aqueous solution of NaCl

FIGURE 9.33

(a) In a mixture of ice and water at 0°C, ice crystals gain and lose water molecules at the same time. The ice and water are in thermal equilibrium. (b) When salt is added to the water, there are fewer water molecules entering the ice because there are fewer of them at the interface.

9.9 Energy and Change of Phase

If you heat a solid sufficiently, it will melt and become a liquid. If you heat the liquid, it will vaporize and become a gas. Energy must be put into a substance to change its phase in the direction from solid to liquid to gas. Conversely, energy must be extracted from a substance to change its phase in the direction from gas to liquid to solid (Figure 9.34).

The cooling cycle of a refrigerator nicely illustrates these concepts. A motor pumps a special fluid through the system, where it undergoes the cyclic process of vaporization and condensation. When the fluid vaporizes, internal energy is drawn from things stored inside the refrigerator. The gas that forms, with its added energy, is directed to and condenses to a liquid in outside coils in the back— appropriately called condensation coils. The next time you're near a refrigerator, place your hand near

Energy is absorbed when change of phase
is in this direction

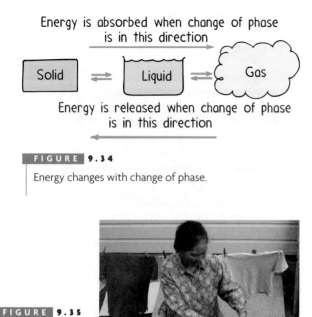

Energy is released when change of phase
is in this direction

Energy changes with change of phase.

The energy of
sunlight simply and
nicely harnessed.

the condensation coils in the back and you'll feel the
heat that has been extracted from the inside.

An air conditioner uses the same principle and
simply pumps heat energy from one part of the unit
to another. If the roles of vaporization and condensation are reversed, the air conditioner becomes a heater.

STOP AND
CHECK YOURSELF

In the process of water vapor condensing in the
air, the slower-moving molecules are the ones
that condense. Does condensation warm or cool
the surrounding air?

CHECK YOUR ANSWER

As slower-moving molecules are removed from the
air, there is an increase in the average kinetic energy
of molecules that remain in the air. Therefore, the
air is warmed. The change of phase is from gas to
liquid, which releases energy (Figure 9.34).

**Heat of vaporization is either the energy required to
separate molecules from the liquid phase or the energy
released when gases condense to the liquid phase.**

The amount of energy needed to change any substance from solid to liquid (and vice versa) is called
the **heat of fusion** for the substance. For water, this is
334 joules per gram. The amount of energy required

to change any substance from liquid to gas (and vice
versa) is called the **heat of vaporization** for the substance. For water, this is a whopping 2256 joules per
gram.

In premodern times, farmers in cold climates prevented jars of food from freezing by taking advantage of water's high heat of fusion. They simply kept
large tubs of water in their cellars. The outside temperature could drop to well below freezing, but not
in the cellars where water was releasing internal
energy while undergoing freezing. Canned food
requires subzero temperatures to freeze because of its
salt or sugar content. So farmers had only to replace
frozen tubs of water with unfrozen ones, and the cellar temperatures wouldn't fall below 0°C.

Water's high heat of vaporization allows you to
briefly touch your wetted finger to a hot skillet on a
hot stove without harm. You can even touch it a few
times in succession as long as your finger remains
wet. Energy that ordinarily would flow into and
burn your finger goes instead into changing the
phase of the moisture on your finger. Similarly, you
are able to judge the hotness of a hot clothes iron.

Paul Ryan, former supervisor in the Department
of Public Works in Malden, Massachusetts, has for
years used molten lead to seal pipes in certain
plumbing operations. He startles onlookers by dragging his finger through molten lead to judge its hotness (Figure 9.36). He is sure that the lead is very hot
and his finger is thoroughly wet before he does this.
(Do not try this on your own: if the lead is not hot
enough, it will stick to your finger—Ouch!)

**Heat of fusion is either the energy needed to separate
molecules from the solid phase or the energy released
when bonds form in a liquid that change it to the solid
phase.**

Paul Ryan tests the hotness
of molten lead by dragging
his wetted finger through it.

SUMMARY OF TERMS

Conduction The transfer of internal energy by molecular and electronic collisions within a substance (especially a solid).

Convection The transfer of internal energy in a gas or liquid by means of currents in the heated fluid. The fluid flows, carrying energy with it.

Radiation The transfer of energy by means of electromagnetic waves.

Terrestrial radiation The radiant energy emitted by Earth.

Newton's law of cooling The rate of loss of internal energy from an object is proportional to the temperature difference between the object and its surroundings.

$$\text{Rate of cooling} \sim \Delta T$$

Evaporation The change of phase at the surface of a liquid as it passes to the gaseous phase.

Sublimation The change of phase directly from solid to gas, bypassing the liquid phase.

Condensation The change of phase from gas to liquid; the opposite of evaporation. Warming of the liquid results.

Boiling A rapid state of evaporation that takes place within the liquid as well as at its surface. As with evaporation, cooling of the liquid results.

Melting The process of changing phase from solid to liquid, as from ice to water.

Freezing The process of changing phase from liquid to solid, as from water to ice.

Heat of fusion The amount of energy needed to change any substance from solid to liquid (and vice versa). For water, this is 334 J/g (or 80 cal/g).

Heat of vaporization The amount of energy required to change any substance from liquid to gas (and vice versa). For water, this is 2256 J/g (or 540 cal/g).

REVIEW QUESTIONS

1. What are the three common ways in which heat is transferred?

9.1 Conduction

2. What is the role of "loose" electrons in heat conductors?
3. Distinguish between a heat conductor and a heat insulator.
4. What is the explanation for a barefoot firewalker being able to walk safely on red-hot wooden coals?
5. Why are such materials as wood, fur, feathers, and even snow good insulators?
6. Does a good insulator prevent heat from getting through it, or does it simply slow its passage?

9.2 Convection

7. By what means is heat transferred by convection?
8. How does buoyancy relate to convection?
9. What happens to the temperature of air when it expands?
10. Why is Millie's hand not burned when she holds it above the escape valve of the pressure cooker (Figure 9.8)?
11. Why does the direction of coastal winds change from day to night?

9.3 Radiation

12. What exactly is radiant energy?
13. How does the frequency of radiant energy relate to the absolute temperature of the radiating source?
14. What is terrestrial radiation? How does it differ from solar radiation?
15. Since all objects emit energy to their surroundings, why don't the temperatures of all objects continuously decrease?
16. What determines whether an object at a given time is a net absorber or a net emitter?

17. Can an object be both a good absorber and a good reflector at the same time?
18. Why does the pupil of the eye appear black?

9.4 Newton's Law of Cooling

19. If you want a room-temperature can of a beverage to cool quickly, should you put it in the freezer compartment or in the main part of your refrigerator? Or does it not matter?
20. Which will undergo the greater rate of cooling, a red-hot poker in a warm oven or a red-hot poker in a cold room? (Or do both cool at the same rate?)
21. Does Newton's law of cooling apply to warming as well as to cooling?

9.5 Global Warming and the Greenhouse Effect

22. What would be the consequence to Earth's temperature if the greenhouse effect were completely eliminated?
23. What is meant by the expression "you can never change only one thing"?

9.6 Heat Transfer and Change of Phase

24. What are the four common phases of matter?
25. Do all the molecules in a liquid have about the same speed, or do they have a wide variety of speeds?
26. What is evaporation, and why is it a cooling process? What is it that cools?
27. What is sublimation?
28. What is condensation, and why is it a warming process? What is it that warms?
29. Why is a steam burn more damaging than a burn from boiling water of the same temperature?

30. Why do you feel uncomfortably warm on a hot and humid day?

9.7 Boiling

31. Distinguish between evaporation and boiling.
32. Why does water not boil at 100°C when it is under greater-than-normal atmospheric pressure?
33. Is it the boiling of the water or the higher temperature of the water that cooks food faster in a pressure cooker?

9.8 Melting and Freezing

34. Why does increasing the temperature of a solid make it melt?
35. Why does decreasing the temperature of a liquid make it freeze?

36. Why will water not freeze at 0°C when foreign ions are present?

9.9 Energy and Change of Phase

37. Does a liquid give off energy or absorb energy when it turns into a gas? When it turns into a solid?
38. Does a gas give off energy or absorb energy when it turns into a liquid? How about a solid when it turns to a liquid?
39. Distinguish between heat of fusion and heat of vaporization.
40. Why is it important that your finger be wet if you intend to touch it briefly to a hot clothes iron to test its temperature?

ACTIVE EXPLORATIONS

1. If you live where there is snow, do as Benjamin Franklin did about two hundred years ago: Lay samples of light and dark cloth on the snow and note the differences in the rate of melting beneath the samples of cloth.

2. Hold the bottom end of a test tube full of cold water in your hand. Heat the top part in a flame until the water boils. The fact that you can still hold the bottom shows that water is a poor conductor of heat. This is even more dramatic when you wedge chunks of ice at the bottom; then the water above can be brought to a boil without melting the ice. Try it and see.

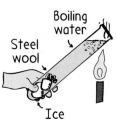

3. Wrap a piece of paper around a thick metal bar and place it in a flame. Note that the paper will not catch fire. Can you figure out why? (Paper generally will not ignite until its temperature reaches 233°C.)

4. Place a Pyrex funnel mouth down in a saucepan full of water so that the straight tube of the funnel sticks above the water. Rest a part of the funnel on a nail or coin so that water can get under it. Place the pan on a stove, and watch the water as it begins to boil. Where do the bubbles form first? Why? As the bubbles rise, they expand rapidly and push water ahead of them. The funnel confines the water, which is forced up the tube and driven out at the top. Now do you know how a geyser and a coffee percolator operate?

5. Watch the spout of a teakettle of boiling water. Notice that you cannot see the steam that issues from the spout. The cloud that you see farther away from the spout is not steam, but condensed water droplets. Now hold the flame of a candle in the cloud of condensed steam. Can you explain your observations?

6. You can make rain in your kitchen. Put a cup of water in a Pyrex saucepan or a Silex coffeemaker and heat it slowly over a low flame. When the water is warm, place a saucer filled with ice cubes on top of the container. As the water below is heated, droplets form at the bottom of the cold saucer and combine until they are large enough to fall, producing a steady "rainfall" as the water below is gently heated. How does this resemble, and how does it differ from, the way in which natural rain is formed?

7. Measure the temperature of boiling water and the temperature of a boiling solution of salt and water. How do the temperatures compare?

8. If you suspend an open-topped container of water in a pan of boiling water, with its top above the surface of the boiling water, water in the inner container will reach 100°C but will not boil. Can you explain why this is so?

EXERCISES

1. Wrap a fur coat around a thermometer. Will the temperature rise?
2. What is the explanation for a feather quilt being so warm on a cold winter night?

3. What is the purpose of the layer of copper or aluminum on the bottom of a piece of stainless-steel cookware?
4. In terms of physics, why do restaurants serve baked potatoes wrapped in aluminum foil?

5. Many tongues have been injured by licking a piece of metal on a very cold day. Why would no harm result if a piece of wood were licked on the same day?

6. Wood is a better insulator than glass. Yet fiberglass is commonly used as an insulator in wooden buildings. Explain.

7. Visit a snow-covered cemetery and note that the snow does not slope upward against the gravestones but, instead, forms depressions around them, as shown. Can you think of a reason for this?

8. You can bring water in a paper cup to a boil by placing it over a hot flame. Why doesn't the paper cup burn?

9. Wood has a very low conductivity. Does it still have a low conductivity if it is very hot—that is, in the stage of smoldering red-hot coals? Could you safely walk across a bed of red-hot wooden coals with bare feet? Although the coals are hot, does much heat conduct from them to your feet if you step quickly? Could you do the same on pieces of red-hot iron? Explain. (**Caution:** coals can stick to your feet, so OUCH—don't try it!)

10. A friend says that, in a mixture of gases in thermal equilibrium, the molecules have the same average kinetic energy. Do you agree or disagree? Defend your answer.

11. A friend says that, in a mixture of gases in thermal equilibrium, the molecules have the same average speed. Do you agree or disagree? Defend your answer.

12. Why would you not expect all of the molecules of air in your room to have the same average speed?

13. In a still room, smoke from a candle will sometimes rise only so far, not reaching the ceiling. Explain why.

14. What does the high specific heat of water have to do with the convection currents in the air at the seashore?

15. How do the average kinetic energies per molecule compare in a mixture of hydrogen and oxygen gases at the same temperature?

16. In a mixture of hydrogen and oxygen gases at the same temperature, which molecules move faster? Why?

17. One container is filled with argon gas and another with krypton gas. If both gases have the same temperature, in which container are the atoms moving faster? Defend your answer.

18. Which atoms have the greatest average speed in a mixture, U-238 or U-235? How would this affect diffusion through a porous membrane of otherwise identical gases made from these isotopes?

19. If we warm a volume of air, it expands. Does it then follow that, if we expand a volume of air, it warms? Defend your answer.

20. Machines used for making snow at ski areas blow a mixture of compressed air and water through a nozzle. The temperature of the mixture may initially be well above the freezing temperature of water, yet crystals of snow are formed as the mixture is ejected from the nozzle. Explain how this happens.

21. Turn an incandescent lamp on and off quickly while you are standing near it. You feel its heat, but you find when you touch the bulb that it is not hot. Explain why you felt heat from the lamp.

22. A number of bodies at different temperatures placed in a closed room share radiant energy and ultimately come to the same temperature. Would this thermal equilibrium be possible if good absorbers were poor emitters and poor absorbers were good emitters? Defend your answer.

23. From the rules that a good absorber of radiation is a good radiator and a good reflector is a poor absorber, state a rule relating the reflecting and radiating properties of a surface.

24. The heat of volcanoes and natural hot springs comes from trace amounts of radioactive minerals in common rock in Earth's interior. Why isn't the same kind of rock at Earth's surface warm to the touch?

25. Suppose that, at a restaurant, you are served coffee before you are ready to drink it. In order that it be hottest when you are ready for it, would you be wiser to add cream to it right away or just before you are ready to drink it?

26. Is it important to convert temperatures to the Kelvin scale when we use Newton's law of cooling? Why or why not?

27. If you wish to save fuel and you're going to leave your warm house for a half hour or so on a very cold day, should you turn your thermostat down a few degrees, turn it off altogether, or let it remain at the room temperature you desire?

28. If you wish to save fuel and you're going to leave your cool house for a half hour or so on a very hot day, should you turn your air conditioning thermostat up a bit, turn it off altogether, or let it remain at the room temperature you desire?

29. Why is whitewash sometimes applied to the glass of florists' greenhouses? Would you expect this practice to be more prevalent in winter or summer months?

30. If the composition of the upper atmosphere were changed so that it permitted a greater amount of terrestrial radiation to escape, what effect would this have on Earth's climate?

31. On a very cold day, you wear both a black coat and a transparent plastic coat. Which should be worn on the outside for maximum warmth? Defend your answer.

32. You can determine wind direction by wetting your finger and holding it up in the air. Explain.

33. If all the molecules in a liquid had the same speed, and some were able to evaporate, would the remaining liquid be cooled? Explain.

34. Where does the energy come from that keeps the dunking bird in Figure 9.26 operating?

35. Why will wrapping a bottle in a wet cloth at a picnic often produce a cooler bottle than placing the bottle in a bucket of cold water?

36. Why does the temperature of boiling water remain the same as long as the heating and boiling continue?

37. Why do vapor bubbles in a pot of boiling water get larger as they rise in the water?

38. Why does the boiling temperature of water decrease when the water is under reduced pressure, such as it is at a higher altitude?

39. Place a jar of water on a small stand within a saucepan of water so that the bottom of the jar is held above the bottom of the pan. When the pan is put on a stove, the water in the pan will boil, but not the water in the jar. Why?

40. Room-temperature water will boil spontaneously in a vacuum—on the Moon, for example. Could you cook an egg in this boiling water? Explain.

41. Your inventor friend proposes a design for cookware that will allow boiling to take place at a temperature of less than 100°C so that food can be cooked with the consumption of less energy. Comment on this idea.

42. When you boil potatoes, will your cooking time be reduced if the water is vigorously boiling instead of gently boiling?

43. Why does putting a lid over a pot of water on a stove shorten the time it takes for the water to come to a boil,

whereas, after the water is boiling, the use of the lid only slightly shortens the cooking time?

44. In the power plant of a nuclear submarine, the temperature of the water in the reactor is above 100°C. How is this possible?

45. A piece of metal and an equal mass of wood are both removed from a hot oven at equal temperatures and dropped onto blocks of ice. The metal has a lower specific heat capacity than the wood. Which will melt more ice before cooling to 0°C?

46. Why is it that, in cold winters, a tub of water placed in a farmer's canning cellar helps prevent canned food from freezing?

47. Why will spraying fruit trees with water before a frost help to protect the fruit from freezing?

48. Why does a hot dog pant?

PROBLEMS

● BEGINNER ■ INTERMEDIATE ◆ EXPERT

1. ● The specific heat capacity of ice is about 0.5 cal/g · °C. Suppose it remains at that value all the way to absolute zero. Show that the heat required to change a 1-gram ice cube at absolute zero (−273°C) to 1 gram of boiling water is 320 calories.

2. ● A small block of ice at 0°C is subjected to 10 g of 100°C steam and melts completely. Show that the mass of the block of ice can be no more than 80 grams.

3. ■ A 10-kg iron ball is dropped onto a pavement from a height of 100 m. Suppose half of the heat generated goes into warming the ball. Show that the temperature increase of the ball is 1.1°C. (In SI units, the specific heat capacity of iron is 450 J/kg · °C.) Why is the answer the same for an iron ball of any mass?

4. ■ A block of ice at 0°C is dropped from a height that causes it to completely melt upon impact. Assume that there is no air resistance and that all the energy goes into

melting the ice. Show that the height necessary for this to occur is at least 34 km. [Hint: Equate the joules of gravitational potential energy to the product of the mass of ice and its heat of fusion (in SI units, 335,000 J/kg). Do you see why the answer doesn't depend on mass?]

5. ■ Fifty grams of hot water at 80°C is poured into a cavity in a very large block of ice at 0°C. The final temperature of the water in the cavity is then 0°C. Show that the mass of ice that melts is 50 grams.

6. ■ A 50-gram chunk of 80°C iron is dropped into a cavity in a very large block of ice at 0°C. Show that the mass of ice that melts is 5.5 grams. (The specific heat capacity of iron is 0.11 cal/g · °C.)

7. ■ The heat of vaporization of ethyl alcohol is about 200 cal/g. Show that if 2 kg of this refrigerant were allowed to vaporize in a refrigerator, it could freeze 5 kg of 0°C water to ice.

CHAPTER 9 ONLINE RESOURCES

The Physics Place

Interactive Figures
9.13

Videos
The Secret to Walking on Hot Coals
Air Is a Poor Conductor
Condensation Is a Warming Process
Boiling Is a Cooling Process

Pressure Cooker: Boiling and Freezing at the Same Time

Quiz

Flashcards

Links

PART THREE

Electricity and Magnetism

How intriguing that this magnet outpulls the whole world when it lifts these nails. The pull between the nails and the Earth I call a **gravitational force,** and the pull between the nails and the magnet I call a **magnetic force.** I can name these forces, but I don't yet understand them. My learning begins by realizing there's a big difference in knowing the names of things and really understanding those things.

CHAPTER 10

Static and Current Electricity

New Zealand physics instructor David Housden constructs a parallel circuit by fastening lamps to extended terminals of a common battery. He asks his class to predict the relative brightnesses of two identical lamps in one wire about to be connected in parallel.

Electricity underlies just about everything around us. It's in the lightning from the sky, it's in the spark when we strike a match, and it's what holds atoms together to form molecules. The control of electricity is evident in technological devices of many kinds, from lamps to computers. More than the physics we've studied thus far, an understanding of electricity requires a step-by-step approach, for one concept is the building block for the next. So please put in extra care in the study of this material. It can be difficult, confusing, and frustrating if you're hasty; but, with careful effort, it can be comprehensible and rewarding. We start with static electricity, electricity at rest, and complete the chapter with current electricity. Let's begin.

10.1 Electric Force and Charge

What if there were a universal force that, like gravity, varies inversely as the square of the distance but that is billions upon billions of times stronger? If there were such a force, and if it were an attractive force like gravity, the universe would be pulled together into a tight ball with all matter pulled as close together as physically possible. But suppose this force were a repelling force, with every bit of matter repelling every other bit of matter. What then? The universe would be an ever-expanding gaseous cloud. Suppose, however, that the universe consisted of two kinds of particles—say, positive and negative. Suppose positives repelled positives but attracted negatives, and that negatives repelled negatives but attracted positives. Like kinds repel and unlike kinds attract (Figure 10.1). Further, suppose there were equal numbers of each so that this strong force was perfectly balanced. What would the universe be like? The answer is simple: it would be like the one we are living in. For there are such particles and there is such a force. We call it *electrical force*.

The Physics Place
Electrostatics

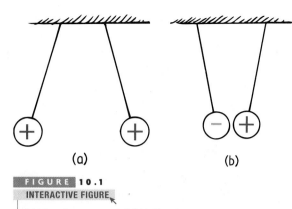

FIGURE 10.1
INTERACTIVE FIGURE

(a) Like charges repel. (b) Unlike charges attract.

FIGURE 10.2
INTERACTIVE FIGURE

Model of a helium atom. The atomic nucleus is made up of two protons and two neutrons. The positively charged protons attract two negatively charged electrons. What is the net charge of this atom?

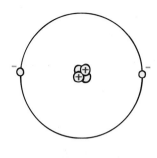

FIGURE 10.3

Electrons are transferred from the fur to the rod. The rod is then negatively charged. Is the fur charged? By how much, compared with the rod? Positively or negatively?

The terms *positive* and *negative* refer to electric charge, the fundamental quantity that underlies all electrical phenomena. The positively charged particles in ordinary matter are protons, and the negatively charged particles are electrons. The attractive force between these particles causes them to clump together into incredibly small units—atoms. (Atoms also contain neutral particles called *neutrons*.) We discussed atoms in Chapter 2. A review of some fundamental facts about atoms:

1. Every atom is composed of a positively charged nucleus surrounded by negatively charged electrons.

2. Each of the electrons in any atom has the same quantity of negative charge and the same mass. Electrons are identical to one another.

3. Protons and neutrons compose the nucleus. (The most common form of hydrogen atom, which has no neutrons, is the only exception.) Protons are about 1800 times more massive than electrons, but each one carries an amount of positive charge equal to the negative charge of electrons. Neutrons have slightly more mass than protons and have no net charge.

4. Atoms usually have as many electrons as protons, so the atom has zero *net* charge.

Which charges are called positive and which negative is the result of a choice made by Benjamin Franklin. It could easily have been the other way around.

When an atom loses one or more electrons, it has a positive net charge, and when it gains one or more electrons, it has a negative net charge. A charged atom is called an *ion*. A *positive ion* has a net positive charge. A *negative ion*, with one or more extra electrons, has a net negative charge.

Material objects are made of atoms, which means they are composed of electrons and protons (and neutrons as well). Although the innermost electrons in an atom are attracted very strongly to the oppositely charged atomic nucleus, the outermost electrons of many atoms are attracted more loosely and can easily be dislodged. The amount of work required to pull an electron away from an atom varies for different substances. Plastic wrap becomes electrically charged as it is drawn from its container, which is why it is attracted to plastic containers. Electrons are held more firmly in rubber or plastic than in your hair, for example. Thus, when you rub a comb against your hair, electrons transfer from the hair to the comb. The comb then has an excess of electrons and is said to be *negatively charged*. Your hair, in turn, has a deficiency of electrons and is said to be positively charged. If you rub a glass or plastic rod with silk, you'll find that the rod becomes positively charged. The silk has a greater affinity for electrons than the glass or plastic rod. Electrons are rubbed off the rod and onto the silk.

FIGURE 10.4

Why will you get a slight shock from the doorknob after scuffing across the carpet?

So protons attract electrons and we have atoms. Electrons repel electrons and we have matter—because atoms don't mesh into one another. This pair of rules is the guts of electricity.

fyi

Static electricity is a problem at gasoline pumps. Even the tiniest of sparks ignites vapors coming from the gasoline and causes fires—frequently lethal. A good rule is to touch metal and discharge static charge from your body before you fuel up. Also, don't use a cell phone when fueling up.

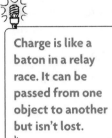

Charge is like a baton in a relay race. It can be passed from one object to another but isn't lost.

Conservation of Charge

Another basic rule is that, whenever something is charged, no electrons are created or destroyed. Electrons are simply transferred from one material to another. Charge is conserved. In every event, whether large scale or at the atomic and nuclear level, the principle of *conservation of charge* has always been found to apply. No case of the creation or destruction of net electric charge has ever been found. The conservation of charge ranks with the conservation of energy and momentum as a significant fundamental principle in physics.

STOP AND CHECK YOURSELF

If you walk across a rug and scuff electrons from your feet, are you negatively or positively charged?

CHECK YOUR ANSWER

You have fewer electrons after you scuff your feet, so you are positively charged (and the rug is negatively charged).

10.2 Coulomb's Law

The electrical force, like gravitational force, decreases inversely as the square of the distance between charges. This relationship, which was discovered by Charles Coulomb in the eighteenth century, is called **Coulomb's law**. It states that, for two charged objects that are much smaller than the distance between them, the force between them varies directly as the product of their charges and inversely as the square of the separation distance. The force acts along a straight line from one charge to the other. Coulomb's law can be expressed as

$$F = k\frac{q_1 q_2}{d^2}$$

SCIENCE AND SOCIETY

■ **ELECTRONICS TECHNOLOGY AND SPARKS**

Electric charge can be dangerous. Two hundred years ago, young boys called powder monkeys ran barefoot below the decks of warships to bring sacks of black gunpowder to the cannons above. It was ship law that this task be done barefoot. Why? Because it was important that no static charge build up on the powder that landed on their bodies as they ran to and fro. Bare feet scuffed the decks much less than shoes and assured no charge accumulation that might produce an igniting spark and an explosion.

Static charge is a danger in many industries today—not because of explosions, but because delicate electronic circuits may be destroyed by static charges. Some circuit components are sensitive enough to be "fried" by sparks of static electricity. Electronics technicians frequently wear clothing of special fabrics with ground wires between their sleeves and their socks. Some wear special wristbands that are connected to a grounded surface so that static charges will not build up—when moving a chair, for example. The smaller the electronic circuit, the more hazardous are sparks that may short-circuit the circuit elements.

■ IONIZED BRACELETS: SCIENCE OR PSEUDOSCIENCE?

Surveys indicate that the vast majority of Americans today believe ionized bracelets can reduce joint or muscle pain. Manufacturers make the claim that ionized bracelets relieve such pain. Are they correct? In 2002, the claim was put to a test by researchers at Mayo Clinic in Jacksonville, Florida, who randomly assigned 305 participants to wear an ionized bracelet for 28 days and another 305 participants to wear a placebo bracelet for the same duration. The study volunteers were men and women 18 and older who had self-reported musculoskeletal pain at the beginning of the study.

Neither the researchers nor the participants knew which volunteers wore an ionized bracelet and which wore a placebo bracelet. Both types of bracelets were identical, were supplied by the manufacturer, and were worn according to the manufacturer's recommendations. Interestingly, both groups reported significant relief from pain. No difference was found in the amount of self-reported pain relief between the group wearing the ionized bracelets and the group wearing the placebo bracelets. Apparently, just believing that the bracelet relieves pain does the trick!

Interestingly, the brain initiates the creation of endorphins (which bind to opiate receptor sites) when the person expects to get relief from pain. The placebo effect is very real and measurable via blood titrations. So there's some merit in the old adage that wishing hard for something will make it come true. But this has nothing to do with the physics, chemistry, or biological interaction with the bracelet. Hence, ionized bracelets join the ranks of pseudoscientific devices.

In any society that thrives more on capturing attention than on informing, pseudoscience is big business.

---

where d is the distance between the charged particles, q_1 represents the quantity of charge of one particle, q_2 represents the quantity of charge of the second particle, and k is the proportionality constant.

The unit of charge is called the **coulomb**, abbreviated C. It turns out that a charge of 1 C is the charge associated with 6.25 billion billion electrons. This might seem like a great number of electrons, but it only represents the amount of charge that flows through a common 100-watt lightbulb in a little more than a second.

> There are about 10^{24} electrons in a penny, all repelling one another. Why don't these electrons fly off the coin?

The proportionality constant k in Coulomb's law is similar to G in Newton's law of gravity. Instead of being a very small number, like G, k is a very large number, approximately

$$k = 9{,}000{,}000{,}000 \text{ N} \cdot \text{m}^2/\text{C}^2$$

In scientific notation, $k = 9.0 \times 10^9 \text{ N} \cdot \text{m}^2/\text{C}^2$. The unit $\text{N} \cdot \text{m}^2/\text{C}^2$ is not central to our interest here; it simply converts the right-hand side of the equation to the unit of force, the newton (N). What is important is the large magnitude of k. If, for example, a pair of like charges of 1 coulomb each were 1 meter apart, the force of repulsion between the two would be 9 billion newtons.* That would be about ten times the weight of a battleship! Obviously, such quantities of net charge do not usually exist in our everyday environment.

So Newton's law of gravitation for masses is similar to Coulomb's law for electrically charged bodies. The most important difference between gravitational and electrical forces is that electrical forces may be either attractive or repulsive, whereas gravitational forces are only attractive. Coulomb's law underlies the bonding forces between molecules that are essential in the field of chemistry.

---

* Contrast this to the gravitational force of attraction between two 1-kg masses 1 m apart: 6.67×10^{-11} N. This is an extremely small force. For the force to be 1 N, the masses at 1 m apart would have to be nearly 123,000 kg each! Gravitational forces between ordinary objects are exceedingly small, and differences in electrical forces between ordinary objects can be exceedingly huge. We don't sense them because the positives and negatives normally balance out, and, even for highly charged objects, the imbalance of electrons to protons is normally less than one part in a trillion trillion.

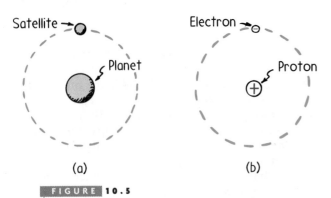

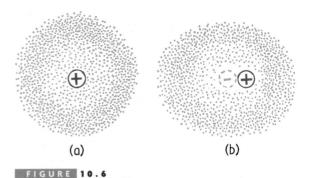

FIGURE 10.5

(a) A gravitational force holds the satellite in orbit about the planet, and (b) an electrical force holds the electron in orbit about the proton. In both cases, there is no contact between the bodies. We say that the orbiting bodies interact with the force fields of the planet and proton and are everywhere in contact with these fields. Thus, the force that one electric charge exerts on another can be described as the interaction between one charge and the field set up by the other.

FIGURE 10.6

(a) The center of the negative "cloud" of electrons coincides with the center of the positive nucleus in an atom. (b) When an external negative charge is brought nearby to the right, as on a charged balloon, the electron cloud is distorted so that the centers of negative and positive charge no longer coincide. The atom is electrically polarized.

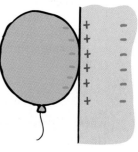

FIGURE 10.7

The negatively charged balloon polarizes molecules in the wooden wall and creates a positively charged surface, so the balloon sticks to the wall.

STOP AND CHECK YOURSELF

1. The proton is the nucleus of the hydrogen atom, and it attracts the electron that orbits it. Relative to this force, does the electron attract the proton with less force, more force, or the same amount of force?

2. If a proton at a particular distance from a charged particle is repelled with a given force, by how much will the force decrease when the proton is three times as distant from the particle? Five times as distant?

3. What is the sign of charge of the particle in this case?

CHECK YOUR ANSWERS

1. The same amount of force, in accord with Newton's third law—basic mechanics! Recall that a force is an interaction between two things—in this case, between the proton and the electron. They pull on each other equally.

2. In accord with the inverse-square law, it decreases to 1/9 its original value. To 1/25 of its original value.

3. Positive.

Charge Polarization

If you charge an inflated balloon by rubbing it on your hair and then place the balloon against a wall, it sticks. This is because the charge on the balloon alters the charge distribution in the atoms or molecules in the wall, effectively inducing an opposite charge on the wall. The molecules cannot move from their relatively stationary positions, but their "centers of charge" are moved. The positive part of the atom or molecule is attracted toward the balloon, whereas the negative part is repelled. This has the effect of distorting the atom or molecule (Figure 10.6). The atom or molecule is said to be **electrically polarized**.

STOP AND CHECK YOURSELF

You know that a balloon rubbed on your hair will stick to a wall. In a humorous vein, does it follow that your oppositely charged head would also stick to the wall?

CHECK YOUR ANSWER

No, unless you're an airhead (having a head mass about the same as that of an air-filled balloon). The force that holds a balloon to the wall cannot support your heavier head.

■ **MICROWAVE OVEN**

Imagine an enclosure filled with Ping-Pong balls among a few batons, all at rest. Now imagine that the batons suddenly rotate backward and forward, striking neighboring Ping-Pong balls. Almost immediately, most of the Ping-Pong balls are energized, vibrating in all directions. A microwave oven works similarly. The batons are water molecules made to rotate to and fro in rhythm with microwaves in the enclosure. The Ping-Pong balls are the other molecules that make up the bulk of material being cooked.

H_2O molecules are electrically polarized, with opposite charges on opposite sides. When an electric field is imposed on them, they align with the field like a compass needle aligns with a magnetic field. When the field is made to oscillate, the H_2O molecules oscillate also—and quite ener-

getically when the frequency of the waves matches the natural rotational frequency of the H_2O. So food is cooked by converting H_2O molecules into flip-flopping energy sources that impart thermal motion to surrounding food molecules. Without polar molecules in the food, a microwave oven wouldn't work. That's why microwaves pass through foam, paper, or ceramic plates and reflect from metals with no effect. However, they do energize water molecules.

A note of caution is due when boiling water in a microwave oven. Water can sometimes heat faster than bubbles can form, and the water then heats beyond its boiling point—it becomes superheated. If the water is bumped or jarred just enough to cause the bubbles to form rapidly, they'll violently expel the hot water from its container. More than one person has had boiling water blast into his or her face.

10.3 Electric Field

Electrical forces, like gravitational forces, can act between things that are not in contact with each other. Both for electricity and gravity, a force field exists that influences distant charges and masses, respectively. The properties of space surrounding any mass are altered such that another mass introduced to this region experiences a force. This "alteration in space" is called its *gravitational field*. We can think of any other mass as interacting with the field and not directly with the mass that produces it. For example, when an apple falls from a tree, we say it is interacting with the mass of Earth, but we can also think of the apple as interacting with the gravitational field of Earth. It is common to think of distant rockets and the like as interacting with gravitational fields rather than bodies responsible for the fields. The field plays an intermediate role in the force between bodies. More important, the field stores energy. So similar to a gravitational field, the space around every electric charge is energized with an **electric field**— an energetic aura that extends through space.*

An electric field is nature's storehouse of electrical energy.

* An electric field is a vector quantity, having both magnitude and direction. The magnitude of the field at any point is simply the force per unit of charge. If a charge q experiences a force F at some point in space, then the electric field E at that point is $E = F/q$.

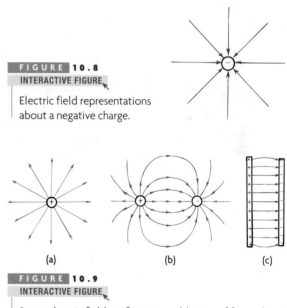

FIGURE 10.8
INTERACTIVE FIGURE

Electric field representations about a negative charge.

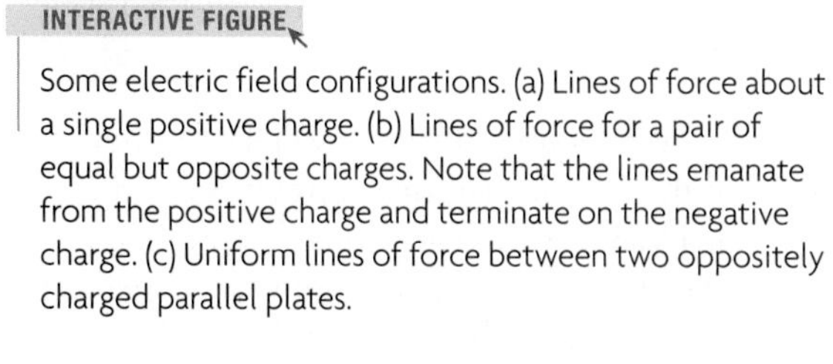

(a)   (b)   (c)

FIGURE 10.9
INTERACTIVE FIGURE

Some electric field configurations. (a) Lines of force about a single positive charge. (b) Lines of force for a pair of equal but opposite charges. Note that the lines emanate from the positive charge and terminate on the negative charge. (c) Uniform lines of force between two oppositely charged parallel plates.

If you place a charged particle in an electric field, it will experience a force. The direction of the force on a positive charge is the same direction as the field. The electric field about a proton extends radially from the proton. About an electron, the field is in the opposite direction (Figure 10.8). As with electric force, the electric field about a particle obeys the inverse-square law. Some electric field configurations are shown in Figure 10.9, and photographs of field patterns are shown in Figure 10.10. In the next chapter, we'll see how bits of iron similarly align with magnetic fields.

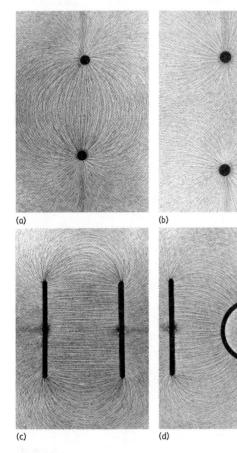

(a)   (b)

(c)   (d)

FIGURE 10.10

Bits of thread suspended in an oil bath line up end-to-end along the direction of the field. (a) Equal and opposite charges. (b) Equal like charges. (c) Oppositely charged plates. (d) Oppositely charged cylinder and plate.

Perhaps your instructor will demonstrate the effects of the electric field that surrounds the charged dome of a Van de Graaff generator (Figure 10.11). Charged objects in the field of the dome are either attracted or repelled, depending on their sign of charge.

Static charge on the surface of any electrically conducting surface will arrange itself such that the electric field inside the conductor will cancel to zero. Note the randomness of threads inside the cylinder of Figure 10.10d, where no field exists.

FIGURE 10.11

Both Lori and the spherical dome of the Van de Graaff generator are electrically charged.

fyi

Whatever the intensity of the electric field about a charged Van de Graaff generator, the electric field inside the dome cancels to zero. This is true for the interiors of all metals that carry static charge.

STOP AND CHECK YOURSELF

Both Lori and the dome of the Van de Graaff generator in Figure 10.11 are charged. Why does Lori's hair stand out?

CHECK YOUR ANSWER

She and her hair are charged. Each hair is repelled by others around it—evidence that *like charges repel*. Even a small charge produces an electrical force greater than the weight of strands of hair. Fortunately, the electrical force is not great enough to make her arms stand out!

10.4 Electric Potential

In our study of energy in Chapter 4, we learned that an object has gravitational potential energy because of its location in a gravitational field. Similarly, a charged object has potential energy by virtue of its location in an electric field. Just as work is required to lift a massive object against the gravitational field of Earth, work is required to push a charged particle against the electric field of a charged body. This work changes the electric potential energy of the

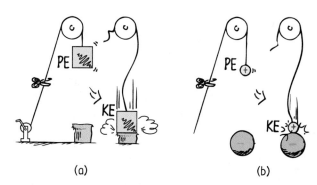

FIGURE 10.12

(a) The PE (gravitational potential energy) of a mass held in a gravitational field. (b) The PE of a charged particle held in an electric field. When the mass and particle are released, how does the KE (kinetic energy) acquired by each compare with the decrease in PE?

FIGURE 10.13

(a) The spring has more elastic PE when compressed. (b) The small charge similarly has more PE when pushed closer to the charged sphere. In both cases, the increased PE is the result of work input.

The Physics Place
Electric Potential

charged particle.* Similarly, work done in compressing a spring increases the potential energy of the spring (Figure 10.13a). Likewise, the work done in pushing a charged particle closer to the charged sphere in Figure 10.13b increases the potential energy of the charged particle. We call the energy possessed by the charged particle that is due to its location electric potential energy. If the particle is released, it accelerates in a direction away from the sphere, and its **electric potential energy** changes to kinetic energy.

If we push a particle with twice the charge, we do twice as much work. Twice the charge in the same location has twice the electric potential energy; with three times the charge, there is three times as much

* This work is positive if it increases the electric potential energy of the charged particle and negative if it decreases it.

potential energy; and so on. When working with electricity, rather than dealing with the total potential energy of a charged body, it is convenient to consider the electric potential energy *per charge*. We simply divide the amount of energy in any case by the amount of charge. The concept of potential energy per charge is called electric potential; that is,

$$\text{electric potential} = \frac{\text{electric potential energy}}{\text{amount of charge}}$$

The unit of measurement for electric potential is the volt, so electric potential is often called *voltage*. A potential of 1 volt (V) equals 1 joule (J) of energy per 1 coulomb (C) of charge.

$$1 \text{ volt} = \frac{1 \text{ joule}}{1 \text{ coulomb}}$$

Thus, a 1.5-volt battery gives 1.5 joules of energy to every 1 coulomb of charge flowing through the battery. *Electric potential* and *voltage* are the same thing, and they are commonly used interchangeably.

In a nutshell: *electric potential* and *potential* mean the same thing—electrical potential energy per unit charge—in units of volts. On the other hand, *potential difference* is the same as *voltage*—the *difference* in electrical potential between two points—also in units of volts.

The significance of voltage is that a definite value for it can be assigned to a location. We can speak about the voltages at different locations in an electric field whether or not charges occupy those locations. The same is true of voltages at various locations in an electric circuit. Later in this chapter, we will see that the location of the positive terminal of a 12-volt battery is maintained at a voltage 12 volts higher than the location of the negative terminal. When a conducting medium connects this voltage difference, any charges in the medium will move between these locations.

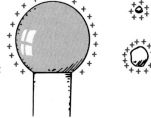

FIGURE 10.14

The larger test charge has more PE in the field of the charged dome, but the electric potential of any amount of charge at the same location is the same.

**STOP AND
CHECK YOURSELF**

1. If there were twice as many coulombs in the test charge near the charged sphere in Figure 10.14, would the *electric potential energy* of the test charge relative to the charged sphere be the same, or would it be twice as great? Would the *electric potential* of the test charge be the same, or would it be twice as great?

2. What does it mean to say that the battery in your car is rated at 12 volts?

CHECK YOUR ANSWERS

1. The result of twice as many coulombs is twice as much *electric potential energy* because it takes twice as much work to put the charge there. But the *electric potential* would be the same. Twice the energy divided by twice the charge gives the same potential as one unit of energy divided by one unit of charge. Electric potential is not the same thing as electric potential energy. Be sure you understand this before you study further.

2. It means that one of the battery terminals is 12 V higher in potential than the other one. We'll soon learn that when a circuit is connected between these terminals, each coulomb of charge in the resulting current will be given 12 J of energy as it passes through the battery (and 12 J of energy "spent" in the circuit).

Rub a balloon on your hair, and the balloon becomes negatively charged—perhaps to several thousand volts! That would be several thousand joules of energy, if the charge were 1 coulomb. However, 1 coulomb is a fairly respectable amount of charge. The charge on a balloon rubbed on hair is typically much less than a millionth of a coulomb. Therefore, the amount of energy associated with the charged balloon is very, very small. A high voltage means a lot of energy only if a lot of charge is involved. Electrical potential energy differs from electric potential (or voltage).

High voltage at low energy is similar to the harmless high-temperature sparks emitted by a fireworks sparkler. Recall that temperature is average kinetic energy per molecule, which means total energy is a lot only for lots of molecules. Similarly, high voltage means a lot of energy only for lots of charge.

FIGURE 10.15

Although the voltage of the charged balloon is high, the electric potential energy is low because of the small amount of charge.

5000 volts?

10.5 Voltage Sources

When the ends of a heat conductor are at different temperatures, heat energy flows from the higher temperature to the lower temperature. The flow ceases when both ends reach the same temperature. Any material having free charged particles that easily flow through it when an electric force acts on them is called an electric **conductor**. Both heat and electric conductors are characterized by electric charges that are free to move. Similar to heat flow, when the ends of an electrical conductor are at different electric potentials—when there is a **potential difference**—charges in the conductor flow from the higher potential to the lower potential. The flow of charge persists until both ends reach the same potential. Without a potential difference, no flow of charge will occur.

The Physics Place
Van de Graff Generator

A battery doesn't supply electrons to a circuit; it instead supplies energy to electrons that already exist in the circuit.

FIGURE 10.16

Although the Wimshurst machine can generate thousands of volts, it puts out no more energy than the work that Jim Stith puts into it by cranking the handle.

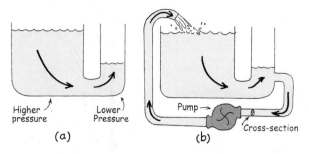

FIGURE 10.17

(a) Water flows from the reservoir of higher pressure to the reservoir of lower pressure. The flow will cease when the difference in pressure ceases. (b) Water continues to flow because a difference in pressure is maintained with the pump.

To attain a sustained flow of charge in a conductor, some arrangement must be provided to maintain a difference in potential while charge flows from one end to the other. The situation is analogous to the flow of water from a higher reservoir to a lower one (Figure 10.17a). Water will flow in a pipe that connects the reservoirs only as long as a difference in water level exists. The flow of water in the pipe, like the flow of charge in a wire, will cease when the pressures at each end are equal. (We imply this phenomenon when we say that water seeks its own level.) A continuous flow is possible if the difference in water levels—hence the difference in water pressures—is maintained with the use of a suitable pump (Figure 10.17b).

A sustained electric current requires a suitable pumping device to maintain a difference in electric potential—to maintain a voltage. Chemical batteries or generators are "electrical pumps" that can maintain a steady flow of charge. These devices do work to pull negative charges apart from positive ones. In chemical batteries, this work is done by the

FIGURE 10.18

An unusual source of voltage. The electric potential between the head and tail of the electric eel (*Electrophorus electricus*) can be up to 650 V.

chemical disintegration of zinc or lead in acid, and the energy stored in the chemical bonds is converted to electric potential energy.

> *fyi*
>
> **Chemical batteries don't respond well to sudden surges of charge. An alternative that does respond well to spurts of energy input is a spinning flywheel. Unlike the ones used by potters for spinning clay, modern flywheels are lightweight composite materials that are stronger and can be spun at high speeds without coming apart. Rotational kinetic energy is then converted to other forms of energy. Watch for flywheels as energy-storing devices.**

Generators separate charge by electromagnetic induction, a process we will describe in the next chapter. The work that is done (by whatever means) in separating the opposite charges is available at the terminals of the battery or generator. This energy per charge provides the difference in potential (voltage) that provides the "electrical pressure" to move electrons through a circuit joined to those terminals.

> *fyi*
>
> **When a common automobile battery provides an electrical pressure of 12 volts to a circuit connected across its terminals, 12 joules of energy are supplied to each coulomb of charge that is made to flow in the circuit.**

10.6 Electric Current

Just as a water current is a flow of H_2O molecules, **electric current** is a flow of charged particles. In circuits of metal wires, electrons make up the flow of charge. One or more electrons from each metal atom are free to move throughout the atomic lattice. These charge carriers are called *conduction electrons*. Protons, on the other hand, do not move in a solid because they are bound within the nuclei of atoms that are more or less locked in fixed positions. In fluids, however, positive ions as well as electrons may constitute the flow of an electric charge.

The Physics Place
Alternating Current
Electric Current

FIGURE 10.19

Each coulomb of charge that is made to flow in a circuit that connects the ends of this 1.5-V flashlight cell is energized with 1.5 J.

An important difference between water flow and electron flow has to do with their conductors. If you purchase a water pipe at a hardware store, the clerk doesn't sell you the water to flow through it. You provide that yourself. By contrast, when you buy an "electron pipe," an electric wire, you also get the electrons. Every bit of matter, wires included, contains enormous numbers of electrons that swarm about in random directions. When a source of voltage sets them moving, we have an electric current.

The *rate* of electrical flow is measured in *amperes*. An **ampere** is the rate of flow of 1 coulomb of charge per second. (That's a flow of 6.25 billion billion electrons per second.) In a wire that carries 4 amperes to a car headlight bulb, for example, 4 coulombs of charge flow past any cross section in the wire each second. In a wire that carries 8 amperes, twice as many coulombs flow past any cross section each second.

It is interesting to note that the speed of electrons as they drift through a wire is surprisingly slow. This is because electrons continually bump into atoms in

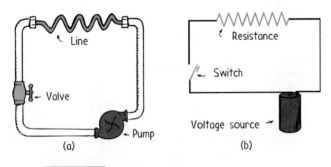

FIGURE 10.20

Analogy between (a) a simple hydraulic circuit and (b) an electrical circuit. Much effort is expended in building particle accelerators that accelerate electrons to speeds approaching the speed of light. If electrons in a common circuit were to travel that fast, one would only have to bend a wire at a sharp angle to cause those high-momentum electrons to fail to make the turn and to fly off into the air. There'd be no need for accelerators! In fact, electrons in circuits move fairly slowly.

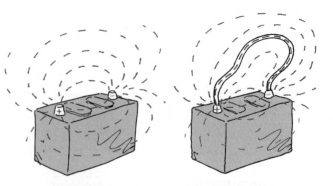

FIGURE 10.21

The electric field lines between the terminals of a battery are directed through a conductor, which joins the terminals. A thick metal wire is shown here, but the path from one terminal to the other is usually an electric circuit. (If you touch this conducting wire, you won't be shocked, but the wire will heat quickly and may burn your hand!)

the wire. The net speed, or *drift speed*, of electrons in a typical circuit is much less than one centimeter per second. The electric signal, however, travels at nearly the speed of light. That's the speed at which the electric *field* in the wire is established.

fyi

> **The danger from car batteries is not so much electrocution as it is explosion. If you touch both terminals with a metal wrench, for instance, you can create a spark that can ignite hydrogen gas in the battery and send pieces of battery and acid flying!**

Also interesting is that a current-carrying wire is not electrically charged. Under ordinary conditions, there are as many conduction electrons swarming through the atomic lattice as there are positively charged atomic nuclei. The numbers of electrons and protons balance, so, whether a wire carries a current or not, the net charge of the wire is normally zero at every moment.

There is often some confusion between charge flowing *through* a circuit and voltage placed, or impressed, *across* a circuit. We can distinguish between these ideas by considering a long pipe filled with water. Water will flow through the pipe if there is a difference in pressure across (or between) its ends. Water flows from the high-pressure end to the low-pressure end. Only the water flows, not the pressure. Similarly, electric charge flows because of the differences in electrical pressure (voltage). You say that *charges* flow through a circuit because of an applied voltage across the circuit. You don't say that *voltage* flows through a circuit. Voltage doesn't go anywhere, for it is the charges that move. Voltage produces current (if there is a complete circuit).

> We often think of current flowing through a circuit, but don't say this around somebody who is picky about grammar, for the expression "current flows" is redundant. More properly, charge flows—which *is* current.

Direct Current and Alternating Current

Electric current may be dc or ac. By dc, we mean **direct current**, which refers to charges flowing in one direction. A battery produces direct current in a circuit because the terminals of the battery always have the same sign. Electrons move from the repelling negative terminal toward the attracting positive terminal, and they always move through the circuit in the same direction.

Alternating current (ac) acts as the name implies. Electrons in the circuit are moved first in one direction and then in the opposite direction, alternating to and fro about relatively fixed positions. This is accomplished in a generator or alternator by periodically switching the sign at the terminals. Nearly all commercial ac circuits involve currents that alternate back and forth at a frequency of 60 cycles per second. This is 60-hertz current (one cycle per second is called a *hertz*). In some countries, 25-hertz, 30-hertz, or 50-hertz current is used. Throughout the world, most residential and

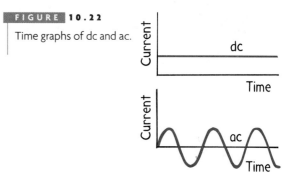

FIGURE 10.22

Time graphs of dc and ac.

commercial circuits are ac because electric energy in the form of ac can easily be stepped up to high voltage to be transmitted great distances with small heat losses, then stepped down to convenient voltages where the energy is consumed. Why this occurs is quite fascinating, and it will be touched on in the next chapter. The rules of electricity in this chapter apply to both dc and ac.

> **fyi**
>
> Conversion from ac to dc is accomplished with an electronic device that allows electron flow in one direction only—a *diode*. A more familiar type diode is the *light-emitting diode* (LED). Photons are emitted when electrons cross a "band gap" in the device. The photon energy most often corresponds to the frequency of red light, which is why most LEDs emit red light. You see these on instrument panels of many kinds, including VCRs and DVD players. Interestingly, when electrical input and light output are reversed, the resulting device is a solar cell!

EVERYDAY APPLICATIONS

■ HISTORY OF 110 VOLTS

In the early days of electrical lighting, high voltages burned out electric light filaments, so low voltages were more practical. The hundreds of power plants built in the United States prior to 1900 adopted 110 volts (or 115 or 120 volts) as their standard. The tradition of 110 volts was decided upon because it made the bulbs of the day glow as brightly as a gas lamp. By the time electrical lighting became popular in Europe, engineers had figured out how to make lightbulbs

that would not burn out so fast at higher voltages. Power transmission is more efficient at higher voltages, so Europe adopted 220 volts as their standard. The United States remained with 110 volts (today, it is officially 120 volts) because of the initial huge expense in the installation of 110-volt equipment. Interestingly, in ac circuits 120 volts is the "root-mean-square" average of the voltage. The actual voltage in a 120-volt ac circuit varies between $+170$ volts and -170 volts, delivering the same power to an iron or a toaster as a 120-volt dc circuit.

10.7 Electric Resistance

How much current is in a circuit depends not only on voltage but also on the **electrical resistance** of the circuit. Just as narrow pipes resist water flow more than wide pipes do, thin wires resist electrical current more than thicker wires do. And length contributes to resistance also. Just as long pipes have more resistance than short ones, long wires offer more electrical resistance. And most important is the material from which the wires were made. Copper has a low electrical resistance, while a strip of rubber has an enormous resistance. Temperature also affects electrical resistance: the greater the jostling of atoms within a conductor (the higher the temperature), the greater its resistance. The resistance of some materials reaches zero at very low temperatures. These materials are referred to as *superconductors*.

Electrical resistance is measured in units called *ohms*. The Greek letter *omega*, Ω, is commonly used as the symbol for the ohm. This unit was

> The unit of electrical resistance is the ohm, Ω. Like the song of old, "Ω, Ω on the Range."

> **fyi**
>
> Some materials, such as germanium or silicon, can be made to alternate between being conductors and insulators. These are *semiconductors*. Between pairs of them the transfer of an electron through their junction can cause emission of light, as in a light-emitting diode (LED). Or, conversely, the absorption of light can lead to an electric current, as in a solar cell.

named after Georg Simon Ohm, a German physicist who, in 1826, discovered a simple and very important relationship among voltage, current, and resistance.

10.8 Ohm's Law

The relationship between voltage, current, and resistance is summarized by a statement called **Ohm's law**. Ohm discovered that the amount of current in a circuit is directly proportional to the voltage established across the circuit and is inversely proportional to the resistance of the circuit:

$$\text{Current} = \frac{\text{voltage}}{\text{resistance}}$$

Or, in units form,

$$\text{Amperes} = \frac{\text{volts}}{\text{ohms}}$$

> **The Physics Place**
> Ohm's Law
> Caution on Handling Electric Wires
> Birds and High-Voltage Wires

So, for a given circuit of constant resistance, current and voltage are proportional to each other.* This means we'll get twice the current for twice the voltage: The greater the voltage, the greater the current. But, if the resistance is doubled for a circuit, the current will be half what it would have been otherwise: The greater the resistance, the smaller the current. Ohm's law makes good sense.

The resistance of a typical lamp cord is much less than 1 ohm, and a typical lightbulb has a resistance of more than 100 ohms. An iron or electric toaster has a resistance of 15 to 20 ohms. The current inside these and all other electrical devices is regulated by circuit elements called resistors (Figure 10.24), whose resistance may be a few ohms or millions of ohms. Resistors heat up when current flows through them, but for small currents the heating is slight.

*Many texts use *V* as the symbol for voltage, *I* for current, *R* for resistance, and express Ohm's law as $V = IR$. It then follows that $I = V/R$, or $R = V/I$, so that, if any two variables are known, the third can be found. (The names of the units are often abbreviated: V for volts, A for amperes, and Ω (the capital Greek letter omega) for ohms.)

FIGURE 10.23

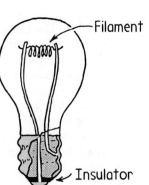

The conduction electrons that surge to and fro in the filament of the lamp do not come from the voltage source. They are within the filament to begin with. The voltage source simply provides them with surges of energy. When switched on, the resistance of the very thin tungsten filament heats up to 3000°C and roughly doubles its resistance.

Filament

Insulator

PROBLEM SOLVING

Problems

1. How much current flows through a lamp with a resistance of 60 Ω when the voltage across the lamp is 12 V?

2. What is the resistance of a toaster that draws a current of 12 A when connected to a 120-V circuit?

3. At 100,000 Ω, how much current will flow through your body if you touch the terminals of a 12-V battery?

4. If your skin is very moist, so that your resistance is only 1000 Ω, and you touch the terminals of a 12-V battery, how much current do you receive?

Solutions

1. From Ohm's law: Current $= \dfrac{\text{voltage}}{\text{resistance}} =$
$\dfrac{12\,\text{V}}{60\,\Omega} = 0.2\text{A}.$

2. Rearranging Ohm's law:
Resistance $= \dfrac{\text{voltage}}{\text{current}} = \dfrac{120\,\text{V}}{12\,\text{A}} = 10\,\Omega.$

3. Current $= \dfrac{\text{voltage}}{\text{resistance}} = \dfrac{12\,\text{V}}{100{,}000\,\Omega} = 0.00012\text{A}.$

4. Current $= \dfrac{\text{voltage}}{\text{resistance}} = \dfrac{12\,\text{V}}{1000\,\Omega} = 0.012\text{A}.$
Ouch!

Superconductors

In common household wiring, flowing electrons collide with atomic nuclei in the wire and convert their kinetic energy to thermal energy in the wire. Early twentieth-century investigators discovered that certain metals in a bath of liquid helium at 4 K lost all electrical resistance. The electrons in these conductors traveled pathways that avoided atomic collisions, permitting them to flow indefinitely. These materials are called **superconductors**, having zero electrical resistance to the flow of charge. No current is lost and no heat is generated in superconductivity. For decades, it was generally thought that zero electrical resistance could occur only in certain metals near absolute zero. Then, in 1986, superconductivity was achieved at 30 K, which spurred hopes of finding superconductivity above 77 K, the point at which nitrogen liquefies. Nitrogen is easier to handle than liquid helium, which is needed for creating colder conditions. The historic leap came in the following year with a nonmetallic compound that lost its resistance at 90 K.

Various ceramic oxides have since been found to be superconducting at temperatures above 100 K. These ceramic materials are "high-temperature" superconductors. High-temperature superconductor (HTS) cables, already in use, carry more current at a lower voltage, which means large power transformers can be located farther away from urban centers—allowing the development of green space. Watch for additional growth of HTS in delivering electric power.

fyi

The air inside a traditional lightbulb is a mixture of nitrogen and argon. As the tungsten filament is heated, minute particles of tungsten evaporate—much like steam leaving boiling water. Over time, these particles are deposited on the inner surface of the glass, causing the bulb to blacken. Losing its tungsten, the filament eventually breaks and the bulb "burns out." A remedy is to replace the air inside the bulb with a halogen gas, such as iodine or bromine. Then the evaporated tungsten combines with the halogen rather than depositing on the glass, which remains clear. Furthermore, the halogen–tungsten combination splits apart when it touches the hot filament, returning halogen as a gas while restoring the filament by depositing tungsten back onto it. This is why halogen lamps have such long lifetimes.

FIGURE 10.24

Resistors. The symbol of resistance in an electric circuit is ‐⋀⋀⋀‐.

Current is a flow of charge, pressured into motion by voltage, and hampered by resistance.

Electric Shock

The damaging effects of shock are the result of current passing through the human body. What causes electric shock in the body—current or voltage? From Ohm's law, we can see that this current depends on the voltage that is applied and also on the electrical resistance of the human body. The resistance of one's body depends on its condition, and it ranges from about 100 ohms, if it is soaked with saltwater, to about 500,000 ohms, if the skin is very dry. If we touch the two electrodes of a battery with dry fingers, completing the circuit from one hand to the other, we offer a resistance of about 100,000 ohms. We usually cannot feel 12 volts, and 24 volts just barely tingles. If our skin is moist, 24 volts can be quite uncomfortable. Table 10.1 describes the effects of different amounts of current on the human body.

To receive a shock, there must be a *difference* in electric potential between one part of your body and another part. Most of the current will pass along the path of least electrical resistance connecting these two points. Suppose you fell from a bridge and managed to grab onto a high-voltage power line, halting your fall. So long as you touch nothing else of different potential, you will receive no shock at all. Even if the wire is a few thousand volts above ground potential and you hang by it with two hands, no appreciable charge will flow from one hand to the other. This is because there is no appreciable difference in electric potential between your hands. If, however, you reach over with one hand and grab onto a wire of different potential . . . *zap*! We have all seen birds perched on high-voltage wires. Every part of their bodies is at the same high potential as the wire, so they feel no ill effects.

Interestingly, the source of electrons in the current that shocks you is your own body. As in all conductors, the electrons are already there. It is the

FIGURE 10.25

The bird can stand harmlessly on one wire of high potential, but it had better not reach over and touch a neighboring wire! Why not?

FIGURE 10.26

The third prong connects the body of the appliance directly to ground. Any charge that builds up on an appliance is therefore conducted to the ground.

energy given to the electrons that you should be wary of. They are energized when a voltage difference exists across different parts of your body.

Most electric plugs and sockets today are wired with three, instead of two, connections. The principal two flat prongs on an electrical plug are for the current-carrying double wire, one part "live" and the other neutral, while the third round prong is grounded—connected directly to the ground (Figure 10.26). Appliances such as irons, stoves, washing machines, and dryers are connected with these three wires. If the live wire accidentally comes in contact with the metal surface of the appliance, and you touch the appliance, you could receive a dangerous shock. This won't occur when the appliance casing is grounded via the ground wire, which assures that the appliance casing is at zero ground potential.

FIGURE 10.27

This table lamp has an insulating body and doesn't need the third (ground) wire.

TABLE 10.1

Effect of Electric Currents on the Body

Current	Effect
0.001 A	Can be felt
0.005 A	Is painful
0.010 A	Causes involuntary muscle contractions (spasms)
0.015 A	Causes loss of muscle control
0.070 A	Goes through the heart; causes serious disruption; probably fatal if current lasts for more than 1 s

■ INJURY BY ELECTRIC SHOCK

Many people are killed each year by current from common 120-volt electric circuits. If your hand touches a faulty 120-volt light fixture while your feet are on the ground, there's likely a 120-volt "electrical pressure" between your hand and the ground. Resistance to current is usually greatest between your feet and the ground, and so the current is usually not enough to do serious harm. But if your feet and the ground are wet, there is a low-resistance electrical path between you and the ground. The 120 volts across this lowered resistance may produce a harmful current in your body.

Pure water is not a good conductor. But the ions that are normally found in water make it a fair conductor. Dissolved materials in water, especially small quantities of salt, lower the resistance even more. There is usually a layer of salt remaining on your skin from perspiration, which, when wet, lowers your skin resistance to a few hundred ohms or less. Handling electrical devices while taking a bath is a definite no-no.

Injury by electric shock occurs in three forms: (1) burning of tissues by heating, (2) contraction of muscles, and (3) disruption of cardiac rhythm. These conditions are caused by the delivery of excessive power for too long a time in critical regions of the body.

Electric shock can upset the nerve center that controls breathing. In rescuing shock victims, the first thing to do is remove them from the source of the electricity. Use a dry wooden stick or some other nonconductor so that you don't get electrocuted yourself. Then apply artificial respiration. It is important to continue artificial respiration. There have been cases of victims of lightning who did not breathe without assistance for several hours, but who were eventually revived and who completely regained good health.

---

fyi

> **Myth:** Lightning never strikes the same place twice.
> **Fact:** Lightning does favor certain spots, mainly high locations. The Empire State Building is struck by lightning about 25 times every year.

STOP AND
CHECK YOURSELF

What causes electric shock—current or voltage?

CHECK YOUR ANSWER

Electric shock *occurs* when current is produced in the body, but the current is *caused* by an impressed voltage.

The Physics Place
Electric Circuits

---

fyi

> **All batteries degrade. The lithium—ion cells, popular in notebook computers, cameras, and cell phones, erode faster when highly charged and warm. So keep yours at about half charge in a cool or cold environment to extend battery life.**

10.9 Electric Circuits

Any path along which electrons can flow is a *circuit*. For a continuous flow of electrons, there must be a complete circuit with no gaps. A gap is usually provided by an electric switch that can be opened or closed to either cut off energy or to allow energy to flow. Most circuits have more than one device that receives electric energy. These devices are commonly connected in a circuit in one of two ways, in *series* or in *parallel*. When connected in series, they form a single pathway for electron flow between the terminals of the battery, generator, or wall outlet (which is simply an extension of these terminals). When connected in parallel, they form branches, each of which is a separate path for the flow of electrons. Both series and parallel connections have their own distinctive characteristics. In the following sections, we shall briefly discuss circuits using these two types of connections.

Series Circuits

A simple **series circuit** is shown in Figure 10.28. Three lamps are connected in series with a battery. The same current exists almost immediately in all three lamps when the switch is closed. The current does not "pile up" or accumulate in any lamp but flows *through* each lamp. Electrons that make up this current leave the negative terminal of the battery, pass through each of the resistive filaments in the

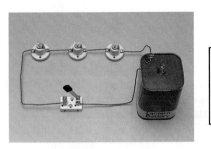 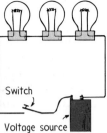

Switch

Voltage source

FIGURE 10.28
INTERACTIVE FIGURE

A simple series circuit. The 6-V battery provides 2 V across each lamp.

lamps in turn, and then return to the positive terminal of the battery. (The same amount of current passes through the battery.) This is the only path of the electrons through the circuit. A break anywhere in the path results in an open circuit, and the flow of electrons ceases. Such a break occurs when the switch is opened, when the wire is accidentally cut, or when one of the lamp filaments burns out. The circuit shown in Figure 10.28 illustrates the following characteristics of series connections:

1. Electric current has but a single pathway through the circuit. This means that the current passing through the resistance of each electrical device along the pathway is the same.

2. This current is resisted by the resistance of the first device, the resistance of the second, and that of the third also, so the total resistance to current in the circuit is the sum of the individual resistances along the circuit path.

3. The current in the circuit is numerically equal to the voltage supplied by the source divided by the total resistance of the circuit. This is in accord with Ohm's law.

4. The total voltage impressed across a series circuit divides among the individual electrical devices in the circuit so that the sum of the "voltage drops" across the resistance of each individual device is equal to the total voltage supplied by the source. This characteristic follows from the fact that the amount of energy given to the total current is equal to the sum of energies given to each device.

5. The voltage drop across each device is proportional to its resistance. This follows from the fact that more energy is dissipated when a current passes through a large resistance than

when the same current passes through a small resistance.

fyi
Solid-state lighting may soon make conventional lightbulbs obsolete. Watch for the progression of light-emitting diodes (LEDs) that are common in flashlights, auto taillights, and novelty gear to widespread and more efficient home and workplace illumination.

STOP AND CHECK YOURSELF

1. What happens to current in other lamps if one lamp in a series circuit burns out?

2. What happens to the brightness of each lamp in a series circuit when more lamps are added to the circuit?

CHECK YOUR ANSWERS

1. If one of the lamp filaments burns out, the path connecting the terminals of the voltage source breaks and current ceases. All lamps go out.

2. Adding more lamps in a series circuit produces a greater circuit resistance. This decreases the current in the circuit and therefore in each lamp, which causes dimming of the lamps. Energy is divided among more lamps, so the voltage drop across each lamp is less.

The rules above hold for ac or dc circuits. It is easy to see the main disadvantage of a series circuit: if one device fails, current in the entire circuit ceases. Some cheap Christmas tree lights are connected in series. When one bulb burns out, it's fun and games (or frustration) trying to locate which one to replace.

Most circuits are wired so that it is possible to operate several electrical devices, each independently of the other. In your home, for example, a lamp can be turned on or off without affecting the operation of other lamps or electrical devices. This is because these devices are connected not in series but in parallel with one another.

fyi
Batteries now deliver power to devices implanted in the human body. Several approaches have been proposed to tap into the power or fuel sources the body already provides. Watch for their implementation in the near future.

Parallel Circuits

A simple **parallel circuit** is shown in Figure 10.29. Three lamps are connected to the same two points, A and B. Electrical devices connected to the same two points of an electrical circuit are said to be *connected in parallel*. Electrons leaving the negative terminal of the battery need travel through only one lamp filament before returning to the positive terminal of the battery. In this case, current branches into three separate pathways from A to B. A break in any one path does not interrupt the flow of charge in the other paths. Each device operates independently of the other devices (whether the circuit is ac or dc).

The circuit shown in Figure 10.29 illustrates the following major characteristics of parallel connections:

1. Each device connects the same two points, A and B, of the circuit. The voltage is therefore the same across each device.

2. The total current in the circuit divides among the parallel branches. Since the voltage across each branch is the same, the amount of current in each branch is inversely proportional to the resistance of the branch.

3. The total current in the circuit equals the sum of the currents in its parallel branches.

4. As the number of parallel branches is increased, the overall resistance of the circuit is *decreased*. Overall resistance is lowered with each added path between any two points of the circuit. This means that the overall resistance of the circuit is less than the resistance of any one of the branches.

STOP AND CHECK YOURSELF

1. What happens to the current in other lamps if one of the lamps in a parallel circuit burns out?

2. What happens to the brightness of each lamp in a parallel circuit when more lamps are added in parallel to the circuit?

CHECK YOUR ANSWERS

1. If one lamp burns out, the other lamps will be unaffected. The current in each branch, according to Ohm's law, is equal to voltage/resistance, and since neither voltage nor resistance is affected in the other branches, the current in those branches is unaffected. The total current in the overall circuit (the current through the battery), however, is decreased by an amount equal to the current drawn by the lamp in question before it burned out. But the current in any other single branch is unchanged.

2. The brightness of each lamp is unchanged as other lamps are introduced (or removed). Only the total resistance and total current in the total circuit changes, which is to say that the current in the battery changes. (There is resistance in a battery also, which we assume is negligible here.) As lamps are introduced, more paths are available between the battery terminals, which effectively decreases total circuit resistance. This decreased resistance is accompanied by an increased current, the same increase that feeds energy to the lamps as they are introduced. Although changes of resistance and current occur for the circuit as a whole, no changes occur in any individual branch in the circuit.

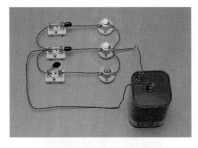
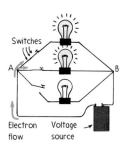

FIGURE 10.29
INTERACTIVE FIGURE

A simple parallel circuit. A 6-V battery provides 6 V across each lamp.

A battery doesn't supply electrons to a circuit; it instead supplies energy to electrons that already exist in the circuit.

FIGURE **10.30**

Just as resistance to checkout is lowered with more open lanes to cash registers in a supermarket, more branches in a parallel circuit lowers total circuit resistance.

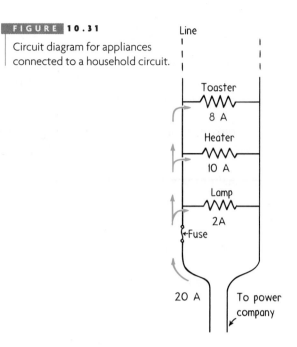

FIGURE **10.31**

Circuit diagram for appliances connected to a household circuit.

Parallel Circuits and Overloading

Electricity is usually fed into a home by way of two wires called *lines*. These lines are very low in resistance and are connected to wall outlets in each room—sometimes through two or more separate circuits. An electric potential of about 110 to 120 volts ac is applied across these lines by a transformer in the neighborhood. (A transformer, as we shall see in the next chapter, is a device that steps down the higher voltage supplied by the power utility.) As more devices are connected to a circuit, more pathways for current result. This lowers the combined resistance of the circuit. Therefore, more current exists in the circuit, which is sometimes a problem. Circuits that carry more than a safe amount of current are said to be *overloaded*.

> In a parallel circuit *most* current travels in the path of least resistance—but not all. *Some* current travels in each path.

We can see how overloading occurs in Figure 10.31. The supply line is connected to a toaster that draws 8 amperes, a heater that draws 10 amperes, and a lamp that draws 2 amperes. When only the toaster is operating and drawing 8 amperes, the total line current is 8 amperes. When the heater is also operating, the total line current increases to 18 amperes (8 amperes to the toaster plus 10 amperes to the heater). If you turn on the lamp, the line current increases to 20 amperes. Connecting additional devices increases the current

still more. Connecting too many devices into the same circuit results in overheating the wires, which can cause a fire.

Safety Fuses

To prevent overloading in circuits, fuses are connected in series along the supply line. In this way the entire line current must pass through the fuse. The fuse shown in Figure 10.32 is constructed with a wire ribbon that will heat up and melt at a given current. If the fuse is rated at 20 amperes, it will pass 20 amperes, but no more. A current above 20 amperes will melt the fuse, which "blows out" and breaks the circuit. Before a blown fuse is replaced, the cause of overloading should be determined and remedied. Sometimes insulation that separates the wires in a circuit wears away and allows the wires to touch. This greatly reduces the resistance in the circuit and is called a *short circuit*.

In modern buildings, fuses have been largely replaced by circuit breakers, which use magnets or bimetallic strips to open a switch when the current

FIGURE **10.32**

A safety fuse.

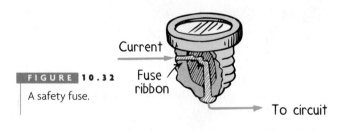

FIGURE 10.33

Electrician Dave Hewitt with a safety fuse and a circuit breaker. He favors the old fuses, which he's found to be more reliable.

is excessive. Utility companies use circuit breakers to protect their lines all the way back to the generators.

> You can prove something to be unsafe, but you can never prove something to be completely safe.

fyi The brightness of the bulb depends on how much power it uses, that is, on how much electricity is converted into heat each second. A tungsten lightbulb that uses 100 watts is brighter than one that uses 60 watts. Because of this, many people mistakenly believe that a watt is a unit of brightness, but it isn't. A 13-watt fluorescent lightbulb (we'll discuss these in Chapter 15) is as bright as a 60-watt conventional (incandescent) bulb. Does that mean that a conventional bulb wastes electricity? Yes. The extra electric power used just heats the bulb. That's why tungsten bulbs are much hotter to touch than equally bright, fluorescent bulbs.

10.10 Electric Power

The moving charges in an electric current do work. This work, for example, can heat a circuit or turn a motor. The rate at which work is done—that is, the rate at which electric energy is converted into another form, such as mechanical energy, heat, or light—is called **electric power.** Electric power is equal to the product of current and voltage.*

$$\text{Power} = \text{current} \times \text{voltage}$$

If the voltage is expressed in volts and the current in amperes, then the power is expressed in watts. So, in units form,

$$\text{Watts} = \text{amperes} \times \text{volts}$$

If a lamp rated at 120 watts operates on a 120-volt line, you can figure that it will draw a current of 1 ampere (120 watts = 1 ampere × 120 volts). A 60-watt lamp draws 1/2 ampere on a 120-volt line. This relationship becomes a practical matter when you wish to know the cost of electrical energy, which is usually a small fraction of a dollar per kilowatt-hour, depending on the locality. A kilowatt is 1000 watts, and a kilowatt-hour represents the amount of energy consumed in 1 hour at the rate of 1 kilowatt.** Therefore, in a locality where electric energy costs 25 cents per kilowatt-hour, a 100-watt electric lightbulb can operate for 10 hours at a cost of 25 cents, or a half nickel for each hour. A toaster or iron, which draws much more current and therefore much more energy, costs about ten times as much to operate.

fyi Your heart uses slightly more than 1 W of power in pumping blood through your body.

FIGURE 10.34

The power and voltage on the lightbulb read "100 W 120 V." Does it *have* 100 W, or does it use 100 W when lit? How many amperes will flow through it when lit?

---

* Recall from Chapter 5 that power = work/time; 1 watt = 1 J/s. Note that the units for mechanical power and electrical power agree (work and energy are both measured in joules):

$$\text{Power} = \frac{\text{charge}}{\text{time}} \times \frac{\text{energy}}{\text{charge}} = \frac{\text{energy}}{\text{time}}$$

** Since power = energy/time, simple rearrangement gives energy = power × time; thus, energy can be expressed in the unit *kilowatt-hours* (kWh).

Roy Unruh harnesses solar energy to produce electricity, which in turn powers demonstration vehicles.

PROBLEM SOLVING

Problems

1. If a 120-V line to a socket is limited to 15 A by a safety fuse, will it operate a 1200-W hair dryer?

2. At 30¢/kWh, what does it cost to operate the 1200-W hair dryer for 1 h?

Solutions

1. Yes. From the expression watts = amperes × volts, we can see that current = 1200 W/120 V = 10 A, so the hair dryer will operate when connected to the circuit. But two hair dryers on the same circuit will blow the fuse.

2. 1200 W = 1.2 kW; 1.2 kW × 1 h × 30¢/1 kWh = 36¢.

EVERYDAY APPLICATIONS

■ ELECTRICAL ENERGY AND TECHNOLOGY

Try to imagine everyday home life before the advent of electrical energy. Think of homes without electric lights, refrigerators, heating and cooling systems, telephones, and radio and TV. We may romanticize a better life without these, but only if we overlook the many hours of daily toil devoted to laundry, cooking, and heating homes. We'd also have to overlook how difficult it was to reach a doctor in times of emergency before the advent of the telephone—when all the doctor had in his bag were laxatives, aspirins, and sugar pills—and when infant death rates were staggering.

We have become so accustomed to the benefits of technology that we are only faintly aware of our dependency on dams, power plants, mass transportation, electrification, modern medicine, and modern agricultural science for our very existence. When we enjoy a good meal, we give little thought to the technology that went into growing, harvesting, and delivering the food on our table. When we turn on a light, we give little thought to the centrally controlled power grid that links the widely separated power stations by long-distance transmission lines. These lines serve as the productive life force of industry, transportation, and the electrification of civilization. Anyone who thinks of science and technology as "inhuman" fails to grasp the ways in which they make our lives more human.

SUMMARY OF TERMS

Coulomb's law The relationship among electrical force, charge, and distance: If the charges are alike in sign, the force is repelling; if the charges are unlike, the force is attractive.

Coulomb The SI unit of electrical charge. One coulomb (symbol C) is equal in magnitude to the total charge of 6.25×10^{18} electrons.

Electrically polarized Term applied to an atom or molecule in which the charges are aligned so that one side has a slight excess of positive charge and the other side a slight excess of negative charge.

Electric field Defined as force per unit charge, it can be considered to be an energetic "aura" surrounding charged objects. About a charged point, the field decreases with distance according to the inverse-square law, like a gravitational field. Between oppositely charged parallel plates, the electric field is uniform.

Electric potential energy The energy a charge possesses by virtue of its location in an electric field.

Conductor Any material having free charged particles that easily flow through it when an electric force acts on them.

Electric potential The electric potential energy per amount of charge, measured in volts, and often called voltage.

Potential difference The difference in potential between two points, measured in volts, and often called voltage difference.

Electric current The flow of electric charge that transports energy from one place to another.

Ampere The unit of electric current; the rate of flow of 1 coulomb of charge per second.

Direct current (dc) An electric current flowing in one direction only.

Alternating current (ac) Electric current that repeatedly reverses its direction; the electric charges vibrate about relatively fixed points. In the United States, the vibrational rate is 60 Hz.

Electrical resistance The property of a material that resists the flow of an electric current through it. It is measured in ohms (Ω).

Superconductor Any material with zero electrical resistance, wherein electrons flow without losing energy and without generating heat.

Ohm's law The statement that the current in a circuit varies in direct proportion to the potential difference or voltage and inversely with the resistance:

$$\text{Current} = \frac{\text{voltage}}{\text{resistance}}$$

A current of 1 A is produced by a potential difference of 1 V across a resistance of 1 Ω.

Series circuit An electric circuit with devices connected in such a way that the same electric current flows through each of them.

Parallel circuit An electric circuit with two or more devices connected in such a way that the same voltage acts across each one, and any single one completes the circuit independently of all the others.

Electric power The rate of energy transfer, or the rate of doing work; the amount of energy per unit time, which can be measured by the product of current and voltage:

$$\text{Power} = \text{current} \times \text{voltage}$$

It is measured in watts (or kilowatts), where

$$1\,\text{A} \times 1\,\text{V} = 1\,\text{W}.$$

REVIEW QUESTIONS

10.1 Electric Force and Charge

1. Which part of an atom is positively charged, and which part is negatively charged?
2. How does the charge of one electron compare with that of another electron?
3. How do the masses of electrons compare with the masses of protons?
4. How does the number of protons in the atomic nucleus normally compare with the number of electrons that orbit the nucleus?
5. What kind of charge does an object acquire when electrons are stripped from it?

6. What is meant by saying that charge is conserved?

10.2 Coulomb's Law

7. How is Coulomb's law similar to Newton's law of gravitation? How is it different?
8. How does a coulomb of charge compare with the charge of a single electron?
9. How does the magnitude of electrical force between a pair of charged particles change when the particles are moved twice as far apart? Three times as far apart?
10. How does an electrically polarized object differ from an electrically charged object?

10.3 Electric Field

11. Give two examples of common force fields.
12. How is the direction of an electric field defined?

10.4 Electric Potential

13. In terms of the units that measure them, distinguish between electric potential energy and electric potential.
14. A balloon may easily be charged to several thousand volts. Does that mean it has several thousand joules of energy? Explain.

10.5 Voltage Sources

15. What condition is necessary for heat energy to flow from one end of a metal bar to the other? For electric charge to flow?
16. What condition is necessary for a sustained flow of electric charge through a conducting medium?
17. How much energy is given to each coulomb of charge passing through a 6-V battery?

10.6 Electric Current

18. Why do electrons, rather than protons, make up the flow of charge in a metal wire?
19. Does electric charge flow *across* a circuit or *through* a circuit? Does voltage flow *across* a circuit or is it *impressed across* a circuit? Explain.
20. Distinguish between dc and ac.
21. Does a battery produce dc or ac? Does the generator at a power station produce dc or ac?

10.7 Electric Resistance

22. Which has the greater resistance, a thick wire or a thin wire of the same length?
23. What is the unit of electrical resistance?

10.8 Ohm's Law

24. What is the effect on current through a circuit of steady resistance when the voltage is doubled? What if both voltage and resistance are doubled?
25. How much current flows through a radio speaker that has a resistance of 8 V when 12 V is impressed across the speaker?

26. Which has the greater electrical resistance, wet skin or dry skin?
27. High voltage by itself does not produce electric shock. What does?
28. What is the function of the third prong on the plug of an electric appliance?
29. What is the source of electrons that makes a shock when you touch a charged conductor?

10.9 Electric Circuits

30. What is an electric circuit, and what is the effect of a gap in such a circuit?
31. In a circuit consisting of two lamps connected in series, if the current in one lamp is 1 A, what is the current in the other lamp?
32. If 6 V were impressed across the circuit in question 31, and the voltage across the first lamp were 2 V, what would be the voltage across the second lamp?
33. In a circuit consisting of two lamps connected in parallel, if there is 6 V across one lamp, what is the voltage across the other lamp?
34. If the current through each of the two branches of a parallel circuit is the same, what does this tell you about the resistance of the two branches?
35. How does the total current through the branches of a parallel circuit compare with the current through the voltage source?
36. As more lines are opened at a fast-food restaurant, the resistance to the motion of people trying to get served is reduced. How is this similar to what happens when more branches are added to a parallel circuit?
37. Are household circuits normally wired in series or in parallel?
38. Why will too many electrical devices operating at one time often blow a fuse?

10.10 Electric Power

39. What is the relationship among electric power, current, and voltage?
40. Which draws more current, a 40-W bulb or a 100-W bulb?

ACTIVE EXPLORATIONS

1. Write a letter to your favorite uncle and bring him up to speed on your progress with physics. Relate the greater number of terms in this chapter, and how learning to distinguish among them contributes to your understanding. Select four of the terms and discuss them. Relate the terms to practical examples.

2. Demonstrate charging by friction and discharging from points with a friend who stands at the far end of a carpeted room. With leather shoes, scuff your way across the rug until your noses are close together. This can be a delightfully tingling experience, depending on how dry the air is and how pointed your noses are.

3. Briskly rub a comb against your hair or a woolen garment and bring it near a small but smooth stream of running water. Is the stream of water charged? (Before you say yes, note the behavior of the stream when an opposite charge is brought nearby.)

4. An electric cell is made by placing two plates of different materials that have different affinities for electrons in a conducting solution. You can make a simple 1.5-V cell by placing a strip of copper and a strip of zinc in a tumbler of saltwater. The voltage of a cell depends on the materials used and the solution they are placed in, not the size of the plates. A battery is actually a series of cells.

An easy cell to construct is the citrus cell. Stick a paper clip and a piece of copper wire into a lemon. Hold the ends of the wire close together, but not touching, and place the ends on your tongue. The slight tingle you feel and the metallic taste you experience result from a slight current of electricity pushed by the citrus cell through the wires when your moist tongue closes the circuit.

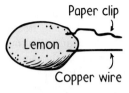

EXERCISES

1. We do not feel the gravitational forces between ourselves and the objects around us because these forces are extremely small. Electrical forces, in comparison, are extremely huge. Since we and the objects around us are composed of charged particles, why don't we usually feel electrical forces?

2. With respect to forces, how are electric charge and mass alike? How are they different?

3. When combing your hair, you scuff electrons from your hair onto the comb. Is your hair then positively or negatively charged? How about the comb?

4. An electroscope is a simple device consisting of a metal ball that is attached by a conductor to two thin leaves of metal foil protected from air disturbances in a jar, as shown. When the ball is touched by a charged body, the leaves that normally hang straight down spread apart. Why? (Electroscopes are useful not only as charge detectors but also for measuring the quantity of charge: the greater the charge transferred to the ball, the more the leaves diverge.)

5. The leaves of a charged electroscope collapse in time. At higher altitudes, they collapse more rapidly. Why is this true? (Hint: The existence of cosmic rays was first indicated by this observation.)

6. Strictly speaking, will a penny be slightly more massive if it has a negative charge or a positive charge? Explain.

7. When one material is rubbed against another, electrons jump readily from one to the other, but protons do not. Why is this? (Think in atomic terms.)

8. If electrons were positive and protons were negative, would Coulomb's law be written the same or differently?

9. The five thousand billion billion freely moving electrons in a penny repel one another. Why don't they fly out of the penny?

10. Two equal charges exert equal forces on each other. What if one charge has twice the magnitude of the other? How do the forces they exert on each other compare?

11. How does the magnitude of electric force compare with the charge between a pair of charged particles when they are brought to half their original distance of separation? To one-quarter their original distance? To four times their original distance? (What law guides your answers?)

12. Suppose that the strength of the electric field about an isolated point charge has a certain value at a distance of 1 m. How will the electric field strength compare at a distance of 2 m from the point charge? What law guides your answer?

13. Why is a good conductor of electricity also a good conductor of heat?

14. When a car is moved into a painting chamber, a mist of paint is sprayed around it. When the body of the car is given a sudden electric charge and the mist of paint is attracted to it, presto—the car is quickly and uniformly painted. What does the phenomenon of polarization have to do with this?

15. If you place a free electron and a free proton in the same electric field, how will the forces acting on them compare? How will their accelerations compare? Their directions of travel?

16. If you put in 10 joules of work to push 1 coulomb of charge against an electric field, what will be its voltage with respect to its starting position? When released, what will be its kinetic energy if it flies past its starting position?

17. What is the voltage at the location of a 0.0001 C charge that has an electric potential energy of 0.5 J (both voltage and potential relative to the same reference point)?

18. What happens to the brightness of light emitted by a lightbulb when the current in it increases?

19. One example of a water system is a garden hose that waters a garden. Another is the cooling system of an automobile. Which of these exhibits behavior more analogous to that of an electric circuit? Why?

20. Is it correct to say that the energy from a car battery ultimately comes from fuel in the gas tank? Defend your answer.

21. Your tutor tells you that an ampere and a volt really measure the same thing, and the different terms only serve to make a simple concept seem confusing. Why should you consider getting a different tutor?

22. In which of the circuits below does a current exist to light the bulb?

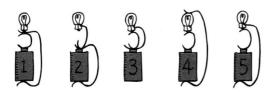

23. Does more current flow out of a battery than into it? Does more current flow into a lightbulb than out of it? Explain.

24. Sometimes you hear someone say that a particular appliance "uses up" electricity. What is it that the appliance actually consumes, and what becomes of it?

25. A simple lie detector consists of an electric circuit, one part of which is part of your body, such as having wires connected to two of your fingers so that your hand is part of the circuit. A sensitive meter shows the current that flows when a small voltage is applied. How does this technique indicate that a person is lying? (And when does this technique not indicate when someone is lying?)

26. Only a small percentage of the electric energy fed into a common lightbulb is transformed into light. What happens to the rest?

27. Will a lamp with a thick filament draw more current or less current than a lamp with a thin filament?

28. A 1-mile-long copper wire has a resistance of 10 ohms. What will be its new resistance when it is shortened by (a) cutting it in half or by (b) doubling it over and using it as if it were one wire of half the length but twice the cross-sectional area?

29. Will the current in a lightbulb connected to a 220-V source be greater or less than that in the same bulb when it is connected to a 110-V source?

30. Which will do less damage—plugging a 110-V appliance into a 220-V circuit or plugging a 220-V appliance into a 110-V circuit? Explain.

31. If a current of one- or two-tenths of an ampere were to flow into one of your hands and out the other, you would probably be electrocuted. But, if the same current were to flow into your hand and out the elbow above the same hand, you could survive, even though the current might be large enough to burn your flesh. Explain.

32. Would you expect to find dc or ac in the filament of a lightbulb in your home? How about in the headlight of an automobile?

33. Are automobile headlights wired in parallel or in series? What is your evidence?

34. A car's headlights dissipate 40 W on low beam and 50 W on high beam. Is there more or less resistance in the high-beam filament?

35. What unit is represented by (a) joule per coulomb, (b) coulomb per second, and (c) watt-second?

36. To connect a pair of resistors so that their equivalent resistance will be greater than the resistance of either one, should you connect them in series or in parallel?

37. To connect a pair of resistors so that their equivalent resistance will be less than the resistance of either one, should you connect them in series or in parallel?

38. A friend says that a battery provides not a source of constant current, but a source of constant voltage. Do you agree or disagree, and why?

39. A friend says that adding bulbs in series to a circuit provides more obstacles to the flow of charge, so there is less current with more bulbs. However, she also says that adding bulbs in parallel provides more paths so more current can flow. Do you agree or disagree, and why?

40. Why might the wingspans of birds be a consideration in determining the spacing between parallel wires on power poles?

41. Estimate the number of electrons that a power company delivers annually to the homes of a typical city of 50,000 people.

42. If electrons flow very slowly through a circuit, why doesn't it take a noticeably long time for a lamp to glow when you turn on a distant switch?

43. Consider a pair of flashlight bulbs connected to a battery. Will they glow brighter if they are connected in series or in parallel? Will the battery run down faster if they are connected in series or in parallel?

44. If several bulbs are connected in series to a battery, they may feel warm to the touch even though they are not visibly glowing. What is your explanation?

45. In the circuit shown, how do the brightnesses of the identical lightbulbs compare? Which lightbulb draws the most current? What will happen if bulb A is unscrewed? If bulb C is unscrewed?

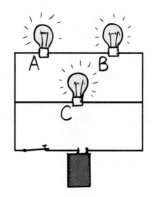

46. As more and more bulbs are connected in series to a flashlight battery, what happens to the brightness of each bulb? Assuming that the heating inside the battery is negligible, what happens to the brightness of each bulb when more and more bulbs are connected in parallel?

47. Are these circuits equivalent to one another? Why or why not?

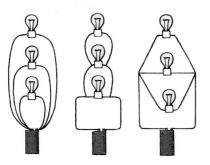

48. A battery has internal resistance, so, if the current it supplies goes up, the voltage it supplies goes down. If too many bulbs are connected in parallel across a battery, will their brightness diminish? Explain.

49. Your friend says that electric current takes the path of least resistance. Why is it more accurate in the case of a parallel circuit to say that *most* current travels in the path of least resistance?

50. If a 60-W bulb and a 100-W bulb are connected in series in a circuit, across which bulb will there be the greater voltage drop? How about if they are connected in parallel?

PROBLEMS

● BEGINNER ■ INTERMEDIATE ◆ EXPERT

1. ● Two pellets, each with a charge of 1 microcoulomb (10^{-6} C), are located 3 cm (0.03 m) apart. Show that the electric force between them is 10 N.

2. ● Two point charges are separated by 6 cm. The attractive force between them is 20 N. Show that when they are separated by 12 cm the force between them is 5 N. (Why can you solve this problem without knowing the magnitudes of the charges?)

3. ● If the charges attracting each other in the preceding problem have equal magnitudes, show that the magnitude of each charge is 2.8 microcoulombs.

4. ● A droplet of ink in an industrial ink-jet printer carries a charge of 1.6×10^{-10} C and is deflected onto paper by a force of 3.2×10^{-4} N. Show that the strength of the electric field required to produce this force is 2×10^{-6} N/C.

5. ● When an electric field does 12 J of work on a charge of 0.0001 C, (a) show that the change in voltage is 120,000 V. (b) When the same electric field does 24 J of work on a charge of 0.0002 C, show that the voltage change is the same.

6. ■ The current driven by voltage V in a circuit of resistance R is given by Ohm's law, $I = V/R$. Show that the resistance of a circuit carrying current I and driven by voltage V is given by the equation $R = V/I$.

7. ■ The same voltage V is impressed on each of the branches of a parallel circuit. The voltage source provides a total current I_{total} to the circuit, and "sees" a total equivalent resistance of R_{eq} in the circuit. That is, $V = I_{total} R_{eq}$ The total current is equal to the sum of the currents through each branch of the parallel circuit. In a circuit with n branches, $I_{total} = I_1 + I_2 + I_3 \ldots I_n$. Use Ohm's law ($I = V/R$) and show that the equivalent resistance of a parallel circuit with n branches is given by

$$\frac{1}{R_{eq}} = \frac{1}{R_1} + \frac{1}{R_2} + \frac{1}{R_3} \ldots + \frac{1}{R_n}$$

8. ● The wattage marked on a light bulb is not an inherent property of the bulb; rather, it depends on the voltage to which it is connected, usually 110 or 120 V. Show that the current in a 60-W bulb connected in a 120-V circuit is 0.5 A.

9. ● Rearrange the equation current = voltage/resistance to express *resistance* in terms of current and voltage. Then consider the following: a certain device in a 120-V circuit has a current rating of 20 A. Show that the resistance of the device is 6 Ω.

10. ● Using the formula Power = current × voltage, show that the current drawn by a 1200-W hair dryer connected to 120 V is 10 A. Then using your same method for the solution to the previous problem, show that the resistance of the hair dryer is 12 Ω.

11. ■ The power in an electric circuit is given by the equation $P = IV$. Use Ohm's law to express V and show that power can be expressed by the equation $P = I^2R$.

12. ■ The total charge that an automobile battery can supply without being recharged is given in terms of ampere-hours. A typical 12-V battery has a rating of 60 ampere-hours (60 A for 1 h, 30 A for 2 h, and so on). Suppose that you forget to turn off the headlights in your parked automobile. If each of the two headlights draws 3 A, show that your battery will be dead in about 10 hours.

13. ■ Suppose you operate a 100-W lamp continuously for 1 week when the power utility rate is 20¢/kWh. Show that this will cost you $3.36.

14. ■ An electric iron connected to a 110-V source draws 9 A of current. Show that the amount of heat generated in 1 minute is almost 60 kJ.

15. ■ For the electric iron of the previous problem, show that the number of coulombs that flow through it in 1 minute is 540 C.

16. ◆ A certain lightbulb with a resistance of 95 ohms is labeled "150 W." Was this bulb designed for use in a 120-V circuit or a 220-V circuit?

17. ◆ In periods of peak demand, power companies lower their voltage. This saves them power (and saves you money)! To see the effect, consider a 1200-W toaster that draws 10 A when connected to 120 V. Suppose the voltage is lowered by 10 percent to 108 V. By how much does the current decrease? By how much does the power decrease? (Caution: The 1200-W label is valid only when 120 V is applied. When the voltage is lowered, it is the resistance of the toaster, not its power, that remains constant.)

CHAPTER 10 ONLINE RESOURCES

The Physics Place

Interactive Figures
10.1, 10.2, 10.8, 10.9, 10.28, 10.29

Tutorials
Electrostatics
Electric Circuits

Videos
Electric Potential
Van de Graff Generator
Alternating Current

Electric Current
Ohm's Law
Caution on Handling Electric Wires
Birds and High-Voltage Wires

Quiz

Flashcards

Links

CHAPTER 11

Magnetism and Electromagnet Induction

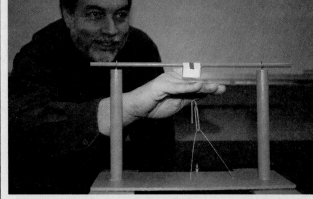

Fred Myers shows that the magnetic field of a ceramic magnet penetrates flesh and the plastic coating on a paper clip.

The term *magnetism* comes from Magnesia, the name of an ancient city in Asia Minor, where the Greeks found certain very unusual stones more than 2000 years ago. These stones, called *lodestones*, possess the unusual property of attracting pieces of iron. Such magnets were first fashioned into compasses and used for navigation by the Chinese in the twelfth century AD.

In the sixteenth century, William Gilbert, Queen Elizabeth's physician, made artificial magnets by rubbing pieces of iron against lodestones. He suggested that a compass always points north and south because Earth itself has magnetic properties. Later, in 1750, John Michell in England found that magnetic poles obey the inverse-square law, and his results were confirmed by Charles Coulomb. The subjects of magnetism and electricity developed almost independently until 1820, when a Danish physicist named Hans Christian Oersted discovered, in a classroom demonstration, that an electric current affects a magnetic compass.* In his demonstration he saw that magnetism was related to electricity. Shortly thereafter, the French physicist André Marie Ampère proposed that electric currents are the source of all magnetic phenomena.

*We can only speculate about how often such relationships become evident when they "aren't supposed to" and are dismissed as "something wrong with the apparatus." Oersted, however, was keen enough to see that nature was revealing another of its secrets.

11.1 Magnetic Poles

Anyone who has played around with magnets knows that magnets exert forces on one another. A **magnetic force** is similar to an electrical force, in that a magnet can both attract and repel without touching (depending on which end of the magnet is held near the another) and the strength of its interaction depends on the distance between magnets. Whereas electric charges produce electrical forces, regions called *magnetic poles* give rise to magnetic forces.

The Physics Place
Oersted's Discovery

If you suspend a bar magnet at its center by a piece of string, you've got a compass. One end, called the *north-seeking pole*, points northward. The opposite end, called the *south-seeking pole*, points southward. More simply, these are called the *north* and *south* poles. All magnets have both a north and a south pole (some have more than one of each). Refrigerator magnets have narrow strips of alternating north and south poles. These magnets are strong enough to hold sheets of paper against a refrigerator door, but they have a very short range because the north and south poles cancel a short distance from the magnet. In a simple bar magnet, the magnetic poles are located at the two ends. A common horseshoe magnet is a bar magnet bent into a U shape. Its poles are also located at its two ends.

> In days gone by, Dick Tracy comic strips, in addition to predicting the advent of cell phones, featured the heading, "He who controls magnetism controls the universe."

If the north pole of one magnet is brought near the north pole of another magnet, they repel. The same is true of a south pole near a south pole. If opposite poles are brought together, however, attraction occurs.*

Like poles repel; opposite poles attract.

This rule is similar to the rule for the forces between electric charges, where like charges repel one another and unlike charges attract. But there is a very important difference between magnetic poles and electric charges. Whereas electric charges can be isolated, magnetic poles cannot. Electrons and protons are entities by themselves. A cluster of electrons need not be accompanied by a cluster of protons, and vice versa. But a north magnetic pole never exists without the presence of a south pole, and vice versa. The north and south poles of a magnet are like the head and tail of the same coin.

If you break a bar magnet in half, each half still behaves as a complete magnet. Break the pieces in half again, and you have four complete magnets. You can continue breaking the pieces in half and never isolate a single pole. Even if your pieces were one atom thick, there would still be two poles on each piece, which suggests that the atoms themselves are magnets.

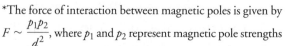

FIGURE 11.2

If you break a magnet in half, you have two magnets. Break these in half, and you have four magnets, each with a north and south pole. Continue breaking the pieces further and further and you find that you always get the same results. Magnetic poles exist in pairs.

---

*The force of interaction between magnetic poles is given by $F \sim \dfrac{p_1 p_2}{d^2}$, where p_1 and p_2 represent magnetic pole strengths and d represents the separation distance between the poles. Note the similarity of this relationship to Coulomb's law and Newton's law of universal gravitation.

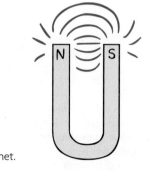

FIGURE 11.1

A horseshoe magnet.

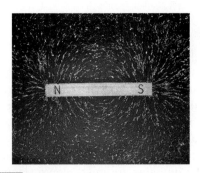

FIGURE 11.3

INTERACTIVE FIGURE

Top view of iron filings sprinkled on a sheet of paper on top of a magnet. The filings trace out a pattern of magnetic field lines in the surrounding space. Interestingly enough, the magnetic field lines continue inside the magnet (not revealed by the filings) and form closed loops.

STOP AND
CHECK YOURSELF

Does every magnet necessarily have a north and a south pole?

CHECK YOUR ANSWER

Yes, just as every coin has two sides, a "head" and a "tail." (Some "trick" magnets have more than two poles, but none has only one.)

fyi

Interestingly, the north pole of a magnet points north because it's attracted to Earth's magnetic *south* pole! Earth's magnetic north pole is in Antarctica. Magnetic and geographic poles don't match.

11.2 Magnetic Fields

If you sprinkle some iron filings on a sheet of paper placed on a magnet, you'll see that the filings trace out an orderly pattern of lines that surround the magnet. The space around the magnet is energized by a **magnetic field**. The shape of the field is revealed by magnetic field lines that spread out from one pole and return to the other pole. It is interesting to compare the field patterns in Figures 11.3 and 11.5 with the electric-field patterns in Figures 10.9 and 10.10 in the previous chapter.

The Physics Place
Magnetic Fields

FIGURE 11.4

When the compass needle is not aligned with the magnetic field, the oppositely directed forces produce a pair of torques (called a couple) that twist the needle into alignment.

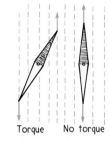

Torque   No torque

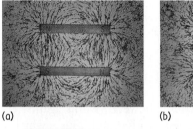

(a)   (b)

FIGURE 11.5

The magnetic field patterns for a pair of magnets. (a) Opposite poles are nearest to each other. (b) Like poles are nearest to each other.

The direction of the field outside the magnet is, by convention, from the north pole to the south pole. Where the lines are closer together, the field is stronger. We can see that the magnetic field strength is greater at the poles. If we place another magnet or a small compass anywhere in the field, its poles will tend to align with the magnetic field.

A magnetic field is produced by the motion of electric charge.* Where, then, is this motion in a common bar magnet? The answer is, in the electrons of the atoms that make up the magnet. These electrons are in constant motion. Two kinds of electron motion produce magnetism: electron spin and electron revolution. A common science model views electrons as spinning about their own axes like tops, while they revolve about the nuclei of their atoms like planets revolving around the Sun. In most common magnets, electron spin is the main contributor to magnetism.

Every spinning electron is a tiny magnet. A pair of electrons spinning in the same direction creates a

*Interestingly, since motion is relative, the magnetic field is relative. For example, when an electron moves by you, there is a definite magnetic field associated with the moving electron. But, if you move along with the electron, so that there is no motion relative to you, you will find no magnetic field associated with the electron. Magnetism is relativistic, as first explained by Albert Einstein when he published his first paper on special relativity, "On the Electrodynamics of Moving Bodies."

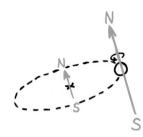

Both the spinning motion and the orbital motion of every electron in an atom produce magnetic fields. These fields combine constructively or destructively to produce the magnetic field of the atom. The resulting field is greatest for iron atoms.

stronger magnet. A pair of electrons spinning in opposite directions, however, works against each other. The magnetic fields cancel. This is why most substances are not magnets. In most atoms, the various fields cancel one another because the electrons spin in opposite directions. In such materials as iron, nickel, and cobalt, however, the fields do not cancel each other entirely. Each iron atom has four electrons whose spin magnetism is uncancelled. Each iron atom, then, is a tiny magnet. The same is true, to a lesser extent, of nickel and cobalt atoms. Most common magnets are therefore made from alloys containing iron, nickel, cobalt, and aluminum in various proportions.

Most of the iron objects around you are magnetized to some degree. A filing cabinet, a refrigerator, or even cans of food on your pantry shelf, have north and south poles induced by Earth's magnetic field. If you pass a compass from their bottoms to their tops, their poles can be easily identified. (See Active Exploration 2 at the end of this chapter, where you are asked to turn cans upside down and note how many days go by for the poles to reverse themselves.)

fyi

Most common magnets are made from alloys containing iron, nickel, cobalt, and aluminum in various proportions. In these, the electron spin contributes virtually all of the magnetic property. (Electrons don't actually spin like a rotating planet, but behave as if they were—the concept of spin is a quantum effect.) In the rare-earth metals, such as gadolinium, the orbital motion is more significant.

11.3 Magnetic Domains

The magnetic field of an individual iron atom is so strong that interactions among adjacent atoms cause large clusters of them to line up with one another. These clusters of aligned atoms are called **magnetic domains**. Each domain is perfectly magnetized and is made up of billions of aligned atoms. The domains are microscopic (Figure 11.7), and there are many of them in a crystal of iron.

Not every piece of iron is a magnet, because the domains in ordinary iron are not aligned. In a common iron nail, for example, the domains are randomly oriented. But when you bring a magnet nearby, they can be induced into alignment. (It is interesting to listen, with an amplified stethoscope, to the clickety-clack of domains aligning in a piece of iron when a strong magnet approaches.) The domains align themselves much as electrical charges in a piece of paper align themselves (become polarized) in the presence of a charged rod. When you remove the nail from the magnet, ordinary

A microscopic view of magnetic domains in a crystal of iron. Each domain consists of billions of aligned iron atoms. In this view, orientation of the domains is random.

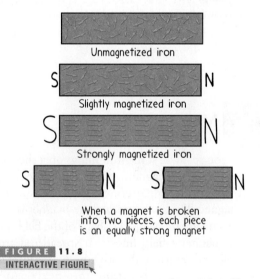

Unmagnetized iron

Slightly magnetized iron

Strongly magnetized iron

When a magnet is broken into two pieces, each piece is an equally strong magnet

Pieces of iron in successive stages of magnetism. The arrows represent domains; the head is a north pole and the tail is a south pole. Poles of neighboring domains neutralize each other's effects, except at the ends.

FIGURE 11.9

Wai Tsan Lee shows iron nails becoming induced magnets.

FIGURE 11.9

Wai Tsan Lee shows iron nails becoming induced magnets.

A magstripe on a credit card contains millions of tiny magnetic domains held together by a resin binder. Data are encoded in binary code, with zeros and ones distinguished by the frequency of domain reversals.

thermal motion causes most or all of the domains in the nail to return to a random arrangement.

Permanent magnets can be made by placing pieces of iron or similar magnetic materials in a strong magnetic field. Alloys of iron differ; soft iron is easier to magnetize than steel. It helps to tap the material to nudge any stubborn domains into alignment. Another way is to stroke the material with a magnet. The stroking motion aligns the domains. If a permanent magnet is dropped or heated outside of the strong magnetic field from which it was made, some of the domains are jostled out of alignment and the magnet becomes weaker.

STOP AND CHECK YOURSELF

1. Why will a magnet not pick up a penny or a piece of wood?

2. How can a magnet attract a piece of iron that is not magnetized?

CHECK YOUR ANSWERS

1. A penny and a piece of wood have no magnetic domains that can be induced into alignment.

2. Like the compass needle in Figure 11.4, domains in the unmagnetized piece of iron are induced into alignment by the magnetic field of the magnet. One domain pole is attracted to the magnet and the other domain pole is repelled. Does this mean the net force is zero? No, because the force is slightly greater on the domain pole closest to the magnet than it is on the farther pole. That's why there is a net attraction. In this way, a magnet attracts unmagnetized pieces of iron (Figure 11.9).

11.4 Electric Currents and Magnetic Fields

A moving charge produces a magnetic field. A current of charges, then, also produces a magnetic field. The magnetic field that surrounds a current-carrying wire can be demonstrated by arranging an assortment of compasses around the wire (Figure 11.10). The magnetic field about the current-carrying wire makes up a pattern of concentric circles. When the current reverses direction, the compass needles turn around, showing that the direction of the magnetic field changes also.*

If the wire is bent into a loop, the magnetic field lines become bunched up inside the loop (Figure 11.11). If the wire is bent into another loop that overlaps the first, the concentration of magnetic field lines inside the loops is doubled. It follows that the magnetic field intensity in this region is increased as the number of loops is increased. The magnetic field intensity is appreciable for a current-carrying coil that has many loops.

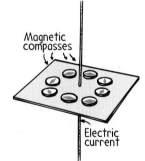

FIGURE 11.10

The compasses show the circular shape of the magnetic field surrounding the current-carrying wire.

*Earth's magnetism is generally accepted as being the result of electric currents that accompany thermal convection in the molten parts of Earth's interior. Earth scientists have found evidence that Earth's poles periodically reverse places—there have been more than 20 reversals in the past 5 million years. This is perhaps the result of changes in the direction of electric currents within Earth.

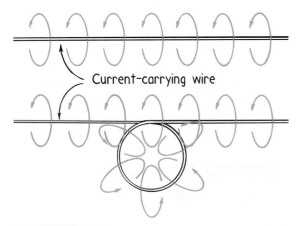

FIGURE 11.11

Magnetic field lines about a current-carrying wire become bunched up when the wire is bent into a loop.

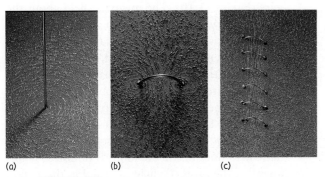

(a)   (b)   (c)

FIGURE 11.12

Iron filings sprinkled on paper reveal the magnetic field configurations about (a) a current-carrying wire, (b) a current-carrying loop, and (c) a coil of loops.

Electromagnets

If a piece of iron is placed in a current-carrying coil of wire, the alignment of magnetic domains in the iron produces a particularly strong magnet known as an **electromagnet**. The strength of an electromagnet can be increased simply by increasing the current through the coil. Strong electromagnets are used to control charged-particle beams in high-energy accelerators. They also levitate and propel prototypes of high-speed trains (Figure 11.13).

Electromagnets powerful enough to lift automobiles are a common sight in junkyards. The strength of these electromagnets is limited mainly by overheating of the current-carrying coils. The most

FIGURE 11.13

A magnetically levitated train—a *magplane*. Whereas conventional trains vibrate as they ride on rails at high speeds, magplanes can travel vibration-free at high speeds because they make no physical contact with the guideway they float above.

powerful electromagnets omit the iron core and employ superconducting coils through which large electrical currents flow with ease.

Superconducting Electromagnets

Ceramic superconductors (Chapter 10) have the interesting property of expelling magnetic fields. Because magnetic fields cannot penetrate the surface of a superconductor, magnets levitate above them. The reasons for this behavior, which are beyond the scope of this book, involve quantum mechanics. One of the more exciting applications of superconducting electromagnets is the levitation of high-speed trains for transportation. Prototype trains have already been demonstrated in the United States, Japan, and Germany. Watch for the growth of this relatively new technology.

FIGURE 11.14

A permanent magnet levitates above a superconductor because its magnetic field cannot penetrate the superconducting material.

11.5 Magnetic Forces on Moving Charges

11.5 Magnetic Forces on Moving Charges

A charged particle at rest will not interact with a static magnetic field. However, if the charged particle moves in a magnetic field, the magnetic character of a charge in motion becomes evident: The charged particle experiences a deflecting force.* The force is greatest when the particle moves in a direction perpendicular to the magnetic field lines. At other angles, the force is less, and it becomes zero when the particle moves parallel to the field lines. In any case, the direction of the force is always perpendicular to the magnetic field lines and the velocity of the charged particle (Figure 11.15). So a moving charge is deflected when it crosses through a magnetic field, but, when it travels parallel to the field, no deflection occurs.

The Physics Place
Magnetic Forces on
Current-Carrying Wires

This deflecting force is very different from the forces that occur in other interactions, such as the gravitational forces between masses, the electric forces between charges, and the magnetic forces between magnetic poles. The force that acts on a moving charged particle, such as an electron in an electron beam, does not act along the line that joins the sources of interaction. Instead, it acts perpendicularly both to the magnetic field and to the electron beam.

We are fortunate that charged particles are deflected by magnetic fields. This fact was employed in guiding electrons onto the inner surface of early television tubes to produce pictures. Also, charged particles from outer space are deflected by Earth's magnetic field. Otherwise the harmful cosmic rays bombarding Earth's surface would be much more intense.

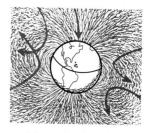

FIGURE **11.16**
The magnetic field of Earth deflects the many charged particles that make up cosmic radiation.

Magnetic Force on Current-Carrying Wires

Simple logic tells you that if a charged particle moving through a magnetic field experiences a deflecting force, then a current of charged particles moving through a magnetic field also experiences a deflecting force. If the particles are deflected while moving inside a wire, the wire is also deflected (Figure 11.17).

If we reverse the direction of current, the deflecting force acts in the opposite direction. The force is strongest when the current is perpendicular to the magnetic field lines. The direction of force is not along the magnetic field lines nor along the direction of current. The force is perpendicular to both field lines and current. It is a sideways force—perpendicular to the wire.

In an advanced course, you'll learn the "simple" right-hand rule!

Insight

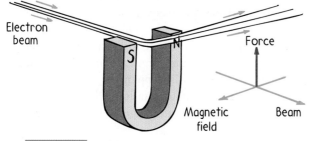

Electron beam

S   N   Force

Magnetic field   Beam

FIGURE **11.15**
A beam of electrons is deflected by a magnetic field.

*When particles of electric charge q and velocity v move perpendicularly into a magnetic field of strength B, the force F on each particle is simply the product of the three variables: $F = qvB$. For nonperpendicular angles, v in this relationship must be the component of velocity perpendicular to B.

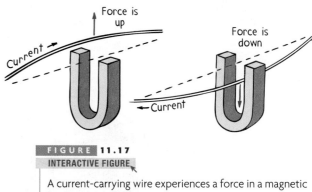

Force is up

Current

Force is down

Current

FIGURE **11.17**
INTERACTIVE FIGURE

A current-carrying wire experiences a force in a magnetic field. (Can you see that this is a simple extension of Figure 11.15?)

We see that, just as a current-carrying wire will deflect a magnet such as a compass needle (as discovered by Oersted in a physics classroom in 1820), a magnet will deflect a current-carrying wire. When discovered, these complementary links between electricity and magnetism created much excitement. Almost immediately, people began harnessing the electromagnetic force for useful purposes—with great sensitivity in electric meters and with great force in electric motors.

STOP AND CHECK YOURSELF

What law of physics tells you that if a current-carrying wire produces a force on a magnet, a magnet must produce a force on a current-carrying wire?

CHECK YOUR ANSWER

Newton's third law, which applies to all forces in nature.

Electric Meters

The simplest meter to detect electric current is a magnetic compass. The next simplest meter is a compass in a coil of wires (Figure 11.18). When an electric current passes through the coil, each loop produces its own effect on the needle, so even a very small current can be detected. Such a current-indicating instrument is called a *galvanometer*.

A more common design is shown in Figure 11.19. It employs more loops of wire and is therefore more sensitive. The coil is mounted for movement, and the magnet is held stationary. The coil turns against a spring, so the greater the current in its windings,

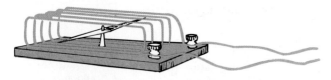

FIGURE **11.18**

A very simple galvanometer.

FIGURE **11.19**

A common galvanometer design.

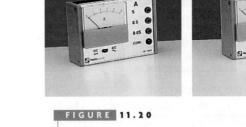

FIGURE **11.20**

Both the ammeter and the voltmeter are basically galvanometers. (The electrical resistance of the instrument is designed to be very low for the ammeter, and very high for the voltmeter.)

the greater its deflection. A galvanometer may be calibrated to measure current (amperes), in which case it is called an *ammeter*. Or it may be calibrated to measure electric potential (volts), in which case it is called a *voltmeter*.*

> **fyi**
>
> The galvanometer is named after Luigi Galvani (1737–1798), who, while dissecting a frog's leg, discovered that dissimilar metals touching the leg caused it to twitch. This chance discovery led to the invention of the chemical cell and the battery. The next time you pick up a galvanized pail, think of Luigi Galvani in his anatomy laboratory.

Electric Motors

If we change the design of the galvanometer slightly, so that deflection makes a complete turn rather than a partial rotation, we have an *electric motor*. The principal difference is that the current in a motor is made to change direction each time the coil makes a half rotation. This happens in a cyclic fashion to produce continuous rotation, which has been used to run clocks, operate gadgets, and lift heavy loads.

*To some degree, measuring instruments change what is being measured—ammeters and voltmeters included. Because an ammeter is connected in series with the circuit it measures, its resistance is made very low. That way, it doesn't appreciably lower the current it measures. Because a voltmeter is connected in parallel, its resistance is made very high, so that it draws very little current for its operation. In the lab part of your course you'll likely learn how to connect these instruments in simple circuits.

FIGURE 11.21
INTERACTIVE FIGURE

A simplified motor.

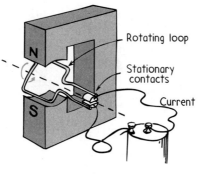

In Figure 11.21 we see the principle of the electric motor in bare outline. A permanent magnet produces a magnetic field in a region where a rectangular loop of wire is mounted to turn about the axis shown by the dashed line. When a current passes through the loop, it flows in opposite directions in the upper and lower sides of the loop. (It must do this because, if charge flows into one end of the loop, it must flow out the other end.) If the upper portion of the loop is forced to the left, then the lower portion is forced to the right, as if it were a galvanometer. But, unlike a galvanometer, the current is reversed during each half revolution by means of stationary contacts on the shaft. The parts of the wire that brush against these contacts are called *brushes*. In this way, the current in the loop alternates so that the forces in the upper and lower regions do not change directions as the loop rotates. The rotation is continuous as long as current is supplied.

We have described here only a very simple dc motor. Larger motors, dc or ac, are usually manufactured by replacing the permanent magnet by an electromagnet that is energized by the power source. Of course, more than a single loop is used. Many loops of wire are wound about an iron cylinder, called an *armature*, which then rotates when the wire carries current.

The advent of electric motors brought to an end much human and animal toil in many parts of the world. Electric motors have greatly changed the way people live.

STOP AND
CHECK YOURSELF

What is the major similarity between a galvanometer and a simple electric motor? What is the major difference?

CHECK YOUR ANSWER

A galvanometer and a motor are similar in that they both use coils positioned in a magnetic field. When a current passes through the coils, forces on the wires rotate the coils. The major difference is that the maximum coil rotation in a galvanometer is one half turn, whereas the coil in a motor (which is wrapped on an armature) rotates through many complete turns. This is accomplished by alternating the direction of the current with each half turn of the armature.

EVERYDAY APPLICATIONS

■ MRI: MAGNETIC RESONANCE IMAGING

Magnetic resonance imaging (MRI) scanners provide high-resolution pictures of the tissues inside a body. Superconducting coils produce a strong magnetic field (up to 60,000 times stronger than the intensity of Earth's magnetic field) that is used to align the protons of hydrogen atoms in the body of the patient.

Like electrons, protons have a "spin" property, so they will align with a magnetic field. Unlike a compass needle that aligns with Earth's magnetic field, the proton's axis wobbles about the applied magnetic field. Wobbling protons are slammed with a burst of radio waves tuned to push the proton's spin axis sideways, perpendicular to the applied magnetic field. When the radio waves pass and the protons quickly return to their wobbling pattern, they emit faint electromagnetic signals whose frequencies depend slightly on the chemical environment in which the proton resides. The signals, which are detected by sensors, are then analyzed by a computer to reveal varying densities of hydrogen atoms in the body and their interactions with surrounding tissue. The images clearly distinguish between fluid and bone, for example.

It is interesting to note that MRI was formerly called NMRI (nuclear magnetic resonance imaging), because hydrogen nuclei resonate with the applied fields. Because of public phobia about anything "nuclear," this diagnostic technique is now called MRI. (Tell your friends that every atom in their bodies contains a nucleus!)

11.6 # Electromagnetic Induction

n the early 1800s, the only current-producing devices were voltaic cells, which produced small currents by dissolving metals in acids. These were the forerunners of our present-day batteries. The question arose as to whether electricity could be produced from magnetism. The answer was provided in 1831 by two physicists, Michael Faraday in England and Joseph Henry in the United States—

The Physics Place
Faraday's Law

each working without knowledge of the other. Their discovery changed the world by making electricity commonplace—powering industries by day and lighting up cities at night.

> **fyi**
> **Multiple loops of wire must be insulated, because bare wire loops touching each other make a short circuit. Interestingly, Joseph Henry's wife tearfully sacrificed part of the silk in her wedding gown to cover the wires of Henry's first electromagnets.**

Faraday and Henry both discovered **electromagnetic induction**—that electric current could be produced in a wire simply by moving a magnet into or out of a coil of wire (Figure 11.22). No battery or other voltage source was needed—only the motion of a magnet in a wire loop. They discovered that voltage is caused, or induced, by the relative motion between a wire and a magnetic field. Whether the magnetic field moves near a stationary conductor or vice versa, voltage is induced either way (Figure 11.23).

The greater the number of loops of wire moving in a magnetic field, the greater the induced voltage (Figure 11.24). Pushing a magnet into a coil with twice as many loops induces twice as much voltage; pushing into a coil with ten times as many loops

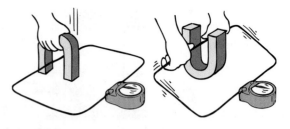

FIGURE 11.23

Voltage is induced in the wire loop whether the magnetic field moves past the wire or the wire moves through the magnetic field.

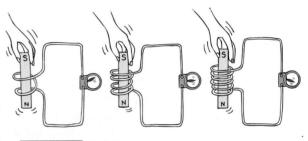

FIGURE 11.24

When a magnet is plunged into a coil with twice as many loops as another, twice as much voltage is induced. If the magnet is plunged into a coil with three times as many loops, three times as much voltage is induced.

induces ten times as much voltage; and so on. It may seem that we get something (energy) for nothing simply by increasing the number of loops in a coil of wire, but we don't. We find that it is more difficult to push the magnet into a coil made up of more loops. This is because the induced voltage produces a current, which makes an electromagnet, which repels the magnet in our hand. So we must do more work against this "back force" to induce more voltage (Figure 11.25).

> **fyi**
> **A long, helically wound coil of insulated wire is called a *solenoid*.**

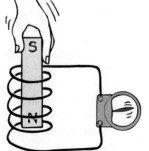

FIGURE 11.22

When the magnet is plunged into the coil, charges in the coil are set in motion, and voltage is induced in the coil.

FIGURE 11.25

It is more difficult to push the magnet into a coil with many loops because the magnetic field of each current loop resists the motion of the magnet.

The amount of voltage induced depends on how fast the magnetic field lines are entering or leaving the coil. Very slow motion produces hardly any voltage at all. Rapid motion induces a greater voltage. This phenomenon of inducing voltage by changing the magnetic field in a coil of wire is called electromagnetic induction.

> Note that a magnetic field does not induce voltage: a *change* in the field over some *time interval* does. If the field changes in a closed loop, and the loop is an electrical conductor, then both voltage and current are induced.

Faraday's Law

Electromagnetic induction is summarized by **Faraday's law:**

The induced voltage in a coil is proportional to the number of loops, multiplied by the rate at which the magnetic field changes within those loops.

The amount of *current* produced by electromagnetic induction depends on the resistance of the coil and the circuit that it connects, as well as the induced voltage.* For example, we can plunge a magnet into and out of a closed rubber loop and into and out of a closed loop of copper. The voltage induced in each is the same, providing the loops are the same size and the magnet moves with the same speed. But the current in each is quite different. The electrons in the rubber sense the same voltage as those in the copper, but their bonding to the fixed atoms prevents the movement of charge that so freely occurs in the copper.

We have mentioned two ways in which voltage can be induced in a loop of wire: by moving the loop near a magnet, or by moving a magnet near the loop. There is a third way—by changing a current in a nearby loop. All three of these cases possess the same essential ingredient—a changing magnetic field in the loop.

We see electromagnetic induction all around us. On the road, we see it operate when a car drives over buried coils of wire to activate a nearby traffic

---

* Current also depends on the "inductance" of the coil. Inductance measures the tendency of a coil to resist a change in current because the magnetism produced by one part of the coil acts to oppose the change of current in other parts of the coil. In ac circuits it is comparable to resistance in dc circuits. To reduce "information overload" we will not treat inductance in this book.

FIGURE 11.26

Guitar pickups are tiny coils with magnets inside them. The magnets magnetize the steel strings. When the strings vibrate, voltage is induced in the coils and boosted by an amplifier, and sound is produced by a speaker.

STOP AND CHECK YOURSELF

If you push a magnet into a coil, as shown in Figure 11.25, you'll feel a resistance to your push. Why is this resistance greater in a coil with more loops?

CHECK YOUR ANSWER

Simply put, more work is required to provide more energy. You can also look at it this way: When you push a magnet into a coil, you induce electric current and cause the coil to become an electromagnet. The more loops in the coil, the stronger the electromagnet that you produce and the stronger it pushes back against you. (If the electromagnetic coil attracted your magnet instead of repelling it, energy would have been created from nothing and the law of energy conservation would have been violated. So the coil must repel the magnet.)

light. When iron parts of a car move over the buried coils, the effect of Earth's magnetic field on the coils is changed, inducing a voltage to trigger the changing of the traffic lights. Similarly, when you walk through the upright coils in the security system at an airport, any metal you carry slightly alters the magnetic field in the coils. This change induces voltage, which sounds an alarm. When the magnetic strip on the back of a credit card is scanned, induced voltage pulses identify the card. Something similar occurs in the recording head of a tape recorder: magnetic domains in the tape are sensed as the tape moves past a current-carrying coil. Electromagnetic induction is at work in computer hard drives, in iPods, and devices galore. As we soon see, it underlies the electromagnetic waves that we call light.

■ MAGNETIC THERAPY

Back in the eighteenth century, a celebrated "magnetizer" from Vienna, Franz Mesmer, brought his magnets to Paris and established himself as a healer in Parisian society. He claimed that he could heal ailing patients simply by waving magnetic wands above their heads.

At that time, Benjamin Franklin, the world's leading authority on electricity, was visiting Paris as a representative of the government of the United States. He suspected that Mesmer's patients did benefit from his ritual—because it kept them away from the bloodletting practices of other physicians. At the urging of the medical establishment, King Louis XVI appointed a royal commission to investigate Mesmer's claims. The commission included Franklin and Antoine Lavoisier, the founder of modern chemistry. The commissioners designed a series of tests in which some subjects thought they were receiving Mesmer's treatment when they weren't, while others received the treatment but were led to believe they had not. The results of these blind experiments established beyond any doubt that Mesmer's success was due solely to the power of suggestion. To this day the report is a model for clarity and reason. Mesmer's reputation was destroyed, and he retired to Austria.

The laws of Faraday and Maxwell are two of the most important statements in physics. They underlie an understanding of the nature of light and of electromagnetic waves in general. In both cases, now, two hundred years later, with all that has been learned about magnetism and physiology, hucksters of magnetism are attracting even larger followings. But there is no government commission of Franklins and Lavoisiers to challenge their claims. Instead, magnetic therapy is another of the untested and unregulated "alternative therapies" given official recognition by Congress in 1992.

Although testimonials about the benefits of magnets are many, there is no scientific evidence whatsoever for magnets boosting body energy or combating aches and pains. Yet millions of "therapeutic" magnets are sold in stores and catalogs. Consumers are buying magnetic bracelets, insoles, wrist and knee bands, back and neck braces, pillows, mattresses, lipstick, and even water (not to mention ionized bracelets, as discussed in the previous chapter). They are told that

magnets have powerful effects on the body, mainly by increasing blood flow to injured areas. The idea that blood is attracted by a magnet is bunk, for the type of iron in blood doesn't respond to a magnet. Furthermore, most therapeutic magnets are the refrigerator type, with a very limited range. To get an idea of how quickly the magnetic field of these magnets drops off, see how many sheets of paper one of these magnets will hold on a refrigerator or any iron surface. The magnet will fall off after a few sheets of paper separate it from the iron surface. The field doesn't extend much more than one millimeter, and it wouldn't penetrate the skin, let alone into muscles. And, even if it did, there is no scientific evidence that magnetism has any beneficial effects on the body at all. But, again, testimonials are another story.

Sometimes an outrageous claim has some truth to it. For example, the practice of bloodletting in previous centuries was in fact beneficial to a small percentage of men. These men suffered the rare genetic disease hemochromatosis, which is characterized mainly by excess iron in the blood. (Women may inherit the disease, but they are generally exempt from its most serious effects because menstruation removes the excess iron from the body.) Although the number of men who benefited from bloodletting was small, testimonials to its success promoted the spread of the practice, which killed many people.

No claim is so outrageous that testimonials can't be found to support it. Claims such as those for Earth being flat or the existence of flying saucers are quite harmless, and may amuse us. Magnetic therapy may likewise be harmless when used to treat many ailments, but not when used to treat a serious disorder in place of modern medicine. Pseudoscience may be intentionally promoted to deceive, or it may merely be the result of flawed and wishful thinking. In either case, pseudoscience is very big business. The market for therapeutic magnets and other such fruits of unreason is enormous.

Scientists must keep their minds open, must be prepared to accept new findings, and must be ready to be challenged by new evidence. But scientists also have a responsibility to inform the public when the public is being deceived—and, in effect, robbed—by pseudoscientists whose claims do not have any substance.

FIGURE **11.27**

When Jean Curtis powers the large coil with ac, an alternating magnetic field is established in the iron bar and thence through the metal ring. Current is therefore induced in the ring, which then establishes its own magnetic field, which always acts in a direction to oppose the field producing it. The result is mutual repulsion—levitation.

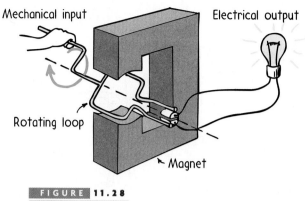

FIGURE **11.28**
INTERACTIVE FIGURE

A simple generator. Voltage is induced in the loop when it is rotated in the magnetic field.

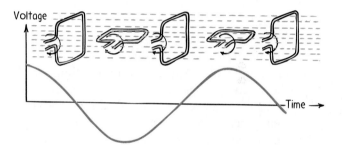

FIGURE **11.29**

As the loop rotates, the magnitude and direction of the induced voltage (and current) changes. One complete rotation of the loop produces one complete cycle in voltage (and current).

fyi
.   **A motor and a generator are actually the same device, with input and output reversed.**

200 years ago, people got light from whale oil. Whales should be glad that humans discovered electricity!

11.7   Generators and Alternating Current

When a magnet is repeatedly plunged into and back out of a coil of wire, the direction of the induced voltage alternates. As the magnetic field strength inside the coil is increased (as the magnet enters), the induced voltage in the coil is directed one way. When the magnetic field strength diminishes (as the magnet leaves), the voltage is induced in the opposite direction. The frequency of the alternating voltage that is induced is equal to the frequency of the changing magnetic field within the loop.

Rather than moving the magnet, it is more practical to move the coil. This is best accomplished by rotating the coil in a stationary magnetic field (Figure 11.28). This arrangement is called a

The Physics Place
Applications of Electro-magnetic Induction

generator. It is essentially the opposite of a motor. Whereas a motor converts electrical energy into mechanical energy, a generator converts mechanical energy into electrical energy.

Because the voltage induced by the generator alternates, the current produced is ac, an alternating current.* The alternating current in our homes is produced by generators standardized so that the current goes through 60 full cycles of change in magnitude and direction each second—60 hertz.

---

* By means such as appropriately designed *brushes* (contacts that brush against the rotating *armature*), the ac in the loop(s) can be converted to dc to make a dc generator.

11.8   Power Production

Fifty years after Faraday and Henry discovered electromagnetic induction, Nikola Tesla and George Westinghouse put those findings to practical use and showed the world that electricity could be generated reliably and in sufficient quantities to light entire cities.

Tesla built generators that were much like those still in use, but quite a bit more complicated than the simple model we have discussed. Tesla's generators had armatures consisting of bundles of copper

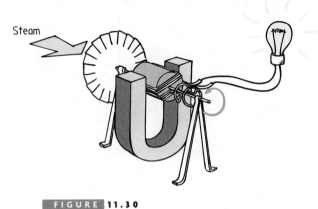

FIGURE **11.30**

Steam drives the turbine, which is connected to the armature of the generator.

wires that were made to spin within strong magnetic fields by means of a turbine, which, in turn, was spun by the energy of steam or falling water. The rotating loops of wire in the armature cut through the magnetic field of the surrounding electromagnets, thereby inducing alternating voltage and current.

We can look at this process from an atomic point of view. When the wires in the spinning armature cut through the magnetic field, oppositely directed electromagnetic forces act on the negative and positive charges. Electrons respond to this force by momentarily swarming relatively freely in one direction throughout the crystalline copper lattice; the copper atoms, which are actually positive ions, are forced in the opposite direction. But the ions are anchored in the lattice, so they barely move at all. Only the electrons move significantly, sloshing back and forth in alternating fashion with each rotation of the armature. The energy of this electronic sloshing is tapped at the electrode terminals of the generator.

fyi **Enormous intergalactic magnetic fields that spread far beyond the galaxies have been recently detected. These giant magnetic fields make up an important part of the cosmic energy store and play a significant role in shaping the evolution of galaxies and large-scale grouping of galaxies.**

It's important to know that generators don't produce energy—they simply convert energy from some other form to electric energy. As we discussed in Chapter 4, energy from a source, whether fossil or nuclear fuel or wind or water, is converted to mechanical energy to drive the turbine. The attached generator converts most of this mechanical energy to electrical energy. Some people think that electricity is a primary source of energy. It is not. It is a carrier of energy that requires a source.

11.9 The Transformer—Boosting or Lowering Voltage

When changes in the magnetic field of a current-carrying coil of wire are intercepted by a second coil of wire, voltage is induced in the second coil. This is the principle of the **transformer**—a simple electromagnetic-induction device consisting of an input coil of wire (the primary) and an output coil of wire (the secondary). The coils need not physically touch each other, but they are normally wound on a common iron core so that the magnetic field of the primary passes through the secondary. The primary is powered by an ac voltage source, and the secondary is connected to some external circuit. Changes in the primary current produce changes in its magnetic field. These changes extend to the secondary, and, by electromagnetic induction, voltage is induced in the secondary. If the number of turns of wire in both coils is the same, voltage input and voltage output will be the same. Nothing is gained. But, if the secondary has more turns than the primary, then greater voltage will be induced in the

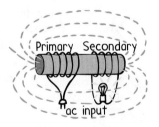

FIGURE **11.31**

A simple transformer.

FIGURE **11.32**

A practical transformer. Both primary and secondary coils are wrapped on the inner part of the iron core (yellow), which guides alternating magnetic field lines (green) produced by ac in the primary. The alternating field induces ac voltage in the secondary. Thus power at one voltage from the primary is transferred to the secondary at a different voltage.

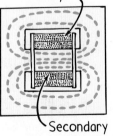

FIGURE 11.33

This common transformer lowers 120 V to 6 V or 9 V. It also converts ac to dc by means of a *diode* inside—a tiny electronic device that acts as a one-way valve.

FIGURE 11.34

A common neighborhood transformer, which typically steps 2400 V down to 240 V for houses and small businesses. Inside the home or business, the 240 V can divide to safer 120 V.

secondary. This is a *step-up transformer*. If the secondary has fewer turns than the primary, the ac voltage induced in the secondary will be lower than that in the primary. This is a *step-down transformer*.

The relationship between primary and secondary voltages relative to the number of turns is:

$$\frac{\text{Primary voltage}}{\text{Number of primary turns}} = \frac{\text{Secondary voltage}}{\text{Number of secondary turns}}$$

It might seem that we get something for nothing with a transformer that steps up the voltage, but we don't. When voltage is stepped up, current in the secondary is less than in the primary. The transformer actually transfers energy from one coil to the other. The rate of transferring energy is *power*. The power used in the secondary is supplied by the primary. The primary gives no more than the secondary uses, in accord with the law of energy conservation. If any slight power losses due to heating of the core can be neglected, then

Power into primary = power out of secondary

Electric power is equal to the product of voltage and current, so we can say that

$$(\text{Voltage} \times \text{current})_{\text{primary}} = (\text{voltage} \times \text{current})_{\text{secondary}}$$

The ease with which voltages can be stepped up or down with a transformer is the principle reason that most electric power is ac rather than dc.

FIGURE 11.35

Voltage generated in power stations is stepped up with transformers prior to being transferred across country by overhead cables. Then other transformers reduce the voltage before supplying it to homes, offices, and factories.

11.10 Field Induction

Electromagnetic induction explains the induction of voltages and currents. Actually, the more basic concept of *fields* is at the root of both voltages and currents. The modern view of electromagnetic induction states that electric and magnetic fields are induced. These, in turn, produce the voltages we have considered. So induction occurs whether or not a conducting wire or any material medium is present. In this more general sense, Faraday's law states:

An electric field is induced in any region of space in which a magnetic field is changing with time.

There is a second effect, an extension of Faraday's law. It is the same except that the roles of electric and magnetic fields are interchanged. It is one of nature's many symmetries. This effect, which was advanced by the British physicist James Clerk Maxwell in about 1860, is known as **Maxwell's counterpart to Faraday's law:**

A magnetic field is induced in any region of space in which an electric field is changing with time.

In each case, the strength of the induced field is proportional to the rates of change of the inducing field. The induced electric and magnetic fields are at right angles to each other.

Maxwell saw the link between electromagnetic waves and light. If electric charges are set into vibration in the range of frequencies that match those of light, waves are produced that are light! Maxwell discovered that light is simply electromagnetic waves in the range of frequencies to which the eye is sensitive.

On the eve of his discovery, Maxwell had a date with the young woman he was later to marry. Story has it that while they were walking in a garden, she remarked about the beauty and wonder of the stars. Maxwell asked her how she would feel if she knew that she was walking with the only person in the world who knew what starlight really was. In fact, at that time, James Clerk Maxwell was the only person in the entire world to know that light of any kind is energy carried in waves of electric and magnetic fields that continually regenerate each other.

The laws of electromagnetic induction were discovered at about the time the American Civil War was being fought. From a long view of human history, there can be little doubt that events such as

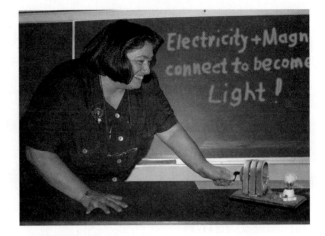

FIGURE 11.36

In turning the crank of the generator, Sheron Snyder does work, which is transformed to voltage and current, which, in turn, is transformed into light.

the American Civil War will pale into provincial insignificance in comparison with the more significant event of the nineteenth century: the discovery of the electromagnetic laws.

Each of us needs a knowledge filter to tell us the difference between what is true and what only pretends to be true. The best knowledge filter ever invented is science.

SUMMARY OF TERMS

Magnetic force (1) Between magnets, it is the attraction of unlike magnetic poles for each other and the repulsion between like magnetic poles. (2) Between a magnetic field and a moving charge, it is a deflecting force due to the motion of the charge: the deflecting force is perpendicular to the velocity of the charge and perpendicular to the magnetic field lines. This force is greatest when the charge moves perpendicular to the field lines and is smallest (zero) when it moves parallel to the field lines.

Magnetic field The region of magnetic influence around a magnetic pole or a moving charged particle.

Magnetic domains Clustered regions of aligned magnetic atoms. When these regions themselves are aligned with one another, the substance containing them is a magnet.

Electromagnet A magnet whose field is produced by an electric current. It is usually in the form of a wire coil with a piece of iron inside the coil.

Electromagnetic induction The induction of voltage when a magnetic field changes with time. If the magnetic field within a closed loop changes in any way, a voltage is induced in the loop:

$$\text{Voltage induced} \sim \text{number of loops} \times \frac{\text{magnetic field change}}{\text{time}}$$

Faraday's law The law of electromagnetic induction, where the induced voltage in a coil is proportional to the number of loops, multiplied by the rate at which the magnetic field changes within those loops. (The induction of voltage is actually the result of a more fundamental phenomenon: the induction of an electric field.)

Generator An electromagnetic induction device that produces electric current by rotating a coil within a stationary magnetic field.

Transformer A device for transferring electric power from one coil of wire to another by means of electromagnetic induction.

Maxwell's counterpart to Faraday's law A magnetic field is induced in any region of space in which an electric field is changing with time. Correspondingly, an electric field is induced in any region of space in which a magnetic field is changing with time.

REVIEW QUESTIONS

1. By whom, and in what setting, was the relationship between electricity and magnetism discovered?

11.1 Magnetic Poles

2. In what way is the rule for the interaction between magnetic poles similar to the rule for the interaction between electric charges?
3. In what way are magnetic poles very different from electric charges?

11.2 Magnetic Fields

4. What produces a magnetic field?
5. What two kinds of motion are exhibited by electrons in an atom?

11.3 Magnetic Domains

6. What is a magnetic domain?
7. Why is iron magnetic and wood not?
8. Why will dropping an iron magnet on a hard floor make it a weaker magnet?

11.4 Electric Currents and Magnetic Fields

9. What is the shape of a magnetic field about a current-carrying wire?
10. What happens to the direction of the magnetic field about an electric current when the direction of the current is reversed?
11. Why is the magnetic field strength inside a current-carrying loop of wire greater than the field strength about a straight section of wire?
12. How is the strength of a magnetic field in a coil affected when a piece of iron is placed inside? Defend your answer.

11.5 Magnetic Forces on Moving Charges

13. In what direction relative to a magnetic field does a charged particle move in order to experience maximum deflecting force? Minimum deflecting force?
14. Both gravitational and electrical forces act along the direction of the force fields. How does the direction of the magnetic force on moving charged particles differ?
15. What effect does Earth's magnetic field have on the intensity of cosmic rays striking Earth's surface?
16. Since a magnetic force acts on a moving charged particle, does it make sense that a magnetic force also acts on a current-carrying wire? Defend your answer.
17. What relative direction between a magnetic field and a current-carrying wire results in the greatest force on the wire? In the smallest force?

18. What happens to the direction of the magnetic force on a wire in a magnetic field when the current in the wire is reversed?
19. What is a galvanometer called when it is calibrated to read current? To read voltage?
20. Is it correct to say that an electric motor is a simple extension of the physics that underlies a galvanometer?

11.6 Electromagnetic Induction

21. What was the important discovery made by physicists Michael Faraday and Joseph Henry?
22. State Faraday's law.
23. What are the three ways in which voltage can be induced in a wire?

11.7 Generators and Alternating Current

24. How does the frequency of induced voltage compare with how frequently a magnet is plunged into and out of a coil of wire?
25. What is the basic difference between a generator and an electric motor?
26. What is the basic similarity between a generator and an electric motor?
27. Why does the voltage induced in a generator alternate?

11.8 Power Production

28. What commonly supplies the energy input to a turbine?
29. Is it correct to say that a generator produces electric energy? Defend your answer.

11.9 The Transformer—Boosting or Lowering Voltage

30. Is it correct to say that a transformer boosts electric energy? Defend your answer.
31. Does a step-up transformer step up the voltage, the current, or the power?
32. Does a step-down transformer step up the voltage, the current, or the power?

11.10 Field Induction

33. What is induced by the rapid alternation of a magnetic field?
34. What is induced by the rapid alternation of an electric field?
35. What important connection did Maxwell discover about electric and magnetic fields?

ACTIVE EXPLORATIONS

1. An iron bar can be magnetized easily by aligning it with the magnetic field lines of Earth and striking it lightly a few times with a hammer. This works best if the bar is tilted down to match the dip of Earth's magnetic field. The hammering jostles the domains so that they can better fall into alignment with Earth's field. The bar can be demagnetized by striking it when it is in an east–west direction.
2. Earth's magnetic field induces some degree of magnetism in most of the iron objects around you. With a compass you can see that cans of food on your pantry shelf have north and south poles. When you pass the compass from their

bottoms to their tops, their poles are easily identified. Mark the poles, either N or S. Then turn the cans upside down and note how many days it takes for the poles to reverse themselves. Explain to your friends why the poles reverse.

EXERCISES

1. Since every iron atom is a tiny magnet, why aren't all iron materials themselves magnets?
2. If you place a chunk of iron near the north pole of a magnet, attraction will occur. Why will attraction also occur if you place the iron near the south pole of the magnet?
3. What is different about the magnetic poles of common refrigerator magnets compared with those of common bar magnets?
4. What surrounds a stationary electric charge? A moving electric charge?
5. "An electron always experiences a force in an electric field, but not always in a magnetic field." Defend this statement.
6. Why will a magnet attract an ordinary nail or paper clip but not a wooden pencil?
7. A friend tells you that a refrigerator door, beneath its layer of white-painted plastic, is made of aluminum. How could you check to see if this is true (without any scraping)?
8. One way to make a compass is to stick a magnetized needle into a piece of cork and to float it in a glass bowl full of water. The needle will align itself with the horizontal component of Earth's magnetic field. Since the north pole of this compass is attracted northward, will the needle float toward the north side of the bowl? Defend your answer.
9. What is the net magnetic force on a compass needle? By what mechanism does a compass needle line up with a magnetic field?
10. Cans of food in your kitchen pantry are likely magnetized. Why?
11. We know a compass points northward because Earth is a giant magnet. Will the northward-pointing needle point northward when the compass is brought to the Southern Hemisphere?

12. When a current-carrying wire is placed in a strong magnetic field, no force acts on the wire. What orientation of the wire is likely?
13. Magnet A has twice the magnetic field strength of magnet B, and, at a certain distance, it pulls on magnet B with a force of 50 N. With how much force, then, does magnet B pull on magnet A?
14. In Figure 11.17, we see a magnet exerting a force on a current-carrying wire. Does a current-carrying wire exert a force on a magnet? Why or why not?
15. A strong magnet attracts a paper clip to itself with a certain force. Does the paper clip exert a force on the strong magnet? If not, why not? If so, does it exert as much force on the magnet as the magnet exerts on it? Defend your answers.
16. When steel naval ships are built, the location of the shipyard and the orientation in the ship while in the shipyard are recorded on a brass plaque permanently fixed to the ship. Why?
17. Can an electron at rest in a magnetic field be set into motion by the magnetic field? What if it were at rest in an electric field?
18. A cyclotron is a device for accelerating charged particles to high speeds as they follow an expanding spiral path. The charged particles are sub- 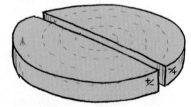 jected to both an electric field and a magnetic field. One of these fields increases the speed of the charged particles, and the other field causes them to follow a curved path. Which field performs which function?
19. A beam of high-energy protons emerges from a cyclotron. Do you suppose there is a magnetic field associated with these particles? Why or why not?

20. A magnetic field can deflect a beam of electrons, but it cannot do work on the electrons to change their speed. Why?

21. Two charged particles are projected into a magnetic field that is perpendicular to their velocities. If the charges are deflected in opposite directions, what does this tell you about the particles?

22. Residents of northern Canada are bombarded by more intense cosmic radiation than are residents of Mexico. Why is this so?

23. What changes in cosmic-ray intensity at Earth's surface would you expect during periods in which Earth's magnetic field is passing through a zero phase while undergoing pole reversals?

24. In a mass spectrometer, ions are directed into a magnetic field, where they curve around in the field and strike a detector. If a variety of singly ionized atoms travel at the same speed through the magnetic field, would you expect them all to be deflected by the same amount? Or would you expect different ions to be bent by different amounts?

25. Historically, replacing dirt roads with paved roads reduced friction between vehicles and the surface of the road. Replacing paved roads with steel rails reduced friction further. What will be the next step in reducing friction between vehicles and the surfaces over which they move? What friction will remain after surface friction has been eliminated?

26. Will a pair of parallel current-carrying wires exert forces on each other?

27. When Tim pushes the wire between the poles of the magnet, the galvanometer registers a pulse. When he lifts the wire, another pulse is registered. How do the pulses differ?

28. Why is a generator armature harder to rotate when it is connected to a circuit and supplying electric current?

29. Will a cyclist coast farther if the lamp connected to his generator is turned off? Explain.

30. If your metal car moves over a wide, closed loop of wire embedded in a road surface, will the magnetic field of Earth within the loop be altered? Will this produce a current pulse? Can you think of a practical application for this at a traffic intersection?

31. At the security area of an airport, you walk through a weak ac magnetic field inside a large coil of wire. What is the result of a small piece of metal on your person that slightly alters the magnetic field in the coil?

32. A piece of plastic tape coated with iron oxide is magnetized more in some parts than in others. When the tape is moved past a small coil of wire, what happens in the coil? What has been a practical application of this?

33. How do the input and output parts of a generator and a motor compare?

34. Your friend says that, if you crank the shaft of a dc motor manually, the motor becomes a dc generator. Do you agree or disagree? Defend your position.

35. If you place a metal ring in a region where a magnetic field is rapidly alternating, the ring may become hot to your touch. Why?

36. A magician places an aluminum ring on a table, underneath which is hidden an electromagnet. When the magician says abracadabra (and pushes a switch that starts current flowing through the coil under the table), the ring jumps into the air. Explain his "trick."

37. How could a lightbulb near, yet not touching, an electromagnet be lit? Is ac or dc required? Defend your answer.

38. Two separate but similar coils of wire are mounted close to each other, as shown. The first coil is connected to a battery and has a direct current flowing through it. The second coil is connected to a galvanometer. How does the galvanometer respond when the switch in the first circuit is closed? After being closed when the current is steady? When the switch is opened?

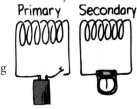

39. Why will more voltage be induced with the apparatus shown above if an iron core is inserted in the coils?

40. Why does a transformer require alternating voltage?

41. How does the current in the secondary of a transformer compare with the current in the primary when the secondary voltage is twice the primary voltage?

42. In what sense can a transformer be thought of as an electrical lever? What does it multiply? What does it not multiply?

43. In the circuit shown, how many volts are impressed across, and how many amperes flow through, the lightbulb?

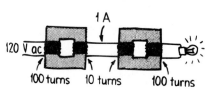

44. In the circuit shown, how many volts are impressed across, and how many amperes flow through, the meter?

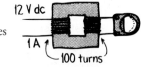

45. How would you answer the previous question if the input were 12-V ac?

46. Can an efficient transformer step up energy? Defend your answer.

47. When a bar magnet is dropped through a vertical length of copper pipe, it falls noticeably more slowly than it does when it is dropped through a vertical length of plastic pipe. If the copper pipe is long enough, the dropped magnet will reach a terminal falling speed. Propose an explanation.

48. What is wrong with the following scheme? To generate electricity without fuel, arrange a motor to run a generator

that will produce electricity that is stepped up with transformers so that the generator can run the motor and simultaneously furnish electricity for other uses.

49. A friend says that changing electric and magnetic fields generate each other, and this gives rise to visible light when the frequency of change matches the frequencies of light. Do you agree? Explain.

50. Would electromagnetic waves exist if changing magnetic fields could produce electric fields but changing electric fields could not in turn produce magnetic fields? Explain.

PROBLEMS

● BEGINNER   ■ INTERMEDIATE   ◆ EXPERT

1. ● An electric doorbell requires 12 volts to operate correctly. A transformer nicely allows it to be powered from a 120-volt outlet. If the primary has 500 turns, show that the secondary should have 50 turns.

2. ● A model electric train requires 6 V to operate. When connected to a 120-V household circuit, a transformer is needed. If the primary coil of the transformer has 240 windings, show that there should be 12 turns on the secondary coil.

3. ● A transformer for a laptop computer converts a 120-V input to a 24-V output. Show that the primary coil has 5 times as many turns as the secondary coil has.

4. ● If the output current for the above transformer is 1.8 A, show that the input current is 0.36 A.

5. ● A transformer has an input of 9 volts and an output of 36 volts. If the input is changed to 12 volts, show that the output would be 48 volts.

6. ● An ideal transformer has 50 turns in its primary and 250 turns in its secondary. 12-V ac is connected to the primary.
   a. Show that there are 60 volts ac available at the secondary.
   b. Show that there are 6 A of current in a 10-ohm device connected to the secondary.
   c. Show that the power supplied to the primary is 360 W.

7. ● Neon signs require about 12,000 V for their operation. Consider a neon-sign transformer that operates off 120-V lines. Show that there should be 100 times as many turns in the secondary as in the primary.

8. ■ 100 kW (10^5 W) of power is delivered to the other side of a city by a pair of power lines, between which the voltage is 12,000 V.
   a. Show that the current in the lines is 8.3 A.
   b. If each of the two lines has a resistance of 10 ohms, show that there is a 83-V change of voltage *along* each line. (Think carefully. This voltage change is along each line, *not between* the lines.)
   c. Show that the power expended as heat in both lines together is 1.38 kW (distinct from power delivered to customers).
   d. How do your calculations support the importance of stepping voltages up with transformers for long-distance transmission?

CHAPTER 11 ONLINE RESOURCES

▞ The Physics Place

Interactive Figures
11.3, 11.8, 11.17, 11.21, 11.24, 11.28

Tutorials
Magnetic Fields

Videos
Oersted's Discovery
Magnetic Forces on Current-Carrying Wires

Faraday's Law
Applications of Electromagnetic Induction

Quiz

Flashcards

Links

PART FOUR

Sound and Light

This CD is the pits—billions of them, inscribed in an array that is scanned by a laser beam at millions of pits per second. It's the sequence of pits detected as light and dark spots that forms a binary code that is converted into a continuous audio waveform. Digitized music! Whoever thought that something as complex as Beethoven's Fifth Symphony could be reduced to a series of ones and zeros? Sound physics!

CHAPTER 12

Waves and Sound

Diane Riendeau uses a classroom wave-generating machine and shows her class how a vibration produces a wave.

Many things in the world about us wiggle and jiggle—the surface of a bell, a string on a guitar, the reed in a clarinet, lips on the mouthpiece of a trumpet, and the vocal cords of your larynx when you speak or sing. All these things *vibrate*. When they vibrate in air, they make the air molecules they touch wiggle and jiggle too, in exactly the same way, and these vibrations spread out in all directions, getting weaker, losing energy as heat, until they die out completely. But if these vibrations were to reach your ear instead, they would be transmitted to a part of your brain, and you would hear sound.

12.1 Vibrations and Waves

In a general sense, anything that moves back and forth, to and fro, from side to side, in and out, or up and down is vibrating. A **vibration** is a wiggle in time. A wiggle in space and time is a **wave**. A wave extends from one location to another. Light and sound are both vibrations that propagate throughout space as waves, but as waves of two very different kinds. Sound is the propagation of vibrations through a material medium—a solid, a liquid, or a gas. If there is no medium to vibrate, then no sound is possible. Sound cannot travel in a vacuum. But light can, because (as we will discuss in the following chapter) light is a vibration of non-material electric and magnetic fields—a vibration of pure energy. Although light can pass through many materials, it needs none. This is evident when it propagates through the vacuum between the Sun and Earth.

The relationship between a vibration and a wave is shown in Figure 12.1. A marking pen on a bob

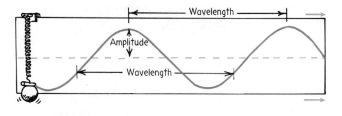

When the bob vibrates up and down, a marking pen traces out a sine curve on the paper, which is moved horizontally at constant speed.

FIGURE 12.2

The source of any wave is something that vibrates. Electrons in the transmitting antenna vibrate 940,000 times each second and produce 940-kHz radio waves. Radio waves can't be seen or heard, but they send a pattern that tells a radio or a TV set what sounds or pictures to make.

attached to a vertical spring vibrates up and down and traces a waveform on a sheet of paper that is moved horizontally at constant speed. The waveform is actually a *sine curve*, a pictorial representation of a wave.

The Physics Place
Waves and Vibrations

Like a water wave, the high points are called *crests*, and the low points are the *troughs*. The straight dashed line represents the "home" position, or midpoint, of the vibration. The word **amplitude** refers to the distance from the midpoint to the crest (or to the trough) of the wave. Therefore, the amplitude equals the maximum displacement from equilibrium.

The **wavelength** of a wave is the distance from the top of one crest to the top of the next one, or, equivalently, the distance between successive identical parts of the wave. The wavelengths of waves at the beach are measured in meters, the wavelengths of ripples in a pond in centimeters, and the wavelengths of light in billionths of a meter (nanometers). All waves have a vibrating source.

How frequently a vibration occurs is described by its frequency. The **frequency** of a vibrating pendulum, or of an object on a spring, specifies the number of to-and-fro vibrations it makes in a given time (usually in one second). A complete to-and-fro oscillation is one vibration. If it occurs in one second, the frequency is one vibration per second. If two vibrations occur in one second, the frequency is two vibrations per second.

The unit of frequency is called the **hertz** (Hz), after Heinrich Hertz, who demonstrated the existence of radio waves in 1886. One vibration per second is 1 hertz, two vibrations per second are 2 hertz, and so on. Higher frequencies are measured in kilohertz (kHz), and still higher frequencies in megahertz (MHz). AM (amplitude modulated)

radio waves are usually measured in kilohertz, while FM (frequency modulated) radio waves are measured in megahertz. A station at 960 kHz on the AM radio dial, for example, broadcasts radio waves that have a frequency of 960,000 vibrations per second. A station at 101.7 MHz on the FM dial broadcasts radio waves with a frequency of 101,700,000 hertz. These radio-wave frequencies are the frequencies at which electrons are forced to vibrate in the antenna of a radio station's transmitting tower. Still higher frequencies are measured in gigahertz (GHz), 1 billion vibrations per second. Cell phones operate in the GHz range, which means electrons inside are jiggling in unison billions of times per second! The frequency of the vibrating electrons and the frequency of the wave produced are the same.

> A bee flaps its wings hundreds of times each second—honey power!

The **period** of a wave or vibration is the time it takes for a complete vibration—for a complete cycle. Period can be calculated from frequency, and vice versa. Suppose, for example, that a pendulum makes two vibrations in one second. Its frequency is 2 Hz. The time needed to complete one vibration—that is, the period of vibration—is 1/2 second. Or if the vibration frequency is 3 Hz, then the period is 1/3 second. The frequency and period are the inverse of each other:

$$\text{Frequency} = \frac{1}{\text{period}}$$

Or, vice versa,

$$\text{Period} = \frac{1}{\text{frequency}}$$

fyi

The frequency of a wave matches the frequency of its vibrating source. This is true not only of sound waves, but, as we'll see in the next chapter, of light waves also. The waves we're learning about, strictly speaking, are *periodic waves*—having distinct periods.

12.2 Wave Motion

If you drop a stone into a calm pond, waves will travel outward in expanding circles. Energy is carried by the wave, traveling from one place to another. The water itself goes nowhere. This can be seen by waves encountering a floating leaf. The leaf bobs up and down, but it doesn't travel with the waves. The waves move along, not the water. The same is true for waves of wind over a field of tall grass on a gusty day. Waves travel across the grass, while the individual grass plants remain in place; instead, stems of grass swing to and fro between definite limits, but they go nowhere. When you speak, molecules in air propagate the disturbance through the air at about 340 meters per second. The disturbance, not the air itself, travels across the room at this speed. In these examples, when the wave motion ceases, the water, the grass, and the air return to their initial positions. It is characteristic of wave motion

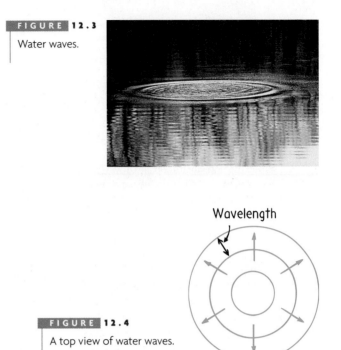

FIGURE 12.3
Water waves.

Wavelength

FIGURE 12.4
A top view of water waves.

that the medium transporting the wave returns to its initial condition after the disturbance has passed.

Wave Speed

The speed of periodic wave motion is related to the frequency and wavelength of the waves. Consider the simple case of water waves (Figures 12.3 and 12.4). Imagine that we fix our eyes on a stationary point on the water's surface and observe the waves passing by that point. We can measure how much time passes between the arrival of one crest and the arrival of the next one (the period), and we can also observe the distance between crests (the wavelength). We know that speed is defined as distance divided by time. In this case, the distance is one wavelength and the time is one period, so the speed of a wave = wavelength/period.

For example, if the wavelength is 10 meters and the time between crests at a point on the surface is 0.5 second, the wave is traveling 10 meters in 0.5 seconds and its speed is 10 meters divided by 0.5 seconds, or 20 meters per second.

Since period is equal to the inverse of frequency, the formula **wave speed** = wavelength/period can also be written,

$$\text{Wave speed} = \text{wavelength} \times \text{frequency}$$

This relationship applies to all kinds of waves, whether they are water waves, sound waves, or light waves.

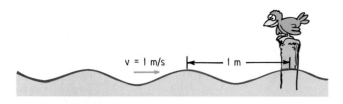

v = 1 m/s | 1 m

FIGURE 12.5
INTERACTIVE FIGURE

If the wavelength is 1 m, and one wavelength per second passes the pole, then the speed of the wave is 1 m/s.

It is customary to express the speed of a wave by the equation $v = f\lambda$, where v is wave speed, f is wave frequency, and λ (the Greek letter lambda) is wavelength.

STOP AND CHECK YOURSELF

1. If a train of freight cars, each 10 m long, rolls by you at the rate of three cars each second, what is the speed of the train?

2. A water wave oscillates up and down three times each second and the distance between wave crests is 2 m.
   a. What is its frequency?
   b. What is its wavelength?
   c. What is its wave speed?

3. The sound from a 60-Hz razor spreads out at 340 meters per second.
   a. What is the frequency of the sound waves?
   b. What is their period?
   c. What is their speed?
   d. What is their wavelength?

CHECK YOUR ANSWERS

1. 30 m/s. We can see this in two ways.
   According to the definition of speed in Chapter 3,

   $$v = \frac{d}{t} = \frac{3 \times 10\,m}{1\,s} = 30\ m/s,$$

   since 30 m of train passes you in 1 s.
   If we compare our train to wave motion, where wavelength corresponds to 10 m and frequency is 3 Hz, then

   Speed = frequency × wavelength
       = 3 Hz × 10 m = 30 m/s

2. a. 3 Hz   b. 2 m
   c. Wave speed = frequency × wavelength
               = 3/s × 2 m = 6 m/s.

3. a. 60 Hz   b. 1/60 second
   c. 340 m/s   d. 5.7 m.

Be clear about the distinction between *frequency* and *speed*. How frequently a wave vibrates is altogether different from how fast it moves from one location to another.

12.3 Transverse and Longitudinal Waves

Fasten one end of a Slinky to a wall and hold the free end in your hand. If you shake the free end up and down, you will produce vibrations that are at right angles to the direction of wave travel. The right-angled, or sideways, motion is called *transverse motion*. This type of wave is called a **transverse wave**. Waves in the stretched strings of musical instruments and upon the surfaces of liquids are transverse waves. We will see later that electromagnetic waves, some of which are radio waves and light waves, are also transverse waves.

The Physics Place
Longitudinal vs. Transverse Waves

A **longitudinal wave** is one in which the direction of wave travel is along the direction in which the source vibrates. You produce a longitudinal wave with your Slinky when you shake it back and forth along the Slinky's axis (Figure 12.6a). The vibrations are then parallel to the direction of energy transfer. Part of the Slinky is compressed, and a wave of **compression** travels along it. In between successive compressions is a stretched region, called a **rarefaction**. Both compressions and

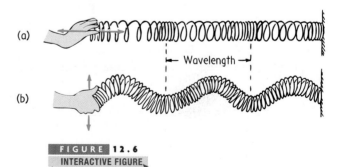

(a)

Wavelength

(b)

FIGURE 12.6
INTERACTIVE FIGURE

Both waves transfer energy from left to right. (a) When the end of the Slinky is pushed and pulled rapidly along its length, a longitudinal wave is produced. (b) When the end of the Slinky is shaken up and down (or side to side), a transverse wave is produced.

rarefactions travel parallel to the Slinky. Together they make up the longitudinal wave.

If you happen to study earthquakes, you'll learn about two types of waves that travel in the ground. One type is transverse (S waves), and the other type is longitudinal (P waves). These travel at different speeds, which provide investigators with a means of determining the source of the waves. Furthermore, the transverse waves cannot travel through liquid matter, while the longitudinal waves can, which provides a means of determining whether matter below ground is molten or solid.

12.4 Sound Waves

Think of the air molecules in a room as tiny, randomly moving Ping-Pong balls. If you vibrate a Ping-Pong paddle in the midst of the balls, you'll set them vibrating to and fro. The balls will vibrate in rhythm with your vibrating paddle. In some regions, they will be momentarily bunched up (compressions), and in other regions in between, they will be momentarily spread out (rarefactions). The vibrating prongs of a tuning fork do the same to air molecules. Vibrations made up of compressions and rarefactions spread from the tuning fork throughout the air, and a *sound wave* is produced.

> Sound requires a medium. It can't travel in a vacuum because there's nothing to compress and stretch.

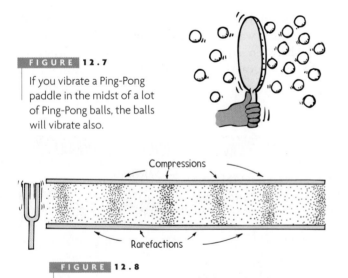

FIGURE 12.7

If you vibrate a Ping-Pong paddle in the midst of a lot of Ping-Pong balls, the balls will vibrate also.

Compressions

Rarefactions

FIGURE 12.8

Compressions and rarefactions travel (both at the same speed and in the same direction) from the tuning fork through the air in the tube. The wavelength is the distance between successive compressions (or rarefactions).

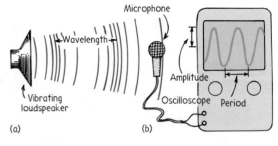

Microphone
Wavelength
Amplitude
Vibrating loudspeaker
Oscilloscope   Period
(a)   (b)

FIGURE 12.9

(a) The radio loudspeaker is a paper cone that vibrates in rhythm with an electric signal. The sound that is produced sets up similar vibrations in the microphone. The vibrations are displayed on an oscilloscope. (b) The waveform on the oscilloscope screen is a graph of pressure versus time, showing how air pressure near the microphone rises and falls as sound waves pass. When the loudness increases, the amplitude of the waveform increases.

The wavelength of a sound wave is the distance between successive compressions or, equivalently, the distance between successive rarefactions. Each molecule in the air vibrates to and fro about some equilibrium position as the waves move by.

Our subjective impression about the frequency of sound is described as **pitch**. A high-pitched sound, like that from a tiny bell, has a high vibration frequency. Sound from a large bell has a low pitch because its vibrations are of a low frequency. Pitch is the high or low we perceive a sound to be, depending on the frequency of the sound wave.

The human ear can normally hear pitches from sound ranging from about 20 hertz to about 20,000 hertz. As we age, this range shrinks. So, by the time you can afford to trade in your old sound system for an expensive hi-fi one, you may not be able to tell the difference. Sound waves of frequencies below 20 hertz are called *infrasonic waves*, and those of frequencies above 20,000 hertz are called *ultrasonic waves*. We cannot hear infrasonic or ultrasonic sound waves.* But dogs and some other animals can.

Most sound is transmitted through air, but any elastic substance—solid, liquid, or gas—can transmit sound.** Air is a poor conductor of sound compared with solids and liquids. You can hear the

---

* In hospitals, concentrated beams of ultrasound are used to break up kidney stones and gallstones, eliminating the need for surgery.

** An elastic substance is "springy," has resilience, and can transmit energy with little loss. Steel, for example, is elastic, whereas lead and putty are not.

■ LOUDSPEAKER

The loudspeaker of your radio or other sound-producing systems changes electrical signals into sound waves. The electrical signals pass through a coil wound around the

neck of a paper cone. This coil, which acts as an electromagnet, is located near a permanent magnet. When current flows one way, magnetic force pushes the electromagnet toward the permanent magnet, pulling the cone inward. When current flows in the opposite direction, the cone is pushed outward. Vibrations in the electric signal cause the cone to vibrate. Vibrations of the cone then produce sound waves in the air.

sound of a distant train clearly by placing your ear against the rail. When swimming, have a friend at a distance click two rocks together beneath the surface of water while you are submerged. Observe how well water conducts the sound. Sound cannot travel in a vacuum because there is nothing to compress and expand. The transmission of sound requires a medium.

fyi   Elephants communicate with one another with infrasonic waves. Their large ears help them to detect these low-frequency sound waves.

Pause to reflect on the physics of sound during a time that you are quietly listening to your radio. The radio loudspeaker is a paper cone that vibrates in rhythm with an electrical signal. Air molecules next to the vibrating cone of the speaker are themselves set into vibration. These, in turn, vibrate against neighboring molecules, which, in turn, do the same, and so on. As a result, rhythmic patterns of compressed and rarefied air emanate from the loudspeaker, showering the entire room with undulating motions. The resulting vibrating air sets your eardrum into vibration, which, in turn, sends cascades of rhythmic electrical impulses along nerves in the cochlea of your inner ear and into the brain. And thus you listen to the sound of music.

FIGURE 12.10

Waves of compressed and rarefied air, generated by the vibrating cone of the loudspeaker, reproduce the sound of music.

fyi   A sound wave traveling through the ear canal vibrates the eardrum, which vibrates three tiny bones, which in turn vibrate the fluid-filled cochlea. Inside the cochlea, tiny hair cells convert the pulse into an electrical signal to the brain.

Speed of Sound

If, from a distance, we watch a person chopping wood or hammering, we can easily see that the blow occurs a noticeable time before its sound reaches our ears. Thunder is often heard seconds after a flash of lightning is seen. These common experiences show that sound requires time to travel from one place to another. The speed of sound depends on wind conditions, temperature, and humidity. It does not depend on the loudness or the frequency of the sound; all sounds travel at the same speed in a given medium. The speed of sound in dry air at 0°C is about 330 meters per second, which is nearly 1200 kilometers per hour. Water vapor in the air increases this speed slightly. Sound travels faster through warm air than through cold air. This is to be expected, because the faster-moving molecules in warm air bump into each other more frequently and, therefore, can transmit a pulse in less time.* For each degree rise in temperature above 0°C, the speed of sound in air increases by 0.6 meter per second. Thus, in air at a normal room temperature of about 20°C, sound travels at about 340 meters per second. In water, sound speed is about four times its speed in air; in steel, it's about fifteen times its speed in air.

* The speed of sound in a gas is about 3/4 the average speed of its molecules.

Your two ears are so sensitive to the differences in sound reaching them that you can detect the direction of incoming sound with almost pinpoint accuracy. With only one ear you would have no idea (and in an emergency might not know which way to move).

STOP AND
CHECK YOURSELF

1. Do compressions and rarefactions in a sound wave travel in the same direction or in opposite directions from one another?

2. What is the approximate distance of a thunderstorm when you note a 3-s delay between the flash of lightning and the sound of thunder?

CHECK YOUR ANSWERS

1. They travel in the same direction.

2. Assuming the speed of sound in air is about 340 m/s, in 3 s it will travel 340 m/s $\times$ 3 s = 1020 m. There is no appreciable time delay for the flash of light, so the storm is slightly more than 1 km away.

12.5   Reflection and Refraction of Sound

Like light, when sound encounters a surface, it can either be returned by the surface, or continue through it. When it is returned, the process is **reflection**. We call the reflection of sound an *echo*. The fraction of sound energy reflected from a surface is large if the surface is rigid and smooth, but it is less if the surface is soft and irregular. The sound energy that is not reflected is transmitted or absorbed.

Sound reflects from a smooth surface in the same way that light does—the angle of incidence (the angle between the direction of the sound and the reflecting surface) is equal to the angle of reflection (Figure 12.11). Sometimes, when sound reflects from the walls, ceiling, and floor of a room, the surfaces are too reflective and the sound becomes garbled. Sound due to multiple reflections is called a **reverberation**. On the other hand, if the reflective surfaces are too absorbent, the sound level is low and the room may sound dull and lifeless. Reflected sound in a room makes it sound lively and full, as you have probably experienced while singing in the shower. In the design of an auditorium or

The Physics Place
Refraction of Sound

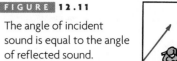
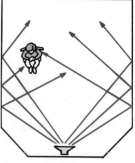

FIGURE 12.11

The angle of incident sound is equal to the angle of reflected sound.

concert hall, a balance must be found between reverberation and absorption. The study of sound properties is called *acoustics*.

It is often advantageous to position highly reflective surfaces behind the stage to direct sound out to the audience. Above the stage, in some concert halls, reflecting surfaces are suspended. The ones in Davies Hall in San Francisco are large shiny plastic surfaces that also reflect light (Figure 12.12). A listener can look up at these reflectors and see the reflected images of the members of the orchestra (the plastic reflectors are somewhat curved, which increases the field of view). Both sound and light obey the same law of reflection. Thus, if a reflector is oriented so that you can see a particular musical instrument, rest assured that you would be able to hear it also. Sound from the instrument will follow the line of sight to the reflector and thence to you.

FIGURE 12.12

The plastic plates above the orchestra reflect both light and sound. Adjusting them is quite simple: what you see is what you hear.

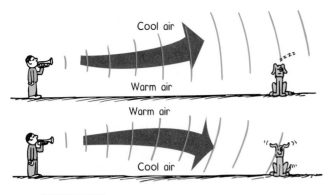

FIGURE 12.13

Sound waves are bent in air of uneven temperatures.

In some halls, absorbers rather than reflectors are used to improve the acoustics.

Refraction occurs when sound continues through a medium and bends. Sound waves bend when parts of the wave fronts travel at different speeds. This may happen when sound waves are affected by uneven winds, or when sound is traveling through air of uneven temperatures. On a warm day, the air near the ground may be appreciably warmer than the air above, so the speed of sound near the ground increases. Sound waves therefore tend to bend away from the ground, resulting in sound that does not seem to transmit well (Figure 12.13).

> **fyi** The direction of travel for both sound and light is always at right angles to their wave fronts.

The refraction of sound occurs underwater, where the speed of sound varies with temperature. This poses a problem for surface vessels that bounce ultrasonic waves off the bottom of the ocean to chart its features. This poses a blessing to submarines that wish to escape detection. Because the ocean has layers of water that are at different temperatures, the refraction of sound leaves gaps or "blind spots" in the water. This is where submarines hide. If it weren't for refraction, submarines would be much easier to detect.

The multiple reflections and refractions of ultrasonic waves are used by physicians in a technique for harmlessly "seeing" the interior of the body without the use of X-rays. When high-frequency sound (ultrasound) enters the body, it is reflected more strongly from the organs' exteriors than from their interiors, and a picture of the outline of the organs is obtained (Figure 12.14). This ultrasound echo technique is nothing new to bats and dolphins, creatures able to emit ultrasonic squeaks and to locate objects by their echoes.

FIGURE 12.14

The 14-week-old fetus that became Megan Hewitt Abrams, who is more recently seen on page 197.

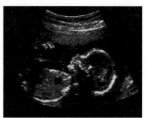

FIGURE 12.15

A dolphin emits ultrahigh-frequency sound to locate and identify objects in its environment. Distance is sensed by the time delay between sending sound and receiving its echo, and direction is sensed by differences in time for the echo to reach the dolphin's two ears. A dolphin's main diet is fish. Because fish hear mainly low frequencies, they are not alerted to the fact they are being hunted.

> **fyi** Owls have extremely sensitive ears. When hunting at night, owls tune into the soft rustles and squeaks of rodents and other small mammals. Like humans, owls locate sound sources by using the fact that sound waves often reach one ear milliseconds before the other. An owl moves its head as it glides toward its prey; when sounds from the target reach both ears at once, the meal is dead ahead. In some owls, one ear is also higher than the other, further sharpening their prey-locating ability.

12.6 Forced Vibrations and Resonance

If you strike an unmounted tuning fork, its sound is rather faint. Repeat with the handle of the fork held against a table after striking it, and the sound is louder. This is because the table is forced to vibrate, and its larger surface sets more air in

▪ DOLPHINS AND ACOUSTICAL IMAGING

The dominant sense of the dolphin is hearing, for vision is not a very useful sense in the often murky and dark depths of the ocean. Whereas sound is a passive sense for us, it is an active sense for the dolphin, which sends out sounds and then perceives its surroundings by means of the echoes that return. The ultrasonic waves emitted by a dolphin enable it to "see" through the bodies of other animals and people. Skin, muscle, and fat are almost transparent to dolphins, so they "see" a thin outline of the body—but the bones, teeth, and gas-filled cavities are clearly apparent. Physical evidence of cancers, tumors, and heart attacks can all be "seen" by the dolphin—as humans have only recently been capable of doing with ultrasound.

What's more fascinating, the dolphin can reproduce the sonic signals that paint the mental image of its surroundings; thus, it is probably able to communicate its experiences to other dolphins by communicating the full acoustic image of what it has "seen," placing it directly in the minds of other dolphins. It needs no word or symbol for "fish," for example, but it is able to communicate an image of the real thing—perhaps with emphasis highlighted by selective filtering, as we similarly communicate a musical concert to others via various means of sound reproduction. Small wonder that the language of the dolphin is very unlike our own!

SAMPLE PROBLEM SOLVING

Problems

1. An oceanic depth-sounding vessel surveys the ocean floor with ultrasonic sound that travels 1530 m/s in seawater. How deep is the water if the time delay of the echo from the ocean floor is 2 s?

2. While sitting at the dock of the bay, Otis notices incoming waves with distance d between crests. The incoming crests lap against the pier pilings at a rate of one every 2 seconds.

a. Find the frequency of the waves.

b. Show that the speed of the waves is given by fd.

c. Suppose the distance d between wave crests is 1.8 m. Show that the speed of the waves is slightly less than 1.0 m/s.

Solutions

1. The round trip is 2 s, meaning 1 s down and 1 s up. Then,

$$d = vt = 1530 \text{ m/s} \times 1 \text{ s} = 1530 \text{ m}$$

(Radar works similarly, where microwaves rather than sound waves are transmitted.)

2. a. The frequency of the waves is given, one per 2 s, or $f = 0.5$ Hz.

   b. $v = f\lambda = fd$.

   c. $v = fd = 0.5 \text{ Hz } (1.8 \text{ m}) = 0.9 \text{ m}$

motion. The table is forced into vibration by a fork of any frequency. This is an example of forced vibration. The vibration of a factory floor caused by the running of heavy machinery is another example of **forced vibration**. A more pleasing example is given by the sounding boards of stringed instruments.

If you drop a wrench and a baseball bat on a concrete floor, you will easily notice the difference in their sounds. This is because each vibrates differently when striking the floor. They are not forced to vibrate at a particular frequency, but, instead, each vibrates at its own characteristic frequency. Any object composed of an elastic material will, when disturbed, vibrate at its own special set of frequencies, which together form its characteristic sound. We speak of an object's **natural frequency**, which depends on such factors as the elasticity and shape of the object. Bells and tuning forks, of course, vibrate at their own characteristic frequencies. Interestingly, most things, from atoms to planets and almost everything else in between, have

The Physics Place
Resonance
Resonance and Bridges

springiness to them, and they vibrate at one or more natural frequencies.

When the frequency of forced vibrations on an object matches the object's natural frequency, a dramatic increase in amplitude occurs. This phenomenon is called **resonance**. Literally, *resonance* means, "resounding," or "sounding again." Putty doesn't resonate, because it isn't elastic, and a dropped handkerchief is too limp to resonate. In order for something to resonate, it needs both a force to pull it back to its starting position and enough energy to maintain its vibration.

A common experience illustrating resonance occurs when you are on a swing. When pumping a swing, you pump in rhythm with the natural frequency of the swing. More important than the force with which you pump is the timing with which you do it. Even small pumps, or small pushes from someone else, if delivered in rhythm with the frequency of the swinging motion, produce large amplitudes.

A common classroom demonstration of resonance is illustrated with a pair of tuning forks adjusted to the same frequency and spaced a meter or so apart (Figure 12.17). When one of the forks is struck, it sets the other fork into vibration. This is a small-scale version of pushing a friend on a swing—it's the timing that's important. When a series of sound waves impinge on the fork, each compression gives the prong of the fork a tiny push. Since the frequency of these pushes corresponds to the natural frequency of the fork, the pushes will successively increase the amplitude of its vibration. This is because the pushes occur at the right time and repeatedly occur in the same direction as the instantaneous motion of the fork. The motion of the second fork is called a *sympathetic vibration*.

fyi
Why does Hollywood persist in playing engine noises whenever a spacecraft in outer space passes by? Wouldn't seeing them float by silently be far more dramatic?

If the forks are not adjusted for matched frequencies, the timing of pushes is off, and resonance doesn't occur. When you tune your radio set, you are similarly adjusting the natural frequency of the electronics in the set to match one of the many surrounding signals. The set then resonates to one station at a time, instead of playing all stations at once.

Resonance is not restricted to wave motion. It occurs whenever successive impulses are applied to

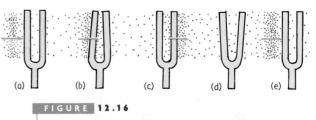

FIGURE 12.16

Stages of resonance. (a) The first compression meets the fork and gives it a tiny and momentary push; (b) the fork bends and then (c) returns to its initial position just at the time a rarefaction arrives and (d) overshoots in the opposite direction. Just when it returns to its initial position, (e) the next compression arrives to repeat the cycle. Now it bends farther because it is moving.

FIGURE 12.17

Ryan demonstrates resonance with a pair of tuning forks with matched frequencies.

a vibrating object in rhythm with its natural frequency. Cavalry troops marching across a footbridge near Manchester, England, in 1831 inadvertently caused the bridge to collapse when they marched in rhythm with the bridge's natural frequency. Since then, it is customary to order troops to "break step" when crossing bridges. A more recent bridge disaster was caused by wind-generated resonance (Figure 12.18).

fyi
Parrots, like humans, use their tongues to craft and shape sound. Tiny changes in tongue position produce big differences in the sound first produced in the syrinx, a voice box organ nestled between the trachea and lungs.

FIGURE 12.18

In 1940, four months after being completed, the Tacoma Narrows Bridge in Washington State was destroyed by wind-generated resonance. The mild gale produced a fluctuating force in resonance with the natural frequency of the bridge, steadily increasing the amplitude until the bridge collapsed.

12.7 Interference

An intriguing property of all waves is **interference**. Consider transverse waves. When the crest of one wave overlaps the crest of another, their individual

The Physics Place
Interference and Beats

effects add together. The result is a wave of increased amplitude. This is *constructive interference* (Figure 12.19). When the crest of one wave overlaps the trough of another, their individual effects are reduced. The high part of one wave simply fills in the low part of another. This is *destructive interference*.

Wave interference is easiest to observe in water. In Figure 12.20, we see the interference pattern produced when two vibrating objects touch the surface of water. We can see the regions in which the crest of one wave overlaps the trough of another to produce a region of zero amplitude. At points along such regions, the waves arrive out of step. We say they are *out of phase* with one another.

Interference is a property of all wave motion, whether the waves are water waves, sound waves, or light waves. We see a comparison of interference for transverse waves and for longitudinal waves in Figure 12.21. In the case of sound, the crest of a wave corresponds to compression and the trough of a wave corresponds to a rarefaction.

Destructive sound interference is at the heart of *antinoise technology*. Some noisy devices such as jackhammers are now equipped with microphones that send the sound of the device to electronic microchips, which create mirror-image wave patterns of the sound signals. This mirror-image sound signal is fed to earphones worn by the operator. In this way, sound compressions (or rarefactions) from the hammer are canceled by mirror image rarefactions (or compressions) in the earphones. The combination of signals cancels the jackhammer noise. Antinoise devices are also common in some aircraft, which are much quieter inside than before this technology was introduced. Are automobiles next, perhaps eliminating the need for mufflers?

Sound interference is dramatically illustrated when monaural sound is played by stereo speakers that are out of phase. Speakers are out of phase when the input wires to one speaker are interchanged

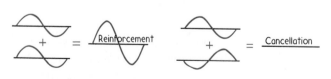

FIGURE 12.19

Constructive and destructive interference in a transverse wave.

FIGURE 12.20

Two sets of overlapping water waves produce an interference pattern.

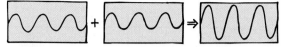

The superposition of two identical transverse waves in phase produces a wave of increased amplitude.

The superposition of two identical longitudinal waves in phase produces a wave of increased intensity.

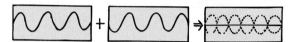

Two identical transverse waves that are out of phase destroy each other when they are superimposed.

Two identical longitudinal waves that are out of phase destroy each other when they are superimposed.

FIGURE 12.21

Constructive (top two panels) and destructive (bottom two panels) wave interference in transverse and longitudinal waves.

FIGURE 12.22

When a mirror image of a sound signal combines with the sound itself, the sound is canceled.

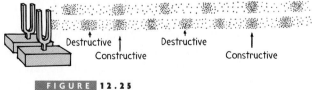

FIGURE 12.23

When the positive and negative wire inputs to one of the stereo speakers have been interchanged, the speakers are then out of phase. When the speakers are far apart, monaural (not stereo) sound is not as loud as it is from properly phased speakers. When they are brought face-to-face, very little sound is heard. Interference is nearly complete, as the compressions of one speaker fill in the rarefactions of the other.

FIGURE 12.24

Ken Ford tows gliders in quiet comfort when he wears his noise-canceling earphones. In larger aircraft, sound from the engines is processed and emitted as antinoise from loudspeakers inside the cabin to provide passengers with a quieter ride.

(positive and negative wire inputs reversed). For a monaural signal, this means that, when one speaker is sending a compression of sound, the other is sending a rarefaction. The sound produced is not as full and not as loud as from speakers properly connected in phase. The longer waves are canceled by interference. Shorter waves are canceled as the speakers are brought closer together, and, when the two speakers are brought face-to-face against each other, very little sound is heard! Only the highest frequencies survive cancellation. You must try this experiment to appreciate it.

Beats

When two tones of slightly different frequencies are sounded together, a fluctuation in the loudness of the combined sounds is heard; the sound is loud, then faint, then loud, then faint, and so on. This periodic variation in the loudness of sound is called **beats**, and it is due to interference. If you strike two

FIGURE 12.25

The interference of two sound sources of slightly different frequencies produces beats.

slightly mismatched tuning forks, one fork vibrates at a different frequency than the other, and the vibrations of the forks will be momentarily in step, then out of step, then in again, and so on. When the combined waves reach our ears in step—say, when a compression from one fork overlaps a compression from the other—the sound is at a maximum. A moment later, when the forks are out of step, a compression from one fork is met with a rarefaction from the other, resulting in a minimum. The sound that reaches our ears throbs between maximum and minimum loudness and produces a tremolo effect.

Beats can occur with any kind of wave, and they can provide a practical way to compare frequencies. To tune a piano, for example, a piano tuner listens for beats produced between a standard tuning fork and those of a particular string on the piano. When the frequencies are identical, the beats disappear. The members of an orchestra tune up their instruments by listening for beats between their instruments and a standard tone produced by a piano or some other instrument.

Standing Waves

Another fascinating effect of interference is *standing waves*. Tie a rope to a wall and shake the free end up and down. The wall is too rigid to shake, so the waves are reflected back along the rope. By shaking the rope just right, you can cause the incident and reflected waves to interfere and form a **standing wave**, where parts of the rope, called the *nodes*, are stationary. You can hold your fingers on either side of the rope at a node, and the rope doesn't touch them. Other parts of the rope, however, would make contact with your fingers. The positions on a standing wave with the largest displacements are known as *antinodes*. Antinodes occur halfway between nodes.

Standing waves are produced when two sets of waves of equal amplitude and wavelength pass through each other in opposite directions. Then the waves are steadily in and out of phase with each other and produce stable regions of constructive and destructive interference (Figure 12.26).

Standing waves are set up in the strings of musical instruments when plucked, bowed, or struck. They are produced in the air in an organ pipe, a flute, or a clarinet—and in the air of a

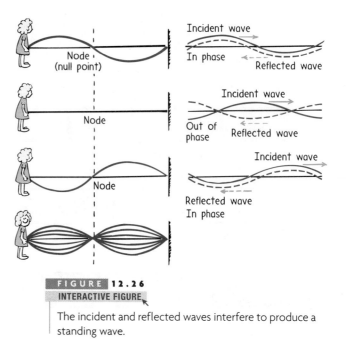

FIGURE 12.26
INTERACTIVE FIGURE

The incident and reflected waves interfere to produce a standing wave.

soda-pop bottle when air is blown over the top. Standing waves appear in a tub of water or a cup of coffee when sloshed back and forth at the appropriate frequency. Standing waves can be produced with either transverse or longitudinal vibrations.

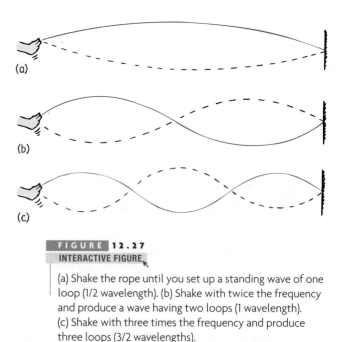

FIGURE 12.27
INTERACTIVE FIGURE

(a) Shake the rope until you set up a standing wave of one loop (1/2 wavelength). (b) Shake with twice the frequency and produce a wave having two loops (1 wavelength). (c) Shake with three times the frequency and produce three loops (3/2 wavelengths).

CHECK YOURSELF

1. Is it possible for one wave to cancel another wave so that there is no amplitude remaining?

2. Suppose you set up a standing wave of three segments, as shown in Figure 12.27c. If you shake with a frequency twice as great, how many wave segments will occur in your new standing wave? How many wavelengths?

CHECK YOUR ANSWERS

1. Yes. This is called destructive interference. When a standing wave is set up in a rope, for example, parts of the rope have no amplitude—the nodes.

2. If you impart twice the frequency to the rope, you'll produce a standing wave with twice as many segments. You'll have six segments. Since a full wavelength has two segments, you'll have three complete wavelengths in your standing wave.

12.8 Doppler Effect

Consider a bug in the middle of a quiet puddle. A pattern of water waves is produced when it jiggles its legs and bobs up and down (Figure 12.28). The bug is not traveling anywhere but merely treads water in a stationary position. The waves it creates are concentric circles because wave speed is the same in all directions. If the bug bobs in the water at a constant frequency, the distance between wave crests (the wavelength) is the same in all directions. Waves encounter point A as frequently as they encounter point B. Therefore, the frequency of wave motion is the same at points A and B, or anywhere in the vicinity of the bug. This wave frequency remains the same as the bobbing frequency of the bug.

The Physics Place
The Doppler Effect

The Physics Place
The Doppler Effect

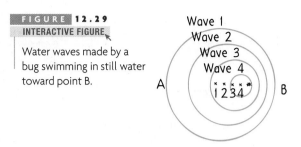

FIGURE 12.29
INTERACTIVE FIGURE

Water waves made by a bug swimming in still water toward point B.

Suppose the jiggling bug moves across the water at a speed less than the wave speed. In effect, the bug chases part of the waves it has produced. The wave pattern is distorted and is no longer that of concentric circles (Figure 12.29). The center of the outer wave originated when the bug was at the center of that circle. The center of the next smaller wave originated when the bug was at the center of that circle, and so forth. The centers of the circular waves move in the direction of the swimming bug. Although the bug maintains the same bobbing frequency as before, an observer at B would see the waves coming more often. The observer would measure a higher frequency. This is because each successive wave has a shorter distance to travel and therefore arrives at B sooner than if the bug weren't moving toward B. An observer at A, on the other hand, measures a lower frequency because of the longer time between wave-crest arrivals. This occurs because each successive wave travels farther to A as a result of the bug's motion. This change in frequency due to the motion of the source (or due to the motion of the receiver) is called the **Doppler effect** (after the Austrian physicist and mathematician Christian Johann Doppler, who lived from 1803–1853).

Water waves spread over the flat surface of the water. Sound and light waves, on the other hand, travel in three-dimensional space in all directions like an expanding balloon. Just as circular waves are closer together in front of the swimming bug, spherical sound or light waves ahead of a moving source are closer together and reach an observer more frequently. The Doppler effect holds for all types of waves.

The Doppler effect is evident when you hear the changing pitch of an ambulance or fire-engine siren. When the siren is approaching you, the crests of the sound waves encounter your ear more frequently, and the pitch is higher than normal. And when the siren passes you and moves away, the

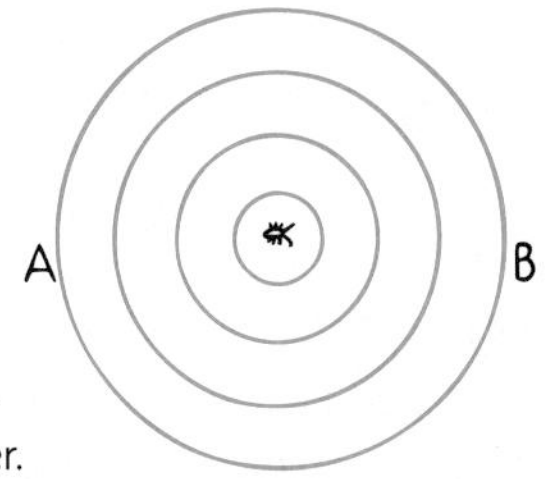

FIGURE 12.28

Top view of water waves made by a stationary bug jiggling in still water.

FIGURE **12.30**
INTERACTIVE FIGURE

The pitch of sound increases when the source moves toward you, and it decreases when the source moves away.

crests of the waves encounter your ear less frequently, and you hear a drop in pitch.

The Doppler effect also occurs for light. When a light source approaches, there is an increase in its measured frequency; when it recedes, there is a decrease in its frequency. An increase in light frequency is called a *blue shift*, because the increase is toward a higher frequency, or toward the blue end of the color spectrum. A decrease in frequency is called a *red shift*, referring to a shift toward a lower frequency, or toward the red end of the color spectrum. The galaxies, for example, show a red shift in the light they emit. A measurement of this shift permits a calculation of their speeds of recession. A rapidly spinning star shows a red shift on the side turning away from us and a relative blue shift on the side turning toward us. This enables us to calculate the star's spin rate.

STOP AND CHECK YOURSELF

When a light or sound source moves toward you, is there an increase or a decrease in the wave speed?

CHECK YOUR ANSWER

Neither! It is the frequency of a wave that undergoes a change when the source is moving, not the wave speed.

12.9 Wave Barriers and Bow Waves

When a source of waves travels as fast as the waves it produces, a "wave barrier" is produced. Consider the bug in our previous example. If it swims as fast as the waves it makes, the bug will keep up with the waves it produces. Instead of the waves going ahead of the bug, they superimpose on one another directly, forming a hump in front of the bug (Figure 12.31). Thus, the bug encounters a wave barrier. Much effort is required of the bug to swim over the hump before it can swim faster than wave speed.

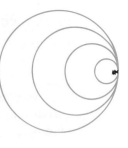

FIGURE **12.31**

The wave pattern made by a bug swimming at wave speed.

The same thing happens when an aircraft travels at the speed of sound. The waves overlap to produce a barrier of compressed air on the leading edges of the wings and on other parts of the aircraft. Considerable thrust is required for the aircraft to push through this barrier (Figure 12.32). Once through, the aircraft can fly faster than the speed of sound without similar opposition. The aircraft is *supersonic*. It is like the bug, which, once it has passed its wave barrier, finds the medium ahead relatively smooth and undisturbed.

When the bug swims faster than wave speed, it produces a pattern of overlapping waves, ideally shown in Figure 12.33. The bug overtakes and outruns the waves it produces. The overlapping waves form a V shape, called a **bow wave**, which appears to be dragging behind the bug. Overlapping waves produce the familiar bow wave generated by a speedboat knifing through the water.

Some wave patterns created by sources moving at various speeds are shown in Figure 12.34. Note

FIGURE **12.32**

Condensation of water vapor by rapid expansion of air can be seen in the rarefied region behind the wall of compressed air.

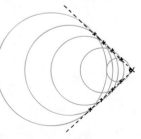

FIGURE **12.33**

Idealized wave pattern made by a bug swimming faster than wave speed.

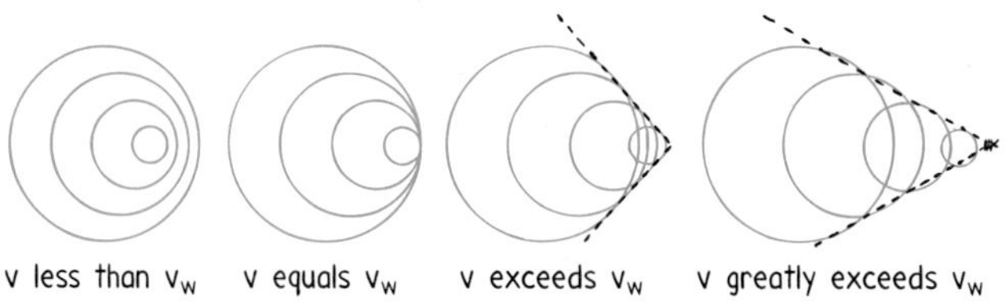

v less than v_w v equals v_w v exceeds v_w v greatly exceeds v_w

FIGURE 12.34

Idealized patterns made by a bug swimming at successively greater speeds. Overlapping at the edges occurs only when the bug swims faster than wave speed.

that, after the speed of the source exceeds wave speed, increased speed produces a narrower V shape.*

12.10 Shock Waves and the Sonic Boom

Whereas a speedboat knifing through the water generates a two-dimensional bow wave at the surface of the water, a supersonic aircraft similarly generates a three-dimensional *shock wave*. Just as a bow wave is produced by overlapping circles that form a V, a **shock wave** is produced by overlapping spheres that form a cone. And just as the bow wave of a speedboat spreads until it reaches the shore of a lake, the conical wake generated by a supersonic aircraft spreads until it reaches the ground.

The bow wave of a speedboat that passes by can splash and douse you if you are at the water's edge. You could say that, in a sense, you are hit by a "water boom." In the same way, when the conical shell of compressed air that sweeps behind a supersonic aircraft reaches listeners on the ground below, the sharp crack they hear is described as a **sonic boom**.

FIGURE 12.35

The shock wave of a bullet piercing a sheet of Plexiglas. Light is deflected as it passes through the compressed air that makes up the shock wave, making it visible. Look carefully and see the second shock wave originating at the tail of the bullet.

* Bow waves generated by boats in water are more complex than is indicated here. Our idealized treatment serves as an analogy for the production of the less complex shock waves in air.

We don't hear a sonic boom from slower-than-sound (subsonic) aircraft because the sound waves reach our ears gradually and are perceived as a continuous tone. Only when the craft moves faster than sound do the waves overlap to reach the listener in a single burst. The sudden increase in pressure is much the same in effect as the sudden expansion of air produced by an explosion. Both processes direct a burst of high-pressure air to the listener. The ear is hard pressed to distinguish between the high pressure caused by an explosion and that produced by many overlapping waves.

A water skier is familiar with the fact that, next to the high hump of the V-shaped bow wave, there is a V-shaped depression. The same is true of a shock wave, which consists of two cones: a high-pressure cone generated at the bow of the supersonic aircraft and a low-pressure cone that follows toward (or at) the tail of the aircraft. The edges of these cones are visible in the photograph of the supersonic bullet in Figure 12.35. Between these two cones, the air pressure rises sharply to above atmospheric pressure, then it falls below atmospheric pressure before sharply returning to normal beyond the inner tail cone (Figure 12.37). This overpressure, suddenly followed by underpressure, intensifies the sonic boom.

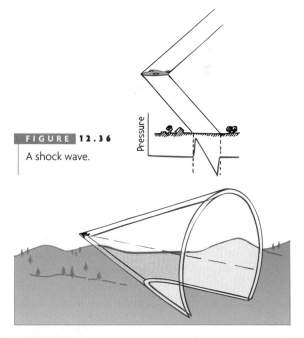

FIGURE 12.36

A shock wave.

FIGURE 12.37

The shock wave actually consists of two cones—a high-pressure cone with its apex at the bow and a low-pressure cone with its apex at the tail. A graph of the air pressure at ground level between the cones takes the shape of the letter N.

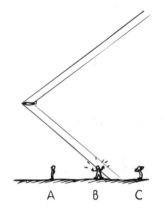

FIGURE **12.38**

The shock wave has not yet reached listener A, but it is now reaching listener B, and it has already reached listener C.

A common misconception is that sonic booms are produced when an aircraft breaks through the sound barrier—that is, just when the aircraft exceeds the speed of sound. This is essentially the same as saying that a boat produces a bow wave when it overtakes its own waves. This is not so. The fact is that a shock wave and its resulting sonic boom are swept continuously behind an aircraft that is traveling faster than sound, just as a bow wave is swept continuously behind a speedboat. In Figure 12.38, listener B is in the process of hearing a sonic boom. Listener C has already heard it, and listener A will hear it shortly. The aircraft that generated this shock wave may have broken through the sound barrier hours ago!

It is not necessary that the moving source be "noisy" to produce a shock wave. Once an object is moving faster than the speed of sound, it will make sound. A supersonic bullet passing overhead produces a crack, which is a small sonic boom. If the bullet were larger and disturbed more air in its path, the crack would be more boomlike. When a lion tamer cracks a circus whip, the cracking sound is actually a sonic boom produced by the tip of the whip traveling faster than the speed of sound. Both the bullet and the whip are not in themselves sound sources, but when they are traveling at supersonic speeds, they produce their own sound as they generate shock waves.

Don't confuse supersonic with ultrasonic. Supersonic has to do with speed—faster than sound. Ultrasonic has to do with frequency—higher than we can hear.

12.11 Musical Sounds

Most of the sounds we hear are noises. The impact of a falling object, the slamming of a door, the roaring of a motorcycle, and most of the sounds from traffic in city streets are noises. Noise corresponds to an irregular vibration of the eardrum produced by an irregularly vibrating source. Graphs that indicate the varying pressure of the air on the eardrum are shown in Figure 12.40a and b. In part a, we see the erratic pattern of noise. In part b, the sound of music has shapes that repeat themselves periodically. These are periodic tones, or musical "notes." (But musical instruments can make noise as well!) Such graphs can be displayed on the screen of an oscilloscope when the electrical signal from a microphone is fed into the input terminal of this useful device.

FIGURE **12.39**

Physics chanteuse Lynda Williams, physics instructor at Santa Rosa Junior College, puts herself fully into the physics of music.

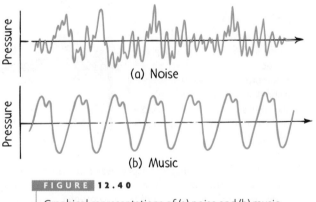

(a) Noise

(b) Music

FIGURE **12.40**

Graphical representations of (a) noise and (b) music.

We have no trouble distinguishing between the tone from a piano and the tone from a clarinet of the same musical pitch (frequency). Each of these tones has a characteristic sound that differs in **quality**, or timbre, a mixture of harmonics of different intensities. Most musical sounds are composed of a superposition of many frequencies called **partial tones**, or simply partials. The lowest frequency, called the **fundamental frequency**, determines the pitch of the note. Partial tones that are whole multiples of the fundamental frequency are called **harmonics**. A tone that has twice the frequency of the fundamental is the second harmonic, a tone with three times the fundamental frequency is the third harmonic, and so on (Figure 12.41).* It is the variety of partial tones that gives a musical note its characteristic quality.

Thus, if we strike middle C on the piano, we produce a fundamental tone with a pitch of about 262 hertz and also a blending of partial tones of two, three, four, five, and so on times the frequency of middle C. The number and relative loudness of the partial tones determine the quality of sound associated with the piano. Sound from practically every musical instrument consists of a fundamental and partials. Pure tones, those having only one frequency, can be produced electronically. Electronic synthesizers, for example, produce pure tones and mixtures of these to produce a vast variety of musical sounds.

The quality of a tone is determined by the presence and relative intensity of the various partials. The ear recognizes the different partials and can therefore differentiate the different sounds produced by a piano and a clarinet. A pair of tones of the same pitch with different qualities has either different partials or a difference in the relative intensity of the partials.

Musical Instruments

Sound production from conventional musical instruments can be grouped into three classes: vibrating strings, vibrating air columns, and percussion.

* Not all partial tones present in a complex tone are integer multiples of the fundamental. Unlike the harmonics of woodwinds and brasses, stringed instruments, such as the piano, produce "stretched" partial tones that are nearly, but not quite, harmonics.

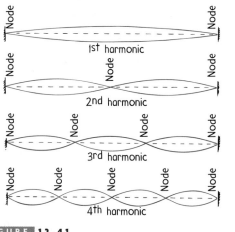

FIGURE 12.41

Modes of vibration of a guitar string.

FIGURE 12.42

A composite vibration of the fundamental mode and the third harmonic.

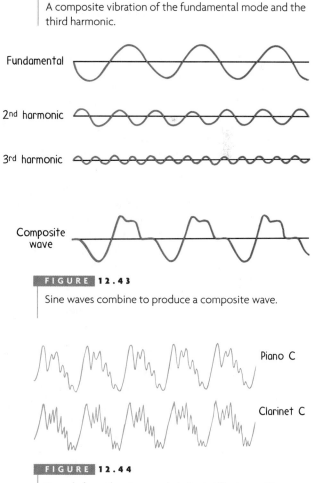

FIGURE 12.43

Sine waves combine to produce a composite wave.

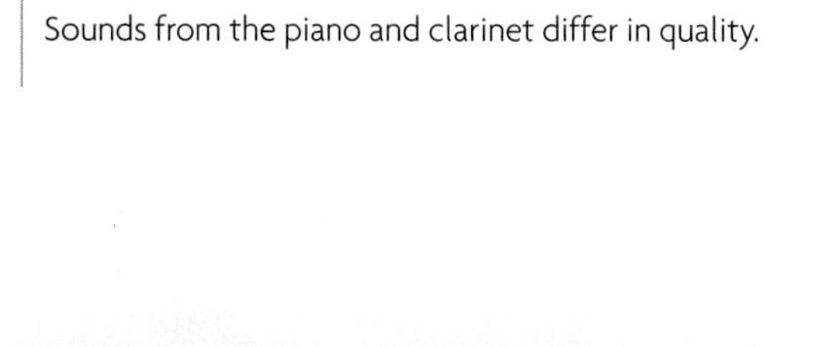

FIGURE 12.44

Sounds from the piano and clarinet differ in quality.

In a stringed instrument, the vibration of the strings is transferred to a sounding board and then to the air, but with low efficiency. To compensate for this effect, we find relatively large string sections in orchestras. A smaller number of the high-efficiency wind instruments sufficiently balance a much larger number of violins.

In a wind instrument, the sound is a vibration of an air column in the instrument. There are various ways to set air columns into vibration. In brass instruments, such as trumpets, French horns, and trombones, vibrations of the player's lips interact with standing waves that are set up by acoustic energy reflected within the instrument by the flared bell. The lengths of the vibrating air columns are manipulated by pushing valves that add or subtract extra segments or by extending or shortening the length of the tube. In woodwinds, such as clarinets, oboes, and saxophones, a stream of air produced by the musician sets a reed vibrating, whereas in fifes, flutes, and piccolos, the musician blows air against the edge of a hole to produce a fluttering stream that sets the air column into vibration.

In percussion instruments, such as drums and cymbals, a two-dimensional membrane or an elastic surface is struck to produce sound. The fundamental tone produced depends on the geometry, the elasticity, and, in some cases, the tension of the surface. Changes in pitch result from changing the tension in the vibrating surface; depressing the edge of a drum membrane with the hand is one way of accomplishing this. Striking the surface in different places can set up different modes of vibration. In the kettledrum, the shape of the kettle changes the frequency of the drum. As in all musical sounds, the quality depends on the number and relative loudness of the partial tones.

Electronic musical instruments differ markedly from conventional musical instruments. Instead of strings that must be bowed, plucked, or struck, or reeds over which air must be blown, or diaphragms that must be tapped to produce sounds, some electronic instruments use electrons to generate the signals that make up musical sounds. Others begin with sound from an acoustical instrument and then modify it. Electronic music demands that the

---

EVERYDAY APPLICATIONS

■ FOURIER ANALYSIS

The French mathematician Joseph Fourier made one of the most interesting discoveries about music in 1822. He discovered that wave motion could be reduced to simple sine waves. A sine wave is the simplest of waves, having a single frequency, as shown in Figure 12.43. All periodic waves, however complicated, can be broken down into constituent sine waves of different amplitudes and frequencies. The mathematical operation for doing this is called **Fourier analysis**. We will not explain the mathematics here, but we will simply point out that, by such analysis, one can find the pure sine tones that constitute the tone of, say, a violin. When these pure tones are sounded together, as by striking a number of tuning forks or by selecting the proper keys on an electric organ, they combine to produce the tone of the violin. The lowest-frequency sine wave is the fundamental, and it determines the pitch of the note. The higher-frequency sine waves are the partials, which give the characteristic quality. Thus, the waveform of any musical sound is no more than a sum of simple sine waves.

Since the waveform of music is a multitude of various sine waves, to duplicate sound accurately by radio, tape recorder, or CD player, we should be able to process as large a range of frequencies as possible. The notes of a piano keyboard range from 27 hertz to 4200 hertz, but to duplicate the music of a piano composition accurately, the sound system must have a range of frequencies up to 20,000 hertz. The greater the range of the frequencies of an electrical sound system, the closer the musical output approximates the original sound, hence the wide range of frequencies that can be produced in a high-fidelity sound system.

Our ear performs a sort of Fourier analysis automatically. It sorts out the complex jumble of air pulsations that reach it, and it transforms them into pure tones. And we recombine various groupings of these pure tones when we listen. What combinations of tones we have learned to focus our attention on determines what we hear when we listen to a concert. We can direct our attention to the sounds of the various instruments and discern the faintest tones from the loudest; we can delight in the intricate interplay of instruments and still detect the extraneous noises of others around us. This is a most incredible feat.

composer and the player demonstrate an expertise beyond the knowledge of musicology. It brings a powerful new tool to the hands of the musician.

> Who better appreciates music—one knowledgeable about it, or the casual listener?

fyi
Scratched CD? Gently wipe a bit of toothpaste on it. The abrasives that polish teeth can also buff scratches from a disc.

FIGURE 12.45
Does each listener hear the same music?

SUMMARY OF TERMS

Vibration A wiggle in time.

Wave A wiggle in both space and time.

Amplitude For a wave or vibration, the maximum displacement on either side of the equilibrium (midpoint) position.

Wavelength The distance between successive crests, troughs, or identical parts of a wave.

Frequency For a vibrating body or medium, the number of vibrations per unit time. For a wave, the number of crests that pass a particular point per unit time.

Hertz The unit of frequency. One hertz (symbol Hz) equals one vibration per second.

Period The time required for a vibration or a wave to make a complete cycle; equal to 1/frequency.

Wave speed The speed with which waves pass a particular point:

$$\text{Wave speed} = \text{frequency} \times \text{wavelength}$$

Transverse wave A wave in which the medium vibrates in a direction perpendicular (transverse) to the direction in which the wave travels. Light consists of transverse waves.

Longitudinal wave A wave in which the medium vibrates in a direction parallel (longitudinal) to the direction in which the wave travels. Sound consists of longitudinal waves.

Compression Condensed region of the medium through which a longitudinal wave travels.

Rarefaction Rarefied region, or region of lessened pressure, of the medium through which a longitudinal wave travels.

Pitch The subjective impression of the frequency of sound.

Reflection The return of a sound wave; an echo.

Reverberation Reechoed sound.

Refraction The bending of a wave, either through a nonuniform medium or from one medium to another, caused by differences in wave speed.

Forced vibration The setting up of vibrations in an object by a vibrating force.

Natural frequency A frequency at which an elastic object naturally tends to vibrate, so that minimum energy is required to produce a forced vibration or to continue vibration at that frequency.

Resonance The response of a body when a forcing frequency matches its natural frequency.

Interference A result of superposing different waves, often of the same wavelength. Constructive interference results from crest-to-crest reinforcement; destructive interference results from crest-to-trough cancellation.

Beats A series of alternate reinforcements and cancellations produced by the interference of two waves of slightly different frequency, heard as a throbbing effect in sound waves.

Standing wave A stationary wave pattern formed in a medium when two sets of identical waves pass through the medium in opposite directions.

Doppler effect The change in frequency of wave motion resulting from motion of the sender or the receiver.

Bow wave The V-shaped wave made by an object moving across a liquid surface at a speed greater than the wave speed.

Shock wave The cone-shaped wave made by an object moving at supersonic speed through a fluid.

Sonic boom The loud sound resulting from a shock wave.

Quality The characteristic timbre of a musical sound, which is governed by the number and relative intensities of partial tones.

Partial tone One of the frequencies present in a complex tone. When a partial tone is an integer multiple of the lowest frequency, it is a harmonic.

Fundamental frequency The lowest frequency of vibration, or the first harmonic. In a string, the vibration makes a single segment.

Harmonics A partial tone that is an integer multiple of the fundamental frequency. The vibration that begins with the fundamental vibrating frequency is the first harmonic, twice the fundamental is the second harmonic, and so on in sequence.

Fourier analysis A mathematical method that disassembles any periodic wave form into a combination of simple sine waves.

SUGGESTED READING

Chiaverina, Chris, and Tom Rossing. *Light Science: Physics for the Visual Arts*. New York: Springer, 1999. Enjoyable reading by two fun physicists!

For more on the acoustics of concert halls, see http://www.concerthalls.org.

See the production of standing waves at http://www2.biglobe.ne.jp/~norimari/science/JavaEd/e-wave4.html.

REVIEW QUESTIONS

12.1 Vibrations and Waves

1. What is a wiggle in time called? A wiggle in space and time?
2. Distinguish between the propagation of sound waves and the propagation of light waves.
3. What is the source of all waves?
4. Distinguish between these different parts of a wave: period, amplitude, wavelength, and frequency.
5. How many vibrations per second are represented in a radio wave of 101.7 MHz?
6. How do frequency and period relate to each other?

12.2 Wave Motion

7. In one word, what is it that moves from source to receiver in wave motion?
8. Does the medium in which a wave travels move with the wave? Give examples to support your answer.
9. What is the relationship among frequency, wavelength, and wave speed?

12.3 Transverse and Longitudinal Waves

10. In what direction are the vibrations in a transverse wave, relative to the direction of wave travel?
11. In what direction are the vibrations in a longitudinal wave, relative to the direction of wave travel?
12. Distinguish between a compression and a rarefaction.

12.4 Sound Waves

13. How does a vibrating tuning fork emit sound?
14. Does sound travel faster in warm air or in cold air? Defend your answer.
15. How does the speed of sound in water compare with the speed of sound in air? How does the speed of sound in steel compare with the speed of sound in air?

12.5 Reflection and Refraction of Sound

16. What is the law of reflection for sound?
17. What is a reverberation?
18. What causes refraction?
19. Does sound tend to bend upward or downward when its speed near the ground is lower than its speed higher up?

20. There is a difference between the way in which we passively see our surroundings in daylight and the way in which we actively probe our surroundings with a searchlight in the darkness. Which of these ways of perceiving our surroundings is more like the way in which a dolphin perceives its environment?

12.6 Forced Vibrations and Resonance

21. Why does a struck tuning fork sound louder when its handle is held against a table?
22. Does a blob of putty have a natural frequency? Explain.
23. Distinguish between forced vibrations and resonance.
24. What is required to make an object resonate?
25. When you listen to a radio, why do you hear only one station at a time instead of all stations at once?
26. Why do troops "break step" when crossing a bridge?

12.7 Interference

27. What kinds of waves exhibit interference?
28. Distinguish between constructive interference and destructive interference.
29. What does it mean to say that one wave is out of phase with another?
30. What physical phenomenon underlies beats?
31. What causes a standing wave?
32. What is a node? What is an antinode?

12.8 Doppler Effect

33. In the Doppler effect, does frequency change? Does wavelength change? Does wave speed change?
34. Can the Doppler effect be observed with longitudinal waves, with transverse waves, or with both?

12.9 Wave Barriers and Bow Waves

35. How do the speed of a wave source and the speed of the waves themselves compare when a wave barrier is being produced? How do they compare when a bow wave is being produced?
36. How does the V shape of a bow wave depend on the speed of the wave source?

12.10 Shock Waves and the Sonic Boom

37. True or false: A sonic boom occurs only when an aircraft is breaking through the sound barrier. Defend your answer.
38. True or false: In order for an object to produce a sonic boom, it must be "noisy." Give two examples to support your answer.

12.11 Musical Sounds

39. Distinguish between a musical sound and noise.
40. Why are there typically more stringed instruments in an orchestra than wind instruments?

ACTIVE EXPLORATIONS

1. Tie a rubber tube, a spring, or a rope to a fixed support and shake it to produce standing waves. See how many nodes you can produce.
2. Test to see which of your ears has the better hearing by covering one ear and finding how far away your open ear can hear the ticking of a clock; repeat for the other ear. Also notice how the sensitivity of your hearing improves when you cup your hands behind your ears.
3. Do the activity suggested in Figure 12.23 with a stereo sound system. Simply reverse the wire inputs to one of the speakers so the two are out of phase. When monaural sound is played and the speakers are brought face-to-face, the lowering of the volume is truly amazing! If the speakers are well insulated, you will hear almost no sound at all.
4. For this activity, you'll need an isolated loudspeaker (bare of its casing) and a sheet of plywood or cardboard—the bigger, the better. Cut a hole in the middle part of the sheet that is about the size of the speaker. Listen to music from the isolated speaker, and then hear the difference when it's placed against the hole. The sheet diminishes the amount of sound from the back of the speaker that interferes with sound coming from the front side, producing a much fuller sound. Now you know why speakers are mounted in enclosures.
5. Wet your finger and rub it slowly around the rim of a thin-rimmed, stemmed glass while you hold the base of the glass firmly to a tabletop with your other hand. The friction of your finger will excite standing waves in the glass, much like the wave made on the strings of a violin by the friction from a violin bow. Try it with a metal bowl.
6. Swing a buzzer of any kind over your head in a circle. You won't hear the Doppler shift, but your friends off to the side will. The pitch will increase as it approaches them, and decrease when it recedes. Then switch places with a friend so you can hear it too.
7. Make the lowest-pitched vocal sound you are capable of; then keep doubling the pitch to see how many octaves your voice can span.
8. Blow over the top of two empty bottles and see if the tone produced is of the same pitch. Then put one in a freezer and try the procedure again. Sound will travel more slowly in the cold denser air of the cold bottle and the note will be lower. Try it and see.

ONE-STEP CALCULATIONS

$$\text{Frequency} = \frac{1}{\text{period}}; f = \frac{1}{T}.$$

$$\text{Period} = \frac{1}{\text{frequency}}; T = \frac{1}{f}.$$

1. What is the frequency, in hertz, that corresponds to each of the following periods?
   a. 0.10 s
   b. 5 s
   c. 1/60 s
2. What is the period, in seconds, that corresponds to each of the following frequencies?
   a. 10 Hz
   b. 0.2 Hz
   c. 60 Hz
3. A weight suspended from a spring is seen to bob up and down over a distance of 20 centimeters twice each second. What is its frequency? Its period? Its amplitude?

Wave speed $v = f\lambda$

4. A skipper on a boat notices wave crests passing his anchor chain every 5 s. He estimates the distance between wave crests to be 15 m. He also correctly estimates the speed of the waves. What is this speed?
5. Radio waves travel at the speed of light—300,000 km/s. What is the wavelength of radio waves received at 101.1 MHz on your FM radio dial?

EXERCISES

1. What is the source of wave motion?
2. If we double the frequency of a vibrating object, what happens to its period?
3. You dip your finger repeatedly into a puddle of water and make waves. What happens to the wavelength if you dip your finger more frequently?
4. How does the frequency of vibration of a small object floating in water compare to the number of waves passing it each second?
5. What kind of motion should you impart to the nozzle of a garden hose so that the resulting stream of water approximates a sine curve?
6. What kind of motion should you impart to a stretched coiled spring (or to a Slinky) to produce a transverse wave? A longitudinal wave?
7. If a gas tap is turned on for a few seconds, someone a couple of meters away will hear the gas escaping long before he or she smells it. What does this indicate about the speed of sound and the motion of molecules in the sound-carrying medium?
8. A cat can hear sound frequencies up to 70,000 Hz. Bats send and receive ultrahigh-frequency squeaks up to 120,000 Hz. Which hears sound of shorter wavelengths, cats or bats?
9. What does it mean to say that a radio station is "at 101.1 on your FM dial"?
10. Sound from Source A has twice the frequency of sound from Source B. Compare the wavelengths of sound from the two sources.
11. Suppose a sound wave and an electromagnetic wave have the same frequency. Which has the longer wavelength?
12. At the stands of a racetrack, you notice smoke from the starter's gun before you hear it fire. Explain.
13. In an Olympic competition, a microphone picks up the sound of the starter's gun and sends it electrically to speakers at every runner's starting block. Why?
14. At the instant that a high-pressure region is created just outside the prongs of a vibrating tuning fork, what is being created inside between the prongs?
15. Why is it so quiet after a snowfall?
16. If a bell is ringing inside a bell jar, we can no longer hear it when the air is pumped out, but we can still see it. What differences in the properties of sound and light does this indicate?
17. Why is the Moon described as a "silent planet"?
18. As you pour water into a glass, you repeatedly tap the glass with a spoon. As the tapped glass is being filled, does the pitch of the sound increase or decrease? (What should you do to answer this question?)
19. If the speed of sound depended on its frequency, would you enjoy a concert sitting in the second balcony?
20. If the frequency of sound is doubled, what change will occur in its speed? What change will occur in its wavelength? Defend your answer.
21. Why does sound travel faster in warm air?

22. Why does sound travel faster in moist air? (Hint: At the same temperature, water-vapor molecules have the same average kinetic energy as the heavier nitrogen and oxygen molecules in the air. How, then, do the average speeds of H_2O molecules compare with those of N_2 and O_2 molecules?)
23. Why is an echo weaker than the original sound?
24. What two physics mistakes occur in a science-fiction movie that shows a distant explosion in outer space, where you see and hear the explosion at the same time?
25. A rule of thumb for estimating the distance in kilometers between an observer and a lightning stroke is to divide the number of seconds in the interval between the flash and the sound by 3. Is this rule correct?
26. If a single disturbance some unknown distance away sends out both transverse and longitudinal waves that travel with distinctly different speeds in the medium, such as in the ground during an earthquake, how could the distance to the disturbance be determined?
27. Why will marchers at the end of a long parade following a band be out of step with marchers near the front?
28. What is the danger posed by people in the balcony of an auditorium stamping their feet in a steady rhythm?
29. Why is the sound of a harp soft in comparison with the sound of a piano?
30. If the handle of a tuning fork is held solidly against a tabletop, the sound from the tuning fork becomes louder. Why? How will this affect the length of time the fork keeps vibrating? Explain.
31. The sitar, an Indian musical instrument, has a set of strings that vibrate and produce music, even though the player never plucks them. These "sympathetic strings" are identical to the plucked strings and are mounted below them. What is your explanation?
32. A special device can transmit sound that is out of phase with the sound of a noisy jackhammer to the jackhammer operator by means of earphones. Over the noise of the jackhammer, the operator can easily hear your voice while you are unable to hear his. Explain.
33. Two sound waves of the same frequency can interfere with each other, but two sound waves must have different frequencies in order to make beats. Why?
34. Walking beside you, your friend takes 50 strides per minute while you take 48 strides per minute. If you start in step, when will you be in step again?
35. Suppose a piano tuner hears three beats per second when listening to the combined sound from his tuning fork and the piano note being tuned. After slightly tightening the string, he hears five beats per second. Should the string be loosened or tightened?
36. A railroad locomotive is at rest with its whistle shrieking, and then it starts moving toward you.
   a. Does the frequency that you hear increase, decrease, or stay the same?
   b. How about the wavelength reaching your ear?

c. How about the speed of sound in the air between you and the locomotive?

37. When you blow your horn while driving toward a stationary listener, an increase in frequency of the horn is heard by the listener. Would the listener hear an increase in the frequency of the horn if he were also in a car traveling at the same speed in the same direction as you are? Explain.

38. How does the Doppler effect aid police in detecting speeding motorists?

39. Astronomers find that light emitted by a particular element at one edge of the Sun has a slightly higher frequency than light from that element at the opposite edge. What do these measurements tell us about the Sun's motion?

40. Would it be correct to say that the Doppler effect is the apparent change in the speed of a wave due to motion of the source? (Why is this question a test of reading comprehension as well as a test of physics knowledge?)

41. Does the conical angle of a shock wave open wider, narrow down, or remain constant as a supersonic aircraft increases its speed?

42. If the sound of an airplane does not originate in the part of the sky where the plane is seen, does this imply that the airplane is traveling faster than the speed of sound? Explain.

43. Does a sonic boom occur at the moment when an aircraft exceeds the speed of sound? Explain.

44. Why is it that a subsonic aircraft, no matter how loud it may be, cannot produce a sonic boom?

45. What physics principle is used by Manuel when he pumps in rhythm with the natural frequency of the swing?

PROBLEMS

● **BEGINNER** ■ **INTERMEDIATE** ◆ **EXPERT**

1. ● A nurse counts 72 heartbeats in 1 minute. Show that the period and frequency of the heartbeats are 0.83 s and 1.2 Hz, respectively.

2. ● We know that speed v = distance/time. Show that when the distance traveled is one wavelength λ and the time of travel is the period T (which equals 1/frequency) you get $v = f\lambda$.

3. ● Microwave ovens typically cook food using microwaves with frequency 2.45 GHz (gigahertz, 10^9 Hz). Show that the wavelength of these microwaves is 12.2 cm.

4. ● For years, marine scientists were mystified by sound waves detected by underwater microphones in the Pacific Ocean. These so-called T waves were among the purest sounds in nature. Eventually, the researchers traced the source to underwater volcanoes whose rising columns of bubbles resonated like organ pipes. A typical T wave has a frequency of 7 Hz. Knowing that the speed of sound in seawater is 1530 m/s, show that the wavelength of a T wave is 219 m.

5. ● An oceanic depth-sounding vessel surveys the ocean bottom with ultrasonic waves that travel 1530 m/s in seawater. Show that when the time delay of an echo to the ocean floor below is 6 seconds, the depth of the water is 4590 m.

6. ● A bat flying in a cave emits a sound and receives its echo 0.1 s later. Show that the distance to the wall of the cave is 17 m.

7. ● Susie hammers on a block of wood when she is 85 m from a large brick wall. Each time she hits the block, she hears an echo 0.5 s later. With this information, show that the speed of sound is 230 m/s.

8. ● Imagine an old hermit type who lives in the mountains. Just before going to sleep, he yells "WAKE UP," and the sound echoes off the nearest mountain and returns 8 hours later. Show that the mountain is almost 5000 km distant.

9. ● On a keyboard, you strike middle C, whose frequency is 256 Hz.
a. Show that the period of one vibration of this tone is 0.00391 s.
b. As the sound leaves the instrument at a speed of 340 m/s, show that its wavelength in air is 1.33 m.

10. ■ a. If you were so foolish as to play your keyboard instrument underwater, where the speed of sound is 1500 m/s, show that the wavelength of the middle-C tone in water would be 5.86 m.
b. Explain why middle C (or any other tone) has a longer wavelength in water than it does in air.

11. ● What beat frequencies are possible with tuning forks of frequencies 256, 259, and 261 Hz?

12. ■ As shown in the drawing, the half-angle of the shock-wave cone generated by a supersonic aircraft is 45°. What is the speed of the plane relative to the speed of sound?

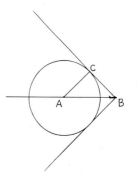

CHAPTER 12 ONLINE RESOURCES

The Physics Place

Interactive Figures
12.1, 12.5, 12.6, 12.26, 12.27, 12.29, 12.30

Tutorials
Waves and Vibrations
The Doppler Effect

Videos
Longitudinal vs. Transverse Waves
Refraction of Sound

Resonance
Resonance and Bridges
Interference and Beats
The Doppler Effect

Quiz

Flashcards

Links

Light Waves

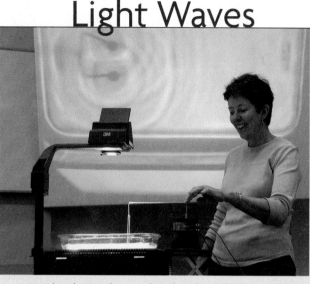

Jennie McKelvie showing that a ripple tank works just fine in New Zealand.

Light is the only thing we can really see. But what is light? We know that, during the day, the primary source of light is the Sun, and a secondary source is the brightness of the sky. Other common sources are white-hot filaments in lightbulbs, glowing gases in glass tubes, and flames. We find that light originates from the accelerated motion of electrons. Light is an electromagnetic phenomenon, and it is only a tiny part of a larger whole—a wide range of electromagnetic waves called the *electromagnetic spectrum*. We begin our study of light by investigating its electromagnetic properties, how it interacts with materials, and its appearance—color. We will see the wave nature of light in the way it diffracts and interferes.

13.1 Electromagnetic Spectrum

If you shake the end of a stick back and forth in still water, you'll create waves on the water's surface. If you similarly shake an electrically charged rod to and fro in empty space, you'll create electromagnetic waves in space. This is because the moving charge is an electric current. Recall from Chapter 11 that a magnetic field surrounds an electric current and that the field changes as the current changes. Recall also that a changing magnetic field induces an electric field—electromagnetic induction. And what does the changing electric field do? It induces a changing magnetic field. The vibrating electric and magnetic fields regenerate each other to make up an **electromagnetic wave**.

In a vacuum, all electromagnetic waves move at the same speed. They differ from one another in their frequency. The classification of electromagnetic waves according to frequency is the **electromagnetic spectrum** (Figure 13.3). Electromagnetic waves have been detected with a frequency as low as 0.01 hertz (Hz). Others, with frequencies of several

FIGURE 13.1
FIGURE 13.1

If you shake an electrically charged object to and fro, you will produce an electromagnetic wave.

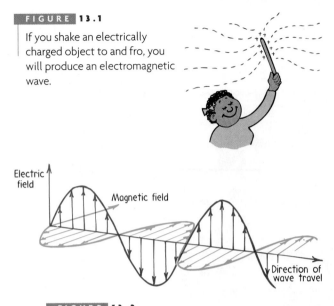

Electric field

Magnetic field

Direction of wave travel

FIGURE 13.2
INTERACTIVE FIGURE

The electric and magnetic fields of an electromagnetic wave in free space are perpendicular to each other and to the direction of motion of the wave.

thousand hertz (kHz), are classified as low-frequency radio waves. One million hertz (1 MHz) lies in the middle of the AM radio band. The very high frequency (VHF) television band of waves begins at about 50 million hertz (MHz) and FM radio frequencies are from 88 to 108 MHz. Then come ultrahigh frequencies (UHF), followed by

FIGURE 13.3
INTERACTIVE FIGURE

The electromagnetic spectrum is a continuous range of waves extending from radio waves to gamma rays. The descriptive names of the sections are merely a historical classification, for all waves are the same in their basic nature, differing principally in frequency and wavelength; all of the waves have the same speed.

microwaves, beyond which are infrared waves, often called "heat waves." Further still is visible light, which makes up less than a millionth of 1 percent of the measured electromagnetic spectrum.

The lowest frequency of light we can see with our eyes appears red. The highest visible frequencies, which are nearly twice the frequency of red light, appear violet. Still higher frequencies in the electromagnetic spectrum are ultraviolet. These higher-frequency waves are more energetic and can cause sunburns. Beyond ultraviolet light are the X-ray and gamma-ray regions. There is no sharp boundary between regions of the spectrum, for they actually grade continuously into one another. The spectrum is divided into these arbitrary regions for the sake of classification.

The frequency of the electromagnetic wave as it vibrates through space is identical with the frequency of the oscillating electric charge that generates it. Different frequencies result in different wavelengths—low frequencies produce long wavelengths and high frequencies produce short wavelengths. The higher the frequency of the vibrating charge, the shorter the wavelength of the radiation.*

The energy carried by light is harvested with solar cells, some of which are now more than 40% efficient. Watch for their greater role in converting the energy of sunlight to electric power.

* The relationship is $c = f\lambda$, where c is the speed of light (constant), f is the frequency, and λ is the wavelength. It is common to describe sound and radio by frequency and light by wavelength. In this book, however, we'll favor the single concept of frequency in describing light.

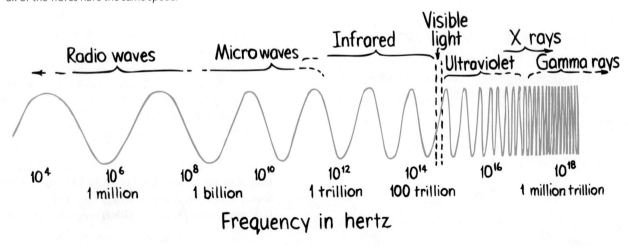

Radio waves

Microwaves

Infrared

Visible light

Ultraviolet

X rays

Gamma rays

10^4 10^6 1 million 10^8 10^{10} 1 billion 10^{12} 1 trillion 10^{14} 100 trillion 10^{16} 10^{18} 1 million trillion

Frequency in hertz

STOP AND
CHECK YOURSELF

Is it correct to say that a radio wave is a low-frequency light wave? Is a radio wave also a sound wave?

CHECK YOUR ANSWER

Yes, both radio waves and light waves are electromagnetic waves that originate in the vibrations of electrons. Radio waves have lower frequencies than light waves, so a radio wave might be considered to be a low-frequency light wave (and a light wave might be considered to be a high-frequency radio wave). A radio wave is definitely not a sound wave. A sound wave is a mechanical vibration of matter, not an electromagnetic vibration. (Don't confuse a radio wave with the sound that a loudspeaker emits.)

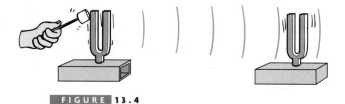

FIGURE 13.4

Just as a sound wave can force a sound receiver into vibration, a light wave can force the electrons in materials into vibration.

FIGURE 13.5

The electrons of atoms have certain natural frequencies of vibration, and these can be modeled as particles connected to the atomic nucleus by springs. As a result, atoms and molecules behave somewhat like optical tuning forks.

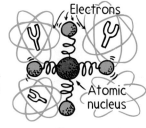

13.2 Transparent and Opaque Materials

Light is energy carried in an electromagnetic wave emitted in most cases by vibrating electrons in atoms. When light is incident upon matter, some of the electrons in the matter are forced into vibration. In this way, vibrations of electrons in the emitter are transmitted to vibrations of electrons in the receiver. This is similar to the way that sound is transmitted (Figure 13.4).

The Physics Place
Light and Transparent Materials

> In air, light travels a million times faster than sound.

Thus, the way a receiving material responds when light is incident upon it depends on the frequency of the light and on the natural frequency of the electrons in the material. Visible light vibrates at a very high rate, some 100 trillion times per second (10^{14} hertz). If a charged object is to respond to these ultrafast vibrations, it must have very, very little inertia. Electrons are light enough to vibrate at this rate.

Materials such as glass and water allow light to pass through without absorption, usually in straight lines. These materials are **transparent** to light. To understand how light penetrates a transparent material, visualize the electrons in an atom as if they were connected to the atomic nucleus by springs (Figure 13.5).* An incident light wave sets the electrons into vibration.

Some materials are springy (elastic) and respond more to vibrations at some frequencies than to others. Bells ring at a particular frequency, tuning forks

vibrate at a particular frequency, and so do the electrons of atoms and molecules. The natural vibration frequencies of an electron depend on how strongly the electron is attached to its atom or molecule. Different atoms and molecules have different "spring strengths." Electrons in glass have a natural vibration frequency in the ultraviolet range. When ultraviolet rays in sunlight shine on glass, resonance occurs as the wave builds and maintains a large amplitude of vibration of the electron, just as pushing someone at the resonant frequency on a swing builds a large amplitude. The energy that atoms in the glass receive may be transferred to neighboring atoms by collisions, or the energy may be reemitted. Resonating atoms in the glass can hold onto the energy of the ultraviolet light for quite a long time (about 100 millionths of a second). During this time, the atom undergoes about 1 million vibrations and collides with neighboring atoms and transfers absorbed energy as heat. Thus, glass is not transparent to ultraviolet. Glass absorbs ultraviolet.

---

* Electrons, of course, are not really connected by springs. We are simply presenting a visual "spring model" of the atom to help us understand the interaction of light with matter. Scientists devise such conceptual models to understand nature, particularly at the submicroscopic level. The worth of a model lies not in whether it is "true" but in whether it is useful—in explaining observations and predicting new ones. If predictions are contrary to new observations, the model is usually either refined or abandoned. The simplified model that we present here—of an atom whose electrons vibrate as if on springs, with a time interval between absorbing energy and reemitting it—is quite useful for understanding how light passes through a transparent material.

3 of many atoms

Glass

FIGURE 13.6

A light wave incident upon a pane of glass sets up vibrations in the molecules that produce a chain of absorptions and reemissions, which pass the light energy through the material and out the other side. Because of the time delay between absorptions and reemissions, the light travels through the glass more slowly than through empty space.

fyi

Materials such as glass are transparent only for those creatures who see in the "visible" part of the spectrum. Other creatures who are tuned to different frequency ranges will see glass as opaque and other materials as transparent.

At lower wave frequencies, such as those of visible light, electrons in the glass are forced into vibration at a lower amplitude. The atoms or molecules in the glass hold the energy for less time, with less chance of collision with neighboring atoms and molecules, and less of the energy transformed to heat. The energy of vibrating electrons is reemitted as light. Glass is transparent to all the frequencies of visible light. The frequency of the reemitted light that is passed from molecule to molecule is identical to the frequency of the light that produced the vibration originally. However, there is a slight time delay between absorption and reemission.

It is this time delay that results in a lower average speed of light through a transparent material (Figure 13.6). Light travels at different average speeds through different materials. We say *average speeds*, for the speed of light in a vacuum, whether in interstellar space or in the space between molecules in a piece of glass, is a constant 300,000 kilometers per second. We call this speed of light c.* The speed of light in the atmosphere is slightly less than it is in a vacuum, but is usually rounded off as c. In water, light travels at 75 percent of its speed in a vacuum, or 0.75 c. In glass, light travels about 0.67 c, depending on the type of glass. In a diamond, light travels at less than half its speed in a

* Today, the accepted value is 299,792 km/s, which is often rounded to 300,000 km/s. (This corresponds to 186,000 mi/s.) We call a corpuscle of light a *photon*.

vacuum, only 0.41 c. Light travels even slower in a silicon carbide crystal called *carborundum*. When light emerges from these materials into the air, it travels at its original speed.

Infrared waves, which have frequencies lower than those of visible light, vibrate not only the electrons but the entire molecules in the structure of the glass and in many other materials. This molecular vibration increases the thermal energy and temperature of the material, which is why infrared waves are often called *heat waves*. Glass is transparent to visible light, but not to ultraviolet and infrared light.

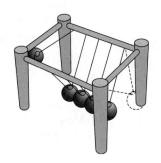

FIGURE 13.7

When the raised ball is released and hits the others, the ball to emerge from the opposite side is not the same ball that initiated the transfer of energy. Likewise, each "photon" that emerges from a pane of glass is not the same photon that was incident upon the glass. Both the emerging ball and emerging photons of light are different, though identical, to the incident ones.

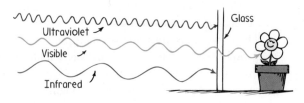

Ultraviolet

Visible

Infrared

Glass

FIGURE 13.8

Clear glass blocks both infrared and ultraviolet, but it is transparent to all the frequencies of visible light.

STOP AND CHECK YOURSELF

1. Why is glass transparent to visible light but opaque to ultraviolet and infrared?

2. Pretend that, while you are at a social gathering, you make several momentary stops across the room to greet people who are "on your wavelength." How is this analogous to light traveling through glass?

3. In what way is it not analogous?

CHECK YOUR ANSWERS

1. The natural frequency of vibration for electrons in glass is the same as the frequency of ultraviolet light, so resonance in the glass occurs when ultraviolet waves shine on it. The absorbed energy is transferred to other atoms as heat, not reemitted as light, so the glass is opaque at ultraviolet frequencies. In the range of visible light, the forced vibrations of electrons in the glass are at smaller amplitudes—vibrations are more subtle, reemission of light (rather than the generation of heat) occurs, and the glass is transparent. Lower-frequency infrared causes whole molecules, rather than electrons, to resonate; again, heat is generated and the glass is opaque.

2. Your average speed across the room would be less because of the time delays associated with your momentary stops. Likewise, the speed of light in glass is less because of the time delays in interactions with atoms along its path.

3. In the case of walking across the room, it is you who begin the walk and you who complete the walk. This is not analogous to the similar case of light, for (according to our model for light passing through a transparent material) the light that is absorbed by an electron that has been made to vibrate is not the same light that is reemitted—even though the two, like identical twins, are indistinguishable.

fyi

 The first to notice a delay in light travel was the Danish astronomer Roemer, who in 1675 saw the effect of light's finite speed "with his own eyes" in eclipses of one of Jupiter's moons due to the increased distance of Earth from Jupiter in six-month intervals. Nearly 300 years later, in 1969, when TV showed astronauts first landing on the Moon, millions of people in their living rooms noticed the time delay between conversations (at the speed of light) between astronauts and earthlings in Mission Control. They noticed the effect of the finite speed of electromagnetic waves "with their own ears."

Most things around us are **opaque**—they absorb light without reemission. Books, desks, chairs, and people are opaque. Energetic vibrations given to the atoms of these materials by light are turned into random kinetic energy—into thermal energy. The materials become slightly warmer.

Metals are opaque to visible light. The outer electrons of atoms in metals are not bound to any particular atom. They are loose and free to wander, with very little restraint, throughout the material (which is why metal conducts electricity and heat so well). When light shines on metal and sets these free electrons into vibration, their energy does not "spring" from atom to atom in the material. It is reflected instead. That's why metals are shiny.

FIGURE 13.9

Metals are shiny because light that shines on them forces free electrons into vibration, which then emit their "own" light waves as reflection.

fyi

 Dark or black skin absorbs ultraviolet radiation before it can penetrate too far. In fair skin, it can travel deeper. Fair skin may develop a tan upon exposure to ultraviolet, which may afford some protection against further exposure. Ultraviolet light is also damaging to the eyes.

Earth's atmosphere is transparent to some ultraviolet light, to all visible light, and to some infrared light. But the atmosphere is opaque to high-frequency ultraviolet light. The small amount of ultraviolet light that does penetrate causes sunburns. If all ultraviolet light penetrated the atmosphere, we would be fried to a crisp. Clouds are semitransparent to ultraviolet light, which is why you can get a sunburn on a cloudy day. Ultraviolet light is not only harmful to your skin, it is also damaging to tar roofs. Now you know why tarred roofs are often covered with gravel.

Have you noticed that things look darker when they are wet than when they are dry? Light incident on a dry surface, such as sand, bounces directly to your eye. But light incident on a wet surface bounces around inside the transparent wet region before it reaches your eye. What happens with each bounce? Absorption! So sand and other things look darker when wet.

STOP AND CHECK YOURSELF

What are two common fates for light shining on a material that isn't absorbed?

CHECK YOUR ANSWER

Transmission and/or reflection. Most light incident on a pane of glass, for example, is transmitted through the pane. But some reflects from its surface. How much transmits and how much reflects varies with different conditions.

13.3 Color

Roses are red and violets are blue; colors intrigue artists and physics types too. To the scientist, the colors of objects are not in the substances of the objects themselves or even in the light they emit or reflect. Color is a physiological experience and is in the eye of the beholder. So, when we say that light from a rose petal is red, in a stricter sense we mean that it *appears* red. Many organisms, including people with defective color vision, will not see the rose as red at all.

The Physics Place
Color

The Physics Place
Colored Shadows
Yellow–Green Peak of Sunlight

FIGURE 13.10

Sunlight passing through a prism separates into a color spectrum. The colors of things depend on the colors of the light that illuminates them.

FIGURE 13.11

The square on the left *reflects* all the colors illuminating it. In sunlight, it is white. When illuminated with blue light, it is blue. The square on the right *absorbs* all the colors illuminating it. In sunlight, it is warmer than the white square.

The colors we see depend on the frequency of the light we see. Different frequencies of light are perceived as different colors; the lowest frequency we detect appears, to most people, as the color red, and the highest appears as violet. Between them range the infinite number of hues that make up the color spectrum of the rainbow. By convention, these hues are grouped into the seven colors: red, orange, yellow, green, blue, indigo, and violet. These colors together appear white. The white light from the Sun is a composite of all the visible frequencies.

Except for such light sources as lamps, lasers, and gas discharge tubes, most of the objects around us reflect rather than emit light. They reflect only part of the light that is incident upon them, the part that provides their color.

> All the colors added together produce white. The absence of all color is black.

Selective Reflection

A rose, for example, doesn't emit light; it reflects light (more about reflection in the next chapter). If we pass sunlight through a prism and then place the petal of a deep-red rose in various parts of the spectrum, the petal will appear brown or black in all regions of the spectrum except in the red region. In the red part of the spectrum, the petal also will appear red, but the green stem and leaves will appear black. This shows that the petal has the ability to reflect red light, but it cannot reflect other colors; the green leaves have the ability to reflect green light and, likewise, cannot reflect other colors. When the rose is held in white light, the petals appear red and the leaves appear green, because the petals reflect the red part of the white light and the leaves reflect the green part of the white light. To understand why objects reflect specific colors of light, we turn our attention to the atom.

Light is reflected from objects in a manner similar to the way sound is "reflected" from a tuning fork when another tuning fork nearby sets it into vibration. A tuning fork can be made to vibrate even when the frequencies are not matched, although at significantly reduced amplitudes. The same is true of atoms and molecules. Electrons can be forced into vibration (oscillation) by the vibrating (oscillating) electric fields of electromagnetic waves.* Once vibrating, these electrons emit their own electromagnetic waves, just as vibrating acoustical tuning forks emit sound waves.

Often a material will reflect light of some frequencies and absorb the rest. If a material absorbs most of the light and reflects red, for example, the material appears red. If it reflects light of all the visible frequencies, like the white part of this page, it will be the same color as the light that shines on it. If a material absorbs light and reflects none, then it is black.

Interestingly, the petals of most yellow flowers, like daffodils, reflect red and green as well as yellow. Yellow daffodils reflect a broad band of frequencies. The reflected colors of most objects are not pure single-frequency colors but are a mixture of frequencies.

An object can reflect only those frequencies present in the illuminating light. The appearance of a colored object therefore depends on the kind of light that illuminates it. An incandescent lamp, for instance, emits light of lower average frequencies than sunlight, enhancing any reds viewed in this light. In a fabric having only a little bit of red in it, the red is more apparent under an incandescent lamp than it is under a fluorescent lamp. Fluorescent lamps are richer in the higher frequencies, and so blues are enhanced in their light. For this reason, it is difficult to tell the true color of objects viewed in artificial light (unless the artificial light is from lamps that match the solar spectrum). How a color appears depends on the light source (Figure 13.12).

Selective Transmission

The color of a transparent object depends on the color of the light it transmits. A red piece of glass appears red because it absorbs cyan or all the colors

* We use the words *oscillate* and *vibrate* interchangeably. Also, the words *oscillators* and *vibrators* have the same meaning.

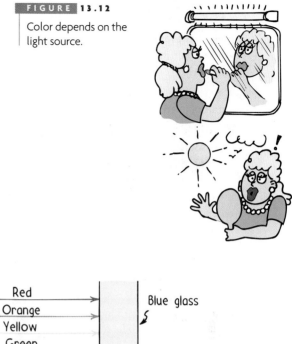

FIGURE 13.12
Color depends on the light source.

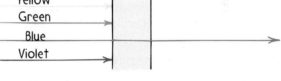

FIGURE 13.13
Only energy having the frequency of blue light is transmitted; energy of the other frequencies, or of the complementary color yellow, is absorbed and warms the glass.

of white light except red, so red light is transmitted. Similarly, a blue piece of glass appears blue because it transmits primarily blue and absorbs the other colors that illuminate it. These pieces of glass contain dyes or *pigments*—fine particles that selectively absorb light of particular frequencies and selectively transmit others. From an atomic point of view, electrons in the pigment molecules are set into vibration by the illuminating light. Light of some of the frequencies is absorbed by the pigments. The rest is reemitted from atom to atom in the glass. The energy of the absorbed light increases the kinetic energy of the atoms, and the glass is warmed. Ordinary window glass doesn't have a color because it transmits light of all visible frequencies equally well.

fyi Carbon is ordinarily black in color, but not when chemically bonded with water in foods such as bread and potatoes. Water is removed when you overheat your toast, which is why burnt toast is black.

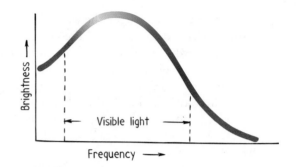

FIGURE 13.14

The radiation curve of sunlight is a graph of brightness versus frequency. Sunlight is brightest in the yellow–green region, which is in the middle of the visible range.

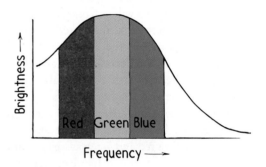

FIGURE 13.15

The radiation curve of sunlight divided into three regions—red, green, and blue. These are the additive primary colors.

Mixing Colored Lights

You can see that white light from the Sun is composed of all the visible frequencies when you pass sunlight through a prism. The white light is dispersed into a rainbow-colored spectrum. The distribution of solar frequencies (Figure 13.14) is uneven, and the light is most intense in the yellow–green part of the spectrum. How fascinating it is that our eyes have evolved to have maximum sensitivity in this range. That's why fire engines and tennis balls are yellow-green for better visibility. Our sensitivity to yellow-green light is also why we see better under the illumination of yellow sodium-vapor lamps at night than we do under incandescent lamps of the same brightness.

All the colors combined produce white. Interestingly, we see white also from the combination of only red, green, and blue light. We can understand this by dividing the solar radiation curve into three regions, as in Figure 13.15. Three types of cone-shaped receptors in our eyes perceive color. Each is stimulated only by certain frequencies of light. Light of lower visible frequencies stimulates the cones sensitive to low frequencies and appears red. Light of middle frequencies stimulates the mid-frequency-sensitive cones and appears green. Light of higher frequencies stimulates the higher-frequency-sensitive cones and appears blue. When all three types of cones are stimulated equally, we see white.

What takes place in the eye seems to be quite complex. Some color sensations depend on intensity, with both rods and cones responding. As intensity increases, orange appears to become yellower and violet seems to get bluer—with no change in frequency. Yellow, green, and blue, however, are independent of intensity, and are called "psychological primaries." The eye is indeed amazing.

All the colors added together produce white. If all the colors are subtracted from white, black is produced.

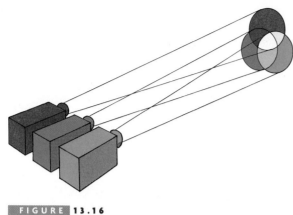

FIGURE 13.16
INTERACTIVE FIGURE

Color addition by the mixing of colored lights. When three projectors shine red, green, and blue light on a white screen, the overlapping parts produce different colors. White is produced where all three overlap.

Project red, green, and blue lights on a screen and, where they all overlap, white is produced. If two of the three colors overlap, or are added, then another color sensation will be produced (Figure 13.16). By adding various amounts of red, green, and blue, the colors to which each of our three types of cones are sensitive, we can produce any color in the spectrum. For this reason, red, green, and blue are called the **additive primary colors**. A close examination of the picture on most color television tubes will reveal that the picture is an assemblage of tiny spots, each less than a millimeter across. When the screen is lit, some of the spots are red, some are green, and some are blue; the mixtures of these primary colors at a distance provide a complete range of colors, plus white.

It's interesting to note that the "black" you see on the darkest scenes on a TV tube is simply the color of the tube face itself, which is more a light gray than black. Because our eyes are sensitive to the contrast with the illuminated parts of the screen, we see this gray as black.

Complementary Colors

Here's what happens when two of the three additive primary colors are combined:

Red + Blue = Magenta

Red + Green = Yellow

Blue + Green = Cyan

We say that magenta is the opposite of green; cyan is opposite red; and yellow is opposite blue. The addition of any color to its opposite color results in white.

Magenta + Green = White (=Red + Blue + Green)

Cyan + Red = White (=Blue + Green + Red)

Yellow + Blue = White (=Red + Green + Blue)

When two colors are added together to produce white, they are called **complementary colors**. Every hue has some complementary color that, when added to it, makes white.

The fact that a color and its complement combine to produce white light is pleasantly used in lighting stage performances. Blue and yellow lights shining on performers, for example, produce the effect of white light—except where one of the two colors is absent, as in the shadows. The shadow of the blue lamp is illuminated by the yellow lamp, and thus it appears yellow. Similarly, the shadow cast by the yellow lamp appears blue. This is a most intriguing effect.

We can see this effect in Figure 13.17, where red, green, and blue lights shine on the golf ball. Note the shadows cast by the ball. The middle shadow is cast by the green spotlight and is not dark because it is illuminated by the red and blue lights, which produces magenta. The shadow cast by the blue light appears yellow because it is illuminated by red and green light. Can you see why the shadow cast by the red light appears cyan?

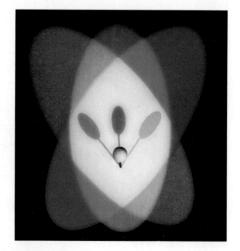

FIGURE 13.17
INTERACTIVE FIGURE

The white golf ball appears white when it is illuminated with red, green, and blue lights of equal intensities. Why are the shadows cast by the ball cyan, magenta, and yellow?

FIGURE 13.18

Paul Robinson displays a variety of colors when he is illuminated by only a red, green, and blue lamp. Can you account for the other resulting colors that appear?

STOP AND CHECK YOURSELF

1. From Figure 13.17, find the complements of cyan, of yellow, and of red.

2. Red + cyan = _____.

3. White − cyan = _____.

4. White − red = _____.

SOLUTIONS

1. Red, blue, cyan

2. White

3. Red

4. Cyan. Interestingly enough, the cyan color of the sea is the result of the removal of red light from white sunlight. The natural frequency of water molecules coincides with the frequency of infrared light, so infrared is strongly absorbed by water. To a lesser extent, red light is also absorbed by water—enough so that it appears a greenish-blue or cyan color.

a   b   c

d   e   f

FIGURE 13.19

Only three colors of ink (plus black) are used to print color photographs—(a) magenta, (b) yellow, (c) cyan, which when combined produce the colors shown in (d). The addition of black (e) produces the finished result (f).

■ COLOR VISION

Light from the world around us focuses upon the retina in our eyes, and we see. The retina is composed of tiny antennae of two kinds that resonate to the incoming light—the rods and the cones. As the names imply, the rods are rod-shaped and the cones are cone-shaped. Rods perceive only intensity of light, while cones perceive color. We see color because of the three types of cones—those sensitive to red, those sensitive to green, and those sensitive to blue. Cones are denser toward the region of distinct vision—the fovea. The rods are sensitive to intensity rather than to frequency, and they predominate away from the fovea, toward the periphery of the retina. Primates and a species of ground squirrel are the only mammals that have the three types of cones and experience full color vision. The retinas of other mammals consist primarily of rods, which are sensitive only to lightness or darkness, so they capture images like those in black-and-white photographs or movies.

Compared with rods, cones require more energy to "fire" an impulse through the nervous system. If the intensity of light is very low, the things we see have no color. We see low intensities with our rods. That's why it's difficult to identify the color of a car by moonlight. Dark-adapted vision is almost entirely due to the rods, while vision in bright light is due to the cones. Stars, for example, look white to us. Yet most stars are actually brightly colored. A time exposure of the stars with a camera reveals reds and red-oranges for the "cooler" stars and blues and blue-violets for the "hotter" stars. The starlight is too faint, however, to fire the color-perceiving cones in the retina. So we see the stars with our rods and perceive them as white or, at most, as only faintly colored. Females have a slightly lower threshold of firing for the cones, however, and they can see more color in stars than males can. So, if she says she sees colored stars and he says she doesn't, she is probably correct!

Mixing Colored Pigments

Every artist knows that, if you mix red, green, and blue paint, the result will not be white but a muddy dark brown. Mixing red and green paint will certainly not produce yellow, so the rule for adding colored lights doesn't apply here. The mixing of pigments in paints and dyes is entirely different from mixing lights. Pigments are tiny particles that absorb specific colors. For example, pigments that produce the color red absorb the complementary color cyan. So something painted red absorbs cyan, which is why it reflects red. In effect, cyan has been subtracted from white light. Something painted blue absorbs yellow, so it reflects all the colors except yellow. Remove yellow from white and you've got blue. The colors magenta, cyan, and yellow are the **subtractive primary colors**. The variety of colors that you see in the colored photographs in this or any book are the result of magenta, cyan, and yellow dots. Light illuminates the book, and light of some frequencies is subtracted from the light reflected. The rules of color subtraction differ from the rules of light addition. We leave this topic to the Suggested Reading.

FIGURE 13.20

Seen through a magnifying glass, the color green on a printed page consists of blue and yellow dots.

FIGURE 13.21

The vivid colors of Sneezlee represent many frequencies of light. The photo, however, is a mixture of only yellow, magenta, cyan, and black.

13.4 # Why the Sky Is Blue, Sunsets Are Red, and Clouds Are White

Why the Sky Is Blue

Not all colors are the result of the addition or subtraction of light. Some colors, like the blue of the sky, are the result of selective scattering.* Consider the analogous case of sound: if a beam of a particular frequency of sound is directed to a tuning fork of a similar frequency, the tuning fork is set into vibration and redirects the beam in multiple directions. The tuning fork *scatters* the sound. A similar process occurs with the scattering of light from atoms and particles that are far apart from one another. This is what happens in the atmosphere.

The Physics Place.
Why the Sky Is Blue and
Sunsets Are Red

We know that atoms behave like tiny optical tuning forks and reemit light waves that shine on them. Very tiny particles act in a similar way: the tinier the particle, the higher the frequency of light it will reemit. This is similar to the way in which small bells ring with higher notes than do larger bells. The nitrogen and oxygen molecules that make up most of the atmosphere are like tiny bells that "ring" with high frequencies when they are energized by sunlight. Like sound from the bells, the reemitted light is sent in all directions. When light is reemitted in all directions, we say that the light is *scattered*.

Of the visible frequencies of sunlight, violet is scattered the most by nitrogen and oxygen in the atmosphere. Then the other colors are scattered in order: blue, green, yellow, orange, and red. Red is scattered only a tenth as much as violet. Although violet light is scattered more than blue, our eyes are not very sensitive to violet light. Therefore, the blue scattered light is what predominates in our vision, so we see a blue sky!

The blue of the sky varies in different locations under various conditions. A main factor is the amount of water vapor in the atmosphere. On clear, dry days, the sky is a much deeper blue than it is on clear, humid days. Places where the upper air is exceptionally dry, such as Italy and Greece, have beautiful blue skies that have inspired painters for centuries. Where the atmosphere contains a lot of particles of dust and other particles larger than oxygen and nitrogen molecules, light of the lower frequencies also undergoes significant scattering. This causes the sky to appear less blue, with a whitish appearance. After a heavy rainstorm, when the airborne particles have been washed away, the sky becomes a deeper blue.

> Doesn't knowing why the sky is blue and why sunsets are red actually add to their beauty? Knowledge doesn't subtract.

The grayish haze in the skies over large cities is the result of particles emitted by automobile and truck engines and by factories. Even when idling, a typical automobile engine emits more than 100 billion particles per second. Most are invisible, but they act as tiny centers to which other particles

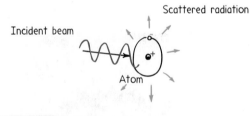

FIGURE 13.22

A beam of light falls on an atom and increases the vibrational motion of electrons in the atom. The vibrating electrons, in turn, reemit light in various directions. Light is scattered.

---

* This type of scattering, which is called Rayleigh scattering, occurs whenever the scattering particles are much smaller than the wavelength of incident light and have resonances at frequencies higher than those of the scattered light.

FIGURE 13.23

In clean air, the scattering of high-frequency light gives us a blue sky. When the air is full of particles larger than oxygen and nitrogen molecules, light of lower frequencies is also scattered, which adds to the high-frequency scattered light to give us a whitish sky.

adhere. These are the primary scatterers of lower-frequency light. With the largest of these particles, absorption rather than scattering occurs, and a brownish haze is produced. Yuck!

Atmospheric soot heats Earth's atmosphere by absorbing light, while cooling local regions by blocking sunlight from reaching the ground. Soot particles in the air can trigger severe rains in one region and droughts and dust storms in another.

Why Sunsets Are Red

Light that isn't scattered is light that is transmitted. Because red, orange, and yellow light are the least scattered by the atmosphere, light of these low frequencies is better transmitted through the air. Red is scattered the least, and it passes through more atmosphere than any other color. So the thicker the atmosphere through which a beam of sunlight travels, the more time there is to scatter all the higher-frequency parts of the light. This means that red light travels through the atmosphere best. As Figure 13.24 shows, sunlight travels through more atmosphere at sunset, which is why sunsets are red.

At noon, sunlight travels through the least amount of atmosphere to reach Earth's surface. Only a small amount of high-frequency light is scattered from the sunlight, enough to make the Sun appear yellowish. As the day progresses and the Sun descends lower in the sky, as Figure 13.24 indicates, the path through the atmosphere is longer, and more violet and blue are scattered from the sunlight. The removal of violet and blue leaves the transmitted light redder. The Sun becomes progressively redder, going from yellow to orange and finally to a red-orange at sunset. Sunsets and sunrises are unusually colorful following volcanic eruptions because particles larger than atmospheric molecules are more abundant in the air.

The colors of the sunset are consistent with our rules for mixing colors. When blue is subtracted from white light, the complementary color that remains is yellow. When higher-frequency violet is subtracted, the resulting complementary color is orange. When medium-frequency green is subtracted, magenta remains. The combinations of resulting colors vary with atmospheric conditions, which change daily, displaying a variety of sunsets.

Why do we see the scattered blue when the background is dark, but not when the background is bright? Because the scattered blue is faint. A faint color will show itself against a dark background, but not against a bright background. For example, when we look from Earth's surface at the atmosphere against the darkness of space, the atmosphere is sky blue. But astronauts above, who look below through the same atmosphere to the bright surface of Earth, do not see the same blueness. They do, of course, see the blueness of the ocean!

Colors in distant landscapes are duller, and color contrasts tend to diminish. That's why a color photograph normally conveys more depth than a black-and-white photograph of the same scene.

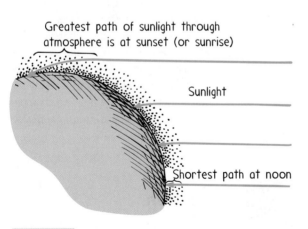

FIGURE 13.24

INTERACTIVE FIGURE.

A sunbeam must travel through more kilometers of atmosphere at sunset than at noon. As a result, more blue is scattered from the beam at sunset than at noon. By the time a beam of initially white light reaches the ground, only light of the lower frequencies survives to produce a red sunset.

Why Clouds Are White

Clouds are made up of clusters of water droplets in a variety of sizes. These clusters of different sizes result in a variety of scattered colors. The tiniest clusters tend to produce blue clouds; slightly larger clusters, green clouds; and still larger clusters, red clouds. The overall result is a white cloud. Electrons close to one another in a cluster vibrate in phase. This results in a greater intensity of scattered light than there would be if the same number of electrons were vibrating separately. Hence, clouds are bright!

FIGURE 13.25

A cloud is composed of water droplets of various sizes. The tiniest droplets scatter blue light, slightly larger ones scatter green light, and still larger ones scatter red light. The overall result is a white cloud.

Larger clusters of droplets absorb much of the light incident upon them, and so the scattered intensity is less. Therefore, clouds composed of larger clusters darken to a deep gray. Further increase in the size of the clusters causes them to fall as raindrops, and we have rain.

The next time you find yourself admiring a crisp blue sky, or delighting in the shapes of bright clouds, or watching a beautiful sunset, think about all those ultra-tiny optical tuning forks vibrating away. You'll appreciate these daily wonders of nature even more!

FIGURE 13.26

The wave appears cyan because seawater absorbs red light. The spray at the crest of the wave appears white because, like clouds, it is composed of a variety of tiny water droplets that scatter all the visible frequencies.

13.5 Diffraction

When you touch your finger to the surface of still water, circular ripples are produced. When you touch the surface with a straightedge, such as a horizontally held meterstick, you produce a plane wave. You can produce a series of plane waves by successively dipping a meterstick into the surface of the water (Figure 13.27).

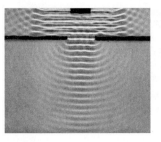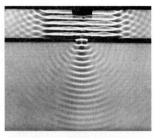

FIGURE 13.27

The oscillating meterstick makes plane waves in the tank of water. Waves diffract through the opening.

The photographs in Figure 13.28 are top views of water ripples in a shallow glass tank (called a ripple tank). A barrier with an adjustable opening is in the tank. When plane waves meet the barrier, they continue through with some distortion. In the left image, where the opening is wide, the waves continue through the opening almost without change. At the two ends of the opening, however, the waves bend. This bending is called **diffraction**. Any bending of light by means other than reflection and refraction is diffraction. As the width of the opening is narrowed, as in the center image in Figure 13.28, the waves spread more. When the opening is small relative to the wavelength of the incident wave, they spread even more. We see that smaller openings will produce greater diffraction. Diffraction is a property of all kinds of waves, including sound and light waves.

Diffraction is not confined to narrow slits or to openings in general but can be seen around the edges of all shadows. On close examination, even the sharpest shadow is blurred slightly at its edges (Figure 13.30).

The amount of diffraction depends on the wavelength of the wave compared with the size of the obstruction that casts the shadow. Long waves are better at filling in shadows, which is why foghorns emit low-frequency sound waves—to fill in any "blind spots." The same is true for radio waves of the standard AM broadcast band, which are very long compared with the sizes of most objects in their path. The wavelength of AM radio waves ranges from 180 to 550 meters, and the waves readily bend around buildings and other objects that might otherwise obstruct them. A long-wavelength radio wave doesn't "see" a relatively small building in its path—but a short-wavelength radio wave does. Because the radio waves of the FM band range from 2.8 to 3.4 meters, they don't bend very well around buildings. This is one of the reasons why FM reception is often poor in localities where AM

FIGURE 13.28

Plane waves passing through openings of various sizes. The smaller the opening, the greater the bending of the waves at the edges.

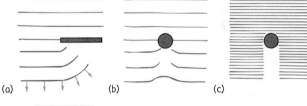

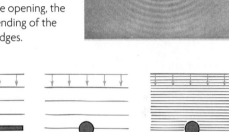

(a) (b) (c)

FIGURE 13.29

(a) Waves tend to spread into the shadow region. (b) When the wavelength is about the size of the object, the shadow is soon filled in. (c) When the wavelength is short compared with the object, a sharp shadow is cast.

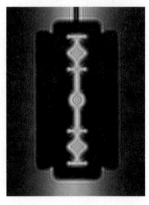

FIGURE 13.30

Diffraction fringes are evident in the shadows of monochromatic (single-frequency) laser light.

reception comes in loud and clear. In the case of radio reception, we don't wish to "see" objects in the path of radio waves, so diffraction is welcome.

Diffraction is not so welcome when you are viewing very small objects with a microscope. If the size of an object is about the same as the wavelength of light, diffraction blurs the image. If the object is

■ SEEING STAR-SHAPED STARS

Have you wondered why stars are represented with spikes? The stars on the American flag have five spikes, and the Jewish Star of David has six spikes. All through the ages, stars have been drawn with spikes. The reason for this has nothing to do with the actual shapes of the stars, which are merely point sources of light in the night sky; rather, it has to do with imperfect eyesight.

The surfaces of our eyes, the corneas, become scratched for a variety of reasons. These scratches make up a diffraction grating of sorts. A scratched cornea is not a very good diffraction grating, but

its effects are evident if you look at a bright point source against a dark background—like a star in the night sky. Instead of seeing a point of light, you see a spiky shape. The spikes will even shimmer and twinkle if there are some temperature differences in the atmosphere to produce some refraction. And, if you live in a windy desert region where sandstorms are frequent, your cornea will be even more scratched and you'll see more vivid star spikes. So stars don't have spikes. They appear spiked because of scratches on the surfaces of our eyes that behave as diffraction gratings. So there's not only physics in all you see, but in how you see!

---

smaller than the wavelength of light, no structure can be seen. The entire image is lost, due to diffraction. No amount of magnification or perfection of microscope design can defeat this fundamental diffraction limit. To get around this problem, microscopists illuminate very tiny objects with electron beams rather than with light. Compared with light waves, electron beams have extremely short wavelengths. *Electron microscopes* take advantage of the fact that all matter has wave properties. A beam of electrons has a wavelength smaller than the wavelengths of visible light. In an electron microscope, electric and magnetic fields, rather than optical lenses, are used to focus and magnify images.

The use of shorter wavelengths to see finer detail is employed by dolphins, who scan their environment with ultrasound. The echoes of long-wavelength sound give the dolphin an overall image of the objects in its surroundings. To examine them in more detail, the dolphin emits sounds of shorter wavelengths. The dolphin has always done naturally what physicians have only recently been capable of doing with ultrasonic imaging devices.

STOP AND
CHECK YOURSELF

Why does a microscopist use blue light rather than white light to illuminate objects being viewed?

CHECK YOUR ANSWER

There is less diffraction with blue light. This allows the microscopist to see more detail (just as a dolphin beautifully investigates fine detail in its environment by means of the echoes of ultra-short wavelengths of sound).

13.6 Interference of Light

Note that the diffracted light in Figure 13.30 shows fringes. These fringes are produced by **interference**, which we discussed in the previous chapter. Constructive and destructive interference is reviewed in Figure 13.31. We see that the addition, or *superposition*, of a pair of identical waves in phase with each other produces a wave of the same frequency but with twice the amplitude. If the waves are exactly one-half wavelength out of phase, their superposition results in complete cancellation. If they are out of phase by other amounts, partial cancellation occurs.

The Physics Place
Soap Bubble Interference

In 1801, the wave nature of light was convincingly demonstrated when the British physicist and physician Thomas Young performed his now famous interference experiment.* Young found that

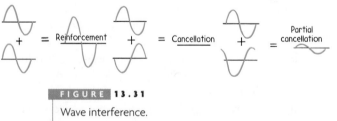

FIGURE 13.31

Wave interference.

---

* Thomas Young read fluently at the age of 2; by 4, he had read the Bible twice; by 14, he knew eight languages. In his adult life, he was a physician and scientist, contributing to an understanding of fluids, work and energy, and the elastic properties of materials. He was the first person to make progress in deciphering Egyptian hieroglyphics. There's no doubt about it—Thomas Young was a bright guy!

light directed through two closely spaced pinholes recombined to produce fringes of brightness and darkness on a screen behind. The bright fringes of light resulted from light waves from the two holes arriving crest to crest, while the dark areas resulted from light waves arriving trough to crest. Figure 13.32 shows Young's drawing of the pattern of superimposed waves from the two sources. His experiment is now done with two closely spaced slits instead of with pinholes, so the fringe patterns are straight lines (Figure 13.33).

We see in Figures 13.34 and 13.35 how the series of bright and dark lines results from the different path lengths from the slits to the screen. For the central bright fringe, the paths from each slit are the same length, and the waves arrive in phase and reinforce each other. The dark fringes on either side of the central fringe result from one path being longer (or shorter) by one-half wavelength, where the waves arrive half a wavelength out of phase. The other sets of dark fringes occur where the paths

FIGURE 13.34

Bright fringes occur when waves from both slits arrive in phase; dark areas result from the overlapping of waves that are out of phase.

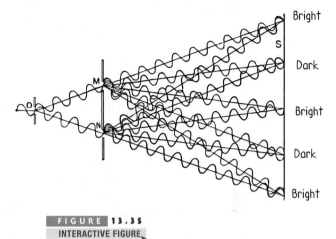

FIGURE 13.35
INTERACTIVE FIGURE

Light from O passes through slits M and N and produces an interference pattern on the screen S.

differ by odd multiples of one-half wavelength: 3/2, 5/2, and so on.

Interference patterns are not limited to one or two slits. A multitude of closely spaced slits makes up a *diffraction grating*. These devices, like prisms, disperse white light into colors. These are used in devices called *spectrometers*, which we will discuss in Chapter 15. The feathers of some birds act as diffraction gratings and disperse colors. The same is true of the microscopic pits on the reflective surface of a compact disc.

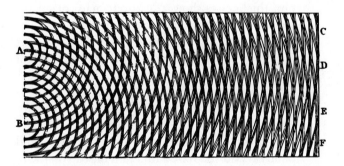

FIGURE 13.32

Thomas Young's original drawing of a two-source interference pattern. Letters C, D, E, and F mark regions of destructive interference.

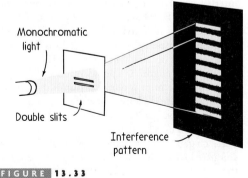

FIGURE 13.33
INTERACTIVE FIGURE

When monochromatic light passes through two closely spaced slits, a striped interference pattern is produced.

FIGURE 13.36

A diffraction grating disperses light into colors by interference. It may be used in place of a prism in a spectrometer.

CHECK YOURSELF

1. If the double slits were illuminated with mono-chromatic (single-frequency) red light, would the fringes be more widely or more closely spaced than if they were illuminated with monochromatic blue light?

2. Why is it important that monochromatic light be used?

CHECK YOUR ANSWERS

1. They would be more widely spaced. Can you see in Figure 13.35 that a slightly longer path—and therefore a slightly more displaced path—from the entrance slit to the screen would result for the longer waves of red light?

2. If light of various wavelengths were diffracted by the slits, dark fringes for one wavelength would be filled in with bright fringes for another, resulting in no distinct fringe pattern. If you haven't seen this, be sure to ask your instructor to demonstrate it.

Interference Colors by Reflection from Thin Films

We have all noticed the beautiful spectrum of colors reflected from a soap bubble or from gasoline on a wet street. These colors are produced by the interference of light waves. This phenomenon, which is often called *iridescence*, is observed in thin transparent films.

A soap bubble appears iridescent in white light when the thickness of the soap film is about the same as the wavelength of light. Light waves reflected from the outer and inner surfaces of the film to your eye travel different distances. When illuminated by white light, the film may be just the right thickness at one place to cause the destructive interference of, say, red light. When red light is subtracted from white light, the mixture remaining appears as the complementary color—cyan. At another place, where the film is thinner, perhaps blue is canceled. Then the light seen is the complement of blue—yellow. Whatever color is canceled by interference, the light seen is its complementary color.

This can be seen when some gasoline has been spilled on a wet street (Figure 13.37). Light reflects from two surfaces: the upper, air–gasoline surface and the lower, gasoline–water surface. If the thick-

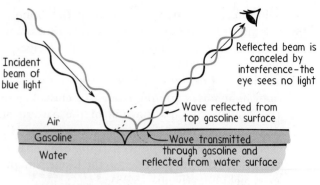

FIGURE 13.37

The thin film of gasoline is just the right thickness to can-cel the reflections of blue light from the top and bottom surfaces. If the film were thinner, perhaps shorter-wavelength violet would be canceled.

ness of the gasoline is such as to cancel blue, as the figure suggests, then the gasoline surface appears yellow to the eye.* As mentioned earlier, blue sub-tracted from white leaves yellow. Why is a variety of colors seen in the thin film of gasoline? The answer is that the film thickness is not uniform. Different film thicknesses show a "contour map" of micro-scopic differences in surface "elevations."

If you view the thin film of gasoline from a lower angle, you'll see different colors. That's because light passing through the film travels a longer path. A longer wave is canceled, and a different color is seen. Different wavelengths of light are canceled for different angles.

Dishes that have been washed in soapy water and poorly rinsed still have a thin film of soap on them. Hold such a dish up to a light source so that *interference colors* can be seen. Then turn the dish to a new position, keeping your eye on the same part

---

* Phase shifts at some reflecting surfaces also contribute to interference. For simplicity and brevity, our concern with this topic will be limited to this footnote: briefly, when light in a medium is reflected at the surface of a second medium in which the speed of transmitted light is less (when there is a greater index of refraction), there is a 180° phase shift (that is, half a wavelength). No phase shift occurs, however, when the second medium is one that transmits light at a higher speed (and there is a lower index of refraction). For example, in a soap bubble, light reflects from the first surface 180° out of phase. Light reflects from the second surface without a phase change. If the thickness of the soap film is very small compared with the wavelength of light, so that the distance through the film is negligible, the parts of the wave reflected from the two surfaces are out of phase and cancel—for all frequencies. This is why parts of a soap film that are extremely thin appear black.

FIGURE 13.38

Bob Greenler shows interference colors with a big bubble. Note that the colors are secondary primaries—magenta, yellow, and cyan.

of the dish. Do you notice a change in color? Light reflecting from the bottom surface of the transparent soap film cancels light reflecting from the top surface.

> Soap-bubble colors result from the interference of reflected light from the inside and outside surfaces of the soap film. When a color is canceled, what you see is its complementary color.

Interference techniques can be used to measure the wavelengths of light and other regions of the electromagnetic spectrum. Interference provides a means of measuring extremely small distances with great accuracy. Instruments called *interferometers*, which use the principle of interference, are the most accurate instruments known for measuring small distances.

STOP AND CHECK YOURSELF

1. What color appears to be reflected from a soap bubble in sunlight when its thickness is such that green light is canceled?

2. In the left column are the colors of certain objects. In the right column are various ways in which colors are produced. Match the right column to the left.

   a. yellow daffodil   1. interference
   b. blue sky   2. diffraction
   c. rainbow   3. selective reflection
   d. peacock feathers   4. refraction
   e. soap bubble   5. scattering

CHECK YOUR ANSWERS

1. The composite of all the visible wavelengths except green is the complementary color, magenta. (Go back and see Figures 13.16 and 13.17.)

2. a–3; b–5; c–4; d–2; e–1

SUMMARY OF TERMS

Electromagnetic wave An energy-carrying wave emitted by vibrating electrical charges (often electrons) and composed of oscillating electric and magnetic fields that regenerate one another.

Electromagnetic spectrum The range of electromagnetic waves that extends in frequency from radio waves to gamma rays.

Transparent The term applied to materials through which light can pass without absorption, usually in straight lines.

Opaque The property of absorbing light without reemission (opposite of transparent).

Additive primary colors The three colors—red, blue, and green—that, when added in certain proportions, will produce any color in the spectrum.

Complementary colors Any two colors that, when added, will produce white light.

Subtractive primary colors The three colors of absorbing pigments—magenta, yellow, and cyan—that, when mixed in certain proportions, will reflect any color in the spectrum.

Diffraction The bending of light that passes around an obstacle or through a narrow opening, causing the light to spread and to produce light and dark fringes.

Interference The result of superposing different waves of the same wavelength. Constructive interference results from crest-to-crest reinforcement; destructive interference results from crest-to-trough cancellation. The interference of selected wavelengths of light produces colors known as interference colors.

SUGGESTED READING

Murphy, Pat, and Paul Doherty. *The Color of Nature*. San Francisco: Chronicle Books, 1996.

REVIEW QUESTIONS

13.1 Electromagnetic Spectrum

1. Does visible light make up a relatively large part or a relatively small part of the electromagnetic spectrum?
2. What is the principal difference between a radio wave and light? Between light and an X-ray?
3. How does the frequency of an electromagnetic wave compare with the frequency of the vibrating electrons that produce it?
4. How is the wavelength of light related to its frequency?

13.2 Transparent and Opaque Materials

5. The sound coming from one tuning fork can force another to vibrate. What is the analogous effect for light?
6. In what region of the electromagnetic spectrum is the resonant frequency of electrons in glass?
7. What is the fate of the energy in ultraviolet light incident on glass?
8. What is the fate of the energy in visible light incident on glass?
9. How does the average speed of light in glass compare with its speed in a vacuum?
10. What part of the electromagnetic spectrum is unable to penetrate Earth's atmosphere?

13.3 Color

11. What is the relationship between the frequency of light and its color?
12. Which has the higher frequency, red light or blue light?
13. Distinguish between the white of this page and the black of this ink, in terms of what happens to the white light that falls on both.
14. How does the color of an object illuminated by an incandescent lamp differ from the color of the same object illuminated by a fluorescent lamp?
15. What is the color of the light that is transmitted through a piece of red glass?
16. Which warms more quickly in sunlight, common window glass or a colored piece of glass? Why?
17. What is the evidence for the statement that white light is a composite of all the colors of the visible part of the electromagnetic spectrum?

18. What is the color of the peak frequency of solar radiation? To what color of light are our eyes most sensitive?
19. What range of frequencies in the radiation curve do red, green, and blue light occupy?
20. Why are red, green, and blue called the additive primary colors?
21. Why are red and cyan called complementary colors?
22. What are the subtractive primary colors? Why are they so called?

13.4 Why the Sky Is Blue, Sunsets Are Red, and Clouds Are White

23. What does it mean to say that light is scattered?
24. Why does the sky sometimes appear whitish?
25. Why does the Sun look reddish at sunrise and sunset but not at noon?
26. What is the evidence for a cloud being composed of particles having a variety of sizes?

13.5 Diffraction

27. Is diffraction more pronounced through a narrow opening or a wide opening?
28. For an opening of a given size, is diffraction more pronounced for a longer or a shorter wavelength?
29. What are some of the ways in which diffraction can be useful or troublesome?

13.6 Interference of Light

30. Is interference restricted only to some types of waves, or does it occur for all types of waves?
31. What is monochromatic light?
32. What produces iridescence?
33. What causes the variety of colors seen in gasoline splotches on a wet street? What two surfaces provide these colors?
34. What accounts for the variety of colors in a soap bubble?
35. If you look at a soap bubble from different angles so that you're viewing different apparent thicknesses of soap film, do you see different colors? Explain.

ACTIVE EXPLORATIONS

1. Which eye do you use more? To test which you favor, hold a finger up at arm's length. With both eyes open, look past it at a distant object. Now close your right eye. If your finger appears to jump to the right, then you use your right eye more.
2. Stare at a piece of colored paper for 45 seconds or so. Then look at a plain white surface. The cones in your retina receptive to the color of the paper become fatigued, so you see an afterimage of the complementary color when you look at a white area. This is because the fatigued cones send a weaker signal to the brain. All the colors produce white, but all the colors minus one produce the color that is complementary to the missing color. Try it and see!
3. Simulate your own sunset: add a few drops of milk to a glass of water and look at a lightbulb through the glass. The bulb appears to be red or pale orange, while light scattered to the side appears blue. Try it and see.
4. With a razor blade, cut a slit in a card and look at a light source through it. You can vary the size of the opening by bending the card slightly. See the interference fringes? Try it with two closely spaced slits.
5. Next time you're in the bathtub, froth up the soapsuds and notice the colors of highlights from the illuminating light overhead on each tiny bubble. Notice that different bubbles reflect different colors, due to the different thicknesses of soap film. If a friend is bathing with you, compare the different colors that you each see reflected from the same bubbles. You'll see that they're different—for what you see depends on your point of view!
6. Do this one at your kitchen sink. Dip a dark-colored coffee cup (dark colors make the best background for viewing interference colors) in dishwashing detergent, and then hold it sideways and look at the reflected light from the detergent film that covers its mouth. Swirling colors appear as the film runs down to form a wedge that grows thicker at the bottom with time. The top becomes thinner, so thin that it appears black. This tells us that its thickness is less than one-fourth the thickness of the shortest waves of visible light. Whatever its wavelength, light reflecting from the inner surface reverses phase, rejoins light reflecting from the outer surface, and cancels. The film soon becomes so thin that it pops.
7. Write a letter to both grandma and grandpa and tell them the reasons for the blueness of the sky, the redness of sunrises and sunsets, and why the clouds are normally white. Explain how knowing the reasons adds, not subtracts, from your appreciation of nature.

EXERCISES

1. What is the fundamental source of electromagnetic radiation?
2. Which have the longest wavelengths: light waves, X-rays, or radio waves?
3. Which has the shorter wavelengths, ultraviolet or infrared? Which has the higher frequencies?
4. We hear people talk of "ultraviolet light" and "infrared light." Why are these terms misleading? Why are we less likely to hear people talk of "radio light" and "X-ray light"?
5. Which requires a physical medium in which to travel, light or sound? Or do both require a physical medium? Explain.
6. Do radio waves travel at the speed of sound, at the speed of light, or at some speed in between?
7. What do radio waves and light have in common? What is different about them?
8. What evidence can you cite to support the idea that light can travel in a vacuum?
9. Short wavelengths of visible light interact more frequently with the atoms in glass than do longer wavelengths. Does this interaction time tend to speed up or to slow down the average speed of light in glass?
10. What determines whether a material is transparent or opaque?
11. You can get a sunburn on a cloudy day, but you can't get a sunburn even on a sunny day if you are behind glass. Explain.
12. Suppose that sunlight falls on both a pair of reading glasses and a pair of dark sunglasses. Which pair of glasses would you expect to become warmer? Defend your answer.
13. In a dress shop with only fluorescent lighting, a customer insists on taking dresses into the daylight at the doorway to check their color. Is she being reasonable? Explain.
14. Fire engines used to be red. Now many of them are yellow-green. Why the change of color?
15. The radiation curve of the Sun (Figure 13.14) shows that the brightest light from the Sun is yellow-green. Why then do we see the Sun as whitish instead of yellow-green?
16. A spotlight is coated so that it won't transmit yellow light from its white-hot filament. What color is the emerging beam of light?
17. How could you use the spotlights at a play to make the yellow clothes of the performers suddenly change to black?
18. Does a color television work by color addition or by color subtraction? Defend your answer.

19. On a TV screen, red, green, and blue spots of fluorescent materials are illuminated at a variety of relative intensities to produce a full spectrum of colors. What dots are activated to produce yellow? To produce magenta? To produce white?

20. What colors of ink do color ink-jet printers use to produce a full range of colors? Do the colors form by color addition or by color subtraction?

21. Below is a photo of science author Suzanne Lyons with her son Tristan wearing red and her daughter Simone wearing green. Below that is the negative of the photo, which shows these colors differently. What is your explanation?

22. Check Figure 13.16 to see if the following three statements are accurate. Then provide the missing word in the last statement. (All colors are combined by the addition of light.)

    Red + green + blue = white.

    Red + green = yellow = white–blue.

    Red + blue = magenta = white–green.

    Green + blue = cyan = white–_____.

23. Under which light will a ripe banana appear black?
    a. red light
    b. yellow light
    c. green light
    d. blue light

24. When white light is shone on red ink that has dried on a clear glass plate, the color that is transmitted is red. But the color that is reflected is not red. What is it?

25. Stare intently for at least a half minute at an American flag. Then turn your gaze to a white wall. What colors do you see in the image of the flag that appears on the wall?

26. Why can't we see stars in the daytime?

27. Why is the sky a darker blue when you are at high altitudes? (Hint: What color is the "sky" on the Moon?)

28. Why does smoke from a campfire look bluish against trees near the ground but yellowish against the sky?

29. Tiny particles, like tiny bells, scatter high-frequency waves more than low-frequency waves. Large particles, like large bells, mostly scatter low frequencies. Intermediate-size particles and bells mostly scatter intermediate frequencies. What does this have to do with the whiteness of clouds?

30. Very big particles, like droplets of water, absorb more radiation than they scatter. What does this have to do with the darkness of rain clouds?

31. The atmosphere of Jupiter is more than 1000 km thick. From the surface of this planet, would you expect to see a white Sun?

32. You're explaining to a youngster at the seashore why the water is cyan colored. The youngster points to the whitecaps of overturning waves and asks why they are white. What is your answer?

33. Why do radio waves diffract around buildings, while light waves do not?

34. Light illuminates two thin, closely spaced slits and produces an interference pattern on a screen behind. How will the distance between the fringes of the pattern differ for red light and blue light?

35. Why is Young's experiment more effective if you use slits rather than the pinholes he first used?

36. A pattern of fringes is produced when monochromatic light passes through a pair of thin slits. Would such a pattern be produced by three thin parallel slits? By thousands of such slits? Give an example to support your answer.

37. Why are interference colors not seen from a thin layer of gasoline on a dry street?

38. If you notice the interference patterns of a thin film of oil or gasoline on water, you'll note that the colors form complete rings. How are these rings similar to the lines of equal elevation on a contour map?

39. Because of wave interference, a film of oil on water is seen as yellow by observers directly above in an airplane. What color does the film appear to a scuba diver looking upward from directly below?

40. Some coated lenses appear bluish when seen by reflected light. What color of light do you suppose they are designed to eliminate?

PROBLEMS

● **BEGINNER**   ■ **INTERMEDIATE**   ◆ **EXPERT**

1. ● Laser pointers emit light waves whose wavelength is 670 nm. What is the frequency for this light? $(1 \text{ nm} = 10^{-9} \text{ m.})$

2. ● Electrons on a radio broadcasting tower are forced to oscillate up and down the tower 535,000 times each second. What is the wavelength of the radio waves that are produced?

3. ● The Hydra galaxy is moving away from Earth at 6.0×10^7 m/s. What fraction of the speed of light is this?

4. ● Consider a pulse of laser light aimed at the Moon that bounces back to Earth. The distance between Earth and the Moon is 3.8×10^8 m. Show that the round-trip time for the light is 2.5 seconds.

5. ● The nearest star beyond the Sun is Alpha Centauri, which is 4.2×10^{16} meters away. If we were to receive a radio message from this star today, show that it would have been sent 4.4 years ago.

6. ● Blue-green light has a frequency of about 6×10^{14} Hz. Using the relationship $c = f\lambda$, show that its wavelength in air is 5×10^{-7} m. How much larger is this wavelength compared to the size of an atom, which is about 10^{-10} m?

7. ● Ultraviolet light has a higher frequency than visible light. Show that the frequency of ultraviolet light of wavelength 360 nm is 8.33×10^{14} Hz.

8. ■ A certain radar installation that is used to track airplanes transmits electromagnetic radiation with a wavelength of 3 cm.
   a. Show that the frequency of this radiation is 10 GHz.
   b. Show that the time required for a pulse of radar waves to reach an airplane 5 km away and return would be 3.3×10^{-5} s.

CHAPTER 13 ONLINE RESOURCES

The Physics Place

Interactive Figures
13.2, 13.3, 13.16, 13.17, 13.24, 13.33, 13.35

Tutorials
Color

Videos
Light and Transparent Materials
Colored Shadows

Yellow–Green Peak of Sunlight
Why the Sky Is Blue and Sunsets Are Red
Soap Bubble Interference

Quiz

Flashcards

Links

Properties of Light

Peter Hopkinson boosts class interest using this zany demonstration of standing astride a large mirror as he lifts his right leg while his unseen left leg provides support behind the mirror.

Most of what we see about us doesn't emit light. Visible things are illuminated by light from a primary source, such as the Sun or a lamp, or from a secondary source, such as a bright sky. Light falling on the surface of a material is usually either reemitted without change in frequency or is absorbed into the material and turned into heat. Usually, both of these processes occur in varying degrees. When the reemitted light returns to the medium from which it came, it is *reflected*. When the reemitted light bends from its original course and proceeds in straight lines from molecule to molecule into a transparent material, it is *refracted*.

14.1 Reflection

When this page is illuminated by sunlight or lamplight, electrons in the atoms of the paper vibrate more energetically in response to the oscillating electric fields of the illuminating light. The energized electrons reemit this light and we see the page. When the page is illuminated by white light, it appears white, which reveals the fact that the electrons reemit all the visible frequencies (recall that all the visible frequencies of light combine to produce white). Very little absorption occurs in the page. The ink on the page is a different story. Except for a bit of reflection, the ink absorbs all the visible frequencies and therefore appears black. Or where a colored dye makes up the ink, we see colored parts of the page.

Law of Reflection

Anyone who has played pool or billiards knows that, when a ball bounces from a surface, the angle of incidence is equal to the angle of rebound. The same is true of light. This is the **law of reflection**, and it holds for all angles:

The angle of reflection equals the angle of incidence.

The law of reflection is illustrated with arrows representing light rays in Figure 14.1. Instead of measuring the angles of incident and reflected rays

 **The Physics Place**
Image Formation in a Mirror

from the reflecting surface, it is customary to measure them from a line perpendicular to the plane of the reflecting surface. This imaginary line is called the *normal*. The incident ray, the normal, and the reflected ray all lie in the same plane.

If you place a candle in front of a mirror, rays of light radiate from the flame in all directions. Figure 14.2 shows only four of the infinite number of rays leaving one of the infinite number of points on the candle. When these rays meet the mirror, they reflect at angles equal to their angles of incidence. The rays diverge from the flame. Note that they also diverge when reflecting from the mirror. These

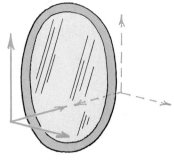

Marjorie's image is as far behind the mirror as she is in front of it. Note that she and her image have the same color of clothing—evidence that light doesn't change frequency upon reflection. Interestingly, her left and right axis is no more reversed than her up and down axis. The axis that is reversed, as shown to the right, is her front and back axis. That's why it appears that her left hand faces the right hand of her image.

divergent rays appear to emanate from behind the mirror (dashed lines). You see an image of the candle at this point. The light rays do not actually come from this point, so the image is called a *virtual image*. The image is as far behind the mirror as the object is in front of the mirror, and image and object have the same size. When you view yourself in a mirror, for example, the size of your image is the same as the size your twin would appear to be, if located as far behind the mirror as you are in front—as long as the mirror is flat. A flat mirror is called a *plane mirror*.

> Your image is as far behind a plane mirror as you are in front of it—as if your twin stood behind a pane of clear glass at a distance as far behind the glass as you are in front of it.

When the mirror is curved, the sizes and distances of object and image are no longer equal. We will not discuss curved mirrors in this text, except to say that the law of reflection still holds. A curved mirror behaves as a succession of flat mirrors, each at a slightly different angular orientation from the one next to it. At each point, the angle of incidence is equal to the angle of reflection (Figure 14.4). Note that in a curved mirror, unlike in a plane mirror, the normals (shown by the dashed black lines) at different points on the surface are not parallel to one another.

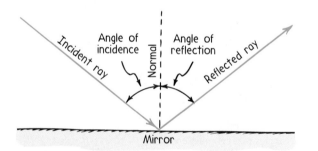

The law of reflection.

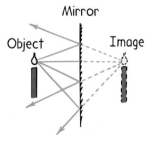

A virtual image is formed behind the mirror and is located at the position where the extended reflected rays (dashed lines) converge.

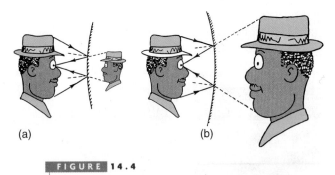

FIGURE 14.4

(a) The virtual image formed by a convex mirror (a mirror that curves outward) is smaller and closer to the mirror than the object. (b) When the object is close to a concave mirror (a mirror that curves inward like a "cave"), the virtual image is larger and farther away than the object. In either case, the law of reflection holds for each ray.

Whether the mirror is plane or curved, the eye–brain system cannot ordinarily distinguish between an object and its reflected image. So the illusion that an object exists behind a mirror (or, in some cases, in front of a concave mirror) is merely due to the fact that the light from the object enters the eye in exactly the same manner, physically, as it would have entered if the object really were at the image location.

STOP AND CHECK YOURSELF

1. What evidence can you cite to support the claim that the frequency of light does not change upon reflection?

2. If you wish to take a picture of your image while standing 5 m in front of a plane mirror, for what distance should you set your camera to provide the sharpest focus?

CHECK YOUR ANSWERS

1. Simply stand in front of a mirror and compare the color of your shirt with the color of its image. The fact that the color is the same is evidence that the frequency of light doesn't change upon reflection.

2. You should set your camera for 10 m; the situation is equivalent to you standing 5 m in front of an open window and viewing your twin standing 5 m in back of the window. Would you like to become rich? Be the first to invent a surface that will reflect 100% of the light incident upon it.

Only part of the light that strikes a surface is reflected. For example, on a surface of clear glass and for normal incidence (light perpendicular to the surface), only about 4% is reflected from each surface. On a clean and polished aluminum or silver surface, however, about 90% of the incident light is reflected.

Diffuse Reflection

In contrast to specula reflection is **diffuse reflection**, which occurs when light is incident on a rough surface and reflected in many directions (Figure 14.5). If the surface is so smooth that the distances between successive elevations on the surface are less than about one-eighth the wavelength of the light, there is very little diffuse reflection, and the surface is said to be *polished*. A surface therefore may be polished for radiation of long wavelengths but rough for light of short wavelengths. The wire-mesh "dish" shown in Figure 14.6 is very rough for light waves and is hardly mirrorlike. But, for long-wavelength radio waves, it is "polished" and is an excellent reflector.

Light reflecting from this page is diffuse. The page may be smooth to a radio wave, but, to a light wave, it is rough. Smoothness is relative to the wavelength of the illuminating waves. Rays of light striking this page encounter millions of tiny flat surfaces facing in all directions. The incident light, therefore, is reflected in all directions. This is a desirable circumstance. It enables us to see objects from any

FIGURE 14.5

Diffuse reflection. Although reflection of each single ray obeys the law of reflection, the many different surface angles that light rays encounter in striking a rough surface produce reflection in many directions.

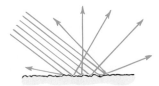

FIGURE 14.6

The open-mesh parabolic dish is a diffuse reflector for short-wavelength light but a polished reflector for long-wavelength radio waves.

FIGURE 14.7
FIGURE 14.7
A magnified view of the surface of ordinary paper.

direction or position. You can see the road ahead of your car at night, for instance, because of diffuse reflection by the rough road surface. When the road is wet, however, it is smoother with less diffuse reflection, and therefore more difficult to see. Most of our environment is seen by diffuse reflection.

An undesirable circumstance related to diffuse reflection is the ghost image that occurs on a TV set when the TV signal bounces off buildings and other obstructions. For antenna reception, this difference in path lengths for the direct signal and the reflected signal produces a slight time delay as well as wave interference. The ghost image is normally displaced to the right, the direction of scanning in the TV tube, because the reflected signal arrives at the receiving antenna later than the direct signal. Multiple reflections contribute to multiple ghosts.

STOP AND CHECK YOURSELF

In terms of the physics of reflection, why is it more dangerous to drive a car on a rainy night?

CHECK YOUR ANSWER

Because the road surface is more mirrored when it is wet, beams from your headlights mostly reflect ahead instead of back to you by diffuse reflection. This makes the road more difficult to see. Furthermore, headlights from oncoming cars reflect from the wet surface full force into your eyes. Glare is much more intense from a mirrored surface.

14.2 Refraction

Recall, from the previous chapter, that light slows down when it enters glass and that it travels at different speeds in different materials.* It travels at 300,000 km/s in a vacuum, at a slightly lower speed in air, and at about three-fourths that speed in water. In a diamond, light travels at about 40% of its speed in a vacuum. As mentioned at the beginning of this chapter, when light passes from one medium to another, we call the process *refraction*. Unless the light is perpendicular to the surface of penetration, bending occurs.

The Physics Place
Models of Refraction

To gain a better understanding of the bending of light in refraction, look at the pair of toy cart wheels in Figure 14.8. The wheels roll from a smooth sidewalk onto a grass lawn. If the wheels meet the grass at an angle, as the figure shows, they are deflected from their straight-line course. Note that the left wheel slows first when it interacts with the grass on the lawn. The right wheel maintains its higher speed while on the sidewalk. It pivots about the slower-moving left wheel because it travels farther in the same time. So the direction of the rolling wheels is bent toward the "normal," the black dashed line perpendicular to the grass–sidewalk border in Figure 14.8.

> A light ray is always at right angles to its wave front.

Figure 14.9 shows how a light wave bends in a similar way. Note the direction of light, indicated by the blue arrow (the light ray). Also note the *wave fronts* drawn at right angles to the ray. (If the light source were close, the wave fronts would appear circular; but, if the distant Sun is the source, the wave fronts are practically straight lines.) The wave fronts are everywhere at right angles to the light rays. In the figure, the wave meets the water surface at an

---

* Just how much the speed of light differs from its speed in a vacuum is given by the index of refraction, n, of the material:

$$n = \frac{\text{speed of light in vacuum}}{\text{speed of light in material}}$$

For example, the speed of light in a diamond is 124,000 km/s, and so the index of refraction for diamond is

$$n = \frac{300,000 \text{ km/s}}{124,000 \text{ km/s}} = 2.42$$

For a vacuum, $n = 1$.

The direction of the rolling wheels changes when one wheel slows down before the other does.

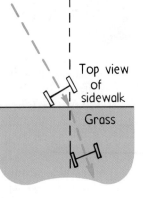

Top view of sidewalk

Grass

angle. This means that the left portion of the wave slows down in the water while the remainder in the air travels at the full speed of light, *c*. The light ray remains perpendicular to the wave front and therefore bends at the surface. It bends like the wheels bend when they roll from the sidewalk onto the grass. In both cases, the bending is caused by a change of speed.*

Figure 14.11 shows a beam of light entering water at the left and exiting at the right. The path would be the same if the light entered from the right and exited at the left. The light paths are reversible for both reflection and refraction. If you

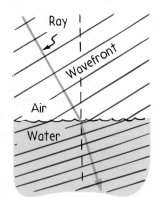

Ray

Wavefront

Air

Water

FIGURE 14.9

The direction of the light waves changes when one part of the wave slows down before the other part.

* The quantitative law of refraction, called Snell's law, is credited to Willebrord Snell, a seventeenth-century Dutch astronomer and mathematician: $n_1 \sin \theta_1 = n_2 \sin \theta_2$, where n_1 and n_2 are the indices of refraction of the media on either side of the surface, and θ_1 and θ_2 are the respective angles of incidence and refraction. If three of these values are known, the fourth can be calculated from this relationship.

For a wave explanation of refraction (and diffraction), read about Huygens' principle in *Conceptual Physics,* 10th edition, pages 558–560.

FIGURE 14.10
INTERACTIVE FIGURE

Refraction.

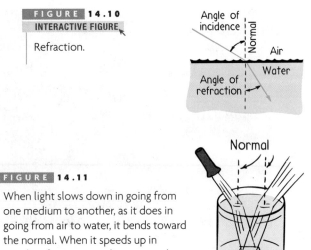

Angle of incidence

Normal

Air

Water

Angle of refraction

Normal

FIGURE 14.11

When light slows down in going from one medium to another, as it does in going from air to water, it bends toward the normal. When it speeds up in traveling from one medium to another, as it does in going from water to air, it bends away from the normal.

Mirror

Although wave speed and wavelength change when undergoing refraction, frequency remains unchanged.

see someone's eyes by way of a reflective or refractive device, such as a mirror or a prism, then that person can see you by way of the device also (unless the device is optically coated to produce a one-way effect).

Refraction causes many illusions. One of them is the apparent bending of a stick that is partially submerged in water. The submerged part appears closer to the surface than it actually is. The same is true when you look at a fish in water. The fish appears nearer to the surface and closer than it really is (Figure 14.12). If we look straight down into water, an object submerged 4 meters beneath the surface appears to be only 3 meters deep. Because of refraction, submerged objects appear to be magnified.

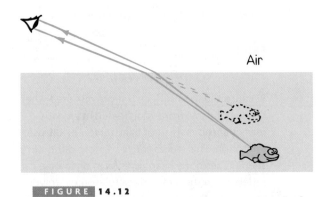

Air

FIGURE 14.12

Because of refraction, a submerged object appears to be nearer to the surface than it actually is.

CHECK YOURSELF

If the speed of light were the same in all media, would refraction still occur when light passes from one medium to another?

CHECK YOUR ANSWER

No.

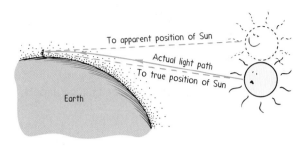

FIGURE 14.13

Because of atmospheric refraction, when the Sun is near the horizon it appears to be higher in the sky.

Refraction occurs in Earth's atmosphere. Whenever we watch a sunset, we see the Sun for several minutes after it has sunk below the horizon (Figure 14.13). Earth's atmosphere is thin at the top and dense at the bottom. Because light travels faster in thin air than it does in dense air, parts of the wave fronts of sunlight at high altitude travel faster than parts closer to the ground. Light rays bend. The density of the atmosphere changes gradually, so light rays bend gradually and follow a curved path. So we gain additional minutes of daylight each day. Furthermore, when the Sun (or Moon) is near the horizon, the rays from the lower edge are bent more than the rays from the upper edge. This shortens the vertical diameter, causing the Sun to appear elliptical (Figure 14.14).

FIGURE 14.14

The Sun is distorted by differential refraction.

A mirage occurs when refracted light appears as if it were reflected light. Mirages are a common sight on a desert when the sky appears to be reflected from water on the distant sand. But when you approach what seems to be water, you find dry sand. Why is this so? The air is very hot close to the sand surface and cooler above the sand. Light travels faster through the thinner hot air near the surface than through the denser cool air above. So wave fronts near the ground travel faster than they do above. The result is upward bending (Figure 14.15). So we see an upside-down view that looks as if reflection were occurring from a water surface. We see a mirage, which is formed by real light and can be photographed (Figure 14.16). A mirage is not, as many people think, a trick of the mind.

FIGURE 14.15

Light from the top of the tree gains speed in the warm and less dense air near the ground. When the light grazes the surface and bends upward, the observer sees a mirage.

When we look at an object over a hot stove or over a hot pavement, we see a wavy, shimmering effect. This is due to varying densities of air caused by changes in temperature. The twinkling of stars results from similar variations in the sky, where light passes through unstable layers in the atmosphere.

FIGURE 14.16

A mirage. The apparent wetness of the road is not a reflection of the sky by water but a refraction of sky light through the warmer and less-dense air near the road surface.

STOP AND
CHECK YOURSELF

If the speed of light were the same in air of various temperatures and densities, would there still be slightly longer daytimes, twinkling stars at night, mirages, and slightly squashed suns at sunset?

CHECK YOUR ANSWER

No.

One of the many beauties of physics is the redness of a fully eclipsed Moon—resulting from the refraction of sunsets and sunrises that completely circle the world.

14.3 Dispersion

Recall, from the previous chapter, that light that resonates with electrons of atoms and molecules in a material is absorbed. Such a material is opaque to light. Also recall that transparency occurs for light of frequencies near (but not at) the resonant frequencies of the material. Light is slowed due to the absorption/reemission sequence, and the closer to the resonant frequencies, the slower the light. This was shown in Figure 12.6. The grand result is

📷 **The Physics Place**
The Rainbow

that high-frequency light in a transparent medium travels slower than low-frequency light. Violet light travels about 1% slower in ordinary glass than red light. Light of colors between red and violet travel at their own respective speeds in glass.

Because light of various frequencies travels at different speeds in transparent materials, different colors of light refract by different amounts. When white light is refracted twice, as in a prism, the separation of light by colors is quite noticeable. This separation of light into colors arranged by frequency is called *dispersion* (Figure 14.17). Because of dispersion, there are rainbows!

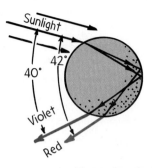

FIGURE 14.18
Dispersion of sunlight by a single raindrop.

Rainbows

For you to see a rainbow, the Sun must shine on drops of water in a cloud or in falling rain. The drops act as prisms that disperse light. When you face a rainbow, the Sun is behind you, in the opposite part of the sky. Seen from an airplane near midday, the bow forms a complete circle. As we will see, all rainbows would be completely round if the ground were not in the way.

You can see how a raindrop disperses light in Figure 14.18. Follow the ray of sunlight as it enters the drop near its top surface. Some of the light here is reflected (not shown), and the remainder is refracted into the water. At this first refraction, the light is dispersed into its spectrum colors, red being deviated the least and violet the most. When the light reaches the opposite side of the drop, each color is partly refracted out into the air (not shown) and partly reflected back into the water. Arriving at the lower surface of the drop, each color is again partly reflected (not shown) and partly refracted back into the air. This refraction at the second surface, like that in a prism, increases the dispersion already produced at the first surface.*

Although each drop disperses a full spectrum of colors, an observer is in a position to see only a single color from any one drop (Figure 14.19). If violet light from a single drop reaches the eye of an observer, red light from the same drop is incident elsewhere toward the feet. To see red light, one must look to a drop higher in the sky. The color red will be seen when the angle between a beam of sunlight and the dispersed light is 42°. The color violet is seen when the angle between the sunbeams and dispersed light is 40°.

---

* We're simplifying when we indicate that the red ray disperses at 42°. Actually, the angle between the incoming and outgoing rays can be anywhere between zero and about 42° (zero degrees corresponding to a full 180° reversal of the light). The strongest concentration of light intensity for red, however, is near the maximum angle of 42°, as shown in Figures 14.18 and 14.19.

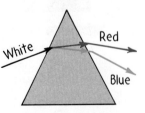

FIGURE 14.17

Dispersion by a prism makes the components of white light visible.

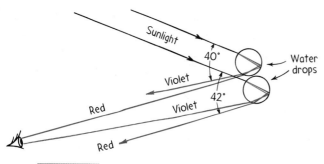

Sunlight incident on two raindrops, as shown, emerges from them as dispersed light. The observer sees the red light from the upper drop and the violet light from the lower drop. Millions of drops produce the entire spectrum of visible light.

Why does the light dispersed by the raindrops form a bow? The answer to this involves a bit of geometry. First of all, a rainbow is not the flat two-dimensional arc it appears to be. The rainbow you see is actually a three-dimensional cone of dispersed light. The apex of this cone is at your eye. To understand this, consider a glass cone, the shape of those paper cones you sometimes see at drinking fountains. If you held the tip of such a glass cone against your eye, what would you see? You'd see the glass as a circle. The same is true with a rainbow. All the drops that disperse the rainbow's light toward you lie in the shape of a cone—a cone of different layers with drops that deflect red to your eye on the outside, orange beneath the red, yellow beneath the orange, and so on, all the way to violet on the inner conical surface (Figure 14.20). The thicker the region containing water drops, the thicker the conical edge you look through and the more vivid the rainbow.

Your cone of vision intersects the cloud of drops and creates your rainbow. It is ever so slightly different from the rainbow seen by a person nearby. So, when a friend says, "Look at the pretty rainbow," you can reply, "Okay, move aside so I can see it, too." Everybody sees his or her own personal rainbow.

Another fact about rainbows: a rainbow always faces you squarely. When you move, your rainbow appears to move with you. So you can never approach the side of a rainbow or see it end-on as in the exaggerated view of Figure 14.20. You *can't* reach its end. Thus the saying "looking for the pot of gold at the end of the rainbow" means pursuing something you can never reach.

Often a larger, secondary bow with its colors reversed can be seen arching at a greater angle around the primary bow. We won't treat this secondary bow except to say that it is formed by similar

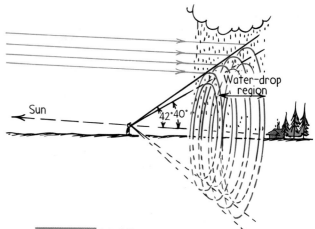

When your eye is located between the Sun (not shown, off to the left) and the water-drop region, the rainbow you see is the edge of a three-dimensional cone that extends through the water-drop region. Violet is dispersed by drops that form a 40° conical surface; red is seen from drops along a 42° conical surface, with other colors in between. (Innumerable layers of drops form innumerable two-dimensional arcs, like the four sets suggested here.)

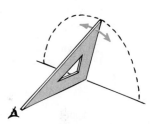

Only the raindrops along the dashed line disperse red light to the observer at a 42° angle; hence, the light forms a bow.

Two refractions and a reflection in water droplets produce light at all angles up to about 42°, with the intensity concentrated where we see the rainbow at 40° to 42°. Light doesn't exit the water droplet at angles greater than 42° unless it undergoes two or more reflections inside the drop. Thus the sky is brighter inside the rainbow than it is outside it. Notice the weak secondary rainbow.

FIGURE **14.23**

Double reflection in a drop produces a secondary bow.

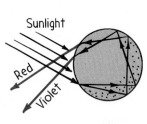

Sunlight

Red

Violet

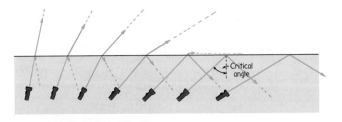

Critical angle

FIGURE **14.24**
INTERACTIVE FIGURE

Light emitted in the water is partly refracted and partly reflected at the surface, as indicated by the lengths of the arrows. At the critical angle, the emerging beam intensity reduces to zero where it tends to graze the surface. Beyond the critical angle, the beam is totally internally reflected.

circumstances and is a result of double reflection within the raindrops (Figure 14.23). Because of this extra reflection (and extra refraction loss), the secondary bow is much dimmer and reversed.

STOP AND
CHECK YOURSELF

1. Suppose you point to a wall with your arm extended. Then you sweep your arm around, making an angle of about 42° to the wall. If you rotate your arm in a full circle while keeping the same angle, what shape does your arm describe? What shape on the wall does your finger sweep out?

2. If light traveled at the same speed in raindrops as it does in air, would we have rainbows?

CHECK YOUR ANSWERS

1. Your arm describes a cone, and your finger sweeps out a circle. The same is true with rainbows.

2. No.

14.4 Total Internal Reflection

Some Saturday night when you're taking your bath, fill the tub extra deep and bring a waterproof flashlight into the tub with you. Turn the bathroom light off. Shine the submerged light straight up and then slowly tip it. Note how the intensity of the emerging beam diminishes and how more light is reflected from the water's surface to the bottom of the tub. When the flashlight is tipped at a certain angle you'll notice that the beam no longer emerges into the air above the surface. This is the **critical angle**. When the flashlight is tipped beyond the critical angle (48° from the normal for water), you'll notice that all the light is reflected back into the tub. This is **total internal reflection**. The light

in water striking the air boundary obeys the law of reflection: the angle of incidence is equal to the angle of reflection. The only light emerging from the water's surface is the light that is diffusely reflected from the bottom of the bathtub. This procedure is shown in Figure 14.24. The proportion of light refracted and the proportion of light internally reflected are indicated by the relative lengths of the arrows.

Interestingly, total internal reflection occurs only for light striking the boundary of a medium wherein the speed of light is greater. Light travels faster in air than it does in water, so total internal reflection can occur when light in water meets an air boundary. But it can't occur for light traveling in air that meets a water boundary.

STOP AND
CHECK YOURSELF

How does the critical angle relate to total internal reflection?

CHECK YOUR ANSWER

Critical angle is the minimum angle of incidence inside a medium for total internal reflection. When a light ray strikes a surface at or beyond the critical angle, total internal reflection occurs.

Your pet goldfish in a large tub looks up to see a compressed view of the outside world (Figure 14.25). The 180° view from horizon to opposite horizon is seen through an angle of 96°—twice the critical angle. A lens that similarly compresses a wide view, called a *fisheye lens*, is used for special-effect photographs.

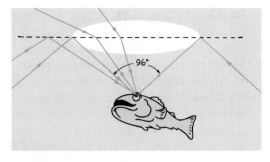

FIGURE 14.25

An observer underwater sees a circle of light at the still surface. Beyond a cone of 96° (twice the critical angle), an observer sees a reflection of the water interior or bottom.

The critical angle for glass is about 43°, depending on the type of glass. This means that, within glass, light that is incident at angles at or greater than 43° will be totally internally reflected. No light will escape beyond this angle; instead, all of it will be reflected back into the glass. Whereas a silvered or aluminized mirror reflects only about 90% of incident light, glass prisms, as shown in Figure 14.26, are more efficient. A little light is lost by reflection before it enters the prism, but, once inside, reflection on the 45°-slanted face is total—100%. Moreover, this light is not marred by any dirt or dust on the outside surface, which is the principal reason for the use of prisms instead of mirrors in many optical instruments.

A pair of prisms each reflecting light through 180° is shown in Figure 14.27. Binoculars use pairs of prisms to lengthen the light path between lenses and thus eliminate the need for long barrels. So a compact set of binoculars is as effective as a longer telescope (Figure 14.28). Another advantage of prisms is that, whereas the image of a straight telescope is upside down, reflection by the prisms in binoculars reinverts the image, so things are seen right side up.

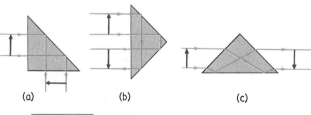

FIGURE 14.26

Total internal reflection in a prism. In (a), the prism changes the direction of the light beam by 90°; in (b), it changes it by 180°; and in (c), it does not change the direction of the beam but instead turns the image upside down.

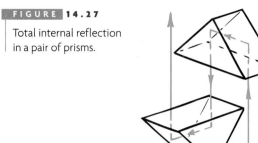

FIGURE 14.27

Total internal reflection in a pair of prisms.

The critical angle for a diamond is about 24.5°, smaller than for any other known substance. The critical angle varies slightly for different colors, because of slight speed variations for different colors. Once light enters a diamond gemstone, most is incident on the sloped backsides at angles greater than 24.5° and is totally internally reflected (Figure 14.30). Because of the great slowdown in speed as light enters a diamond, refraction is pronounced, and, because of the frequency-dependence of the speed, there is great dispersion. Further dispersion occurs as the light exits through the many facets at its face. Hence we see unexpected flashes of a wide array of colors. Interestingly, when these flashes are narrow enough to be seen by only one eye at a time, the diamond "sparkles."

Total internal reflection also underlies the operation of optical fibers, or light pipes (Figure 14.31). An optical fiber "pipes" light from one place to another by a series of total internal reflections, much as a bullet ricochets down a steel pipe. Light rays bounce along the inner walls, following the twists and turns of the fiber. Optical fibers are used to illuminate instrument displays on automobile dashboards from a single bulb. Dentists use them with flashlights to get light where they want it. Bundles of thin flexible glass or plastic fibers are used to see

FIGURE 14.28

Prism binoculars.

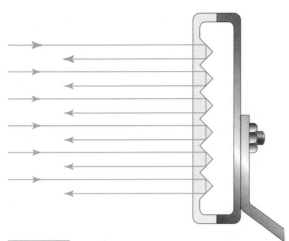

FIGURE 14.29

The rear reflectors on cars, bicycles, and other vehicles contain arrays of tiny prisms that use total internal reflection to send light back in the opposite direction.

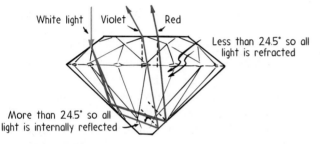

FIGURE 14.30

Paths of light in a diamond. Rays that strike the inner surface of a diamond at angles greater than the critical angle (about 24.5°, depending on the color of the light) are internally reflected and exit via refraction at the top surface.

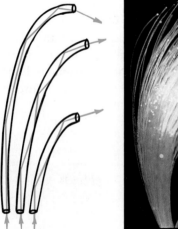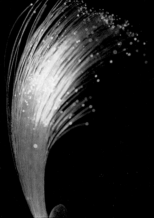

FIGURE 14.31

The light in optical fibers is "piped" from below by a succession of total internal reflections until it emerges at the top ends of the fibers.

what is occurring in inaccessible regions, such as in the interior of a motor or in a patient's stomach. They can be made small enough to snake through blood vessels or through narrow passages in the body, such as the urethra. Light shines down some of the fibers to illuminate the scene and is reflected back along others.

Optical fibers are important in communications because they offer a practical alternative to copper wires and cables. In many places, thin glass fibers now replace thick, bulky, and expensive copper cables to carry thousands of simultaneous telephone messages among the major switching centers. In many aircraft, control signals are fed from the pilot to the control surfaces by means of optical fibers. Signals are carried in the modulations of laser light. Unlike electricity, light is indifferent to temperature and fluctuations in surrounding magnetic fields, so the signal is clearer. Also, it is much less likely to be tapped by eavesdroppers.

14.5 Lenses

When you think of lenses, think of sets of glass prisms arranged as shown in Figure 14.32. They refract incoming parallel light rays so that the rays converge to (or diverge from) a point. The arrangement shown in Figure 14.32a converges the light, and we have a **converging lens**. Notice that it is thicker in the middle. In the arrangement shown in Figure 14.32b, the middle is thinner than the edges. Because this lens diverges the light, we have a **diverging lens**. Note that the prisms in part b diverge the incident rays in a way that makes them appear to originate from a single point in front of the lens.

In both lenses, the greatest deviation of rays occurs at the outermost prisms, because they have the greatest angle between the two refracting surfaces. No deviation occurs exactly in the middle, for in that region the two surfaces of the glass are parallel to each other (light doesn't deviate when going through glass with parallel surfaces, like window glass). A real lens is not made of prisms, of course. It is made of a solid piece of glass with surfaces usually ground to a circular curve. In Figure 14.33, we see how smooth lenses refract waves.

> Learning about lenses is a hands-on activity. Not fiddling with lenses while learning about them is like taking swimming lessons away from water.

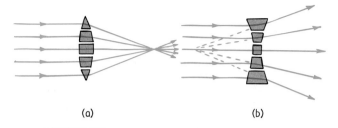

(a)   (b)

FIGURE 14.32

A lens may be thought of as a set of prisms.

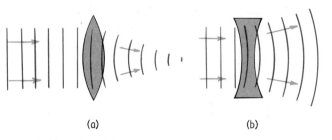

(a)   (b)

FIGURE 14.33

Wave fronts travel more slowly in glass than in air. In (a), the waves are retarded more through the center of the lens, and convergence results. In (b), the waves are retarded more at the edges, and divergence results.

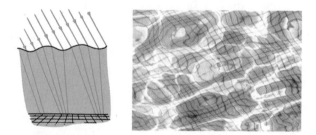

FIGURE 14.34

The moving patterns of bright and dark areas at the bottom of the pool result from the uneven surface of the water, which behaves like a blanket of undulating lenses. Just as we see the pool bottom shimmering, a fish looking upward at the Sun would see it shimmering, too. Because of similar irregularities in the atmosphere, we see the stars twinkle.

Some key features of lenses are shown for a converging lens in Figure 14.35. The *principal axis* is the line joining the centers of curvature of the two lens surfaces. The *focal point* is the point of convergence for light parallel to the principal axis. Incident beams not parallel to the principal axis focus at points above or below the focal point. All such possible points make up a *focal plane* (not shown).

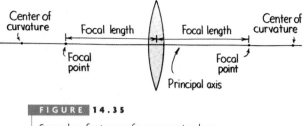

FIGURE 14.35

Some key features of a converging lens.

Because a lens has two surfaces, it has two focal points and two focal planes. The *focal length* of the lens is the distance between the center of the lens and either focal point.

In a diverging lens, an incident beam of light parallel to the principal axis is not converged to a point; it is diverged, so the light appears to emerge from a point in front of the lens.

Image Formation by a Lens

At this moment, light is reflecting from your face onto this page. Light that reflects from your forehead, for example, strikes every part of the page. The same is true of the light that reflects from your chin. Every part of the page is illuminated with reflected light from your forehead, your nose, your chin, and every other part of your face. You don't see an image of your face on the page because there is too much overlapping of light. But place a barrier with a pinhole in it between your face and the page, and the light that reaches the page from your forehead does not overlap the light from your chin. The same is true for the rest of your face. Without this overlapping, an image of your face is formed on the page. It will be very dim, for very little light reflected from your face passes through the pinhole. To see the image, you'd have to shield the page from other light sources. The same is true of the vase and flower in Figure 14.36b.

The first cameras had no lenses and admitted light through a small pinhole. Long exposure times were required because of the small amount of light admitted by the pinhole. This meant that subjects being photographed had to remain very still. Motion would produce a blur. If the hole were a bit larger, exposure time would be shorter, but overlapping rays would produce a blurry image. Too large a hole would allow too much overlapping, resulting in no image. That's where a converging lens plays a role (Figure 14.36). The lens converges light onto

■ YOUR EYE

With all of today's technology, the most remarkable
optical instrument known
is your eye. Light enters
through your cornea, which
does about 70% of the necessary bending of the light
before it passes through

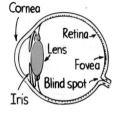

your pupil (the aperture, or opening, in the iris).
Light then passes through your lens, which provides the extra bending power needed to focus
images of nearby objects on your extremely sensitive retina. (Only recently have artificial detectors
been made with greater sensitivity to light than
the human eye.) An image of the visual field outside your eye is spread over the retina. The retina
is not uniform. There is a spot in the center of
your field of the retina called the fovea, which is
the region of most acute vision. You see greater
detail here than at any other part of your retina.
There is also a spot in your retina where the
nerves carrying all the information exit the eye on
their way to the brain. This is your blind spot.

You can demonstrate that you have a blind
spot in each eye. Simply hold this book at
arm's length, close your left eye, and look at
the round dot and X to its right with your
right eye only. You can see both the dot and
the X at this distance. Now move the book
slowly toward your face, with your right eye

fixed upon the dot, and you'll reach a position
about 20–25 centimeters from your eye where the
X disappears. When both eyes are open, one eye
"fills in" the part to which your other eye is blind.
Now repeat with only the left eye open, looking
this time at the X, and the dot will disappear. But
note that your brain fills in the two intersecting
lines. Amazingly, your brain fills in the "expected"
view even with one eye closed. Instead of seeing
nothing, your brain graciously fills in the appropriate background. Repeat this for small objects
on various backgrounds. You not only see what's
there—you see what's not there!

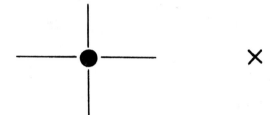

The light receptors in your retina do not connect
directly to your optic nerve but are instead inter-
connected with many other cells. Through these
interconnections, a certain amount of information
is combined and "digested" in your retina. In this
way, the light signal is "thought about" before it
goes to the optic nerve and then to the main body
of your brain. So some brain functioning occurs in
your eye. Amazingly, your eye does some of your
"thinking."

FIGURE 14.36
INTERACTIVE FIGURE

Image formation. (a) No image appears on the wall
because rays from all parts of the object overlap the
entire wall. (b) A single small opening in a barrier prevents
overlapping rays from reaching the wall; a dim upside-
down image is formed. (c) A lens converges the rays upon
the wall without overlapping; more light makes a brighter
image.

Can you see why the image in Figure 14.36b is upside
down? And is it true that for photographs you have
developed and printed, whether via chip or film,
they're all upside down?

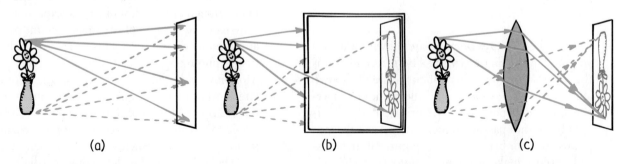

(a)   (b)   (c)

the film without any overlapping of rays. Moving objects can be taken with the lens camera because of the short exposure time. As mentioned earlier, that's why early photographs taken with lens cameras were called snapshots.

The simplest application of a converging lens is as a magnifying glass. To understand how it works, think about how you examine objects near and far. With unaided vision, you see a distant object through a relatively narrow angle of view, and you see a close object through a wider angle of view (Figure 14.37). To see the details of a small object, you want to get as close to it as possible for the widest-angle view. But your eye can't focus when it's too close to the object. That's where the magnifying glass is useful. When close to the object, the magnifying glass gives you a clear image that would be blurry without it.

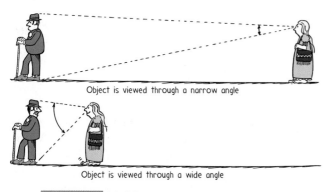

Object is viewed through a narrow angle

Object is viewed through a wide angle

FIGURE 14.37

When you use a magnifying glass, you hold it close to the object you wish to examine. This is because a converging lens provides an enlarged image that is right side up only when the object is inside the focal point. If a screen is placed at the image distance, no image appears on it because no light is directed to the image position. The rays that reach your eye, however, behave virtually as if they originated at the image position. This is called a **virtual image**—one formed by light rays that do not converge at the image location (Figure 14.38).

When the object is distant enough to be outside the focal point of a converging lens, a **real image** is formed instead of a virtual image; Figure 14.39 shows this case. Light rays converge to form a real image that can be displayed on a screen. Real images formed with a single lens are always upside down.

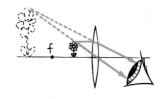

FIGURE 14.38

When an object is near a converging lens (inside its focal point *f*), the lens acts as a magnifying glass to produce a virtual image. The image appears larger and farther from the lens than the object is.

That's why, for correct viewing, slides are inserted upside down in a slide projector. The frames of motion pictures are likewise upside down. The same is true for the image in a camera.

A diverging lens used alone produces a reduced virtual image. It makes no difference how far or how near the object is. The image is always virtual, right side up, and smaller than the object. That's why a diverging lens is often used as a "finder" on a camera. When you view the object to be photographed through such a lens, you see a virtual image that approximates the same proportions as the photograph.

> Poke a hole in a piece of paper, hold it in sunlight so the solar image is the same size as a coin on the ground, then determine how many coins would fit between the ground and the pinhole. That's the same number of solar diameters that would occupy the distance from Earth to the Sun. (See this exercise in *Practice Book for Conceptual Physics Fundamentals*.)

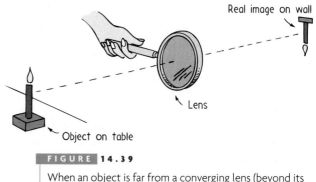

Real image on wall

Lens

Object on table

FIGURE 14.39

When an object is far from a converging lens (beyond its focal point), a real upside-down image is formed.

FIGURE 14.40

A diverging lens forms a virtual, right-side-up image of Jamie and his cat.

STOP AND
CHECK YOURSELF

Why is the greater part of the photograph in Figure 14.40 out of focus?

CHECK YOUR ANSWER

Both Jamie and his cat and the virtual image of Jamie and his cat are "objects" for the lens of the camera that took this photograph. Since the objects are at different distances from the lens, images are at different distances relative to the film in the camera. So only one can be brought into focus. The same is true of your eyes. You cannot focus on near and far objects at the same time.

Lens Defects

No lens provides a perfect image. A distortion in an image is called an **aberration**. By combining lenses in certain ways, you can minimize aberrations. For this reason, most optical instruments use compound lenses, each consisting of several simple lenses, instead of single lenses.

Spherical aberration results from light passing through the edges of a lens and focusing at a slightly different place from where light passing through the center of the lens focuses (Figure 14.41). This can be remedied by covering the edges of a lens, as with

FIGURE 14.41

Spherical aberration.

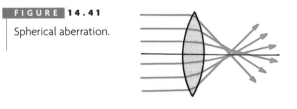

a diaphragm in a camera. Spherical aberration is corrected in good optical instruments by a combination of lenses.

Chromatic aberration is the result of various colors having different speeds and hence different refractions in the lens (Figure 14.42). In a simple lens (as in a prism), red light and blue light do not come to focus in the same place. Achromatic lenses, which combine simple lenses of different kinds of glass, correct this defect.

FIGURE 14.42

Chromatic aberration.

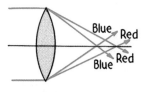

The pupil of the eye regulates the amount of light that enters the eye by changing its size. Vision is sharpest when the pupil is smallest because light then passes through only the center of the eye's lens, where spherical and chromatic aberrations are minimal. Also, the eye then acts more like a pinhole camera, so minimal focusing is required for a sharp image. You see better in bright light because your pupils are smaller.

Astigmatism of the eye is a defect that results when the cornea is curved more in one direction than in another, somewhat like the side of a barrel. Because of this defect, the eye does not form sharp images. The remedy is eyeglasses with cylindrical lenses that have more curvature in one direction than in another.

If you wear glasses and ever misplace them, or if you find it difficult to read small print as in a telephone book, squint, or, even better, hold a pinhole (in a piece of paper or whatever) in front of your eye, close to the page. You'll see the print clearly, and because you're close, it is magnified. Try it and see!

STOP AND CHECK YOURSELF

1. If light traveled at the same speed in glass and in air, would glass lenses change the direction of light rays?

2. Why is chromatic aberration associated with a lens but not with a mirror?

3. How can chromatic aberration be corrected?

4. There have been reports of round fishbowls starting fires by focusing the Sun's rays coming in a window. Can you cite a possible explanation for this occurrence?

CHECK YOUR ANSWERS

1. No.

2. Light of different frequencies travels at different speeds in a transparent medium and therefore refracts at different angles. This is the cause of chromatic aberration. The angles of reflected light, however, have no relation to frequency. One color reflects the same as any other color. In telescopes, therefore, mirrors are preferred over lenses because of the absence of chromatic aberration for reflection.

3. This defect can be corrected with a combination of lenses. These are called achromatic lenses.

4. This can certainly happen. The water-filled bowl acts as a converging lens and, like a lens made entirely of glass, can converge rays of sunlight to a focus. If the point of focus is flammable, a fire can occur.

EVERYDAY APPLICATIONS

■ LATERAL INHIBITION

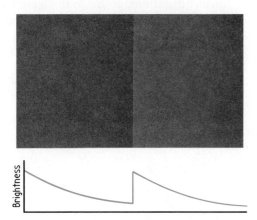

The human eye can do what no camera film can do: it can perceive degrees of brightness that range from about 500 million to 1. The difference in brightness between the Sun and Moon, for example, is about 1 million to 1. But, because of an effect called lateral inhibition, we don't perceive the actual differences in brightness. The brightest places in our visual field are prevented from outshining the rest, for whenever a receptor cell on our retina sends a strong brightness signal to our brain, it also signals neighboring cells to dim their responses. In this way, we even out our visual field, which allows us to discern detail in very bright areas and in dark areas as well.

Lateral inhibition exaggerates the difference in brightness at the edges of places in our visual field. Edges, by definition, separate one thing from another. So we accentuate differences rather than similarities. This is illustrated in the pair of shaded rectangles to the right. They look to be different shades of brightness because of the edge that separates them. But cover the edge with your pencil or your finger, and they look equally bright (try it now)! That's because both rectangles are equally bright; each rectangle is shaded from lighter to darker, moving from left to right. Our eye concentrates on the boundary where the dark edge of the left rectangle joins the light edge of the right rectangle, and our eye–brain system assumes that the rest of the rectangle is the same. We pay attention to the boundary and ignore the rest.

Questions to ponder: Is the way the eye picks out edges and makes assumptions about what lies beyond similar to the way in which we sometimes make judgments about other cultures and other people? Don't we, in the same way, tend to exaggerate the differences on the surface while ignoring the similarities and subtle differences within?

The advent of eyeglasses probably occurred in Italy late in the thirteenth century. (Curiously, the telescope wasn't invented until some 300 years later. If, in the meantime, anyone viewed objects through a pair of lenses separated along their axes, such as fixed at the ends of a tube, there is no record of it.) An option to wearing eyeglasses for correcting vision is contact lenses. A more recent option is LASIK (acronym for laser-assisted in situ keratomileusis), the procedure of reshaping the cornea using pulses from a laser. Another recent procedure is PRK (photorefractive keratectomy), and still another is IntraLase, where intraocular lenses are implanted in the eye like a contact lens, a procedure of choice for extremely nearsighted or farsighted people and for those who can't have laser surgery. The wearing of eyeglasses and contact lenses may soon be a thing of the past.

14.6 Polarization

Interference and diffraction provide the best evidence that light is wavelike. As we learned in Chapter 12, waves can be either longitudinal or transverse. Sound waves are longitudinal, which means the vibratory motion of the medium is along the direction of wave travel. The fact that light waves exhibit **polarization** demonstrates that they are transverse.

The Physics Place
Polarized Light and 3-D Viewing

If you shake a rope either up and down or from side to side as shown in Figure 14.43, you'll produce a transverse wave along the rope. The plane of vibration is the same as the plane of the wave. If we shake it up and down, the wave vibrates in a vertical plane. If we shake it back and forth, the wave vibrates in a horizontal plane. We say that such a wave is *plane-polarized*—that the waves traveling along the rope are confined to a single plane. Polarization is a property of transverse waves. (Polarization does not occur among longitudinal waves—there is no such thing as polarized sound.)

FIGURE 14.43

A vertically plane-polarized plane wave and a horizontally plane-polarized plane wave.

(a)   (b)

FIGURE 14.44

(a) A vertically plane-polarized wave from a charge vibrating vertically. (b) A horizontally plane-polarized wave from a charge vibrating horizontally.

A single vibrating electron can emit an electromagnetic wave that is plane-polarized. The plane of polarization matches the vibrational direction of the electron. That means that a vertically accelerating electron emits light that is vertically polarized. A horizontally accelerating electron emits light that is horizontally polarized (Figure 14.44)*.

A common light source, such as an incandescent lamp, a fluorescent lamp, or a candle flame, emits light that is unpolarized. This is because the electrons that emit the light are vibrating in many random directions. There are as many planes of vibration as the vibrating electrons producing them. A few planes are represented in Figure 14.45a. We can represent all these planes by radial lines, shown in Figure 14.45b. (Or, more simply, the planes can be represented by vectors in two mutually perpendicular directions, as shown in Figure 14.45c.) The vertical vector represents all the components of vibration in the vertical direction. The horizontal vector represents all the components of vibration horizontally. The simple model of Figure 14.45c represents unpolarized light. Polarized light would be represented by a single vector.

All transparent crystals having a noncubic natural shape have the property of polarizing light.

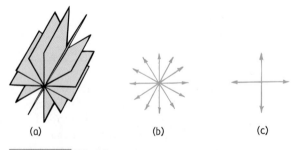

(a)   (b)   (c)

FIGURE 14.45

Representations of plane-polarized waves. Electric vectors, a and b, show the electric part of the electromagnetic wave.

* Light may also be circularly polarized and elliptically polarized, which are also transverse polarizations. But we will not study these cases.

FIGURE 14.46

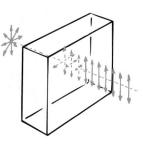

FIGURE 14.46

One component of the incident unpolarized light is absorbed, resulting in emerging polarized light.

These crystals divide unpolarized light into two internal beams polarized at right angles to each other. Some crystals strongly absorb one beam while transmitting the other (Figure 14.46). This makes them excellent polarizers. Herapathite is such a crystal. Microscopic herapathite crystals are aligned and embedded between cellulose sheets. They make up Polaroid filters, popular in sunglasses. Other Polaroid sheets consist of certain aligned molecules rather than tiny crystals.

> **Polarization occurs only for transverse waves. In fact, it is an important way of telling whether a wave is transverse or longitudinal.**

If you look at unpolarized light through a Polaroid filter, you can rotate the filter in any direction and the light appears unchanged. But, if the light is polarized, rotating the filter allows you to block out more and more of the light until it is completely blocked out. An ideal Polaroid filter transmits a full 50% of incident unpolarized light. That 50% is polarized. When two Polaroid filters are arranged so that their polarization axes are aligned, light can pass through both, as shown in the rope analogy (Figure 14.47a). If their axes are at right angles to each other (in this case, we say the filters are crossed), almost no light penetrates the pair (Figure 14.47b). (A small amount of shorter wavelengths do get through.) When Polaroid filters are used in pairs like this, the first one is called the *polarizer* and the second one is called the *analyzer*.

STOP AND
CHECK YOURSELF

Which pair of glasses is best suited for automobile drivers? (The polarization axes are shown by the straight lines.)

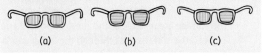

(a)   (b)   (c)

CHECK YOUR ANSWER

Glasses A are best suited because the vertical axis blocks horizontally polarized light, which makes up much of the glare from horizontal surfaces. Glasses C are suited for viewing 3-D movies.

Much of the light reflected from nonmetallic surfaces is polarized. The glare from glass or water is a good example. Except for light that hits vertically, the reflected ray has more vibrations parallel to the reflecting surface. The part of the ray that penetrates the surface has more vibrations at right angles to the surface (Figure 14.48). Skipping flat rocks off the surface of a pond provides an appropriate analogy. When the rocks hit parallel to the surface, they are easily reflected by the surface. But when they hit with their faces at right angles to the surface, they "refract" into the water. The glare from reflecting surfaces can be dimmed a lot with the use of Polaroid sunglasses. The polarization axes of the lenses are vertical because most of the glare reflects from horizontal surfaces.

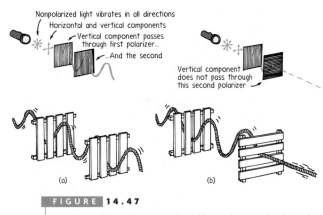

Nonpolarized light vibrates in all directions
Horizontal and vertical components
Vertical component passes through first polarizer...
...And the second

Vertical component does not pass through this second polarizer

(a)   (b)

FIGURE 14.47

A rope analogy illustrates the effect of crossed Polaroids.

FIGURE 14.48

Polaroid sunglasses block out horizontally vibrating light. When the lenses overlap at right angles, no light gets through.

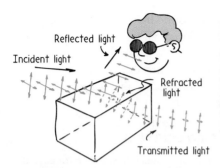

FIGURE 14.49

Most glare from nonmetallic surfaces is polarized. Here we see that the components of incident light parallel to the surface are reflected, and the components perpendicular to the surface pass through the surface into the medium. Because most of the glare we encounter is from horizontal surfaces, the polarization axes of Polaroid sunglasses are vertical.

FIGURE 14.50

Light is transmitted when the axes of the Polaroids are aligned (a), but absorbed when Ludmila rotates one so that the axes are at right angles to each other (b). When she inserts a third Polaroid at an angle between the crossed Polaroids, light is again transmitted (c). Why? (For the answer, after you have given this some thought see Appendix C, "More About Vectors.")

Beautiful colors similar to interference colors can be seen when certain materials are placed between crossed Polaroid filters. Cellophane works wonderfully. Why these colors are produced is another story—one that is left to the Suggested Reading and Web sites for this chapter.

fyi

Cosmic microwave background (CMB) fills all of space and approaches us from every direction. It is an echo of the Big Bang that got our universe started some 14 billion years ago. Recent findings show this radiation to be polarized. Polarization observations are unaffected by gravity and provide a clear and detailed look at the early cosmos.

SUMMARY OF TERMS

Reflection The return of light rays from a surface in such a way that the angle at which a given ray is returned is equal to the angle at which it strikes the surface (also called *specular reflection*).

Refraction The bending of an oblique ray of light when it passes from one transparent medium to another. This is caused by a difference in the speed of light in the transparent media. When the change in medium is abrupt (say, from air to water), the bending is abrupt; when the change in medium is gradual (say, from cool air to warm air), the bending is gradual, which accounts for mirages.

Law of reflection The angle of incidence equals the angle of reflection. The incident and reflected rays lie in a plane that is normal to the reflecting surface.

Diffuse reflection Reflection in irregular directions from an irregular surface.

Critical angle The minimum angle of incidence inside a medium at which a light ray is totally reflected.

Total internal reflection The total reflection of light traveling within a medium that strikes the boundary of another medium at an angle at, or greater than, the critical angle.

Converging lens A lens that is thicker in the middle than at the edges and that refracts parallel rays passing through it to a focus.

Diverging lens A lens that is thinner in the middle than at the edges, causing parallel rays passing through it to diverge as if from a point.

Virtual image An image formed by light rays that do not converge at the location of the image. Mirrors, converging lenses used as magnifying glasses, and diverging lenses all produce virtual images.

Real image An image formed by light rays that converge at the location of the image. A real image can be displayed on a screen.

Aberrations Distortions in the formation of perfect images, which are inherent, to some degree, in all optical systems.

Polarization The alignment of the transverse electric vectors that make up electromagnetic radiation. Such waves of aligned vibrations are said to be *polarized*.

REVIEW QUESTIONS

1. Distinguish between reflection and refraction.

14.1 Reflection

2. What does incident light that falls on an object do to the electrons in the atoms of the object?
3. What do the electrons in an illuminated object do when they are made to oscillate with greater energy?
4. What is the law of reflection?
5. Relative to the distance of an object in front of a plane mirror, how far behind the mirror is the image?
6. Does the law of reflection hold for curved mirrors? Explain.
7. Does the law of reflection hold for diffuse reflection? Explain.
8. How can a surface be polished for some waves and not for others?

14.2 Refraction

9. What is the angle between a light ray and its wave front?
10. When a wheel rolls from a smooth sidewalk onto grass, the interaction of the wheel with the blades of grass slows the wheel. What slows light when it passes from air into glass or water?
11. What causes the bending of light in refraction?
12. Does light travel faster in thin air or in dense air? What does this difference in speed have to do with the length of daylight?
13. What is a mirage?
14. Why do stars twinkle?

14.3 Dispersion

15. What happens to light of a certain frequency when it is incident on a material whose natural frequency is the same as the frequency of the light?
16. Which travels more slowly in glass, red light or violet light?
17. What is dispersion? Cite a common example of dispersion.
18. What prevents rainbows from being seen as complete circles?

19. Does a single raindrop illuminated by sunlight disperse a spectrum of colors? Does a viewer see a spectrum from a single faraway drop?
20. Is a rainbow flat, or is it three-dimensional?
21. Why is a secondary rainbow dimmer than a primary bow?

14.4 Total Internal Reflection

22. What is meant by critical angle?
23. When is light totally reflected in water or glass?
24. When is light totally reflected in a diamond?
25. Light normally travels in straight lines, but it "bends" in an optical fiber. Explain.

14.5 Lenses

26. Distinguish between a converging lens and a diverging lens.
27. What is the focal length of a lens?
28. Distinguish between a virtual image and a real image.
29. Is a converging lens or a diverging lens used to produce a real image? To produce a virtual image?
30. Distinguish between spherical aberration and chromatic aberration.
31. What is astigmatism? What is a remedy for astigmatism?

14.6 Polarization

32. What phenomenon distinguishes between longitudinal and transverse waves?
33. How does the direction of polarization of light compare with the direction of vibration of the electrons that produced it?
34. Why does light pass through a pair of Polaroid filters when the axes are aligned but not when the axes are at right angles to each other?
35. How much unpolarized light does an ideal Polaroid filter transmit?
36. When unpolarized light is incident at a grazing angle upon water, what can you say about the reflected light?

ACTIVE EXPLORATIONS

1. Make a pinhole camera, as illustrated on the next page. Cut out one end of a small cardboard box, and cover the end with tissue or wax paper. Make a clean-cut pinhole at the other end. (If the cardboard is thick, make the pinhole through a piece of aluminum foil placed over an opening in the cardboard.) Aim the camera at a bright object in a darkened room, and you will see an upside-down image on the tissue paper. If, in a dark room, you replace the tissue paper with unexposed photographic film, cover the back so it is light-tight, and cover the pinhole with a removable flap. Now you are ready to take a picture. Exposure times differ, depending principally on the kind

of film and the amount of light. Try different exposure times, starting with about 3 seconds. Also try boxes of various lengths. You'll find everything in focus in your photographs, but the pictures will not have clear-cut, sharp outlines. The lens on a commercial camera is much bigger than the pinhole and therefore admits more light in less time—hence the name *snapshots*.

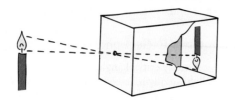

2. Stand a pair of mirrors on edge with the faces parallel to each other. Place an object, such as a coin, between the mirrors, and look at the reflections in each mirror. Nice?

3. Set up two pocket mirrors at right angles, and place a coin between them. You'll see four coins. Change the angle of the mirrors, and see how many images of the coin you can see. With the mirrors at right angles, look at your face. Then wink. What do you see? You now see yourself as others see you. Hold a printed page up to the double mirrors and compare its appearance with the reflection of a single mirror.

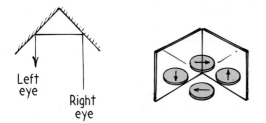

4. Rotate a pair of mirrors, keeping them at right angles to each other. Does your image rotate also? Then place the mirrors 60 degrees apart so that you can see your face. Again rotate the mirrors, and see if your image rotates also. Amazing?

5. Determine the magnifying power of a lens by focusing on the lines of a ruled piece of paper. Count the spaces between the lines that fit into one magnified space, and you have the magnifying power of the lens. You can do the same with binoculars and a distant brick wall. Hold the binoculars so that only one eye looks at the bricks through the eyepiece while the other eye looks directly at the bricks. The number of bricks seen with the unaided eye that will fit into one magnified brick reveals the magnification of the instrument.

6. Look at the reflections of overhead lights from the inner and outer surfaces of eyeglasses, and you will see two fascinatingly different images. Why are they different?

7. When you're wearing Polaroid sunglasses, view the glare from a nonmetallic surface, such as a road or a body of water. Tip your head from side to side, and see how the glare intensity changes as you vary the magnitude of the electric vector component aligned with the polarization axis of the glasses. Also notice the polarization of different parts of the sky when you hold the sunglasses in your hand and rotate them.

8. Place a bottle of corn syrup between two Polaroid filters. Place a white light source at the back. Then look through the filters and the syrup to view spectacular colors as you rotate one of the filters.

9. See spectacular interference colors with a polarized-light microscope. Any microscope, including an inexpensive toy microscope, can be converted into a polarized-light microscope by fitting a piece of Polaroid filter inside the eyepiece and taping another onto the stage of the microscope. Stretch various pieces of plastic wrap across a slide, rotate the eyepiece, and view impressive colors.

EXERCISES

1. Her eye at point *P* looks into the mirror. Which of the numbered cards can she see reflected in the mirror?

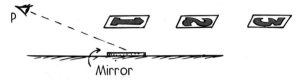

2. Cowboy Joe wishes to shoot his assailant by ricocheting a bullet off a mirrored metal plate. To do so, should he simply aim at the mirrored image of his assailant? Explain.
3. Trucks often have signs on their backsides that say, "If you can't see my mirrors, I can't see you." Explain the physics here.
4. Why is the lettering on the front of some vehicles, for example ambulances, "backward"?

AMBULANCE

5. When you look at yourself in the mirror and wave your right hand, your beautiful image waves its left hand. Then why don't the feet of your image wiggle when you shake your head?
6. Car mirrors are uncoated on the front surface and silvered on the back surface. When the mirror is properly adjusted, light from behind reflects from the silvered surface into the driver's eyes. Good. But this is not so good at nighttime with the glare of headlights behind. This problem is solved by the wedge shape of the mirror (see sketch). When the mirror is tilted slightly upward to the "nighttime" position, glare is directed upward toward the ceiling, away from the driver's eyes. Explain how the driver can still see cars behind in the mirror.

7. To reduce the glare of the surroundings, the windows of some department stores slant inward at the bottom, rather than being vertical. How does this reduce glare?
8. A person in a dark room looking through a window can clearly see a person outside in the daylight, whereas the person outside cannot see the person inside. Explain.
9. Which kind of road surface is easier to see when driving at night, an uneven pebbled surface or a mirror-smooth surface? Explain.
10. Why is it difficult to see the roadway in front of you when driving on a rainy night?

11. We see the bird and its reflection. Why do we not see the bird's feet in the reflection?
12. What must be the minimum length of a plane mirror in order for you to see a full view of yourself?
13. What effect does your distance from the plane mirror have on your answer to the preceding question? (Try it and see!)
14. Hold a pocket mirror almost at arm's length from your face and note the amount of your face you can see. To see more of your face, should you hold the mirror closer or farther, or would you have to have a larger mirror? (Try it and see!)
15. From a steamy mirror, wipe away just enough steam to allow you to see your full face. How tall will the wiped area be compared with the vertical dimension of your face?
16. The diagram shows a person and her twin at equal distances on opposite sides of a thin wall. Suppose that a window were to be cut in the wall so each twin can see a complete view of the other. Show the size and location of the smallest window that could be cut in the wall to do the job. (Hint: Draw rays from the top of each twin's head to the other twin's eyes. Do the same from the feet of each to the eyes of the other.)

17. Why does reflected light from the Sun or Moon appear as a column in the body of water as shown? How would it appear if the water's surface were perfectly smooth?
18. What is wrong with the cartoon of the man looking at himself in the mirror? (Have a friend face a mirror as shown, and you'll see.)
19. A pair of toy cart wheels is rolled obliquely from a smooth surface onto two plots of grass, a rectangular plot and a triangular plot, as shown. The ground is on a slight incline, so that, after slowing down in the grass, the wheels will speed up again when emerging onto the smooth surface. Finish the sketches by showing some positions of the wheels inside each plot and on the other side of each plot, thereby indicating their direction of travel.

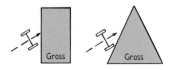

20. A pulse of red light and a pulse of blue light enter a glass block normal to its surface at the same time. Strictly speaking, after passing through the block, which pulse exits first?

21. During a lunar eclipse, the Moon is not completely dark, but it is often a deep red in color. Explain this in terms of the refraction of all the sunsets and sunrises around the world.

22. If you place a glass test tube in water, you will be able to see the tube. If you place it in clear soybean oil, you may not be able to see it. What does this tell you about the speed of light in the oil and in the glass?

23. If, while standing on the bank of a stream, you wished to spear a fish swimming in the water out in front of you, would you aim above, below, or directly at the observed fish to make a direct hit? If you decided instead to zap the fish with a laser, would you aim above, below, or directly at the observed fish? Defend your answers.

24. If the fish in the previous exercise were small and blue, and your laser light were red, what corrections should you make? Explain.

25. When a fish looks upward at an angle of 45°, does it see the sky or only the reflection of the bottom? Defend your answer.

26. If you were to send a beam of laser light to a space station above the atmosphere and just above the horizon, would you aim the laser above, below, or directly at the visible space station? Defend your answer.

27. Rays of light in water that shine up to the water–air boundary at angles of more than 48° to the normal are totally reflected. No rays beyond 48° refract outside. How about the other way around? Is there an angle at which light rays in air meeting the air–water boundary will reflect totally? Or will some light be refracted at all angles?

28. When your eye is submerged in water, is the bending of light rays from water to your eyes greater than, less than, or the same as it is in air?

29. When you stand with your back to the Sun, you see a rainbow as a circular arc. Could you move off to one side and then see the rainbow as the segment of an ellipse rather than the segment of a circle (such as Figure 14.20 suggests)? Defend your answer.

30. Two observers standing apart from one another do not see the "same" rainbow. Explain.

31. A rainbow viewed from an airplane may form a complete circle. Where will the shadow of the airplane appear? Explain.

32. How is a rainbow similar to the halo sometimes seen around the Moon on a frosty night? How are rainbows and halos different?

33. What is responsible for the rainbow-colored fringe commonly seen at the edges of a spot of white light from the beam of a lantern or slide projector?

34. Transparent plastic swimming-pool covers called solar heat sheets have thousands of small lenses made up of air-filled bubbles. The lenses in these sheets are advertised to focus heat from the Sun into the water and raise its temperature. Do you think the lenses of such sheets direct more solar energy into the water? Defend your answer.

35. Would the average intensity of sunlight measured by a light meter at the bottom of the pool in Figure 14.34 be different if the water were still?

36. What accounts for the large shadows cast by the ends of the thin legs of the water strider? What accounts for the ring of bright light around the shadows on the bottom?

37. Why will goggles allow a swimmer underwater to focus more clearly on what he is looking at?

38. Cover the top half of a camera lens. What effect does this have on the pictures taken?

39. Would refracting telescopes and microscopes magnify if light had the same speed in glass as it does in air? Explain.

40. Consider a simple magnifying glass underwater. Will it magnify more or less? Explain why.

41. Can you take a photograph of your image in a plane mirror and focus the camera both on your image and on the mirror frame? Explain.

42. For correct viewing, why are slides put into a slide projector upside down?

43. Maps of the Moon are upside down. Why?

44. What does polarization tell you about the nature of light waves?

45. The digital displays of watches and other devices are normally polarized. What problem occurs with these displays when one is wearing polarized sunglasses?

46. Why will an ideal Polaroid filter transmit 50% of incident nonpolarized light?

47. Why may an ideal Polaroid filter transmit anything from zero to 100% of incident polarized light?

48. What percentage of light is transmitted by two ideal Polaroid filters, one on top of the other with their polarization axes aligned? With their polarization axes at right angles to each other?

49. How can a single Polaroid filter be used to show that the sky is partially polarized? (Interestingly enough, unlike humans, bees and many insects can discern polarized light, and they use this ability for navigation.)

50. Light will not pass through a pair of Polaroid filters when they are aligned perpendicularly. But, if a third Polaroid filter is sandwiched between the other two with its alignment halfway between the alignments of the others (that is, with its axis making a 45° angle with each of the other two alignment axes), some light does get through. Why?

PROBLEMS

1. ● If you take a photograph of your image in a plane mirror, how many meters away should you set your focus if you are 3 m in front of the mirror?

2. ● A spider hangs by a strand of silk at an eye level 20 cm in front of a plane mirror. You are behind the spider, 50 cm from the mirror. Show that the distance between your eye and the image of the spider in the mirror is 70 cm.

3. ● Suppose that you walk toward a mirror at 2 m/s. How rapidly do you and your image approach each other? (The answer is not 2 m/s.)

4. ■ When light strikes glass perpendicularly, about 4% of the light is reflected at each surface. Show that the amount of light transmitted through a pane of window glass is approximately 92%.

5. ■ Show with a simple diagram that, when a mirror with a fixed beam incident on it is rotated through a certain angle, the reflected beam is rotated through an angle twice as large. (This doubling of displacement makes irregularities in ordinary window glass more evident.)

6. ■ The average speed of light slows to $0.75c$ when it refracts through a particular piece of plastic.
   a. What change is there in the light's frequency in the plastic?
   b. In its wavelength?

7. ◆ The diameter of the Sun makes an angle of 0.53° with its apex at the surface of Earth. Show that 2.1 minutes elapses as the Sun moves one solar diameter in an overhead sky. (Remember that it takes 24 hours, or 1440 minutes, for the Sun to move through 360°.) How does your answer compare with the time it takes for the Sun to disappear once its lower edge meets the horizon at sunset? (Does refraction affect your answer?)

8. ◆ A quantitative way of relating the object distance d_o with image distance d_i for a lens is given by the *thin-lens equation* $\dfrac{1}{d_o} + \dfrac{1}{d_i} = \dfrac{1}{f}$. Rearrange this to show that $d_i = \dfrac{d_o f}{d_o - f}$.

CHAPTER 14 ONLINE RESOURCES

The Physics Place

Interactive Figures
14.1, 14.10, 14.24, 14.36

Videos
Image Formation in a Mirror
Models of Refraction

The Rainbow
Polarized Light and 3-D Viewing

Quiz

Flashcards

Links

Is the slanted line really broken?

Are the dashes on the right really shorter?

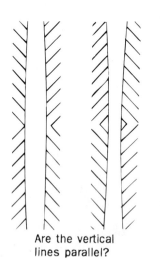

Can you count the black dots?

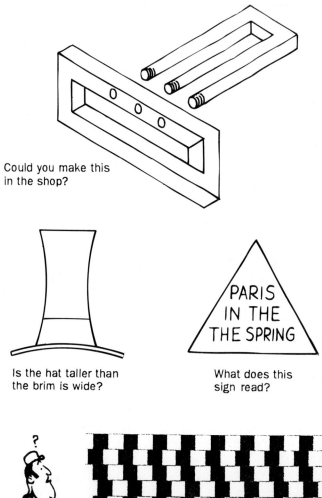

Could you make this in the shop?

Is the hat taller than the brim is wide?

What does this sign read?

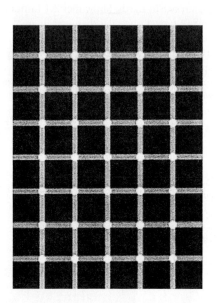

Are the vertical lines parallel?

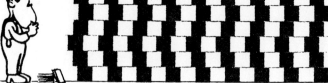

Are the tiles really crooked?

Optical illusions.

PART FIVE

Atomic and Nuclear Physics

"Know nukes!" The Earth's natural heat that warms this natural hot spring, or that powers a geyser or a volcano, comes from nuclear power—the radioactivity of minerals in the Earth's interior. Power from atomic nuclei is as old as the Earth itself, and isn't restricted to today's nuclear reactors, or "nukes," as they are called. How about that!

Quantum Theory

David Kagan models an orbiting electron with a strip of corrugated plastic. The stacked wood blocks model energy levels.

Back in Chapter 2 we discussed the atom as a building block of matter and touched upon the structure of the atom. We know that the atom is composed of a central nucleus surrounded by a sea of electrons. The study of the atom above the atomic nucleus, the arrangements of the surrounding electrons, is called *atomic physics*— how this chapter begins. We will outline some of the developments that led from atomic physics to *quantum physics*. We begin at the dawn of the twentieth century when physicists were greatly puzzled by several phenomena that begged explanations. All the puzzles had to do with the interaction between light and matter.

15.1 The Photoelectric Effect

In 1900, the German theoretical physicist Max Planck was trying to explain why the higher frequencies of light are only emitted by objects at higher temperatures. Why, for example, does a red-hot lightbulb not emit any violet light? Classical models of radiating bodies predicted that objects should radiate more of their energy at higher frequencies. That these high frequencies didn't appear was called the "ultraviolet catastrophe" and required a new model for how matter radiates. Planck hypothesized that warm bodies emit radiant energy (light) in individualized bundles. Each of these bundles he called a **quantum** (plural, *quanta*). According to Planck, the energy in each quantum is proportional to the frequency of radiation. Thus, more energy is associated with a quantum of violet light than it is of red light. Hence, a body that is red hot won't emit higher-energy violet light quanta until its temperature is much higher.

Physicists were reluctant to adopt Planck's revolutionary quantum notion. To be taken seriously,

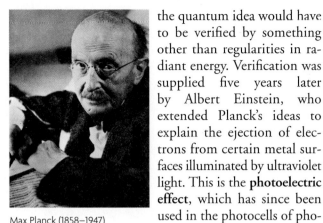

Max Planck (1858–1947)

the quantum idea would have to be verified by something other than regularities in radiant energy. Verification was supplied five years later by Albert Einstein, who extended Planck's ideas to explain the ejection of electrons from certain metal surfaces illuminated by ultraviolet light. This is the **photoelectric effect**, which has since been used in the photocells of photographer's light meters and automatic door openers.*

Einstein viewed light not as a continuous wave, but as a stream of particles, bundles of energy (later called *photons*). He stated that the energy E of any single photon is proportional to the frequency f of the corresponding light wave. This energy is given by

$$E = hf$$

where h is a number called **Planck's constant.**** So a photon of violet light (high frequency) carries more energy than does a photon of red light (low frequency). Although a bright beam of red light has more photons, and therefore more energy, than does a dim beam of violet light, the violet beam carries *more energy per photon*. The photoelectric effect shows that photons interact with matter one at a time.

The equation $E = hf$ tells us why ultraviolet light and X-rays cause damage to molecules in living cells, whereas microwave radiation doesn't. The relatively low frequency of microwaves ensures low energy per photon. Photons of ultraviolet radiation, on the other hand, can deliver about a million times more energy to a molecule because the

* Einstein's paper on the photoelectric effect won him the 1921 Nobel Prize in Physics.

** Planck's constant, h, has the numerical value 6.6×10^{-34} J·s. We shall see that Planck's constant is a fundamental constant of nature that serves to set a lower limit on the smallness of things. It ranks with the velocity of light and Newton's gravitational constant as a basic constant of nature, and it appears again and again in quantum physics.

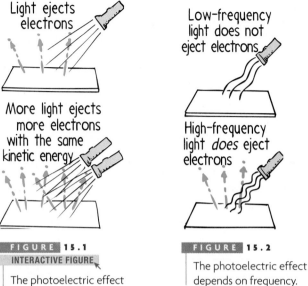

FIGURE 15.1
INTERACTIVE FIGURE

The photoelectric effect depends on intensity.

FIGURE 15.2

The photoelectric effect depends on frequency.

frequency of ultraviolet radiation is about a million times greater than the frequency of microwaves. X-ray photons, with even higher frequencies, can deliver even more energy.

The concept of the photon exemplifies the revolution in physics—*quantization*. Quantization is the idea that the natural world is granular rather than smoothly continuous. Matter is quantized; water is made up of individual water molecules. Electric charge is quantized; it is always some whole-number multiple of the charge of a single

FIGURE 15.3

Phil Wolf, co-author of *Problem Solving in Conceptual Physics*, employs the photoelectric effect by directing light of various frequencies onto a photosensitive surface, which allows him to measure the energies of ejected electrons.

electron. Any elemental particle that makes up matter or carries energy is called a *quantum*.

Consider the photograph of Max Planck on the previous page. The blending areas of black, white, and gray in the photograph do not look smooth at all when viewed through a magnifying glass. With magnification, you can see that the photograph consists of many tiny dots. In a similar way, we live in a world that is a blurred image of the grainy world of atoms. The "common-sense" world described by classical physics seems smooth and continuous because quantum graininess is on a very small scale compared with the sizes of things in the familiar world.

If a vending machine accepts only quarters, you can't fool it by inserting five nickels taped together. Just as the machine accepts coins only one at a time, electrons interact with photons one at a time. Either the photon has enough energy to eject the electron or it doesn't.

STOP AND CHECK YOURSELF

1. What does the word *quantum* mean?

2. How much total energy is in a monochromatic beam composed of *n* photons of frequency *f* ?

CHECK YOUR ANSWERS

1. A *quantum* is the smallest elemental unit of a quantity. Radiant energy, for example, is composed of many quanta, each of which is called a *photon*. So the more photons there are in a beam of light, the more energy there is in that beam.

2. The total energy would be *nhf*.

15.2 Emission Spectra

Another puzzle of early investigators was finding nature's rules for the emission of light from a glowing gas. Physicists and chemists discovered that each glowing element emits its own characteristic pattern of frequencies and produces its own **emission spectrum.** Such spectra can be seen when light is passed through a prism—or, better, when it is first passed through a thin slit and

The Physics Place
Light and Spectroscopy
Bohr's Shell Model

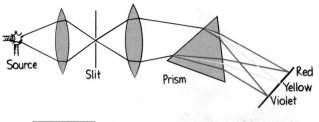

FIGURE 15.4

A simple spectroscope. Images of the illuminated slit are cast on a screen and make up a pattern of lines. The spectral pattern is characteristic of the light used to illuminate the slit.

then focused through a prism onto a viewing screen behind. Such an arrangement of slit, focusing lenses, and prism (or diffraction grating) is called a **spectroscope** (Figure 15.4).

Each component color is focused at a definite position according to its frequency and forms an image of the slit on the screen, photographic film, or appropriate detector. The different-colored images of the slit are called *spectral lines*. Some typical spectral patterns labeled by wavelengths are shown in Figure 15.6 (it is customary to refer to colors in terms of their wavelengths rather than their frequencies). A given frequency corresponds to a definite wavelength.*

FIGURE 15.5

George Curtis separates light from an argon source into its component frequencies with a typical student-lab spectroscope.

---

* Recall, from Chapter 12, that $v = f\lambda$, where v is the wave speed, f is the wave frequency, and $\lambda \approx f\lambda$ (lambda) is the wavelength. For light, v is the constant c, so we see from $c = f\lambda$ the relationship between frequency and wavelength—namely, $f = c/\lambda$ and $\lambda = c/f$.

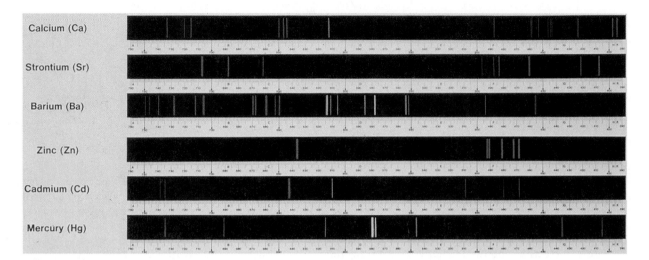

Spectral patterns of some elements.

If, for example, the light given off by a sodium-vapor lamp is analyzed in a spectroscope, a single yellow line predominates. If we narrow the width of the slit, we find that this line is really composed of two very close lines. These lines correspond to the two predominant frequencies of light emitted by glowing sodium atoms.

The same happens with all glowing vapors. The light from a mercury-vapor lamp shows a pair of bright yellow lines close together (but in different positions from those of sodium), a very intense green line, and several blue and violet lines. A neon tube produces a more complicated pattern of lines. The light emitted by each element in the gaseous phase produces its own characteristic pattern of lines. These lines are as characteristic of each element as are fingerprints of people. The spectroscope, therefore, is widely used in chemical analysis.

Atomic spectra are the fingerprints of atoms.

While chemists were using the spectroscope for chemical analysis, physicists were busy trying to find order in the confusing arrays of spectral lines. It had long been known that the lightest element, hydrogen, has a far more orderly spectrum than the other elements. An important sequence of lines in the hydrogen spectrum starts with a line in the red region, followed by one in the blue, then by several lines in the violet, and many in the ultraviolet (Figure 15.7). Successive lines become closer from the first in the red to the last in the ultraviolet, until the lines seem to merge. In 1884 a Swiss school-teacher, Johann Jakob Balmer, first expressed the wavelengths of these lines in a single mathematical formula. This is the basis of science—collecting data and developing a formula to organize them. Balmer, however, could give no reason why his formula worked. He predicted that other elements might follow a similar formula, which proved to be correct. His discovery led to the prediction of lines that had not yet been measured.

Another regularity in atomic spectra was found by the Swedish physicist and mathematician Johannes Rydberg. He noticed that the sum of the frequencies of two lines in the spectrum of hydrogen often equals the frequency of a third line. This relationship was later advanced as a general principle by the Swiss physicist Walter Ritz and is called the

fyi
The mystery of the chemical makeup of stars was solved with the spectroscope. Identity of stellar elements is in their starlight.

A portion of the hydrogen spectrum. Each line, an image of the slit in the spectroscope, represents light of a specific frequency emitted by hydrogen gas when excited (higher frequency is to the right).

Ritz combination principle. It states that the spectral lines of any element include frequencies that are either the sum or the difference of the frequencies of two other lines. Like Balmer, Ritz was unable to offer an explanation for this regularity.

The puzzle: When elements are made to glow, why are only certain colors of light emitted but not others? And why do different elements emit light with different sets of frequencies?

Emission Spectra Explained

Electrons occupy shells about the atomic nucleus. An electron in a shell farther from the nucleus has a greater electric potential energy relative to the nucleus than an electron in a shell nearer the nucleus. We say that the more distant electron is in a higher energy state, or, equivalently, at a higher energy level.

When an electron is in any way raised to a higher energy level, the atom is said to be *excited*. The electron's higher position is only momentary—the atom soon loses its temporarily acquired energy when the electron returns to a lower level and emits radiant energy. The atom has undergone the process of **excitation** and *de-excitation.*

Each electrically neutral element has its own characteristic set of energy levels. Each electron that makes a transition from a higher to a lower energy level in an excited atom emits a photon of electromagnetic radiation. The frequency of the emitted photon is related to the electron's energy transition. The change in the transition energy is the same as the amount of energy in the photon. In this way energy from the atom is given to the photon, the frequency of which is directly proportional to its energy: $E = hf$.

If many atoms in a material are excited, many photons with many frequencies are emitted, each

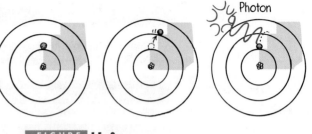

FIGURE 15.9

INTERACTIVE FIGURE

When an electron in an atom is boosted to a higher orbit, the atom is excited. When the electron returns to its original orbit, the atom de-excites and gives off a photon of light.

exactly matching a transition from a higher to a lower energy level. These frequencies correspond to the characteristic colors of light from each chemical element.

A gas sample that is given no energy emits no light. If the amount of energy given to the gas is strictly controlled, so just enough energy is imparted to boost the electrons of atoms from the ground state to the first excited state but not beyond, the sample will emit only one frequency of light. Then there is only one pathway by which the electron can de-excite. The emission spectrum of the sample will have only a single line. As the energy given to the gas increases, more lines appear in the emission spectrum.

This model solves the mystery of the Ritz combination principle. You can see from Figure 15.10 that an electron transition from the third energy level to the ground state corresponds to a particular energy difference. A direct transition produces a single photon with this amount of energy, while a "two-step" transition (third energy level to second, and then second to first) produces two photons, whose energies add up to the energy of the single

FIGURE 15.8

Excitation and de-excitation.

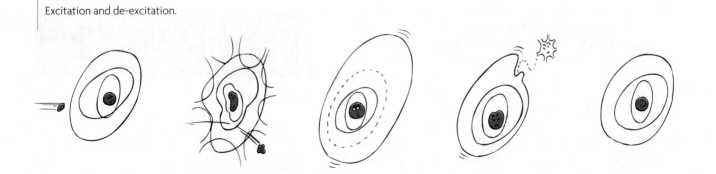

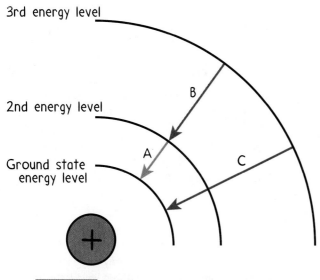

3rd energy level

2nd energy level

Ground state
energy level

B

A

C

+

FIGURE 15.10

Three of many energy levels in an atom. An electron making a transition from the third level to the second level is shown in red (B), and on jumping from the second level to the ground state is shown in green (for infrared) (A). The sum of the energies (and the frequencies) for these two transitions equals the energy (and the frequency) of the single jump from the third to the ground state, shown in blue (C).

direct-transition photon. Since a photon's energy is proportional to its frequency, then it follows that

$$E_1 + E_2 = E_3 \Rightarrow hf_1 + hf_2 = hf_3$$
$$\Rightarrow f_1 + f_2 = f_3.$$

Niels Bohr, a Danish physicist, was the first to use these regularities in atomic behavior to develop the familiar planetary model of the atom.* Bohr was the first to explain the excitation–de-exitation process. Just as each element is characterized by the number of electrons occupying shells surrounding its atomic nucleus, each element also possesses its own characteristic *pattern* of electron shells, or *energy states*. These energy states are found only in shells of certain radii from the nucleus. Because these states can have only certain energies, we say they are *discrete*. We call these discrete states *quantum states*.

* This model, like most models, has major defects because the electrons do not revolve in planes as planets do. Later, the model was revised; "orbits" became "shells" and "clouds." We use *orbit* because it was, and still is, commonly used. Electrons are not merely bodies, like planets, but rather behave like waves concentrated in certain parts of the atom.

Yet Bohr's planetary model of the atom begged a major question. Accelerated electrons, according to Maxwell's theory, radiate energy in the form of electromagnetic waves. So an electron accelerating around a nucleus should radiate energy continuously. This radiating away of energy should cause the electron to spiral into the nucleus (Figure 15.11). Bohr boldly deviated from classical physics by stating that the electron doesn't radiate light while it accelerates around the nucleus in a single orbit. Bohr stated, in effect, it just ain't so!

Bohr was able to account for X-rays in heavier elements, showing that they are emitted when electrons drop from outer to innermost orbits. He predicted X-ray frequencies that were later experimentally confirmed. Bohr was also able to calculate the "ionization energy" of a hydrogen atom—the energy needed to knock the electron out of the atom completely. This also was verified by experiment.

Using measured frequencies of X-rays as well as visible, infrared, and ultraviolet light, scientists could map energy levels of all the atomic elements. Bohr's model had electrons orbiting in neat circles (or ellipses) arranged in groups or shells. This model of the atom accounted for the general chemical properties of the elements. It also predicted a missing element, which led to the discovery of hafnium.

Bohr solved the mystery of atomic spectra while providing an extremely useful model of the atom. He was quick to point out that his model was to be interpreted as a crude beginning, and the picture of electrons whirling about the nucleus like planets about the Sun was not to be taken literally (to which popularizers of science paid no heed). His sharply defined orbits were conceptual representations of an atom whose later description involved waves—*quantum mechanics*. His ideas of quantum jumps and frequencies being proportional to energy differences remain part of twenty-first century physics.

FIGURE 15.11

According to classical theory, an electron accelerating around its orbit should continuously emit radiation. This loss of energy should cause it to spiral rapidly into the nucleus. But this does not happen.

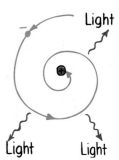

1. According to Figure 15.10, how many transitions can an electron in quantum level 3 make in returning to the ground state?

2. Two predominant spectral lines in the hydrogen spectrum, an infrared one and a red one, have frequencies 2.7×10^{14} Hz and 4.6×10^{14} Hz, respectively. Can you predict a higher- frequency line in the hydrogen spectrum?

CHECK YOUR ANSWERS

1. Three (transitions A, B, and C), as shown in Figure 15.10.

2. The sum of the frequencies is 2.7×10^{14} Hz + 4.6×10^{14} Hz = 7.3×10^{14} Hz, which happens to be the frequency of a violet line in the hydrogen spectrum. Using Figure 15.10 as a model, can you see that, if the infrared line is produced by a transition corresponding to path A and the red line corresponds to path B, then the blue line corresponds to path C?

The light emitted in the glass tubes of advertising signs is a familiar consequence of excitation. The different colors correspond to the excitation of different gases, although it is common to refer to any of these as "neon." Only the red light is that of neon. At the ends of each glass tube that contains the neon gas are electrodes. Electrons are "boiled off" these electrodes and are jostled back and forth at high speeds by a high ac voltage. Millions of high-speed electrons vibrate back and forth inside the glass tube and collide with millions of target atoms, boosting orbital electrons into higher energy levels by an amount of energy equal to the decrease in kinetic energy of the bombarding electron. As the electrons fall back to their stable orbits, this energy is radiated as the characteristic red light of neon. The process occurs and recurs many times, as neon atoms continually undergo a cycle of excitation and de-excitation. The overall result of this process is the transformation of electrical energy into radiant energy.

The colors of various flames are due to excitation. Different atoms in the flame emit colors characteristic of their energy-level spacings. Common table salt placed in a flame, for example, produces the characteristic yellow of sodium. Mercury-vapor street lamps emit light rich in blues and violet, producing a white different from that of incandescent lamps. Every element, excited in a flame or otherwise, emits its own characteristic color or colors.

Suppose a friend suggests that, for a first-rate operation, the gaseous neon atoms in a neon tube should be periodically replaced with fresh atoms because the energy of the atoms tends to be used up with continued excitation, producing dimmer and dimmer light. What do you say to this?

CHECK YOUR ANSWER

The neon atoms don't release any energy that is not given to them by the electric current in the tube and therefore don't get "used up." Any single atom may be excited and re-excited without limit. If the light is, in fact, becoming dimmer and dimmer, it is probably because a leak exists. Otherwise, there is no advantage whatsoever to changing the gas in the tube, because a "fresh" atom is indistinguishable from a "used" one. Both are ageless, in that they are older than the solar system.

Excitation is illustrated in the aurora borealis. High-speed electrons that originate in the solar wind strike atoms and molecules in the upper atmosphere. They emit light exactly as occurs in a neon tube. The different colors in the aurora correspond to the excitation of different gases—oxygen atoms produce a greenish-white color, nitrogen molecules produce a red-violet color, and nitrogen ions produce a blue-violet color. Auroral emissions are not restricted to visible light; they also include infrared, ultraviolet, and X-ray radiation.

> **Exciting an atom is like trying to kick a ball out of a ditch. Many short kicks won't do the job, because the ball keeps falling back. A kick of just the right energy is enough to get the ball out of the ditch. The same is true with the excitation of atoms.**

15.3 Absorption Spectra

When we view white light from an incandescent source with a spectroscope, we see a continuous spectrum over the whole rainbow of colors. If a gas is placed between the source and the spectroscope,

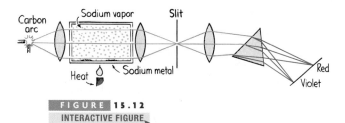

Experimental arrangement for demonstrating the absorption spectrum of a gas.

however, careful inspection will show that the spectrum is not quite continuous. This is an **absorption spectrum,** and there are dark lines distributed throughout it; these dark lines against a rainbow-colored background are *absorption lines.*

An atom may be excited by absorbing a photon of light. An atom will most strongly absorb light having the frequencies to which it is tuned—the same frequencies that it ordinarily would emit when excited. When a beam of white light passes through a gas, the photons that get absorbed are those whose energies are just right to excite an electron to a higher energy level. When the electron de-excites, the absorbed energy is reradiated, but in *all* directions instead of only in the direction of the incident beam. When the light remaining in the beam is spread out into a spectrum, the frequencies that were absorbed show up as dark lines in the otherwise continuous spectrum. The absorption spectrum is much like the corresponding emission spectrum in reverse (Figure 15.13).

Although the Sun is a source of incandescent light, the spectrum it produces, upon close examination, is not continuous. There are many absorption lines, called *Fraunhofer lines* in honor of the Bavarian optician and physicist Joseph von Fraunhofer, who first observed and accurately mapped them. Similar lines are found in the spectra produced by the other stars. These lines indicate that the Sun and stars are each surrounded by an atmosphere of gases that absorb some of the light coming from the main body. Analysis of these lines reveals the chemical composition of the atmospheres of such sources. We find from these analyses that the stellar elements are the same elements that exist on Earth. An interesting sidelight is that, in 1868, spectroscopic analysis of sunlight showed some spectral lines different from any known on Earth. These lines identified a new element, which was named *helium,* after Helios, the Greek god of the Sun. Helium was discovered in the Sun before it was discovered on Earth. How about that!

We can determine the speed of stars by studying the spectra they emit. Just as a moving sound source produces a Doppler shift in its pitch (Chapter 12), a moving light source produces a Doppler shift in its light frequency. Compared with the frequency of light from a stationary source, the frequency (not the speed!) of light emitted by an approaching source is higher, while the frequency of light from a receding source is lower. The corresponding spectral lines are displaced toward the red end of the spectrum (called a *red shift*) for receding sources, and toward the blue end (called a *blue shift*) for approaching sources. Since the universe is expanding, almost all the galaxies show a red shift in their spectra.

15.4 Fluorescence

Many materials that are excited by ultraviolet light emit visible light upon de-excitation. The action of these materials is called **fluorescence.** In these materials, a photon of ultraviolet light excites the atom, boosting an electron to a higher energy state. In this upward quantum jump, the atom is likely to leapfrog over several intermediate energy states. So when the atom de-excites, it may make smaller jumps, emitting photons with less energy.

This excitation and de-excitation process is like leaping up a small staircase in a single bound, and then descending one or two steps at a time. Hence,

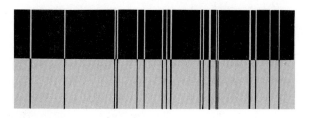

FIGURE 15.13

Emission and absorption spectra.

FIGURE 15.14

Crayons fluorescing in various colors under ultraviolet light.

ultraviolet light shining on the material causes it to glow an overall red, yellow, or whatever color is characteristic of the material. Fluorescent dyes are used in paints and fabrics to make them glow when bombarded with ultraviolet photons in sunlight.

When a time delay occurs between excitation and de-excitation, **phosphorescence** occurs. The element phosphorus is a good example. Phosphorus and other materials are used for items that are made to glow in the dark. The delay depends on the material, and can last for several hours. When the source of excitation is removed (such as when the room lights are turned off) an afterglow occurs while millions of atoms spontaneously undergo gradual de-excitation.

> STOP AND
> ## CHECK YOURSELF
>
> Why would it be impossible for a fluorescent material to emit ultraviolet light when illuminated by infrared light?
>
> ### CHECK YOUR ANSWER
>
> Photon energy output would be greater than photon energy input, which would violate the law of conservation of energy.

The next time you visit a natural-science museum, go to the geology section and take in the exhibit of minerals illuminated with ultraviolet light (Figure 15.15). You'll notice that different minerals radiate various colors. High-energy ultraviolet photons strike the minerals, causing the excitation of atoms in the mineral structure. The light of the several frequencies that you see corresponds to the electron cascading down through series of tiny energy-level spacings. Every excited atom emits light of its characteristic frequencies, with no two different minerals giving off light of exactly the same color. Beauty is in both the eye and the mind of the beholder.

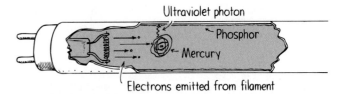

FIGURE 15.16

A fluorescent tube. Ultraviolet (UV) light is emitted by gas in the tube excited by an alternating electric current. The UV light, in turn, excites phosphors on the inner surface of the glass tube, which emit white light.

Fluorescent Lamps

The common fluorescent lamp consists of a cylindrical glass tube with electrodes at each end (Figure 15.16). In the lamp, as in the tube of a neon sign, electrons are "boiled" from one of the electrodes and forced to vibrate to and fro at high speeds within the tube by the ac voltage. The tube is filled with very low-pressure mercury vapor, which is excited by the impact of the high-speed electrons. Much of the emitted light is in the ultraviolet region. This is the primary excitation process. The secondary process occurs when the ultraviolet light strikes *phosphors,* a powdery material on the inner surface of the tube. The phosphors are excited by the absorption of the ultraviolet photons and fluoresce, giving off a multitude of lower-frequency photons that combine to produce white light. Different phosphors can be used to produce different colors or "textures" of light.

15.5 Incandescence

The phenomenon of light produced due to a high temperature is **incandescence**. If you connect a lightbulb to an adjustable voltage source and slowly turn up the voltage, the filament appears first red-hot, then orange-hot, then yellow-hot, and finally white-hot. This is incandescent light (*incandescent* is from a Latin word meaning "to grow hot"). As we learned in Chapter 9, as the temperature of a light source increases, the predominant frequency of emitted radiation (that is, the brightest part of the spectrum), the *peak frequency,* is directly proportional to the absolute temperature of the emitter:

$$\bar{f} \sim T$$

As discussed in Chapter 9, the bar above the f indicates the peak frequency, for radiations of many

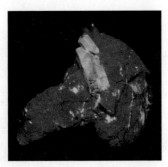

FIGURE 15.15

The rock contains the fluorescing minerals calcite and willemite, which, under ultraviolet light, are clearly seen as red and green, respectively.

frequencies are emitted from an incandescent source. The temperature of incandescent bodies, whether they be stars or blast-furnace interiors, can be determined by measuring the peak frequency (or color) of the radiant energy they emit.

The main difference between light from an incandescent lamp and light from neon or any other gas-discharge tube is that light from the incandescent source contains an infinite number of frequencies, spread smoothly across the spectrum. This does not mean that an infinite number of energy levels characterize the tungsten atoms in the lamp filament. If the filament were vaporized and then excited, the tungsten gas would emit light with a finite number of frequencies and produce an overall bluish color. In the gaseous phase, light is emitted by atoms spaced far from one another, much different from the light emitted by the same atoms closely packed into a solid. Similarly, sound from an isolated ringing bell is different from the sound of a box crammed with ringing bells (Figure 15.18). When billions of atoms are crammed together in a solid, their electron energy levels are smudged to make a *band* of energy, where the levels are so close together that they appear to be continuous. Outer electrons in the atoms make transitions not only with the energy levels of their "parent" atoms, but also between the levels of neighboring atoms. Electrons bounce around over dimensions larger than a single atom, resulting in an infinite variety of transitions.

> Humans glow at infrared frequencies. Infrared thermometers utilize $\bar{f} \sim T$ when they measure our radiant energy and convert it to a temperature reading.

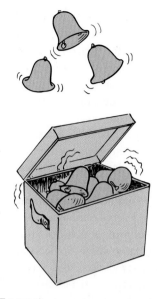

FIGURE 15.18

The sound of an isolated bell rings with a clear and distinct frequency, whereas the sound emanating from a box of bells crowded together is discordant. The same is true with the difference in light emitted from atoms in a gaseous state compared with light emitted from atoms in a solid state.

> The color of light emitted by an incandescent solid depends on the temperature of the solid. The color of light emitted by an exited gas does not depend on the gas temperature—but does depend on the energy levels of its atoms.

15.6 Lasers

The phenomena of excitation, fluorescence, and phosphorescence underlie the operation of a most intriguing instrument, the **laser** (**l**ight **a**mplification by **s**timulated **e**mission of **r**adiation).* Although the first laser was invented in 1958, the concept of stimulated emission was predicted by Albert Einstein in 1917. To understand how a laser operates, we first discuss coherent light.

Light emitted by incandescence is incoherent; that is, photons of many frequencies and in many

* A word constructed from the initials of a phrase is called an *acronym*.

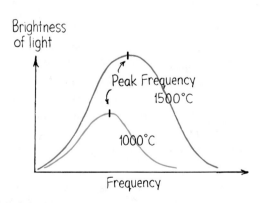

FIGURE 15.17

Radiation curves for an incandescent solid.

phases of vibration are emitted. The light is as incoherent as the footsteps on an auditorium floor when a mob of people are chaotically rushing about. Incoherent light is chaotic. A beam of incoherent light spreads out after a short distance, becoming wider and wider and less intense with increased distance.

Even if the beam is filtered so that it consists of single-frequency waves (monochromatic light), it is still incoherent, because the waves are out of phase with one another.

A laser is a device that produces a beam of coherent light. Every laser has a source of atoms called an active medium, which can be a gas, liquid, or solid (the first laser was a ruby crystal). The atoms in the medium are excited to delayed (metastable) states by an external source of energy. When most of the atoms in the medium are excited, a single photon from an atom that undergoes de-excitation can start a chain reaction. This photon strikes another atom, stimulating it into emission, and so on, producing coherent light. Most of this light is initially moving in random directions. Light traveling along the laser axis, however, is selectively reflected from mirrors

FIGURE 15.19
INTERACTIVE FIGURE

Incoherent white light contains waves of many frequencies (and of many wavelengths) that are out of phase with one another.

FIGURE 15.20

Light of a single frequency and wavelength still contains a mixture of phases.

FIGURE 15.21

Coherent light: all the waves are identical and in phase.

coated to reflect light of the desired wavelength. One mirror is totally reflecting, while the other is partially reflecting. The reflected waves reinforce each other after each round-trip reflection between the mirrors, thereby setting up a to-and-fro resonance condition wherein the light builds up to an appreciable intensity. The light that escapes through the more transparent-mirrored end makes up the laser beam.*

In addition to gas and crystal lasers, other types have joined the laser family: glass, chemical, liquid, and semiconductor lasers. Present models produce light beams ranging in frequency from infrared through ultraviolet. Some models can be tuned to various frequency ranges. Most exciting is the prospect of an X-ray laser beam.

The laser is not a source of energy. It is simply a converter of energy that takes advantage of the process of stimulated emission to concentrate a certain fraction of its energy (commonly 1%) into radiant energy of a single frequency moving in a single direction. Like all devices, a laser can put out no more energy than is put into it.

> A laser beam is not seen unless it scatters off something in the air. Like sunbeams or moonbeams, what you see are the particles in the scattering medium, not the beam itself. When the beam lands on a diffuse surface, part of it is scattered toward your eye as a dot.

15.7 Wave–Particle Duality

The wave and particle nature of light is evident in the formation of optical images. We understand the photographic image produced by a camera in terms of light waves, which spread from each point of the object, refract as they pass through the lens system, and converge to focus on photographic film or some other detector. The path of light from the object through the lens system and to the focal plane can be calculated using methods developed from the wave theory of light.

* The narrowness of a laser beam is evident when you see a lecturer produce a tiny red or green spot on a screen using a laser "pointer." Light from an intense laser pointed at the Moon has been reflected and detected back on Earth.

But now consider carefully the way in which an image is formed on photographic film. Film consists of an emulsion that contains grains of silver halide crystal, each grain containing about 10^{10} silver atoms. Each photon that is absorbed gives up its energy hf to a single grain in the emulsion. This energy activates surrounding crystals in the entire grain and is used in development to complete the photochemical process. Many photons activating many grains produce the usual photographic exposure. When a photograph is taken with exceedingly feeble light, we find that the image is built up by individual photons that arrive independently and are seemingly random in their distribution. We see this strikingly illustrated in Figure 15.22, which shows how an exposure progresses photon by photon.

Double-Slit Experiment

Let's return to Thomas Young's double-slit experiment, which we discussed in terms of waves in Chapter 13. Recall that, when we pass monochro-

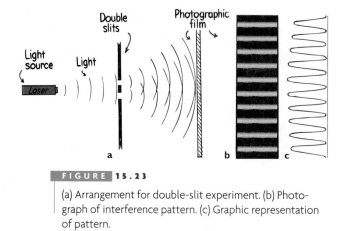

FIGURE 15.23

(a) Arrangement for double-slit experiment. (b) Photograph of interference pattern. (c) Graphic representation of pattern.

matic light through a pair of closely spaced thin slits, we produce an interference pattern (Figure 15.23). Now let's consider the experiment in terms of photons. Suppose we dim our light source so that, in effect, only one photon at a time reaches the barrier with the thin slits. If film behind the barrier is exposed to the light for a very short time, the film gets exposed as simulated in Figure 15.23a. Each spot represents the place where the film has been exposed by a photon. If the light is allowed to expose the film for a longer time, a

FIGURE 15.22

Stages of film exposure reveal the photon-by-photon production of a photograph. The approximate numbers of photons at each stage are (a) 3×10^3, (b) 1.2×10^4, (c) 9.3×10^4, (d) 7.6×10^5, (e) 3.6×10^6, and (f) 2.8×10^7.

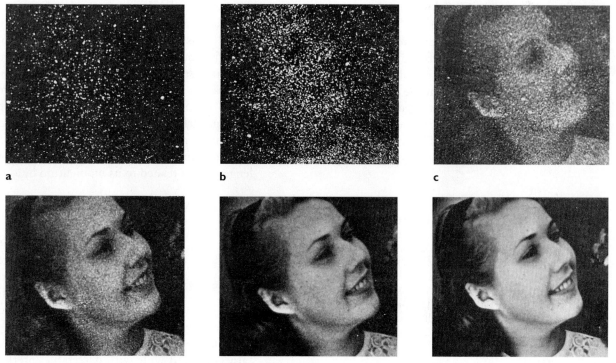

a   b   c

d   e   f

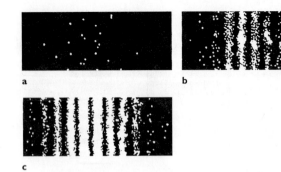

FIGURE **15.24**
INTERACTIVE FIGURE

Stages of two-slit interference pattern. The pattern of individually exposed grains progresses from (a) 28 photons to (b) 1000 photons to (c) 10,000 photons. As more photons hit the screen, a pattern of interference fringes appears.

pattern of fringes begins to emerge, as in Figures 15.23b and 15.23c. This is quite amazing! Spots on the film are seen to progress, photon by photon, to form the same interference pattern characterized by waves!

If we cover one slit so that photons striking the photographic film can pass only through a single slit, the tiny spots on the film accumulate to form a single-slit diffraction pattern (Figure 15.24). Photons hit the film at places they would not hit if both slits were open! If we think about this classically, we are perplexed and may ask how photons passing through the single slit "know" that the other slit is covered and therefore fan out to produce the wide single-slit diffraction pattern. Or, if both slits are open, how do photons traveling through one slit "know" that the other slit is open and avoid certain regions, proceeding only to areas that will ultimately fill to form the fringed double-slit interference pattern?* The modern answer is the photon interferes with itself and passes

---

* From a prequantum point of view, this wave–particle duality is indeed mysterious. This leads some people to believe that quanta have some sort of consciousness, with each photon or electron having "a mind of its own." The mystery, however, is like beauty. It is in the mind of the beholder rather than in nature itself. We conjure models to understand nature and, when inconsistencies arise, we sharpen or change our models. The wave–particle duality of light doesn't fit a model built on classical ideas. One model is that quanta have minds of their own. Another model is quantum physics. In this book, we subscribe to the latter.

FIGURE **15.25**

Single-slit diffraction pattern.

> **Light travels as a wave and hits like a particle.**

through both slits. Each single photon has wave properties as well as particle properties. The photon displays different aspects at different times. *A photon behaves as a particle when it is being emitted by an atom or absorbed by photographic film or other detectors and behaves as a wave in traveling from a source to the place where it is detected.* The fact that light exhibits both wave and particle behavior was one of the interesting surprises of the early twentieth century.

15.8 Particles as Waves: Electron Diffraction

If a photon of light has both wave and particle properties, why shouldn't a material particle (one with mass) also have both wave and particle properties? This question was posed by the French physicist Louis de Broglie in 1924 while he was still a graduate student. His answer constituted his doctoral thesis in physics and later earned him the Nobel Prize in Physics. According to de Broglie, every particle of matter is associated with a corresponding wave. Each body—whether an electron, a proton, an atom, a mouse, or you—has a wavelength that is related to its momentum by

$$\text{Wavelength} = \frac{h}{\text{momentum}}$$

where *h* is Planck's constant. Under the proper conditions, every particle will produce an interference or diffraction pattern. A body of large mass and ordinary speed has such a small wavelength that interference and diffraction are negligible; rifle bullets fly straight and do not pepper their targets far and wide with detectable interference

The Physics Place
Electron Waves

Louis de Broglie (1892–1987)

patches.* But for smaller particles, such as electrons, diffraction can be appreciable.

A beam of electrons can be diffracted in the same way that a beam of photons can be diffracted, as is evident in Figure 15.26. Beams of electrons directed through double slits exhibit interference patterns, just as photons do. The apparatus is more complex for electrons than it is for photons, but the procedure is essentially the same. The intensity of the source can be reduced to direct electrons one at a time through a double-slit arrangement, producing the same remarkable results as with photons. Like photons, electrons strike the screen as particles, but the *pattern* of arrival is wavelike. The angular deflection of electrons to form the interference pattern agrees perfectly with calculations using de Broglie's equation for the wavelength of an electron.

In Figure 15.29, we see displayed on a TV monitor the step-by-step buildup of an interference pattern produced by individual electrons in a standard electron microscope. The image is gradually completed as the electrons produce the interference pattern customarily associated with waves. Neutrons, protons, whole atoms, and, to an immeasurable degree, even high-speed rifle bullets exhibit a duality of particle and wave behavior.

Electron Waves

The idea that electrons in an atom may occupy only certain energy levels was very puzzling to early investigators and to Bohr himself. Why the electron occupies only discrete levels became understood by considering the electron to be not a particle, but as de Broglie hypothesized, a *wave*.

* A bullet of mass 0.02 kg traveling at 330 m/s, for example, has a de Broglie wavelength of

$$\frac{h}{mv} = \frac{6.6 \times 10^{-34}\,\text{J} \cdot \text{s}}{(0.02\ \text{kg})(330\ \text{m/s})} = 10^{-34}\text{m},$$

an incredibly small size a million million million millionth the size of a hydrogen atom. An electron traveling at 2% the speed of light, on the other hand, has a wavelength 10^{-10} m, which is equal to the diameter of the hydrogen atom. Diffraction effects for electrons are measurable, whereas diffraction effects for bullets are not.

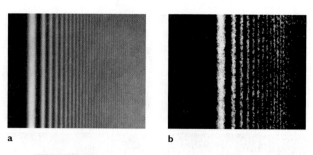

a   b

FIGURE **15.26**

Fringes produced by the diffraction of (a) light and (b) an electron beam.

FIGURE **15.27**

An electron microscope makes practical use of the wave nature of electrons. The wavelength of an electron beam is typically thousands of times shorter than the wavelength of visible light, so the electron microscope is able to distinguish detail not visible with optical microscopes.

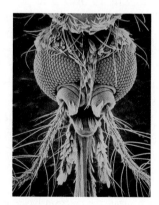

FIGURE **15.28**

Detail of the head of a female mosquito as seen with a scanning electron microscope at a "low" magnification of 200 times.

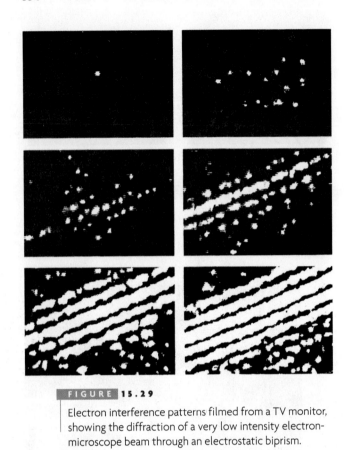

FIGURE 15.29

Electron interference patterns filmed from a TV monitor, showing the diffraction of a very low intensity electron-microscope beam through an electrostatic biprism.

Using the idea of interference, de Broglie showed that the discrete values of radii of Bohr's orbits are a natural consequence of standing electron waves. A Bohr orbit exists where an electron wave can close on itself constructively. The electron wave becomes a standing wave, like a wave on a musical string. The standing wave has a whole number of wavelengths fitting evenly into the circumferences of the orbits (Figure 15.30). The circumference of the innermost orbit, according to this picture, is equal to one wavelength. The second orbit has a circumference of two electron wavelengths, the third has three, and so forth (Figure 15.31). This is similar to a chain necklace made of paper clips. No matter what size necklace is made, its circumference is equal to some multiple of the length of a single paper clip.* Since the circumferences of electron orbits are discrete, it follows that the radii of these orbits, and hence the energy levels, are also discrete.

* For each orbit, the electron has a unique speed, which determines its wavelength. For higher energy, larger radius orbits, the electron speeds are less, and wavelengths are longer. So for our analogy to be accurate, we'd have to use not only more paper clips to make increasingly longer necklaces but increasingly larger paper clips as well.

STOP AND CHECK YOURSELF

1. If electrons behaved only like particles, what pattern would you expect on the screen after the electron passed through the double slits?

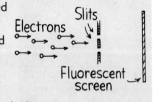

2. We don't notice the de Broglie wavelength for a pitched baseball. Is this because the wavelength is very large or because it is very small?

3. If an electron and a proton have the same de Broglie wavelength, which particle has the greater speed?

CHECK YOUR ANSWERS

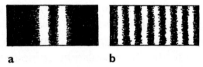

a   b

1. If electrons behaved only like particles, they would form two bands, as indicated in the figure labeled a. Because of their wave nature, they actually produce the pattern shown in the figure labeled b.

2. We don't notice the wavelength of a pitched baseball because it is extremely small—on the order of 10^{-20} times smaller than the atomic nucleus.

3. The same wavelength means that the two particles have the same momentum. This means that the less massive electron must travel faster than the heavier proton.

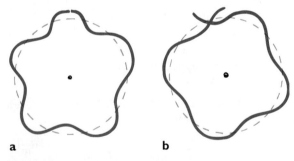

a   b

FIGURE 15.30

(a) An orbiting electron forms a standing wave only when the circumference of its orbit is equal to a whole-number multiple of the wavelength. (b) When the wave does not close in on itself in phase, it undergoes destructive interference. Hence, orbits exist only where waves close in on themselves in phase.

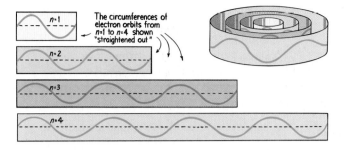

Erwin Schrödinger (1887–1961)

FIGURE 15.31

The electron orbits in an atom have discrete radii because the circumferences of the orbits are whole-number multiples of the electron wavelength. This results in a discrete energy state for each orbit. (The figure is greatly oversimplified, as the standing waves make up spherical and ellipsoidal shells rather than flat, circular ones.)

This model explains why electrons don't spiral closer and closer to the nucleus, causing atoms to shrink to the size of the tiny nucleus. If each electron orbit is described by a standing wave, the circumference of the smallest orbit can be no smaller than one wavelength—no fraction of a wavelength is possible in a circular (or elliptical) standing wave. As long as an electron carries the momentum necessary for wave behavior, atoms don't shrink in on themselves.

In the still more modern wave model of the atom, electron waves move not only around the nucleus but also in and out, toward and away from the nucleus. The electron wave is spread out in three dimensions, leading to the picture of an electron "cloud." As we shall see, this is a cloud of *probability,* not a cloud made up of a pulverized electron scattered over space. The electron, when detected, remains a point particle.

15.9 Quantum Mechanics

Many changes in physics occurred in the mid-1920s. Starting with de Broglie's matter waves, the Austrian-German physicist Erwin Schrödinger formulated an equation that describes how matter waves change under the influence of external forces. Schrödinger's equation plays the same role in **quantum mechanics** that Newton's equation (acceleration = force/mass) plays in classical physics.*

The matter waves in Schrödinger's equation are mathematical entities that are not directly observable, so the equation provides us with a purely mathematical rather than a visual model of the atom—which places it beyond the scope of this book. So our discussion of it will be brief.**

In **Schrödinger's wave equation**, the thing that "waves" is the nonmaterial *matter wave amplitude*—a mathematical entity called a *wave function,* represented by the symbol ψ (the Greek letter psi). All of the information about the matter waves is contained in the wave function. One derives answers for the probable values of, say, the momentum or energy or position of the particle by operating mathematically upon the wave function. For example, the electron in a hydrogen atom may be located anywhere between the center of the nucleus and some radial distance far away. A physicist can calculate the probability of it being in a certain volume of space by multiplying the wave function by itself ($|\psi|^2$). This produces a second mathematical entity called a *probability density function,* which at a given time indicates the probability per unit volume for each of the possibilities represented by ψ.

Experimentally, there is a finite probability (chance) of finding an electron in a particular region at any instant. The value of this probability lies between the limits 0 and 1, where 0 indicates never and 1 indicates always. For example, if the probability is 0.40 for finding the electron within a certain range of radii, this signifies a 40% chance that the electron will be found there. The Schrödinger equation cannot tell a physicist where an electron can be found in an atom at any

---

* Schrödinger's wave equation, strictly for math types, is

$$\left(-\frac{h^2}{2m}\nabla^2 + U\right)\psi = ih\,\frac{\partial\psi}{\partial t}.$$

** Our short treatment of this complex subject is hardly conducive to any real understanding of quantum mechanics. At best, it serves as a brief overview and possible introduction to further study. The readings suggested at the end of the chapter may be quite useful.

FIGURE **15.32**

Probability distribution of an electron cloud.

moment, but only the *likelihood* of finding it there—or, for a large number of measurements, what fraction of measurements will find the electron in each region. When the position of an electron in its Bohr energy level (quantum state) is repeatedly measured and each of its locations is plotted as a dot, the resulting pattern resembles a sort of electron cloud (Figure 15.32). An individual electron may, at various times, be detected anywhere in this probability cloud; it even has an extremely small but finite probability of momentarily existing inside the nucleus. It is detected most of the time, however, close to an average distance from the nucleus, which fits the orbital radius described by Niels Bohr.

> Considering something to be impossible may reflect a *lack* of understanding, as when scientists thought a single atom could never be seen. Or it may represent a *deep* understanding, as when scientists (and the patent office!) reject perpetual motion machines.

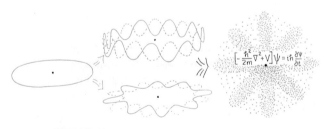

FIGURE **15.33**

From the Bohr model of the atom, to the modified model with de Broglie waves, to a wave model with the electrons distributed in a "cloud" throughout the atomic volume.

STOP AND CHECK YOURSELF

1. Consider 100 photons diffracting through a thin slit to form a diffraction pattern. If we detect five of the photons in a certain region in the pattern, what is the probability (between 0 and 1) of detecting a photon in this region?

2. Suppose that you open a second identical slit and that the diffraction pattern is one of bright and dark bands. Suppose the region where five photons hit before now has none. A wave theory says that waves that hit before are now canceled by waves from the other slit—that crests and troughs combine to zero. But our measurement is of photons that either make a hit or don't. How does quantum mechanics reconcile this?

CHECK YOUR ANSWERS

1. We have approximately a 0.05 probability of detecting a photon at this location. In quantum mechanics we say $|\psi|^2 \approx 0.05$. The true probability could be somewhat more or less than 0.05. Put the other way around, if the true probability is 0.05, then the number of photons detected could be somewhat more or less than 5.

2. Quantum mechanics says that photons propagate as waves and are absorbed as particles, with the probability of absorption governed by the maxima and minima of wave interference. Where the combined wave from the two slits has zero amplitude, the probability of a particle being absorbed is zero.

Most physicists, but not all, view quantum mechanics as a fundamental theory of nature. Interestingly enough, Albert Einstein, one of the founders of **quantum physics**, never accepted it as fundamental; he considered the probabilistic nature of quantum phenomena as the outcome of a deeper, as yet undiscovered, physics. He stated, "Quantum mechanics is certainly imposing. But an inner voice tells me it is not yet the real thing. The theory says a lot, but does not really bring us closer to the secret of 'the Old One.'"*

* Although Einstein practiced no religion, he often invoked God as the "Old One" in his statements about the mysteries of nature.

15.10 Uncertainty Principle

The wave–particle duality of quanta has inspired interesting discussions about the limits of our ability to accurately measure the properties of small objects. The discussions center on the idea that the act of measuring something affects the quantity being measured.

For example, we know that, if we place a cool glass thermometer in a cup of hot coffee, the temperature of the coffee is altered as it gives heat to the thermometer. The measuring device alters the quantity being measured. But we can correct for these errors in measurement if we know the initial temperature of the thermometer, the masses and specific heats involved, and so forth. Such corrections fall well within the domain of classical physics—these are *not* the uncertainties of quantum physics. Quantum uncertainties stem from the wave nature of matter. A wave, by its very nature, occupies some space and lasts for some time. It cannot be squeezed to a point in space or limited to a single instant of time, for then it would not be a wave. This inherent "fuzziness" of a wave gives a fuzziness to measurements at the quantum level. Innumerable experiments have shown that any measurement that in any way probes a system necessarily disturbs the system by at least one quantum of action, *h*—Planck's constant. So any measurement that involves interaction between the measurer and what is being measured is subject to this minimum uncertainty.

We distinguish between observing and probing. Consider a cup of coffee on the other side of a room. If you passively glance at it and see steam rising from it, this act of "measuring" involves no physical interaction between your eyes and the coffee. Your glance neither adds nor subtracts energy from the coffee. You can assert that it's hot with no *probing*. Placing a thermometer in it is a different story. You physically interact with the coffee and thereby subject it to alteration. The quantum contribution to this alteration, however, is completely dwarfed by classical uncertainties and is negligible. Quantum uncertainties are significant only in the atomic and subatomic realm.

Compare the acts of making measurements of a pitched baseball and of an electron. You can measure the speed of a pitched baseball by having it fly through a pair of photogates that are a known distance apart (Figure 15.34). The ball is timed as it

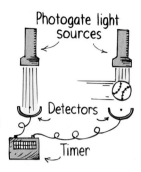

The ball's speed is measured by dividing the distance between the photogates by the time difference between crossing the two light paths. Photons hitting the ball alter its motion much less than the motion of an oil supertanker would be altered by a few fleas bumping into it.

interrupts beams of light in the gates. The accuracy of the ball's measured speed has to do with uncertainties in the measured distance between the gates and in the timing mechanisms. Interactions between the macroscopic ball and the photons it encounters are insignificant. But not so in the case of measuring the electron. Even a single photon bouncing off an electron appreciably alters the motion of the electron—and in an unpredictable way. If you wish to observe an electron and determine its whereabouts with light, the wavelength of the light would have to be very short (long waves go by without interaction). You fall into a dilemma. Light of a short wavelength, which can "see" the tiny electron better, corresponds to a large quantum of energy, which greatly alters the electron's state of motion. If instead you use a long wavelength that corresponds to a smaller quantum of energy, the change you induce to the electron's state of motion will be smaller, but the determination of its position by means of the coarser wave will be less precise. The act of observing something as tiny as an electron probes the electron and, in so doing, produces a considerable uncertainty in either its position or its motion. Although this uncertainty is completely negligible for measurements of position and motion regarding everyday (macroscopic) objects, it is a predominant fact of life in the atomic domain.

The uncertainty of measurement in the atomic domain, first stated mathematically by the German physicist Werner Heisenberg, is called the **uncertainty principle**. It is a fundamental principle in quantum mechanics. Heisenberg found that when the uncertainties in the measurement of momentum and position for a particle are multiplied together, the product

Werner Heisenberg (1901–1976)

must be equal to or greater than Planck's constant, h, divided by 2π, which is represented as $\hbar$ (called *h-bar*). We can state the uncertainty principle in a simple formula:

$$\Delta p \Delta x \geq \hbar$$

The Δ here means "uncertainty in our measurement of": Δp is the uncertainty in our measurement of momentum (the symbol for momentum is conventionally p), and Δx is the uncertainty in position. The product of these two uncertainties must be equal to or greater than ($\geq$) the size of $\hbar$. This means that if we wish to know the momentum of an electron with great precision (small Δp), the corresponding uncertainty in position will be large. Or if we wish to know the position with great accuracy (small Δx), the corresponding uncertainty in momentum will be large. The sharper one of these quantities is, the less sharp is the other.*

The uncertainty principle operates similarly with energy and time. We cannot measure a particle's energy with complete precision in an infinitesimally short span of time. The uncertainty in our knowledge of energy, ΔE, and the duration taken to measure the energy, Δt, are related by the expression**

$$\Delta E \Delta t \geq \hbar$$

The greatest precision we can ever hope to attain is that case in which the product of the energy and time uncertainties equals $\hbar$. The more precisely we determine the energy of a photon, an electron, or a particle of whatever kind, the more uncertain we will be of the time during which it has that energy.

The uncertainty principle is relevant only to quantum phenomena. The inaccuracies in measuring the position and momentum of a baseball due to the interactions of observation, for example, are completely negligible. But the inaccuracies in measuring the position and momentum of an electron are far from negligible. This is because the uncertainties in the measurements of these subatomic quantities are comparable to the magnitudes of the quantities themselves.†

It is reckless to apply the uncertainty principle to areas outside of quantum mechanics. Some people conclude from statements about the interaction between the observer and the observed that the universe does not exist "out there," independent of all acts of observation, and that reality is created by the observer. Others interpret the uncertainty principle as nature's shield of forbidden secrets. Some critics of science use the uncertainty principle as evidence that science itself is uncertain. The reality of the universe (whether observed or not), nature's secrets, and the uncertainties of science have very little to do with Heisenberg's uncertainty principle. The profundity of the uncertainty principle has to do with the unavoidable interaction between nature at the atomic level and the means by which we probe it.

15.11 Correspondence Principle

If a new theory is valid, it must account for the verified results of the old theory. This is the **correspondence principle**, first articulated by Bohr. New theory and old must correspond; that is, they must overlap and agree in the region where the results of the old theory have been fully verified.

When the techniques of quantum mechanics are applied to macroscopic systems rather than atomic systems, the results are essentially identical with those of classical mechanics. For a large system, such as the solar system, where classical physics is successful, the Schrödinger equation leads to results that differ from classical theory only by infinitesimal amounts. The two domains blend when the

---

* Only in the classical limit where $\hbar$ becomes zero could the uncertainties of both position and momentum be arbitrarily small. Planck's constant is greater than zero, and we cannot, in principle, simultaneously know both quantities with absolute certainty.

** We can see that this is consistent with the uncertainty in momentum and position. Recall that Δ momentum = force $\times$ Δ time and that Δenergy = force $\times$ Δdistance. Then

$$\begin{aligned} \hbar &= \Delta\text{momentum} \times \Delta\text{distance} \\ &= (\text{force} \times \Delta\text{distance}) \times \Delta\text{time} \\ &= \Delta\text{energy} \times \Delta\text{time} \end{aligned}$$

† The uncertainties in measurements of momentum, position, energy, or time that are related to the uncertainty principle for a pitched baseball are only 1 part in about 10 million billion billion billion (10^{-34}). Quantum effects are negligible even for the swiftest bacterium, where the uncertainties are about 1 part in a billion (10^{-9}). Quantum effects become evident for atoms, where the uncertainties can be as large as 100%. For electrons moving in an atom, quantum uncertainties dominate, and we are in the full-scale quantum realm.

CHECK YOURSELF

1. Is Heisenberg's uncertainty principle applicable to the practical case of using a thermometer to measure the temperature of a glass of water?

2. A Geiger counter measures radioactive decay by registering the electrical pulses produced in a gas tube when high-energy particles pass through it. The particles emanate from a radioactive source—say, radium. Does the act of measuring the decay rate of radium alter the radium or its decay rate?

3. Can the quantum principle that we cannot observe something without changing it be reasonably extrapolated to support the claim that you can make a stranger turn around and look at you by staring intently at his back?

CHECK YOUR ANSWERS

1. No. Although we probably subject the temperature of water to a change by the act of probing it with a thermometer, especially one appreciably colder or hotter than the water, the uncertainties that relate to the precision of the thermometer are quite within the domain of classical physics. The role of uncertainties at the subatomic level is inapplicable here.

2. Not at all, because the interaction involved is between the Geiger counter and the particles, not between the Geiger counter and the radium.

It's the behavior of the high-energy particles that is altered by measurement, not the behavior of the radium from which they emanate. See how this ties into the next question.

3. No. Here we must be careful in defining what we mean by *observing*. If our observation involves probing (giving or extracting energy), we indeed change to some degree that which we observe. For example, if we shine a light source onto the person's back, our observation consists of probing, which, however slight, physically alters the configuration of atoms on his back. If he senses this, he may turn around. But simply staring intently at his back is observing in the passive sense. The light you receive (or block by blinking, for example) has already left his back. So whether you stare, squint, or close your eyes completely, you in no physical way alter the atomic configuration on his back. Shining a light or otherwise probing something is not the same thing as passively looking at something. A failure to make the simple distinction between *probing* and *passive observation* is at the root of much nonsense that is claimed by some to be supported by quantum physics. Better support for the above claim would be positive results from a simple and practical test, rather than on the assertion that it rides on the hard-earned reputation of quantum theory.

de Broglie wavelength is small compared with the dimensions of the system or of the pieces of matter in the system. It is, in fact, impractical to use quantum mechanics in the domains in which classical physics is successful; but, at the atomic level, quantum physics reigns and is the only theory that gives results consistent with what is observed.

> The correspondence principle is a general rule not only for good science but for all good theory—even in areas as far removed from science as government, religion, and ethics.

Complementarity

The realm of quantum physics can seem confusing. Light waves that interfere and diffract deliver

their energy in packages of quanta—particles. Electrons that move through space in straight lines and experience collisions as if they were particles distribute themselves spatially in interference patterns as if they were waves. In this confusion, there is an underlying order. The behavior of light and electrons is confusing in the same way! Light and electrons both exhibit wave and particle characteristics.

Niels Bohr, a founder of quantum physics, formulated an explicit expression of the wholeness inherent in this dualism. He called it **complementarity**. As Bohr expressed it, quantum phenomena exhibit complementary (mutually exclusive) properties—appearing either as particles or as waves—depending on the type of experiment conducted. Experiments designed to examine individual exchanges of energy and momentum bring out particle-like properties, while experiments designed to examine spatial distribution of energy bring out wavelike properties. The wavelike properties of light and the particle-like properties of light complement one another—both are necessary for the understanding of "light." Which part is emphasized depends on what question one puts to nature.

Complementarity is not a compromise, and it doesn't mean that the whole truth about light lies somewhere in between particles and waves. It's rather like viewing the sides of a crystal. What you see depends on what facet you look at, which is why light, energy, and matter appear to be behaving as quanta in some experiments and as waves in others.

The idea that opposites are components of a wholeness is not new. Ancient Eastern cultures incorporated it as an integral part of their worldview. This is demonstrated in the yin–yang diagram of T'ai Chi Tu (Figure 15.35). One side of the circle is called *yin,* and the other side is called *yang.* Where there is yin, there is yang. Only the union of yin and yang forms a whole. Where there is low, there is also high. Where there is night, there is also day. Where there is birth, there is also death. A whole person integrates yin (feminine traits, right

FIGURE 15.35

Opposites are seen to complement one another in the yin–yang symbol of Eastern cultures.

brain, emotion, intuition, darkness, cold, wetness) with yang (masculine traits, left brain, reason, logic, light, heat, dryness). Each has aspects of the other. For Niels Bohr, the yin–yang diagram symbolized the principle of complementarity. In later life, Bohr wrote broadly on the implications of complementarity. In 1947, when he was knighted for his contributions to physics, he chose for his coat of arms the yin–yang symbol.

fyi   **You won't fully appreciate the frontiers of physics unless you're familiar with its foothills.**

SUMMARY OF TERMS

Quantum (*pl.* quanta) From the Latin word *quantus,* meaning "how much," a quantum is the smallest elemental unit of a quantity, the smallest discrete amount of something. One quantum of electromagnetic energy is called a photon.

Photoelectric effect The emission of electrons from a metal surface when light shines upon it.

Planck's constant A fundamental constant, h, that relates the energy of light quanta to their frequency:

$$h = 6.6 \times 10^{-34} \text{ joule} \cdot \text{second}$$

Emission spectrum The distribution of wavelengths in the light from a luminous source.

Spectroscope An optical instrument that separates light into its constituent wavelengths in the form of spectral lines.

Ritz combination principle The statement that the frequencies of some spectral lines of the elements are either the sums or the differences of the frequencies of two other lines.

Excitation The process of boosting one or more electrons in an atom or molecule from a lower to a higher energy level. An atom in an excited state will usually decay (de-excite) rapidly to a lower state by the emission of a photon. The energy of the photon is proportional to its frequency: $E = hf$.

Absorption spectrum A continuous spectrum, like that of white light, interrupted by dark lines or bands that result from the absorption of light of certain frequencies by a substance through which the radiant energy passes.

Fluorescence The property of certain substances to absorb radiation of one frequency and to reemit radiation of lower frequency. It occurs when an atom is boosted up to an excited state and loses its energy in two or more downward jumps to a lower energy state.

Phosphorescence A type of light emission that is the same as fluorescence except for a delay between excitation and de-excitation, which provides an afterglow. The delay is caused by atoms being excited to energy states that do not decay rapidly. The afterglow may last from fractions of a second to hours or even days, depending on the type of material, temperature, and other factors.

Incandescence The state of glowing while at a high temperature, caused by electrons bouncing around over dimensions larger than the size of an atom, emitting radiant energy in the process. The peak frequency of radiant energy is proportional to the absolute temperature of the heated substance:

$$\bar{f} \sim T$$

Laser (**l**ight **a**mplification by **s**timulated **e**mission of **r**adiation) An optical instrument that produces a beam of coherent monochromatic light.

Quantum mechanics The theory of the microworld based on wave functions and probabilities developed especially by Werner Heisenberg (1925) and Erwin Schrödinger (1926).

Schrödinger's wave equation A fundamental equation of quantum mechanics, which relates probability wave amplitudes to the forces acting on a system. It is as basic to quantum mechanics as Newton's laws of motion are to classical mechanics.

Quantum physics The physics that describes the microworld, where many quantities are granular (in units called *quanta*), not continuous, and where particles of light (*photons*) and particles of matter (such as electrons) exhibit wave as well as particle properties.

Uncertainty principle The principle, formulated by Werner Heisenberg, stating that Planck's constant, *h*, sets a limit on the precision of measurement. According to the uncertainty principle, it is not possible to measure exactly both the position and the momentum of a particle at the same time, nor the energy and the time during which the particle has that energy.

Correspondence principle The rule that a new theory must give the same results as the old theory where the old theory is known to be valid.

Complementarity The principle, enunciated by Niels Bohr, stating that the wave and particle aspects of both matter and radiation are necessary, complementary parts of the whole. Which part is emphasized depends on what experiment is conducted (i.e., on what question one puts to nature).

SUGGESTED READING

Cole, K. C. *The Hole in the Universe: How Scientists Peered over the Edge of Emptiness and Found Everything.* New York: Harcourt, 2001.

Ford, K. W. *The Quantum World: Quantum Physics for Everyone.* Cambridge, MA: Harvard University Press, 2004. An intriguing account of the development of quantum physics, with emphasis on the participating physicists.

Rigden, J. S. *Hydrogen—The Essential Element.* Cambridge, MA: Harvard University Press, 2002. An enjoyable biography of nature's most abundant element.

Trefil, J. *Atoms to Quarks.* New York: Scribner's, 1980. A nice development of quantum theory in the early chapters, which lead on to particle physics.

REVIEW QUESTIONS

1. Distinguish between *atomic physics* and *nuclear physics*.

15.1 The Photoelectric Effect

2. What is the photoelectric effect?
3. Which would be more successful in dislodging electrons from a metal surface—photons of violet light or photons of red light? Why?
4. Why won't a very bright beam of red light impart more energy to an ejected electron than a feeble beam of violet light?
5. What is meant by *quantization*?

15.2 Emission Spectra

6. What is an emission spectrum?
7. What is a *spectroscope,* and what does it accomplish?
8. What is predicted in the Ritz combination principle?
9. What does it mean to say an atom is excited?
10. Is light emitted when an atom is excited, or when it de-excites?
11. What does it mean to say that energy states are *discrete*?

15.3 Absorption Spectra

12. How does an absorption spectrum differ in appearance from an emission spectrum?
13. What are Fraunhofer lines?
14. How can astrophysicists tell whether a star is receding from or approaching Earth?

15.4 Fluorescence

15. Why is ultraviolet light, but not infrared light, effective in making certain materials fluoresce?
16. Distinguish between the primary and secondary excitation processes that occur in a fluorescent lamp.
17. What is responsible for the afterglow of phosphorescent materials?

15.5 Incandescence

18. How is the peak frequency of emitted light related to the temperature of its incandescent source?
19. When a gas glows, discrete colors are emitted. When a solid glows, the colors are smudged. Why?

15.6 Lasers

20. Distinguish between *monochromatic light* and *coherent light*.
21. How does the avalanche of photons in a laser beam differ from the hordes of photons emitted by an incandescent lamp?

15.7 Wave–Particle Duality

22. Does light behave primarily as a wave or as a particle when it interacts with the crystals of matter in photographic film?
23. When does light behave as a wave? When does it behave as a particle?

15.8 Particles as Waves: Electron Diffraction

24. What hypothesis was made by Louis de Broglie in 1924?
25. When electrons are diffracted through a double slit, do they arrive at the screen in a wavelike way or in a particle-like way? Is the pattern of hits wavelike or particle-like?
26. Why does each element have its own pattern of spectral lines?
27. How does treating the electron as a wave rather than as a particle solve the riddle of why electron orbits are discrete?
28. According to the simple de Broglie model, how many wavelengths are there in an electron wave in the first orbit? In the second orbit? In the *n*th orbit?
29. How can we explain why electrons don't spiral into the attracting nucleus?

15.9 Quantum Mechanics

30. What does the wave function psi (ψ) represent?
31. Distinguish between a wave function and a probability density function.
32. How does the probability cloud of the electron in a hydrogen atom relate to the orbit described by Niels Bohr?

15.10 Uncertainty Principle

33. In which of the following are quantum uncertainties significant? Measuring simultaneously the speed and location of a baseball; of a pebble; of an electron.
34. What is the uncertainty principle with respect to motion and position?
35. If measurements show a precise position for an electron, can those measurements show precise momentum also? Explain.
36. If measurement shows a precise value for the energy radiated by an electron, can that measurement show a precise time for this event as well? Explain.

15.11 Correspondence Principle

37. Exactly what is it that "corresponds" in the correspondence principle?
38. How does Schrödinger's equation fare when applied to the solar system?
39. What is the principle of complementarity?
40. Cite evidence that the idea of opposites as components of a wholeness preceded Bohr's principle of complementarity.

ACTIVE EXPLORATIONS

1. Write a letter to a 12-year-old youngster and explain how light is emitted from lamps, flames, and lasers. Explain why fluorescent dyes and paints are so impressively vivid when illuminated with an ultraviolet lamp. Go on to discuss the similarities and differences between fluorescence and phosphorescence.
2. Borrow a diffraction grating from your physics instructor. The common kind looks like a photographic slide, and light passing through it or reflecting from it is diffracted into its component colors by thousands of finely ruled lines. Look through the grating at the light from a sodium-vapor street lamp. If it's a low-pressure lamp, you'll see the nice yellow spectral "line" that dominates sodium light (actually, it's two closely spaced lines). If the street lamp is round, you'll see circles instead of lines; if you look through a slit cut in cardboard or some similar material, you'll see lines. What happens with the now-common high-pressure sodium lamps is more interesting. Because of the collisions of excited atoms, you'll see a smeared-out spectrum that is nearly continuous, almost like that of an incandescent lamp. Right at the yellow location, where you'd expect to see the sodium line, is a dark area. This is the sodium absorption band. It is due to the cooler sodium that surrounds the high-pressure emission region. You should view this from a block or so away so that the line, or circle, is small enough to allow the resolution to be maintained. Try this. It is very easy to see!

EXERCISES

1. Which has more energy—a photon of visible light or a photon of ultraviolet light?
2. We speak of photons of red light and photons of green light. Can we speak of photons of white light? Why or why not?
3. Which laser beam carries more energy per photon—a red beam or a green beam?
4. If a beam of red light and a beam of blue light have exactly the same energy, which beam contains the greater number of photons?

5. If we double the frequency of light, we double the energy of each of its photons. If we instead double the wavelength of light, what happens to the photon energy?

6. Do phosphors on the inside of fluorescent lamps convert ultraviolet light to visible light, or visible light to ultraviolet light? Defend your answer.

7. Silver bromide (AgBr) is a light-sensitive substance used in some types of photographic film. To cause exposure of the film, it must be illuminated with light having sufficient energy to break apart the molecules. Why do you suppose this film may be handled without exposure in a darkroom illuminated with only red light? How about blue light? How about very bright red light relative to very dim blue light?

8. Sunburn produces cell damage in the skin. Why is ultraviolet radiation capable of producing this damage, while visible radiation, even if more intense, is not?

9. In the photoelectric effect, does brightness or frequency determine the kinetic energy of the ejected electrons? Which determines the number of the ejected electrons?

10. A very bright source of red light has much more energy than a dim source of blue light, but the red light has no effect in ejecting electrons from a certain photosensitive surface. Why is this so?

11. Why are ultraviolet photons more effective at inducing the photoelectric effect than photons of visible light?

12. Why does light striking a metal surface eject only electrons, not protons?

13. Does the photoelectric effect depend on the wave nature or the particle nature of light?

14. What is the evidence for the claim that iron exists in the relatively cool outer layer of the Sun?

15. What difference does an astronomer see between the emission spectrum of an element in a receding star and a spectrum of the same element in the lab? (Hint: This relates to information in Chapter 12.)

16. A blue-hot star is about twice as hot as a red-hot star. But the temperatures of the gases in advertising signs are about the same, whether they emit red or blue light. What is your explanation?

17. Does atomic excitation occur in solids as well as in gases? How does the radiant energy from an incandescent solid differ from the radiant energy emitted by an excited gas?

18. If atoms of a substance absorb ultraviolet light and emit blue light, what becomes of the "missing" energy?

19. When an electron makes a transition from its first quantum level to ground level, the energy difference is carried by the emitted photon. In comparison, how much energy is involved to return an electron at ground level to the first quantum level?

20. Your friend reasons that, if ultraviolet light can activate the process of *fluorescence*, infrared light ought to also. Your friend looks to you for approval or disapproval of this idea. What is your position?

21. The forerunner to the laser involved microwaves rather than visible light. What does *maser* mean?

22. The first laser consisted of a red ruby rod activated by a photoflash tube that emitted green light. Why would a laser composed of a green crystal rod and a photoflash tube that emits red light not work?

23. A laboratory laser has a power of only 0.8 mW (8×10^{-4} W). Why does it seem more powerful than light from a 100-W lamp?

24. How do the avalanches of photons in a laser beam differ from the hordes of photons emitted by an incandescent lamp?

25. A friend speculates that scientists in a certain country have developed a laser that produces far more energy than is put into it. Your friend asks for your response to this speculation. What is your response?

26. Since every object has some temperature, every object radiates energy. Why, then, can't we see objects in the dark?

27. If we continue heating a piece of initially room-temperature metal in a dark room, it will begin to glow visibly. What will be its first visible color, and why?

28. We can heat a piece of metal to red hot and then to white hot. Can we heat it until the metal glows blue hot? And would it be solid at this temperature?

29. How do the surface temperatures of reddish, bluish, and whitish stars compare?

30. Part a in the sketch on the next page shows a radiation curve of an incandescent solid and its spectral pattern as produced with a spectroscope. Part b shows the "radiation curve" of an excited gas and its emission spectral pattern. Part c shows the curve produced when a cool gas is between an incandescent source and the viewer; the corresponding spectral pattern is left as an exercise for you to construct. Part d shows the spectral pattern of an incandescent source as seen through a piece of green glass; you are to sketch in the corresponding radiation curve.

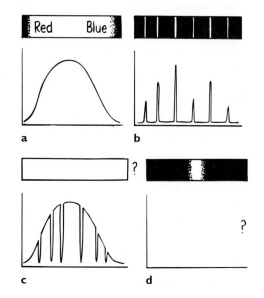

31. Consider just four of the energy levels in a certain atom, as shown in the diagram. How many spectral lines will result from all possible transitions among these levels? Which transition corresponds to the

highest-frequency light emitted? To the lowest-frequency light emitted?

n = 4 ————
n = 3 ————
n = 2 ————

n = 1 ————

32. An electron de-excites from the fourth quantum level in the diagram above to the third and then directly to the ground state. Two photons are emitted. How does the sum of their frequencies compare with the frequency of the single photon that would be emitted by de-excitation from the fourth level directly to the ground state?

33. When do photons behave like a wave? When do they behave like particles?

34. Light has been argued to be a wave and then a particle, and then back again. Does this indicate that light's true nature probably lies somewhere between these two models?

35. What laboratory device utilizes the wave nature of electrons?

36. When a photon hits an electron and gives it energy, the photon has less energy after it bounces from the electron. What happens to the frequency of the photon after it bounces from the electron? (This phenomenon is called the *Compton effect.*)

37. An electron and a proton travel at the same speed. Which has more momentum? Which has the longer wavelength?

38. One electron travels twice as fast as another. Which has the longer wavelength?

39. Does the de Broglie wavelength of a proton become longer or shorter as its velocity increases?

40. We don't notice the wavelength of moving matter in our common experience. Is this because the wavelength is extraordinarily large or extraordinarily small?

41. What principal advantage does an electron microscope have over an optical microscope?

42. A friend says, "If an electron is not a particle, then it must be a wave." What is your response? (Do you hear "either–or" statements like this often?)

43. Consider one of the many electrons on the tip of your nose. If somebody looks at it, will its motion be altered?

How about if it is looked at with one eye closed? With two eyes open, but crossed? Does Heisenberg's uncertainty principle apply here?

44. Does the uncertainty principle tell us that we can never know anything for certain?

45. Do we inadvertently alter the realities that we attempt to measure in a public opinion survey? Does Heisenberg's uncertainty principle apply here?

46. If the behavior of a system is measured exactly for some period of time and is understood, does it follow that the future behavior of that system can be exactly predicted? (Is there a distinction between properties that are *measurable* and properties that are *predictable*?)

47. If a butterfly causes a tornado, does it make sense to eradicate butterflies? Defend your answer.

48. We hear the expression "taking a quantum leap" to describe large changes. Is the expression appropriate? Defend your answer.

49. What is it that waves in the Schrödinger wave equation?

50. If the world of the atom is so uncertain and subject to the laws of probabilities, how can we accurately measure such things as light intensity, electric current, and temperature?

51. What evidence supports the notion that light has wave properties? What evidence supports the view that light has particle properties?

52. When and where do Newton's laws of motion and quantum mechanics overlap?

53. What does Bohr's correspondence principle say about quantum mechanics versus classical mechanics?

54. Richard Feynman in his book *The Character of Physical Law* states: "A philosopher once said, 'It is necessary for the very existence of science that the same conditions always produce the same results.' Well, they don't!" Who was speaking of classical physics, and who was speaking of quantum physics?

55. To measure the exact age of Old Methuselah, the oldest living tree in the world, a Nevada professor of dendrology, aided by an employee of the U.S. Bureau of Land Management, cut the tree down in 1965 and counted its rings. Is this an extreme example of changing that which you measure or an example of arrogant and criminal stupidity?

PROBLEMS

● **BEGINNER**   ■ **INTERMEDIATE**   ◆ **EXPERT**

1. ■ In the accompanying diagram, the energy difference between states A and B is twice the energy difference between states B and C. In a transition (quantum jump) from C to B, an electron emits a photon of wavelength 600 nm.
   a. What is the wavelength emitted when the photon jumps from B to A?
   b. What is the wavelength emitted when the photon jumps from C to A?

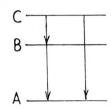

2. ■ A typical wavelength of infrared radiation emitted by your body is 25 mm (2.5×10^{-5} m). Show that the energy per photon of this radiation is 7.9×10^{-21} J.

3. ■ An electron strikes the back of the face of an older-model TV screen at 1/10 the speed of light. Show that the de Broglie wavelength of the electron is 2.4×10^{-11} m, less than the diameter of a single atom.

4. ■ You decide to roll a 0.1-kg ball across the floor so slowly that it will have a small momentum and a large de Broglie wavelength. If you roll it at 0.001 m/s, show that its wavelength would be 6.6×10^{-30} m (incredibly small compared with the electron wavelength in the previous problem).

CHAPTER 15 ONLINE RESOURCES

The Physics Place

Interactive Figures
15.1, 15.6, 15.9, 15.12, 15.19, 15.24

Tutorials
Light and Spectroscopy
Bohr's Shell Model

Videos
Electron Waves

Quiz

Flashcards

Links

CHAPTER 16

The Atomic Nucleus and Radioactivity

Dean Zollman investigates nuclear properties with a modern version of Rutherford's scattering experiment.

The atomic nucleus and its processes are perhaps the most misunderstood and controversial areas of physics. Distrust of anything *nuclear*, or anything *radioactive*, is much like the fears of electricity more than a century ago. The distrust of electricity in households stemmed from ignorance. Indeed, electricity can be most dangerous and even lethal when improperly handled. But with safeguards and well-informed consumers, society has determined that the benefits of electricity outweigh its risks. Today, we are making similar decisions about nuclear technology's risks versus its benefits—made with an adequate understanding of the atomic nucleus and its inner processes.

16.1 Radioactivity

Elements with unstable nuclei are said to be *radioactive*. They sooner or later break down and eject energetic particles or emit high-frequency electromagnetic radiation. This process is **radioactivity**, which, because it involves the decay of the atomic nucleus, is often called *radioactive decay*.

The Physics Place
Nuclear Physics

The Physics Place
Radioactive Decay

A common misconception is that radioactivity is something new in the environment, but it has been around far longer than the human race. It is as much a part of our environment as the Sun and the rain. It has always been in the soil we walk on and in the air we breathe, and it is what warms the interior of Earth and makes it molten. In fact, radioactive decay in Earth's interior is what heats the water that spurts from a geyser or wells up from a natural hot spring. Even the helium in a child's balloon is nothing more than the product of radioactive decay. Radioactivity is as natural as sunshine and rain.

FIGURE 16.1

Origins of radiation exposure for an average individual in the United States.

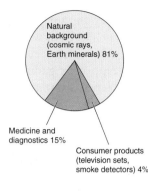

Natural background (cosmic rays, Earth minerals) 81%

Medicine and diagnostics 15%

Consumer products (television sets, smoke detectors) 4%

16.2 Alpha, Beta, and Gamma Rays

All elements having an atomic number greater than 82 (lead) are radioactive. These elements, and others, emit three distinct types of radiation, named by the first three letters of the Greek alphabet, α, β, γ—*alpha*, *beta*, and *gamma*. Alpha rays carry a positive electrical charge, beta rays carry a negative charge, and gamma rays carry no charge. The three rays can be separated by placing a magnetic field across their paths (Figure 16.2).

The Physics Place
Atoms and Isotopes

An **alpha particle** is the combination of two protons and two neutrons (in other words, it is the nucleus of the helium atom, atomic number 2). Alpha particles are easy to shield against because of their relatively large mass and their double positive charge (+2). For example, they do not normally penetrate through light materials such as paper or clothing. Because of their great kinetic energies, however, alpha particles can cause significant damage to the surface of a material, especially living tissue. When traveling only a tiny distance through rock beneath Earth's surface, alpha particles pick up electrons and become nothing more than harmless helium. As a matter of fact, that's where the helium in a child's balloon comes from—practically all Earth's helium atoms were at one time energetic alpha particles.

A **beta particle** is an electron ejected from a nucleus. Once ejected, it is indistinguishable from an electron in a cathode ray tube, electrical circuit, or one orbiting the atomic nucleus. The difference is that a beta particle originates inside the nucleus—where it is created when a neutron transforms into a proton. A beta particle is normally faster than an alpha particle and carries only a single negative charge (−1). Beta particles are not as easy to stop as alpha particles are, and they are able to penetrate light materials such as paper or clothing. They can penetrate fairly deeply into skin, where they have the potential for harming or killing living cells. But they are not able to penetrate deeply into denser materials such as aluminum. Beta particles, once stopped, simply become a part of the material they are in, like any other electron.

> Once alpha and beta particles slow by collisions, they become harmless. Alpha particles become helium nuclei and electrons join other atoms.

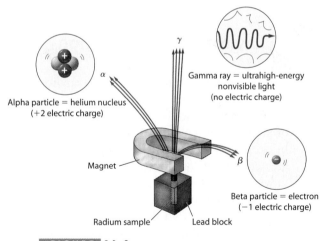

Alpha particle = helium nucleus (+2 electric charge)

γ

α

Gamma ray = ultrahigh-energy nonvisible light (no electric charge)

Magnet

β

Beta particle = electron (−1 electric charge)

Radium sample

Lead block

FIGURE 16.2
INTERACTIVE FIGURE

In a magnetic field, alpha rays bend one way, beta rays bend the other way, and gamma rays don't bend at all. Note that the alpha rays bend less than do the beta rays. This occurs because alpha particles have more inertia (mass) than beta particles. The combined beam comes from a radioactive material placed at the bottom of a hole drilled in a lead block.

FIGURE 16.3
INTERACTIVE FIGURE

A gamma ray is simply electromagnetic radiation, much higher in frequency and energy than light and X-rays.

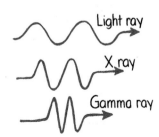

Light ray

X ray

Gamma ray

Gamma rays are the high-frequency electromagnetic radiation emitted by radioactive elements. Like visible light, a gamma ray is pure energy. It is made up of photons, each with an energy much greater than the photons that make up visible light, ultraviolet light, or even X-rays. Because they have no mass or electric charge and because of their high energies, gamma rays are able to penetrate through most materials. However,

they cannot penetrate unusually dense materials such as lead, which absorbs them. Delicate molecules inside cells throughout our bodies that are zapped by gamma rays suffer structural damage. Hence, gamma rays are generally more harmful to us than alpha or beta particles (unless the alphas or betas are ingested).

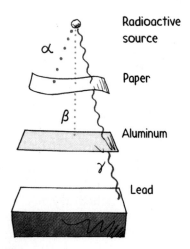

FIGURE 16.4
INTERACTIVE FIGURE

Alpha particles are the least penetrating and can be stopped by a few sheets of paper. Beta particles will readily pass through paper, but not through a sheet of aluminum. Gamma rays are absorbed by solid lead.

FIGURE 16.5

The shelf life of fresh strawberries and other perishables is markedly increased when the food is subjected to gamma rays from a radioactive source. The strawberries on the right were treated with gamma radiation, which kills the microorganisms that normally lead to spoilage. The food is only a receiver of radiation and is in no way transformed into an emitter of radiation as can be confirmed with a radiation detector.

CHECK YOURSELF

Pretend you are given three radioactive rocks—one an alpha emitter, one a beta emitter, and one a gamma emitter. You can throw away one, but of the remaining two, you must hold one in your hand and the other you must place in your pocket. What can you do to minimize your exposure to radiation?

CHECK YOUR ANSWER

Hold the alpha emitter in your hand because the skin on your hand will shield you. Put the beta emitter in your pocket because beta particles will likely be stopped by the combined thickness of your clothing and skin. Throw away the gamma emitter because it would penetrate your body from any of these locations. Ideally, of course, you should distance yourself as much as possible from all the rocks.

16.3 Environmental Radiation

Common rock and minerals in our environment contain significant quantities of radioactive isotopes because most of them contain trace amounts of uranium. As a matter of fact, people who live in brick, concrete, or stone buildings are exposed to greater amounts of radiation than people who live in wooden buildings.

The leading source of naturally occurring radiation is radon-222, an inert gas arising from uranium deposits. Radon is a heavy gas that tends to accumulate in basements after it seeps up through cracks in the floor. Levels of radon vary from region to region, depending upon local geology. You can check the radon level in your home with a radon detector kit (Figure 16.6). If levels are abnormally high, corrective measures, such as sealing the basement floor and walls and maintaining adequate ventilation, are recommended.

FIGURE 16.6

A commercially available radon test kit for the home.

About one-fifth of our annual exposure to radiation comes from nonnatural sources, primarily medical procedures. Television sets, fallout from long-ago nuclear testing, and the coal and nuclear power industries are also contributors. The coal industry far outranks the nuclear power industry as a source of radiation. Globally, the combustion of coal annually releases about 13,000 tons of radioactive thorium and uranium into the atmosphere. Both these minerals are found naturally in coal deposits so that their release is a natural consequence of burning coal. Worldwide, the nuclear power industries generate about 10,000 tons of radioactive waste each year. Most all of this waste, however, is contained and *not* released into the environment.

Units of Radiation

> Radioactivity has been around since Earth's beginning.

Radiation dosage is commonly measured in *rads* (*r*adiation *a*bsorbed *d*ose), a unit of absorbed energy. One **rad** is equal to 0.01 joule of radiant energy absorbed per kilogram of tissue.

The capacity for nuclear radiation to cause damage is not just a function of its level of energy, however. Some forms of radiation are more harmful than others. For example, suppose you have two arrows, one with a pointed tip and one with a suction cup at its tip. Shoot both arrows at an apple at the same speed and both have the same kinetic energy. The one with the pointed tip, however, will invariably do more damage to the apple than the one with the suction cup. Similarly, some forms of radiation cause greater harm than other forms even when we receive the same number of rads from both forms.

The unit of measure for radiation dosage based on potential damage is the **rem** (*r*oentgen *e*quivalent *m*an).* In calculating the dosage in rems, we multiply the number of rads by a factor that corresponds to different health effects of different types of radiation determined by clinical studies. For example, 1 rad of alpha particles has the same biological effect as 10 rads of beta particles.** We call both of these dosages 10 rems:

Particle	Radiation dosage		Factor		Health effect
alpha	1 rad	×	10	=	10 rems
beta	10 rads	×	1	=	10 rems

STOP AND CHECK YOURSELF

Would you rather be exposed to 1 rad of alpha particles or 1 rad of beta particles?

CHECK YOUR ANSWER

Multiply these quantities of radiation by the appropriate factor to get the dosages in rems. Alpha: 1 rad × 10 = 10 rems; beta: 1 rad × 1 = 1 rem. The factors show us that, physiologically speaking, alpha particles are 10 times more damaging than beta particles.

Doses of Radiation

Lethal doses of radiation begin at 500 rems. A person has about a 50% chance of surviving a dose of this magnitude delivered to the whole body over a short period of time. During radiation therapy, a patient may receive localized doses in excess of 200 rems each day for a period of weeks (Figure 16.7).

All the radiation we receive from natural sources and from diagnostic medical procedures is only a fraction of 1 rem. For convenience, the smaller unit *millirem* is used, where 1 millirem (mrem) is 1/1000th of a rem.

The average person in the United States is exposed to about 360 mrems a year, as Table 16.1 indicates. About 80% of this radiation comes from natural sources, such as cosmic rays and Earth itself. A typical chest X-ray exposes a person to 5 to 30 mrems (0.005 to 0.030 rem), less than one ten-thousandth of the lethal dose. Interestingly, the human body is a

---

* This unit is named for the discoverer of X-rays, Wilhelm Roentgen.
** This is true even though beta particles have more penetrating power, as previously discussed.

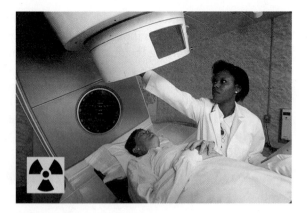

FIGURE 16.7

Nuclear radiation is focused on harmful tissue, such as a cancerous tumor, to selectively kill or shrink the tissue in a technique known as *radiation therapy*. This application of nuclear radiation has saved millions of lives—a clear-cut example of the benefits of nuclear technology. The inset shows the internationally used symbol indicating an area where radioactive material is being handled or produced.

FIGURE 16.8

The film badges worn by Tammy and Larry contain audible alerts for both radiation surge and accumulated exposure. Information from the individualized badges is periodically downloaded to a database for analysis and storage.

significant source of natural radiation, primarily from the potassium we ingest. Our bodies contain about 200 grams of potassium. Of this quantity, about 20 milligrams is the radioactive isotope potassium-40, which is a gamma ray emitter. Between every heartbeat about 60,000 potassium-40 isotopes in the average human body undergo spontaneous radioactive decay. Radiation is indeed everywhere.

When radiation encounters the intricately structured molecules in the watery, ion-rich brine that makes up our cells, the radiation can create chaos on the atomic scale. Some molecules are broken, and this change alters other molecules, which can be harmful to life processes.

Cells are able to repair most kinds of molecular damage caused by radiation if the radiation is not too severe. A cell can survive an otherwise lethal dose of radiation if the dose is spread over a long period of time to allow intervals for healing. When radiation is sufficient to kill cells, the dead cells can be replaced by new ones (except for most nerve cells, which are irreplaceable). Sometimes a radiated cell will survive with a damaged DNA molecule. New cells arising from the damaged cell retain the altered genetic information, producing a *mutation*. Usually the effects of a mutation are insignificant, but occasionally the mutation results in cells that do not function as well as unaffected ones, sometimes leading to a cancer. If the damaged DNA is in an individual's reproductive cells, the genetic code of the individual's offspring may retain the mutation.

Radioactive Tracers

In scientific laboratories radioactive samples of all the elements have been made. This is accomplished by bombardment with neutrons or other particles. Radioactive materials are extremely useful in scientific research and industry. To check the action of a fertilizer, for example, researchers combine a small amount of radioactive material with the fertilizer and then apply the combination to a few plants. The amount of radioactive fertilizer taken up by the plants can be easily measured with radiation detectors. From such measurements, scientists can inform farmers of the proper amount of fertilizer to

TABLE 16.1

Annual Radiation Exposure

Source	Typical Dose (mrem) Received Annually
Natural Origin	
Cosmic radiation	26
Ground	33
Air (radon-222)	198
Human tissues (K-40; Ra-226)	35
Human Origin	
Medical procedures	
Diagnostic X-rays	40
Nuclear diagnostics	15
TV tubes, other consumer products	11
Weapons-test fallout	1
Commercial fossil fuel power plants	<1
Commercial nuclear power plants	≪1

FIGURE   16.9

Tracking fertilizer uptake with a radioactive isotope.

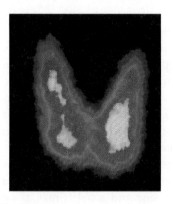

FIGURE   16.10

The thyroid gland, located in the neck, absorbs much of the iodine that enters the body through food and drink. Images of the thyroid gland, such as the one shown here, can be obtained by giving a patient a small amount of the radioactive isotope iodine-131. These images are useful in diagnosing metabolic disorders.

use. Radioactive isotopes used to trace such pathways are called *tracers*.

In a technique known as medical imaging, tracers are used for the diagnosis of internal disorders. This technique works because the path the tracer takes is influenced only by its physical and chemical properties, not by its radioactivity. The tracer may be introduced alone or along with some other chemical that helps target the tracer to a particular type of tissue in the body.

16.4 The Atomic Nucleus and the Strong Force

The atomic nucleus occupies only a few quadrillionths the volume of the atom, leaving most of the atom as empty space. The nucleus is composed of **nucleons**, which is the collective name for protons and neutrons. (Each nucleon is composed of three smaller particles called quarks—believed to be fundamental, not made of smaller parts.)

Just as there are energy levels for the orbital electrons of an atom, so there are energy levels within the nucleus. Whereas orbiting electrons emit photons when making transitions to lower orbits, similar changes of energy states in radioactive nuclei result in the emission of gamma-ray photons. This is gamma radiation.

We know that electrical charges of like sign repel one another. So how is it possible that positively charged protons in the nucleus stay clumped together? This question led to the discovery of an attraction called the **strong force**, which acts between all nucleons. This force is very strong but only over extremely short distances (about 10^{-15} meters, the approximate diameter of a proton or neutron). Repulsive electrical interactions, on the other hand, are relatively long-ranged. Figure 16.11 suggests a comparison of the strengths of these two forces over distance. For protons that are close together, as in small nuclei, the attractive strong nuclear force easily overcomes the repulsive electrical force. But for protons that are far apart, like those on opposite edges of a large nucleus, the attractive strong nuclear force may be weaker than the repulsive electrical force.

> **fyi**
> **Without the nuclear strong force—strong interaction—there would be no atoms beyond hydrogen.**

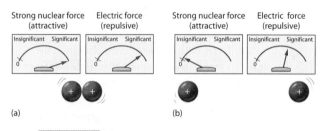

(a)   (b)

FIGURE   16.11

(a) Two protons near each other experience both an attractive strong nuclear force and a repulsive electric force. At this tiny separation distance, the strong nuclear force overcomes the electric force, resulting in them staying together. (b) When the two protons are relatively far from each other, the electric force is more significant. The protons repel each other. This proton–proton repulsion in large atomic nuclei reduces nuclear stability.

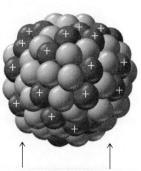

(a) Nucleons close together   (b) Nucleons far apart

FIGURE 16.12

(a) All nucleons in a small atomic nucleus are close to one another; hence, they experience an attractive strong nuclear force. (b) Nucleons on opposite sides of a larger nucleus are not as close to one another, and so the attractive strong nuclear forces holding them together are much weaker. The result is that the large nucleus is less stable.

A large nucleus is not as stable as a small one. In a helium nucleus, for example, each of the two protons feels the repulsive effect of the other. In a uranium nucleus, each of the 92 protons feels the repulsive effects of the other 91 protons! The nucleus is unstable. We see that there is a limit to the size of the atomic nucleus. It is for this reason that all nuclei having more than 82 protons are radioactive.

STOP AND
CHECK YOURSELF

Two protons in the atomic nucleus repel each other, but they are also attracted to each other. Why?

CHECK YOUR ANSWER

While two protons repel each other by the electric force, they also attract each other by the strong nuclear force. Both of these forces act simultaneously. So long as the attractive strong nuclear force is stronger than the repulsive electric force, the protons will remain together. Under conditions where the electric force overcomes the strong nuclear force, however, the protons fly apart from each other.

Neutrons serve as "nuclear cement," holding the atomic nucleus together. Protons attract both protons and neutrons by the strong nuclear force. Protons also repel other protons by the electric force. Neutrons, on the other hand, have no electric charge and so only attract other protons and neutrons by the strong nuclear force. The presence of neutrons therefore adds to the attraction among nucleons and helps hold the nucleus together (Figure 16.13).

The more protons there are in a nucleus, the more neutrons are needed to help balance the repulsive electric forces. For light elements, it is sufficient to have about as many neutrons as protons. The most common isotope of carbon, C-12, for instance, has equal numbers of each—six protons and six neutrons. For large nuclei, more neutrons than protons are needed. Because the strong nuclear force diminishes rapidly over distance, nucleons must be practically touching in order for the strong nuclear force to be effective. Nucleons on opposite sides of a large atomic nucleus are not as attracted to one another. The electric force, however, does not diminish by much across the diameter of a large nucleus and so begins to win out over the strong nuclear force. To compensate for the weakening of the strong nuclear force across the diameter of the nucleus, large nuclei have more neutrons than protons. Lead, for example, has about one and a half times as many neutrons as protons.

So we see that neutrons are stabilizing and large nuclei require an abundance of them. But neutrons are not always successful in keeping a nucleus intact. Interestingly, neutrons are not stable when they are by themselves. A lone neutron is radioactive, and spontaneously transforms to a proton and an electron (Figure 16.14a). A neutron needs protons around to keep this from happening. The alpha particles emitted in alpha decay are literally nuclear "chunks," and only heavy nuclei emit

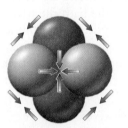
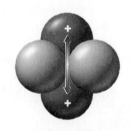

All nucleons, both protons and neutrons, attract one another by the strong nuclear force.

Only protons repel one another by the electric force.

FIGURE 16.13

The presence of neutrons helps hold the nucleus together by increasing the effect of the strong nuclear force, represented by the single-headed arrows.

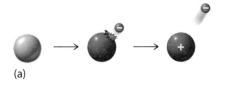

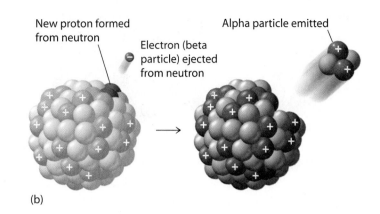

New proton formed from neutron

Electron (beta particle) ejected from neutron

Alpha particle emitted

FIGURE 16.14

(a) A neutron near a proton is stable, but a neutron by itself is unstable and decays to a proton by emitting an electron. (b) Destabilized by an increase in the number of protons, the nucleus begins to shed fragments, such as alpha particles.

(b)

them.* Beta and gamma particles, on the other hand, can be emitted by radioactive nuclei both heavy and light. The beta decay of a single neutron and the alpha decay of a heavy nucleus are shown in Figure 16.14b.

STOP AND
CHECK YOURSELF

What role do neutrons serve in the atomic nucleus? What is the fate of a neutron when alone or distant from one or more protons?

CHECK YOUR ANSWER

Neutrons serve as nuclear cement in nuclei, and add to nuclear stability. But when alone, a neutron is radioactive and spontaneously transforms to a proton and an electron.

16.5 Radioactive Half-Life

The rate of decay for a radioactive isotope is measured in terms of a characteristic time, the **half-life**. This is the time it takes for half of an original quantity of an element to decay. For example, radium-226 has a half-life of 1620 years, which means that half of a radium-226 sample will be converted to other elements by the end of 1620 years. In the next 1620 years, half of the remaining

radium will decay, leaving only one-fourth the original amount of radium. (After 20 half-lives, the initial quantity of radium-226 will be diminished by a factor of about one million.)

Half-lives are remarkably constant and not affected by external conditions. Some radioactive isotopes have half-lives that are less than a millionth of a second, while others have half-lives of more than a billion years. Uranium-238 has a half-life of 4.5 billion years. All uranium eventually decays to lead in a series of steps. In 4.5 billion years, half the uranium that is in Earth today will be lead.

It is not necessary to wait through the duration of a half-life in order to measure it. The half-life of an element can be calculated at any given moment by measuring the rate of decay of a known quantity. This is easily

The Physics Place
Half-Life

The radioactive half-life of a material is also the time for its decay rate to reduce to half.

* An exception to the rule that alpha decay is limited to heavy nuclei is the highly radioactive nucleus of beryllium 8, with four protons and four neutrons, which splits into two alpha particles—a form of nuclear fission.

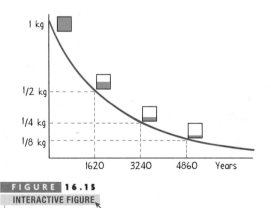

FIGURE 16.15
INTERACTIVE FIGURE

Every 1620 years the amount of radium decreases by half.

(a)   (b)

FIGURE 16.16

Some radiation detectors. (a) A Geiger counter detects incoming radiation by its ionizing effect on enclosed gas in the tube. (b) A scintillation counter detects incoming radiation by flashes of light that are produced when charged particles or gamma rays pass through it.

done using a radiation detector (Figure 16.16). In general, the shorter the half-life of a substance, the faster it disintegrates, and the more radioactivity per amount is detected.

STOP AND CHECK YOURSELF

1. If a sample of radioactive isotopes has a half-life of 1 day, how much of the original sample will be left at the end of the second day? The third day?

2. Which will give a higher counting rate on a radiation detector, radioactive material that has a short half-life or radioactive material that has a long half-life?

CHECK YOUR ANSWERS

1. One-fourth of the original sample will be left— the three-fourths that underwent decay is then a different element altogether. At the end of 3 days, one-eighth of the original sample will remain.

2. The material with the shorter half-life is more active and will show a higher counting rate on a radiation detector.

16.6 Transmutation of the Elements

When a radioactive nucleus emits an alpha or a beta particle, there is a change in atomic number—a different element is formed. The changing of one chemical element to another is called **transmutation**. Transmutation occurs in natural events, and is also initiated artificially in the laboratory.

Natural Transmutation

Consider uranium-238, the nucleus of which contains 92 protons and 146 neutrons. When an alpha particle is ejected, the nucleus loses two protons and two neutrons. Because an element is defined by the number of protons in its nucleus, the 90 protons and 144 neutrons left behind are no longer identified as being uranium. What we have is the nucleus of a different element—thorium. This transmutation can be written as a nuclear equation:

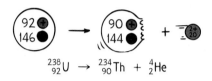

$$^{238}_{92}U \rightarrow {}^{234}_{90}Th + {}^{4}_{2}He$$

We see that $^{238}_{92}U$ transmutes to the two elements written to the right of the arrow. When this transmutation occurs, energy is released, partly in the form of kinetic energy of the alpha particle ($^{4}_{2}He$), partly in the kinetic energy of the thorium nucleus, and partly in the form of gamma radiation. In this and all such equations, the mass numbers at the top balance ($238 = 234 + 4$) and the atomic numbers at the bottom also balance ($92 = 90 + 2$).

Thorium-234, the product of this reaction, is also radioactive. When it decays, it emits a beta particle.* Since a beta particle is an electron, the atomic number of the resulting nucleus is *increased* by 1. So after beta emission by thorium with 90 protons, the resulting element has 91 protons. It is no longer thorium, but the element protactinium. Although the atomic number has increased by 1 in this

---

* Beta emission is always accompanied by the emission of a neutrino (actually an antineutrino), a neutral particle with nearly zero mass that travels at about the speed of light. The neutrino ("little neutral one") was postulated by Wolfgang Pauli in 1930 and detected in 1956. Neutrinos are hard to detect because they interact very weakly with matter. Whereas a piece of solid lead a few centimeters thick will stop most gamma rays from a radium source, a piece of lead about 8 light years thick would be needed to stop half the neutrinos produced in typical nuclear decays. Thousands of neutrinos are flying through you every second of every day, because the universe is filled with them. Only occasionally, one or two times a year or so, does a neutrino interact with the matter of your body.

At this writing, the mass of neutrinos is unknown. Neutrinos are so numerous in the universe that if they have even the tiniest mass, they might make up most of the mass of the universe. Neutrinos may be the "glue" that holds the universe together.

process, the mass number (protons + neutrons) remains the same. The nuclear equation is

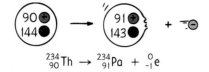

$$^{234}_{90}\text{Th} \rightarrow {}^{234}_{91}\text{Pa} + {}^{0}_{-1}\text{e}$$

We write an electron as ${}^{0}_{1}e$. The superscript 0 indicates that the electron's mass is insignificant relative to that of protons and neutrons. The subscript -1 is the electric charge of the electron.

So we see that when an element ejects an alpha particle from its nucleus, the mass number of the resulting atom is decreased by 4, and its atomic number is decreased by 2. The resulting atom is an element two spaces back in the periodic table of the elements. When an element ejects a beta particle from its nucleus, the mass of the atom is practically unaffected, meaning there is no change in mass number, but its atomic number increases by 1. The resulting atom belongs to an element one place forward in the periodic table. Gamma emission results in no change in either the mass number or the atomic number. So we see that radioactive elements can decay backward or forward in the periodic table.*

The successions of radioactive decays of ${}^{238}_{92}\text{U}$ to ${}^{206}_{82}\text{Pb}$, an isotope of lead, is shown in Figure 16.17. Each gray arrow shows an alpha decay, and each red arrow shows a beta decay. Notice that some of the nuclei in the series can decay in both ways. This is one of several similar radioactive series that occur in nature.

Artificial Transmutation

Ernest Rutherford, in 1919, was the first of many investigators to succeed in transmuting a chemical element. He bombarded nitrogen gas with alpha particles from a piece of radioactive ore. The impact of an alpha particle on a nitrogen nucleus transmutes nitrogen into oxygen:

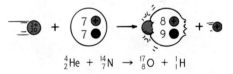

$$^{4}_{2}\text{He} + {}^{14}_{7}\text{N} \rightarrow {}^{17}_{8}\text{O} + {}^{1}_{1}\text{H}$$

Rutherford used a device called a *cloud chamber* to record this event (Figure 16.18). In a cloud chamber, moving charged particles show a trail of ions along their path in a way similar to the ice crystals that show the trail of jet planes high in the sky. From a quarter-of-a-million cloud-chamber tracks photographed on movie film, Rutherford showed seven examples of atomic transmutation. Analysis of tracks bent by a strong external magnetic field showed that when an alpha particle collided with a nitrogen atom, a proton bounced out and the heavy atom recoiled a short distance. The alpha particle disappeared. The alpha particle was absorbed in the process, transforming nitrogen to oxygen.

Since Rutherford's announcement in 1919, experimenters have carried out many other nuclear reactions, first with natural bombarding projectiles

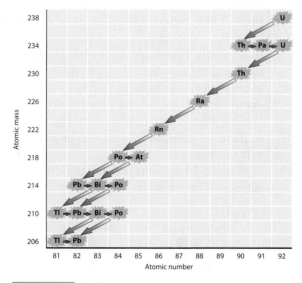

U-238 decays to Pb-206 through a series of alpha and beta decays.

* Sometimes a nucleus emits a positron, which is the "antiparticle" of an electron. In this case, a proton becomes a neutron, and the atomic number is decreased.

FIGURE **16.18**
A cloud chamber. Charged particles moving through supersaturated vapor leave trails. When the chamber is in a strong magnetic field, bending of the tracks provides information about the charge, mass, and momentum of the particles.

Radioactive sample

Vapor trails

Piston

FIGURE **16.19**

Walter Steiger, pioneer of telescopes in Hawaii, examines vapor trails in a small cloud chamber.

FIGURE **16.20**

Tracks of elementary particles in a bubble chamber, a device similar to, but more complicated than, a cloud chamber. Two particles have been destroyed at the points where the spirals emanate, and four others created in the collision.

from radioactive ores and then with still more energetic projectiles—protons and electrons hurled by huge particle accelerators. Artificial transmutation is what produces the hitherto unknown synthetic elements from atomic number 93 to 118. All these artificially made elements have short half-lives. If they ever existed naturally when Earth was formed, they have long since decayed.

16.7 Radiometric Dating

Earth's atmosphere is continuously bombarded by cosmic rays and this bombardment causes many atoms in the upper atmosphere to transmute. These transmutations result in many protons and neutrons being "sprayed out" into the environment. Most of the protons are stopped as they collide with the atoms of the upper atmosphere, stripping electrons from these atoms to become hydrogen atoms.

The Physics Place

Carbon Dating
Plutonium

The neutrons, however, keep going for longer distances because they have no electrical charge and therefore do not interact electrically with matter. Eventually, many of them collide with the nuclei in the denser lower atmosphere. A nitrogen that captures a neutron, for instance, becomes an isotope of carbon by emitting a proton:

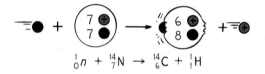

$${}_{0}^{1}n + {}_{7}^{14}N \rightarrow {}_{6}^{14}C + {}_{1}^{1}H$$

This carbon-14 isotope, which makes up less than one-millionth of 1% of the carbon in the atmosphere, is radioactive and has eight neutrons. (The most common isotope, carbon-12, has six neutrons and is not radioactive.) Because both carbon-12 and carbon-14 are forms of carbon, they have the same chemical properties. Both these isotopes can chemically react with oxygen to form carbon dioxide, which is taken in by plants. This means that all plants contain a tiny bit of radioactive carbon-14. All animals eat plants (or other animals that ate plants), and therefore have a little carbon-14 in them. In short, all living things on Earth contain some carbon-14.

Carbon-14 is a beta emitter and decays back to nitrogen by the following reaction:

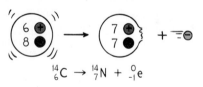

$${}_{6}^{14}C \rightarrow {}_{7}^{14}N + {}_{-1}^{0}e$$

Because plants continue to take in carbon dioxide as long as they live, any carbon-14 lost by decay is immediately replenished with fresh carbon-14 from the atmosphere. In this way, a radioactive equilibrium is reached where there is a constant ratio of about one carbon-14 atom to every 100 billion carbon-12 atoms. When a plant dies, replenishment of carbon-14 stops. Then the percentage of carbon-14 decreases at a constant rate given by its half-life.* The longer a plant or other organism is dead, therefore, the less carbon-14 it contains relative to the constant amount of carbon-12.

The half-life of carbon-14 is about 5730 years. This means that half of the carbon-14 atoms that are now present in a plant or animal that dies today will decay in the next 5730 years. Half of the remaining carbon-14 atoms will then decay in the following 5730 years, and so forth.

With this knowledge, scientists are able to calculate the age of carbon-containing artifacts, such as wooden tools or skeletons, by measuring their current level of radioactivity. This process, known as **carbon dating**, enables us to probe as much as 50,000 years into the past. Beyond this time span, there is too little carbon-14 remaining to permit accurate analysis.

Carbon-14 dating would be an extremely simple and accurate dating method if the amount of radioactive carbon in the atmosphere had been constant over the ages. But it hasn't been. Fluctuations in the Sun's magnetic field as well as changes in the strength of Earth's magnetic field affect cosmic-ray intensities in Earth's atmosphere, which in turn produce fluctuations in the production of C-14. In addition, changes in Earth's climate affect the amount of carbon dioxide in the atmosphere. The oceans are great reservoirs of carbon dioxide. When

the oceans are cold, they release less carbon dioxide into the atmosphere than when they are warm.

The dating of older, but nonliving, things is accomplished with radioactive minerals, such as uranium. The naturally occurring isotopes U-238 and U-235 decay very slowly and ultimately become isotopes of lead—but not the common lead isotope Pb-208. For example, U-238 decays through several stages to finally become Pb-206, whereas U-235 finally becomes the isotope Pb-207. Lead isotopes 206 and 207 that now exist were at one time uranium. The older the uranium-bearing rock is, the higher the percentage of these remnant isotopes will be.

From the half-lives of uranium isotopes, and the percentage of lead isotopes in uranium-bearing rock, it is possible to calculate the date at which the rock was formed.

fyi

> One ton of ordinary granite contains about 9 grams of uranium and 20 grams of thorium. Basalt rocks contain 3.5 and 7.7 grams of the same, respectively.

STOP AND
CHECK YOURSELF

Suppose an archeologist extracts a gram of carbon from an ancient ax handle and finds it one-fourth as radioactive as a gram of carbon extracted from a freshly cut tree branch. About how old is the ax handle?

CHECK YOUR ANSWER

Assuming the ratio of C-14/C-12 was the same when the ax was made, the ax handle is two half-lives of C-14, about 11,500 years old.

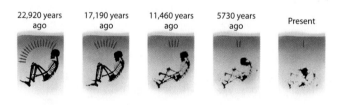

22,920 years ago   17,190 years ago   11,460 years ago   5730 years ago   Present

FIGURE 16.21

The amount of radioactive carbon-14 in the skeleton diminishes by one-half every 5730 years, with the result that today the skeleton contains only a fraction of the carbon-14 it had originally. The red arrows symbolize relative amounts of carbon-14.

* A 1-g sample of contemporary carbon contains about 5×10^{22} atoms, 6.5×10^{10} of which are C-14 atoms, and has a beta disintegration rate of about 13.5 decays per minute.

16.8 Nuclear Fission

In 1938, two German scientists, Otto Hahn and Fritz Strassmann, made an accidental discovery that was to change the world. While bombarding a sample of uranium with neutrons in the hope of creating new, heavier elements, they were astonished to find chemical evidence for the production of barium, an element having a little more than half the mass of uranium. Hahn wrote of this news to his former colleague Lise Meitner, who had fled from Nazi Germany to Sweden because of her Jewish

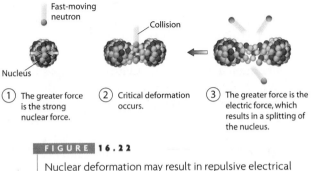

Fast-moving neutron

Collision

Nucleus

① The greater force is the strong nuclear force.

② Critical deformation occurs.

③ The greater force is the electric force, which results in a splitting of the nucleus.

FIGURE 16.22

Nuclear deformation may result in repulsive electrical forces overcoming attractive nuclear forces, in which case fission occurs.

ancestry. From the evidence given to her by Hahn, Meitner concluded that the uranium nucleus, activated by neutron bombardment, had split in two. Soon thereafter, Meitner, working with her nephew, Otto Frisch, also a physicist, published a paper in which the term *nuclear fission* was first coined.*

In the nucleus of every atom there exists a delicate balance between attractive nuclear forces and repulsive electric forces between protons. In all known nuclei, the nuclear forces dominate. In uranium, however, this domination is tenuous. If a uranium nucleus stretches into an elongated shape (Figure 16.22), the electrical forces may push it into an even more elongated shape. If the elongation passes a certain point, electrical forces overwhelm strong nuclear forces, and the nucleus splits. This is **nuclear fission.**

The energy released by the fission of one U-235 nucleus is relatively enormous—about seven million times the energy released by the explosion of one TNT molecule. This energy is mainly in the form of kinetic energy of the fission fragments that fly apart from one another, with some energy given to ejected neutrons and the rest to gamma radiation.

A typical uranium fission reaction is

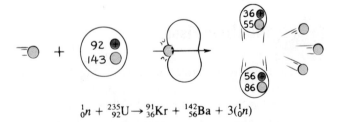

$$\,_0^1 n + \,_{92}^{235}U \rightarrow \,_{36}^{91}Kr + \,_{56}^{142}Ba + 3(\,_0^1 n)$$

* Otto Hahn, rather than Lise Meitner, received the Nobel Prize for the work on nuclear fission. Notoriously, Hahn didn't even acknowledge Meitner's role, although it was recognized by other physicists, including Niels Bohr. See more about this in the readable book $E = mc^2$, by David Bodanis.

Note in this example that one neutron starts the fission of a uranium nucleus and that the fission produces three neutrons. (A fission reaction may produce fewer or more than three neutrons.) These product neutrons can cause the fissioning of three other uranium atoms, releasing nine more neutrons. If each of these 9 neutrons succeeds in splitting a uranium atom, the next step in the reaction produces 27 neutrons, and so on. Such a sequence, illustrated in Figure 16.23, is called a **chain reaction**—a self-sustaining reaction in which the products of one reaction event stimulate further reaction events.

Why do chain reactions not occur in naturally occurring uranium ore deposits? They would if all uranium atoms fissioned so easily. Fission occurs mainly for the rare isotope U-235, which makes up only 0.7% of the uranium in pure uranium metal. When the more abundant isotope U-238 absorbs neutrons created by fission of U-235, the U-238 typically does not undergo fission. So any chain reaction is snuffed out by the neutron-absorbing U-238, as well as by the rock in which the ore is imbedded.

fyi There is evidence, found in a mine in Gabon, that some 2 billion years ago when the percentage of U-235 in ore was much greater than it is today, there *were*, in fact, some natural reactors on Earth.

If a chain reaction occurred in a baseball-size chunk of pure U-235, an enormous explosion would result. If the chain reaction were started in a smaller chunk of pure U-235, however, no explosion would occur. This is because of geometry: the ratio of surface area to mass is larger in a small piece than in a large one (just as there is more skin on six small potatoes having a combined

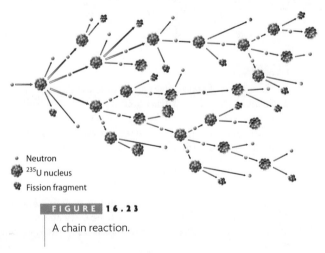

• Neutron

🔵 ^{235}U nucleus

🔵 Fission fragment

FIGURE 16.23

A chain reaction.

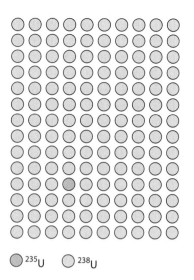

FIGURE 16.24

Only 1 part in 140 of naturally occurring uranium is U-235.

^{235}U ^{238}U

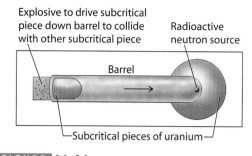

Explosive to drive subcritical piece down barrel to collide with other subcritical piece

Radioactive neutron source

Barrel

Subcritical pieces of uranium

FIGURE 16.26

Simplified diagram of a uranium fission bomb.

mass of 1 kilogram than there is on a single 1-kilogram potato). So there is more surface area on a bunch of small pieces of uranium than on a large piece. In a small piece of U-235, neutrons leak through the surface before an explosion can occur. In a bigger piece, the chain reaction builds up to enormous energies before the neutrons get to the surface and escape (Figure 16.25). For masses greater than a certain amount, called the **critical mass**, an explosion of enormous magnitude may take place.

Consider a large quantity of U-235 divided into two pieces, each having a mass less than critical. The units are *subcritical*. Neutrons in either piece readily reach a surface and escape before a sizable chain reaction builds up. But if the pieces are suddenly driven together, the total surface area decreases. If the timing is right and the combined mass is greater than critical, a violent explosion takes place. This is

what can happen in a nuclear fission bomb (Figure 16.26). A bomb in which pieces of uranium are driven together is a so-called "gun-type" weapon, as opposed to the now more common "implosion weapon."

Constructing a fission bomb is a formidable task. The difficulty is separating enough U-235 from the more abundant U-238. Scientists took more than two years to extract enough U-235 from uranium ore to make the bomb that was detonated at Hiroshima in 1945. To this day uranium isotope separation remains a difficult process, although advanced centrifuges have made it less formidable than it was in World War II.

STOP AND CHECK YOURSELF

A 1-kilogram ball of U-235 is critical, but the same ball broken up into small chunks is not. Explain.

CHECK YOUR ANSWER

The small chunks have more combined surface area than the ball from which they came (just as the combined surface area of gravel is greater than the surface area of a boulder of the same mass). Neutrons escape via the surface before a sustained chain reaction can build up.

Nuclear Fission Reactors

The awesome energy of nuclear fission was introduced to the world in the form of nuclear bombs, and this violent image still colors our thinking about nuclear power, making it difficult for many people to recognize its potential usefulness. Currently, about 20% of electric energy in the United States is generated by *nuclear fission reactors* (it is

FIGURE 16.25

The exaggerated view shows that a chain reaction in a small piece of pure U-235 runs its course before it can cause a large explosion because neutrons leak from the surface too soon. The surface area of the small piece is large relative to the mass. In a larger piece, more uranium and less surface are presented to the neutrons.

Neutrons escape surface

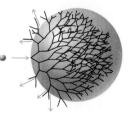

Neutrons trigger more reactions

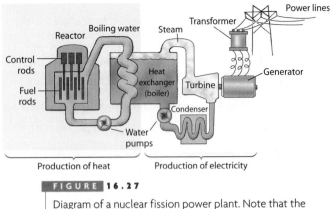

FIGURE 16.27

Diagram of a nuclear fission power plant. Note that the water in contact with the fuel rods is completely contained, and radioactive materials are not involved directly in the generation of electricity.

more in some other countries—about 75% is generated that way in France). These reactors are simply nuclear furnaces. They, like fossil fuel furnaces, do nothing more elegant than boil water to produce steam for a turbine (Figure 16.27). The greatest practical difference is the amount of fuel involved: a mere 1 kilogram of uranium fuel, less than the size of a baseball, yields more energy than 30 freight-car loads of coal.

A fission reactor contains four components: nuclear fuel, control rods, moderator (to slow neutrons, which is required for fission),* and liquid (usually water) to transfer heat from the reactor to the turbine and generator. The nuclear fuel is primarily U-238 plus about 3% U-235. Because the U-235 is so highly diluted with U-238, an explosion like that of a nuclear bomb is not possible.** The reaction rate, which depends on the number of neutrons that initiate the fission of other U-235

nuclei, is controlled by rods inserted into the reactor. The control rods are made of a neutron-absorbing material, usually the metal cadmium or boron.

Heated water around the nuclear fuel is kept under high pressure to keep it at a high temperature without boiling. It transfers heat to a second lower-pressure water system, which operates the turbine and electric generator in a conventional fashion. In this design, two separate water systems are used so that no radioactivity reaches the turbine or the outside environment.

One disadvantage of fission power is the generation of radioactive waste products. Light atomic nuclei are most stable when composed of equal numbers of protons and neutrons, as discussed earlier, and mainly heavy nuclei need more neutrons than protons for stability. For example, in U-235 there are 143 neutrons but only 92 protons. When uranium fissions into two medium-weight elements, the extra neutrons in their nuclei make them unstable. They are radioactive, with a wide range of half-lives averaging about ten years. Among the longer-lived isotopes are cesium-137 and strontium-90, both with half-lives of roughly 30 years. Trace amounts have half-lives of thousands of years. Safely disposing of these waste products as well as materials made radioactive in the production of nuclear fuels

Know nukes before you say "No nukes"!

requires special storage casks and procedures. Although fission has been successfully producing electricity for more than a half century, disposing of radioactive wastes in the United States remains problematic.[†]

* Moderators are substances such as graphite and heavy water that decrease the speeds of neutrons so they can be captured by the fissionable isotope. Interestingly, although slow neutrons in a reactor keep the fission process going, in a detonated nuclear bomb, slow neutrons couldn't keep up with the explosion and it would fizzle out. So one of the safeguards of commercial reactors is that slow neutrons can't sustain a substantial explosion. Even the 1986 Chernobyl accident was an incomplete explosion in a primitive reactor that is no longer manufactured.

** In a worst-case scenario, however, heat sufficient to melt the reactor core is possible—and, if the reactor building is not strong enough, this will scatter radioactivity into the environment. Such an accident occurred at the Chernobyl reactor.

† American policy has been to look for ways to deeply bury radioactive wastes, but many nuclear scientists argue that "spent" nuclear fuel should first be treated in ways to derive value from it or make it less hazardous before what is left over is finally buried. A concept called the *Integral Fast Reactor* (IFR), studied in the 1990s (but never built) would derive additional energy from what is now waste and reduce the chance of diversion of spent fuel to weapons. Other devices are being researched that convert long-life isotopes to ones of shorter half-life. Rather than deeply burying nuclear wastes, for many years the French have been tending and monitoring them in underground storage facilities. Just as the tailings of gold mines and other mines were considered worthless a century ago, but are today being reworked for their commercial value, so it may well be for today's radioactive wastes. If these wastes are kept where they are accessible, it may turn out that they can be modified to be less of a plague to future generations than is commonly thought.

FIGURE 16.28

The nuclear reactor is housed within a dome-shaped containment building that is designed to prevent the release of radioactive isotopes in the event of an accident.

Containment building for nuclear reactor

> *fyi*
> **Weapons-grade plutonium is 90% pure Pu-239.**

The benefits of fission power are plentiful electricity, conservation of the many billions of tons of fossil fuels that every year are literally converted to heat and smoke (which in the long run may be far more precious as sources of organic molecules than as sources of heat), and the elimination of the megatons of carbon dioxide, sulfur oxides, and other deleterious substances that are put into the air each year by the burning of these fossil fuels.

> *fyi*
> **An average ton of coal contains 1.3 ppm (parts per million) of uranium and 3.2 ppm of thorium. That's why the average coal-burning power plant is a far greater source of airborne radioactive material than a nuclear power plant.**

STOP AND CHECK YOURSELF

Coal contains tiny quantities of radioactive materials, enough so there is more environmental radiation surrounding a typical coal-fired power plant than surrounding a fission power plant. What does this indicate about the shielding typically surrounding the two types of power plants?

CHECK YOUR ANSWER

Coal-fired power plants are as American as apple pie, with, so far, no required (and expensive) technology to restrict the emissions of radioactive particles. Nukes, on the other hand, are required to have shielding to ensure strictly low levels of radioactive emissions.

The Breeder Reactor

One of the fascinating features of fission power is the breeding of fission fuel from nonfissionable U-238. This breeding occurs when small amounts of fissionable isotopes are mixed with U-238 in a reactor. Fission liberates neutrons that convert the relatively abundant nonfissionable U-238 to U-239, which beta decays to Np-239, which in turn beta decays to fissionable plutonium—Pu-239 (Figure 16.29). So in addition to the energy produced, fission fuel is bred from the relatively abundant U-238 in the process.

Breeding occurs to some extent in all fission reactors, but a reactor specifically designed to breed more fissionable fuel than what is put into it is called a **breeder reactor**. Using a breeder reactor is like filling your car's gas tank with water, adding some gasoline, then driving the car and having more gasoline after the trip than you did at the beginning! The basic principle of the breeder reactor is very attractive, for after a few years of operation a breeder-reactor power plant can produce vast amounts of power while at the same time breeding twice as much fuel as its original fuel.

The downside is the enormous complexity of successful and safe operation. The United States gave up on breeders about two decades ago, and only Russia, France, Japan, and India are still investing in them. Officials in these countries point out that the supplies of naturally occurring U-235 are limited. At present rates of consumption, all natural sources of U-235 may be depleted within a century. If countries then decide to turn to breeder reactors, they may well find themselves digging up the radioactive wastes they once buried.

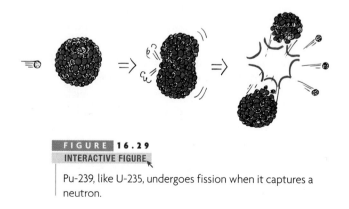

FIGURE 16.29
INTERACTIVE FIGURE

Pu-239, like U-235, undergoes fission when it captures a neutron.

■ PLUTONIUM

Early in the nineteenth century, the farthest planet known in the solar system was Uranus. The first planet to be discovered beyond Uranus was named Neptune. In 1930 a planet beyond Neptune was discovered, and was named Pluto. During this time the heaviest element known was uranium. Appropriately enough, the first transuranic element to be discovered was named *neptunium* and the second transuranic element was named *plutonium*.

Neptunium is produced when a neutron is absorbed by a U-238 nucleus. Rather than undergoing fission, the nucleus emits a beta particle and becomes neptunium, the first synthetic element beyond uranium. The half-life of neptunium is only 2.3 days, so it isn't around very long. Neptunium is a beta emitter, and very soon becomes plutonium. The half-life of plutonium is about 24,000 years, so it lasts a considerable time. The isotope plutonium-239, like U-235, will undergo fission when it captures a neutron. Whereas the separation of fissionable U-235 from uranium metal is a very difficult process (because U-235 and U-238 have the same chemistry), the separation of plutonium from uranium metal is

relatively easy. This is because plutonium is an element distinct from uranium, with its own chemical properties.

The element plutonium is chemically poisonous in the same sense as are lead and arsenic. It attacks the nervous system and can cause paralysis. Death can follow if the dose is sufficiently large. Fortunately, plutonium does not remain in its elemental form for long because it rapidly combines with oxygen to form three compounds: PuO, PuO_2, and Pu_2O_3, all of which chemically are relatively benign. They will not dissolve in water or in biological systems. These plutonium compounds do not attack the nervous system and have been found to be chemically harmless.

Plutonium in any form, however, is radioactively toxic. It is more toxic than uranium, although less toxic than radium. Plutonium emits high-energy alpha particles, which kill cells rather than simply disrupting them and leading to mutations. Interestingly, damaged cells rather than dead cells contribute to cancer. This is why plutonium ranks relatively low as a cancer-producing substance. The greatest danger that plutonium presents to humans is its potential for use in nuclear fission bombs. Its usefulness is in fission reactors—particularly breeder reactors.

16.9 Mass–Energy Equivalence— $E = mc^2$

Early in the early 1900s, Albert Einstein discovered that mass is actually "congealed" energy. Mass and energy are two sides of the same coin, as stated in his celebrated equation $E = mc^2$. In this equation E stands for the energy that any mass has at rest, m stands for mass, and c is the speed of light. The quantity c^2 is the proportionality constant of energy and mass. This relationship between energy and mass is the key to understanding why and how energy is released in nuclear reactions.

The more energy that is associated with a particle, the greater is the mass of the particle. Is the mass of a nucleon inside a nucleus the same as that of the same nucleon outside a nucleus? This question can be answered by considering the work that would be required to separate nucleons from a nucleus. From physics we know that work, which is expended energy, is equal to *force* × *distance*.

Think of the amount of force required to pull a nucleon out of the nucleus through a sufficient distance to overcome the attractive strong nuclear force, comically indicated in Figure 16.30. Enormous work would be required. This work is energy added to the nucleon that is pulled out.

According to Einstein's equation, this newly acquired energy reveals itself as an increase in the nucleon's mass. The mass of a nucleon outside a nucleus is greater than the mass of the same nucleon locked inside a nucleus. A carbon-12 atom—the nucleus of which is made up of six protons and six neutrons—has a mass of exactly 12.00000 atomic

FIGURE 16.30

Work is required to pull a nucleon from an atomic nucleus. This work increases the energy and hence the mass of the nucleon outside the nucleus.

mass units (amu). Therefore, on average, each nucleon contributes a mass of 1 amu. However, outside the nucleus, a proton has a mass of 1.00728 amu and a neutron has a mass of 1.00867 amu. Thus, we see that the combined mass of six free protons and six free neutrons—(6 × 1.00728) + (6 × 1.00867) = 12.09570—is greater than the mass of one carbon-12 nucleus. The greater mass reflects the energy required to pull the nucleons apart from one another. Thus, what mass a nucleon has depends on where the nucleon is.

The masses of the isotopes of various elements can be very accurately measured with a mass spectrometer (Figure 16.31). This important device uses a magnetic field to deflect ions of these isotopes into circular arcs. The greater the inertia (mass) of the ion, the more it resists deflection, and the greater the radius of its curved path. The magnetic force sweeps lighter ions into shorter arcs and heavier ions into larger arcs.*

> **Is something amiss here? We live in a nuclear-powered universe, yet we get most of our electric energy from burning coal.**

A graph of the nuclear masses for the elements from hydrogen through uranium is shown in Figure 16.32. The graph slopes upward with increasing atomic number as expected: Elements are more massive as atomic number increases. The slope steepens because there are proportionally more neutrons in the more massive atoms.

A more important graph results from the plot of nuclear mass *per nucleon* from hydrogen through uranium (Figure 16.33). This is perhaps the most important graph in this book, for it is the key to understanding the energy associated with nuclear processes. To obtain the average mass per nucleon, you divide the total mass of a nucleus by the number of nucleons in the nucleus. (Similarly, if you divide the total mass of a roomful of people by the number of people in the room, you get the average mass per person.)

Note that the masses per nucleon vary, almost as if the individual nucleons had different masses in

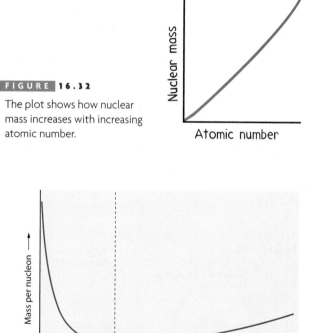

FIGURE 16.32

The plot shows how nuclear mass increases with increasing atomic number.

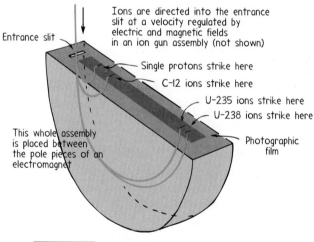

FIGURE 16.31

The mass spectrometer. Electrically charged isotopes are directed into the semicircular "drum," where they are forced into curved paths by a strong magnetic field. Lighter isotopes have less inertia (mass) and so they easily change direction and are pulled into curves of smaller radii. Heavier isotopes have greater inertia (mass) and so they are pulled into curves of larger radii. The mass of an isotope, therefore, is directly related to how far away from the slit it lands.

---

* Interestingly, miniature mass spectrometers are used for detecting molecules associated with explosives at airport security stations. The security agent swabs luggage with a soft cloth that is placed in the device. Molecules on the swab are ionized and scrutinized.

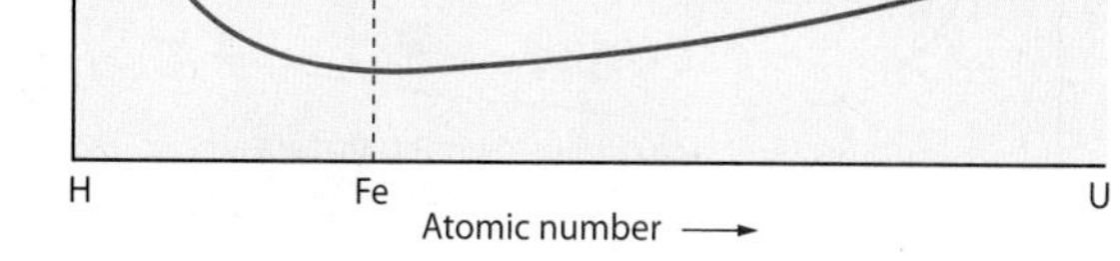

FIGURE 16.33

INTERACTIVE FIGURE

This graph shows the average mass per nucleon in atomic nuclei. This quantity is greatest for the lightest nuclei, where the nucleons are held together least tightly; least for iron, whose nucleons are held together most tightly; and of an intermediate value for the heaviest nuclei.

different nuclei. The greatest mass per nucleon occurs for the proton alone, hydrogen, because it has no binding energy to pull its mass down. Progressing beyond hydrogen, the mass per nucleon gets smaller, and is least in one of the isotopes of the iron atom. Beyond iron, the process reverses itself as the average binding energy per nucleon decreases, thanks to the repulsive electric force between protons. This trend continues all the way to uranium and the transuranic elements.

From Figure 16.33 we can see how energy is released when a uranium nucleus splits into two nuclei of lower atomic number. Uranium, being toward the right-hand side of the graph, is shown to have a relatively large amount of mass per nucleon. When the uranium nucleus splits in half, however, smaller nuclei of lower atomic numbers are formed. As shown in Figure 16.34, these nuclei are lower on the graph than uranium, which means that they have a smaller amount of mass per nucleon. When this decrease in mass is multiplied by the speed of light squared (c^2 in Einstein's equation), the product is equal to the energy yielded by each uranium nucleus as it undergoes fission.

We can think of the mass-per-nucleon graph as an energy valley that starts at hydrogen (the highest point) and slopes steeply to the lowest point (iron), then slopes gradually up to uranium. Iron is at the bottom of the energy valley and is the most stable nucleus. It is also the most tightly bound nucleus; more energy per nucleon is required to separate a nucleon from its nucleus than from any other nucleus.

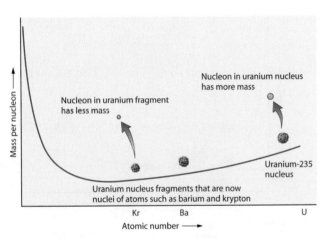

FIGURE 16.34

The mass per nucleon in a uranium nucleus is greater than in any one of its nuclear fission fragments. This lost mass is mass that has been transformed into energy, which is why nuclear fission is an energy-releasing process.

All nuclear power today is by way of nuclear fission. A more promising long-range source of energy is to be found on the left side of the energy valley.

STOP AND CHECK YOURSELF

Correct the following incorrect statement: When a heavy element such as uranium undergoes fission, there are fewer nucleons after the reaction than before.

CHECK YOUR ANSWER

When a heavy element such as uranium undergoes fission, there aren't fewer nucleons after the reaction. Instead, there's *less mass* in the same number of nucleons.

The graph of Figure 16.33 (and Figures 16.34 and 16.35) reveals the energy of the atomic nucleus, a primary source of energy in the universe—which is why it can be considered the most important graph in this book.

16.10 Nuclear Fusion

Notice in the graphs of Figures 16.33 and Figure 16.35 that the steepest part of the energy valley goes from hydrogen to iron. Mass is decreased and energy is released as light nuclei combine. This combining of nuclei is **nuclear fusion**—the opposite of nuclear fission. We see from Figure 16.35 that, as we move along the list of elements from hydrogen to iron, the average mass per nucleon decreases. Thus, when two small nuclei fuse—say, two hydrogen isotopes—the mass of the resulting helium-4 nucleus is less than the mass of the two small nuclei before fusion. Energy is released as smaller nuclei fuse.

For a fusion reaction to occur, the nuclei must collide at a very high speed in order to overcome their mutual electric repulsion. The required speeds correspond to the extremely high temperatures found in the Sun and other stars. Fusion initiated by high temperatures is called **thermonuclear fusion**. In the high temperatures of the Sun, approximately 657 million tons of hydrogen are converted into 653 million tons of helium *each second*. The missing 4 million tons of mass are discharged as radiant energy.

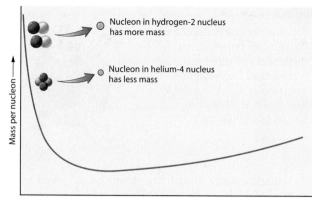

Such reactions are, quite literally, nuclear burning. Thermonuclear fusion is analogous to ordinary chemical combustion. In both chemical and nuclear burning, a high temperature starts the reaction; the release of energy by the reaction maintains a high enough temperature to spread the fire. The net result of the chemical reaction is a combination of atoms into more tightly bound molecules. In nuclear fusion reactions, the net result is more tightly bound nuclei. In both cases mass decreases as energy is given off.

FIGURE 16.35

The mass per nucleon for a hydrogen-2 nucleus is greater than for helium-4. Mass converts to energy. Nuclear fusion for light nuclei is an energy-releasing process.

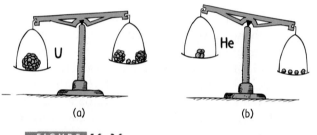

(a)   (b)

FIGURE 16.36
INTERACTIVE FIGURE

The mass of a nucleus is not equal to the sum of the masses of its parts. (a) The fission fragments of a heavy nucleus like uranium are less massive than the uranium nucleus. (b) Two protons and two neutrons are more massive in their free states than when combined to form a helium nucleus.

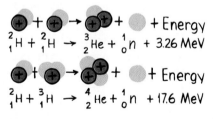

$$^2_1H + ^2_1H \rightarrow ^3_2He + ^1_0n + 3.26 \text{ MeV}$$

$$^2_1H + ^3_1H \rightarrow ^4_2He + ^1_0n + 17.6 \text{ MeV}$$

FIGURE 16.37

Fusion reactions of two hydrogen isotopes. Most of the energy released is carried by the neutrons, which are ejected at high speeds.

fyi   Thermonuclear reactors on Earth will require temperatures even higher than in the Sun. The Sun's slow pace—10 billion years to burn out—is too leisurely for application on Earth.

1. Fission and fusion are opposite processes, yet each releases energy. Isn't this contradictory?

2. To get nuclear energy release from the element iron, should iron be fissioned or fused?

3. Predict whether the temperature of the core of a star increases or decreases when iron and elements of higher atomic number than iron in the core are fused.

CHECK YOUR ANSWERS

1. No, no, no! This is contradictory only if the same element is said to release energy by both the processes of fission and fusion. Only the fusion of light elements and the fission of heavy elements result in a decrease in nucleon mass and a release of energy.

2. Neither, because iron is at the very bottom of the "energy valley." Fusing a pair of iron nuclei produces an element to the right of iron on the curve, where mass per nucleon is higher. If you split an iron nucleus, the products lie to the left of iron on the curve—also a higher mass per nucleon. So no energy is released. For energy release, "Decrease Mass" is the name of the game—any game, chemical or nuclear.

3. In the fusion of iron and any nuclei beyond, energy is absorbed and the star core cools at this late stage of its evolution. This, however, leads to the star's collapse, which then greatly increases it temperature. Interestingly, elements beyond iron are not manufactured in normal fusion cycles in stellar sources, but are manufactured when stars violently explode—supernovae.

Controlling Fusion

Carrying out fusion reactions under controlled conditions requires temperatures of millions of degrees. There are a variety of techniques for attaining high temperatures. No matter how the temperature is produced, a problem is that all materials melt and vaporize at the temperatures required for fusion. One solution to this problem is to confine the reaction in a nonmaterial container.

A nonmaterial container is a magnetic field, which can exist at any temperature and can exert powerful forces on charged particles in motion. "Magnetic walls" of sufficient strength provide a kind of magnetic straightjacket for hot gases called plasmas. Magnetic compression further heats the plasma to fusion temperatures. At this writing, fusion by magnetic confinement has only been partially successful—a sustained and controlled reaction has been out of reach.

Another approach bypasses magnetic confinement altogether with high-energy lasers. An intriguing technique is aiming an array of laser beams at a common point and dropping solid pellets composed of hydrogen isotopes through the synchronous cross fire (Figure 16.38). The energy of the multiple beams should crush pellets to densities 20 times that of lead. Such a fusion "burn" could produce several hundred times more energy than is delivered by the laser beams that compress and ignite the pellets. Like the succession of small fuel/air explosions in an automobile engine's cylinders that convert into a smooth flow of mechanical power, the successive ignition of dropping pellets in a fusion power plant may similarly produce a steady stream of electric power. The success of this technique requires precise timing, for the necessary compression must occur before a shock wave causes the pellet to disperse. Reliable high power lasers are vital. However, break-even (where energy output

FIGURE 16.38

Fusion with multiple laser beams. Pellets of deuterium are rhythmically dropped into synchronized laser cross fire in this planned device. The resulting heat is carried off by molten lithium to produce steam.

equals energy input) has yet to be achieved with laser fusion.

Other fusion schemes involve the bombardment of fuel pellets not by laser light but by beams of electrons and ions. We await "break-even day," when one of the techniques for controlled nuclear fusion produces sustained energy.

If people are one day to dart about the universe in the same way we jet about Earth today, their supply of fuel is assured. The fuel for fusion—hydrogen—is found in every part of the universe, not only in the stars but also in the space between them. About 91% of the atoms in the universe are estimated to be hydrogen. For people of the future, the supply of raw materials is also assured because all the elements known to exist result from the fusing of more and more hydrogen nuclei. Future humans might synthesize their own elements and produce energy in the process, just as the stars have always done.

fyi Research on nuclear fusion is headquartered in France at the forthcoming International Thermonuclear Experimental Reactor (ITER), a joint fusion project by the European Union and the United States. In addition, a projected mini-ITER in Japan will focus on superconducting magnets. Similar fusion research is occurring in India, Korea, and China. Watch (but don't hold your breath) for the advent of fusion power.

SUMMARY OF TERMS

Radioactivity The process wherein unstable atomic nuclei break down and emit radiation.

Alpha particle The nucleus of a helium atom, which consists of two neutrons and two protons, ejected by certain radioactive elements.

Beta particle An electron (or positron) emitted during the radioactive decay of certain nuclei.

Gamma ray High-frequency electromagnetic radiation emitted by the nuclei of radioactive atoms.

Rad Acronym (*radiation absorbed dose*), a unit of absorbed energy. One rad is equal to 0.01 joule of radiant energy absorbed per kilogram of tissue.

Rem Acronym (*roentgen equivalent man*), a unit used to measure the effect of ionizing radiation on humans.

Nucleon The collective name for a nuclear proton or neutron.

Strong force Force that attracts nucleons to each other within the nucleus (also called the strong interaction).

Half-life The time required for half the atoms in a sample of a radioactive isotope to decay.

Transmutation The conversion of an atomic nucleus of one element into an atomic nucleus of another element through a loss or gain in the number of protons.

Carbon dating Process of determining the elapsed time since death by measuring the radioactivity of the remaining carbon-14 atoms.

Nuclear fission The splitting of the nucleus of a heavy atom, such as uranium-235, into two main parts, accompanied by the release of much energy.

Chain reaction A self-sustaining reaction in which the products of one reaction event stimulate further reaction events.

Critical mass The minimum mass of fissionable material in a reactor or nuclear bomb that will sustain a chain reaction.

Breeder reactor Nuclear fission reactor that not only produces power but produces more nuclear fuel than it consumes by converting a nonfissionable uranium isotope into a fissionable plutonium isotope.

Nuclear fusion The combination of the nuclei of light atoms to form heavier nuclei, with the release of much energy.

Thermonuclear fusion Nuclear fusion produced by high temperature.

SUGGESTED READING

Bodansky, D. *Nuclear Energy: Principles, Practices, and Prospects*, 2nd ed. New York: Springer, 2004.

Hannum, W. H., G. E. Marsh, and G. S. Stanford. *Physics and Society* 33(3), 8 (July 2004); see http://www.aps.org/units/fps.

Vandenbosch, R., and S. E. Vandenbosch. *Physics and Society* 35(3), 7 (July 2006); see http://www.aps.org/units/fps.

REVIEW QUESTIONS

16.1 Radioactivity

1. Where does most of the radiation you encounter originate?
2. What are cosmic rays, and where do they originate?

16.2 Alpha, Beta, and Gamma Rays

3. How do the electric charges of alpha, beta, and gamma rays differ?
4. Why are alpha and beta rays deflected in opposite directions in a magnetic field? Why are gamma rays undeflected?
5. What is the origin of gamma rays?

16.3 Environmental Radiation

6. What is the leading source of naturally occurring radiation?
7. What type of power plant most exposes us to radioactivity, coal or nuclear?
8. Distinguish between a *rad* and a *rem*.
9. Do humans receive more radiation from artificial or natural sources of radiation?
10. What is the lethal dose of radiation? What is the average radiation dosage per year for the average person in the United States? What is an average dose delivered by a typical X-ray?
11. Is the human body radioactive?

12. What kinds of cells are in most danger when they are irradiated?
13. What is a radioactive tracer?

16.4 The Atomic Nucleus and the Strong Force

14. Name the two different nucleons.
15. Why doesn't the repulsive electric force of protons in the atomic nucleus cause the protons to fly apart?
16. Why is a larger nucleus generally less stable than a smaller nucleus?
17. What is the role of neutrons in the atomic nucleus?
18. Which have more neutrons than protons, large nuclei or small nuclei?

16.5 Radioactive Half-Life

19. What is meant by *radioactive half-life*?
20. What is the half-life of Ra-226?
21. For any radioactive isotope, what is the relationship between decay rate and half-life?

16.6 Transmutation of the Elements

22. What is transmutation?
23. When thorium, atomic number 90, decays by emitting an alpha particle, what is the atomic number of the resulting nucleus?

24. When thorium decays by emitting a beta particle, what is the atomic number of the resulting nucleus?
25. Does atomic mass increase or decrease or stay the same in each of the reactions referred to in questions 23 and 24?
26. What change in atomic number occurs when a nucleus emits an alpha particle? When it emits a beta particle? When it emits a gamma ray?
27. What is the long-range fate of all the uranium that exists in the world?
28. When and by whom did the first successful intentional transmutation of an element occur?
29. Why are the elements beyond uranium not common in Earth's crust?

16.7 Radiometric Dating

30. How are radioactive isotopes produced?
31. What occurs when a nitrogen nucleus captures an extra neutron?
32. What does the proportion of lead and uranium in a rock tell us about the age of the rock?

16.8 Nuclear Fission

33. When a nucleus undergoes fission, what role can the ejected neutrons play?
34. Why is a chain reaction more likely in a big piece of uranium than it is in a small piece?
35. What is a critical mass?
36. Which will leak more neutrons, two separate pieces of uranium or the same two pieces stuck together?
37. What are the four main components of a fission reactor?

38. Why can a reactor not explode like a fission bomb?
39. What is the effect of putting small amounts of fissionable isotopes with large amounts of U-238?
40. How does a breeder reactor breed nuclear fuel?

16.9 Mass–Energy Equivalence —$E = mc^2$

41. Is work required to pull a nucleon out of an atomic nucleus? Does the nucleon, once outside, then have more energy than it did when it was inside the nucleus? In what form is this energy?
42. Which ions are least deflected in a mass spectrometer?
43. What is the basic difference between the graphs of Figure 16.32 and Figure 16.33?
44. In which atomic nucleus is the mass per nucleon greatest? In which is it least?
45. What becomes of the missing mass when a uranium nucleus undergoes fission?

16.10 Nuclear Fusion

46. If the graph in Figure 16.35 is seen as an energy valley, what can be said of nuclear transformations that progress toward iron?
47. When a pair of hydrogen isotopes is fused, is the mass of the product nucleus more or less than the sum of the masses of the two hydrogen nuclei?
48. For helium to release energy, should it be fissioned or fused?
49. What kind of containers are used to contain plasmas at temperatures of millions of degrees?
50. In what form is energy initially released in nuclear fusion?

ACTIVE EXPLORATIONS

1. Write a letter to a grandparent to dispel any notion his or her friends might have about radioactivity being something new in the world. Tie this to the idea that many people have the strongest views on that which they least understand.

2. Write a letter to an uncle or aunt discussing nuclear power. Cite both the ups and downs of it, and explain how the comparison affects your personal view of nuclear power. Also explain how nuclear fission and nuclear fusion differ.

EXERCISES

1. Is radioactivity in the world something relatively new? Defend your answer.
2. Can it be truthfully said that, whenever a nucleus emits an alpha or beta particle, it necessarily becomes the nucleus of a different element?
3. Why is a sample of radium always a little warmer than its surroundings?
4. Some people say that all things are possible. Is it at all possible for a hydrogen nucleus to emit an alpha particle? Defend your answer.

5. Why are alpha and beta rays deflected in opposite directions in a magnetic field? Why are gamma rays undeflected?
6. The alpha particle has twice the electric charge of the beta particle, but it is deflected less than the beta particle in a magnetic field. Why is this so?
7. How would the paths of alpha, beta, and gamma radiations compare in an electric field?
8. In what way is the emission of gamma radiation from a nucleus similar to the emission of light from an atom?

9. Which type of radiation—alpha, beta, or gamma—results in the greatest change in atomic number? In the least change in atomic number?

10. Which type of radiation—alpha, beta, or gamma—produces the greatest change in mass number? The least change in mass number?

11. In bombarding atomic nuclei with proton "bullets," why must the protons be accelerated to high energies to make contact with the target nuclei?

12. Just after an alpha particle leaves the nucleus, would you expect it to speed up? Defend your answer.

13. Within the atomic nucleus, which interaction tends to hold it together and which interaction tends to push it apart?

14. What evidence supports the contention that the strong nuclear force is stronger than the electrical force at short distances within the nucleus?

15. A friend asks if a radioactive substance with a half-life of 1 day will be entirely gone at the end of 2 days. What is your answer?

16. When the isotope bismuth-213 emits an alpha particle, what new element results? What new element results if it instead emits a beta particle?

17. When $^{226}_{84}$Ra decays by emitting an alpha particle, what is the atomic number of the resulting nucleus? What is the resulting atomic mass?

18. When $^{218}_{84}$Po emits a beta particle, it transforms into a new element. What are the atomic number and atomic mass of this new element? What are they if the polonium instead emits an alpha particle?

19. State the number of neutrons and protons in each of the following nuclei: $^2_1$H, $^{12}_6$C, $^{56}_{26}$Fe, $^{197}_{79}$Au, $^{90}_{38}$Sr, and $^{238}_{92}$U.

20. How is it possible for an element to decay "forward in the periodic table"—that is, to decay to an element of higher atomic number?

21. Elements with atomic numbers greater than that of uranium do not exist in any appreciable amounts in nature because they have short half-lives. Yet there are several elements with atomic numbers smaller than that of uranium that have equally short half-lives and that do exist in appreciable amounts in nature. How can you account for this?

22. You and your friend journey to the mountain foothills to get closer to nature and to escape such things as radioactivity. While bathing in the warmth of a natural hot spring, she wonders aloud how the spring gets its heat. What do you tell her?

23. Coal contains minute quantities of radioactive materials, yet there is more environmental radiation surrounding a coal-fired power plant than a fission power plant. What does this indicate about the shielding that typically surrounds these power plants?

24. When we speak of dangerous radiation exposure, are we customarily speaking of alpha radiation, beta radiation, or gamma radiation? Discuss.

25. People who work around radioactivity wear film badges to monitor the amount of radiation that reaches their bodies. These badges consist of small pieces of photographic film enclosed in a light-proof wrapper (Figure 16.8). What kind of radiation do these devices monitor?

26. A friend produces a Geiger counter to check the local background radiation. It ticks. Another friend, who normally fears most that which is understood least, makes an effort to keep away from the region of the Geiger counter and looks to you for advice. What do you say?

27. When food is irradiated with gamma rays from a cobalt-60 source, does the food become radioactive? Defend your answer.

28. If it were known that cosmic ray intensity was much greater thousands of years ago, how would this affect the ages assigned to ancient samples of once-living matter?

29. The age of the Dead Sea Scrolls was found by carbon dating. Could this technique have worked if they were carved in stone tablets? Explain.

30. Why will nuclear fission probably not be used directly for powering automobiles? How could it be used indirectly?

31. Why does a neutron make a better nuclear bullet than a proton or an electron when all three have low energy?

32. Does the average distance that a neutron travels through fissionable material before escaping increase or decrease when two pieces of fissionable material are assembled into one piece? Does this assembly increase or decrease the probability of an explosion?

33. U-235 releases an average of 2.5 neutrons per fission, while Pu-239 releases an average of 2.7 neutrons per fission. Which of these elements might you therefore expect to have the smaller critical mass?

34. Why is lead found in all deposits of uranium ores?

35. Why does plutonium not occur in appreciable amounts in natural ore deposits?

36. Why does a chain reaction not occur in uranium mines?

37. A friend makes the claim that the explosive power of a nuclear bomb is due to static electricity. Do you agree or disagree? Defend your answer.

38. If a nucleus of $^{232}_{90}$Th absorbs a neutron and the resulting nucleus undergoes two successive beta decays (emitting electrons), what nucleus results?

39. The energy release of nuclear fission is tied to the fact that the heaviest nuclei have about 0.1% more mass per nucleon than nuclei near the middle of the periodic table of the elements. What would be the effect on energy release if the 0.1% figure were instead 1%?

40. How does the mass per nucleon in uranium compare with the mass per nucleon in the fission fragments of uranium?

41. How is chemical burning similar to nuclear fusion?

42. To predict the approximate energy release of either a fission or a fusion reaction, explain how a physicist makes use of the curve of Figure 16.33 or a table of nuclear masses and the equation $E = mc^2$.

43. Which process would release energy from gold, fission or fusion? From carbon? From iron?

44. If uranium were to split into three segments of equal size instead of two, would more energy or less energy be released? Defend your answer in terms of Figure 16.33.

45. Explain how radioactive decay has always warmed Earth from the inside, and nuclear fusion has always warmed Earth from the outside.

46. What effect on the mining industry can you foresee for a future in which the synthesis of elements is widespread?

47. The world has never been the same since the discovery of electromagnetic induction and its applications to electric motors and generators. Speculate on and list some of the worldwide changes that are likely to follow the advent of successful fusion reactors.
48. Ordinary hydrogen is sometimes called a perfect fuel because there is an almost unlimited supply on Earth, and, when it burns (oxidizes), harmless water is the combustion product. So why don't we abandon fission energy and fusion energy, not to mention fossil fuel energy, and just use hydrogen?
49. Referring to the previous exercise, why may it take fusion power or the equivalent to see hydrogen taking the place of gasoline?
50. Discuss and make a comparison of pollution by conventional fossil fuel power plants and nuclear fission power plants. Consider thermal pollution, chemical pollution, and radioactive pollution.

PROBLEMS

● BEGINNER   ■ INTERMEDIATE   ◆ EXPERT

1. ● Radiation from a point source obeys the inverse-square law. If a Geiger counter 1 meter from a small sample reads 360 counts per minute, what will be its counting rate at 2 meters from the source? At 3 meters from the source?
2. ● If a sample of a radioactive isotope has a half-life of 1 year, how much of the original sample will be left at the end of the second year? At the end of the third year? At the end of the fourth year?
3. ● A certain radioactive substance has a half-life of 1 hour. If you start with 1 gram of the material at noon, how much will be left at 3 p.m.? at 6 p.m.? at 10 p.m.?
4. ● A sample of a particular radioisotope is placed near a Geiger counter, which is observed to register 160 counts per minute. Eight hours later, the detector counts at a rate of 10 counts per minute. What is the half-life of the material?
5. ● The isotope cesium-137, which has a half-life of 30 years, is a product of nuclear power plants. Show that it take 120 years for this isotope to decay to about one-sixteenth its original amount.

6. ■ Suppose that you want to find out how much gasoline is in an underground storage tank. You pour in one gallon of gasoline that contains some radioactive material with a long half-life that gives off 5000 counts per minute. The next day, you remove a gallon from the underground tank and measure its radioactivity to be 10 counts per minute. How much gasoline is in the tank?
7. ■ Suppose that you measure the intensity of radiation from carbon-14 in an ancient piece of wood to be 6% of what it would be in a freshly cut piece of wood. Show that the age of this artifact is about 23,000 years old.
8. ■ The kiloton, which is used to measure the energy released in an atomic explosion, is equal to 4.2×10^{12} J (approximately the energy released in the explosion of 1000 tons of TNT). Recall that 1 kilocalorie of energy raises the temperature of 1 kilogram of water by 1°C and that 4184 joules is equal to 1 kilocalorie. Show that the temperature of 4.0×10^8 kilograms of water can be increased 50°C by a 20-kiloton atomic bomb.

CHAPTER 16 ONLINE RESOURCES

The Physics Place

Interactive Figures
16.2, 16.3, 16.4, 16.15, 16.29, 16.33, 16.36

Tutorials
Nuclear Physics
Atoms and Isotopes

Videos
Radioactive Decay
Half-Life

Carbon Dating
Plutonium

Quiz

Flashcards

Links

I hope you've enjoyed *Conceptual Physics Fundamentals* and will value your knowledge of physics as a worthwhile component of your general education. Viewing physics as a study of the rules of nature enhances your sense of wonder and the way you see the physical world—knowing that so much in nature is connected, with seemingly diverse phenomena often following the same basic rules. How intriguing that the rules governing a falling apple also apply to a space station orbiting the Earth, that a sky's redness at sunset is connected to its blueness at midday, that the rules discovered by Faraday and Maxwell show how electricity and magnetism connect to become light.

The value of science is more than its applications to fast cars, computers, iPods, and other products. Its greatest value lies in its methods of understanding and investigating nature—that hypotheses are framed so that they are capable of being disproved, and experiments are designed so that their results can be reproduced by others. Science is more than a body of knowledge; it is a way of thinking.

And then there are the purveyors of junk science who dress up their claims in the language of science but intentionally ignore its methods. The several boxes on pseudoscience throughout this book are an attempt to expose this. Being able to distinguish between scientific experiments and unsupported claims is particularly important because so much misinformation and hype are used by charlatans to peddle their bogus wares and bogus ideas. Pseudoscience cheapens science. Its purveyors wish to topple the scientific way of viewing the world and disavow skeptical thinking.

Skeptical thinking, in addition to sharpening common sense, is an essential ingredient in formulating a hypothesis that requires a test for wrongness: If I'm wrong, how would I know? This key question can accompany any important idea, scientific or otherwise. When it is applied to social, political, and religious questions, you become stronger for it. Socially, you see others' points of view more clearly. Politically, you see all social movements as experiments. Religiously, you see that conflicts between science and religion stem mainly from misapplications of one or both of them. Properly applied, science is not only compatible with spirituality but can be a profound source of spirituality.

Contemplating the immensity of the universe and the geologic time scale of our planet evokes a soaring feeling that is surely spiritual. We've learned that four hundred million years ago, long before mammals, there were fish; then came amphibians and then reptiles. In the struggle of species survival, trillions upon trillions of life forms passed their genetic traits on to their offspring, sometimes here and there making adaptive changes. After a long and prodigious ascent, humans emerged. Innumerable lives brought us to where we are. Rather than sweeping this long and astounding journey aside, we all should celebrate it—for we are the benefactors.

Science offers a modern-day means of establishing our origins, how we can survive, and even who we can become. We're at a present vantage point where science can proceed from "how" to "why"— ironically at a time when the potential for world calamity has never been greater. Overpopulation denial, energy greed, and other socioeconomic and political problems beset our age. Yet science provides us with the physical and intellectual tools to improve our lives and our relationships with one another and our environment. Our hope lies with those with open scientific minds who understand and can sensibly address the global issues that threaten our survival. Earth is the only home we all share, deserving our utmost care. As I see it, and hopefully you too, caring, knowledgeable people applying the methods of science are humanity's best hope.

ON MEASUREMENT AND UNIT CONVERSION

Two major systems of measurement prevail in the world today: the *United States Customary System* (USCS, formerly called the British system of units), used in the United States of America and in Burma, and the *Système International* (SI) (known also as the international system and as the metric

The Physics Place
Significant Figures

system), used everywhere else. Each system has its own standards of length, mass, and time. The units of length, mass, and time are sometimes called the *fundamental units* because, once they are selected, other quantities can be measured in terms of them.

United States Customary System

Based on the British Imperial System, the USCS is familiar to everyone in the United States. It uses the foot as the unit of length, the pound as the unit of weight or force, and the second as the unit of time. The USCS is presently being replaced by the international system—rapidly in science and technology (all 1988 Department of Defense contracts) and some sports (track and swimming), but so slowly in other areas and in some specialties it seems the change may never come. For example, we will continue to buy seats on the 50-yard line. Camera film is in millimeters but computer disks are in inches.

For measuring time, there is no difference between the two systems except that in pure SI the only unit is the second (s, not sec) with prefixes; but in general, minute, hour, day, year, and so on, with two or more lettered abbreviations (h, not hr), are accepted in the USCS.

TABLE A.1

SI Units

Quantity	Unit	Symbol
Length	meter	m
Mass	kilogram	kg
Time	second	s
Force	newton	N
Energy	joule	J
Current	ampere	A
Temperature	kelvin	K

Systéme International

During the 1960 International Conference on Weights and Measures held in Paris, the SI units were defined and given status. Table A.1 shows SI units and their symbols. SI is based on the *metric system,* originated by French scientists after the French Revolution in 1791. The orderliness of this system makes it useful for scientific work, and it is used by scientists all over the world. The metric system branches into two systems of units. In one of these the unit of length is the meter, the unit of mass is the kilogram, and the unit of time is the second. This is called the *meter-kilogram-second* (mks) system and is preferred in physics. The other branch is the *centimeter-gram-second* (cgs) system, which,

The Physics Place
The Metric System

because of its smaller values, is favored in chemistry. The cgs and mks units are related to each other as follows: 100 centimeters equal 1 meter; 1000 grams equal 1 kilogram. Table A.2 shows several units of length related to each other.

One major advantage of a metric system is that it uses the decimal system, in which all units are

TABLE A.2

Table Conversions Between Different Units of Length

Unit of Length	Kilometer	Meter	Centimeter	Inch	Foot	Mile
1 kilometer	= 1	1000	100,000	39,370	3280.84	0.62140
1 meter	= 0.00100	1	100	39.370	3.28084	6.21×10^{-4}
1 centimeter	= 1.0×10^{-5}	0.0100	1	0.39370	0.032808	6.21×10^{-6}
1 inch	= 2.54×10^{-5}	0.02540	2.5400	1	0.08333	1.58×10^{-5}
1 foot	= 3.05×10^{-4}	0.30480	30.480	12	1	1.89×10^{-4}
1 mile	= 1.60934	1609.34	160,934	63,360	5280	1

TABLE A.3

Some Prefixes

Prefix	Definition
micro-	One-millionth: a microsecond is one-millionth of a second
milli-	One-thousandth: a milligram is one-thousandth of a gram
centi-	One-hundredth: a centimeter is one-hundredth of a meter
kilo-	One thousand: a kilogram is 1000 grams
mega-	One million: a megahertz is 1 million hertz
giga-	One billion: a gigahertz is 1 billion hertz

related to smaller or larger units by dividing or multiplying by 10. The prefixes shown in Table A.3 are commonly used to show the relationship among units.

METER

The standard of length of the metric system orginally was defined in terms of the distance from the north pole to the equator. This distance was thought at the time to be close to 10,000 kilometers. One ten-millionth of this, the meter, was carefully determined and marked off by means of scratches on a bar of platinum-iridium alloy. This bar is kept at the International Bureau of Weights and Measures in France. The standard meter in France has since been calibrated in terms of the wavelength of light—it is 1,650,763.73 times the wavelength of orange light emitted by the atoms of the gas krypton-86. The meter is now defined as being the length of the path traveled by light in a vacuum during a time interval of 1/299,792,458 of a second.

KILOGRAM

The standard unit of mass, the kilogram, is a block of platinum–iridium alloy, also preserved at the International Bureau of Weights and Measures located in France (Figure A.1). The kilogram equals 1000 grams. A gram is the mass of 1 cubic centimeter

FIGURE A.1

The standard kilogram.

(cc) of water at a temperature of 4°C. (The standard pound is defined in terms of the standard kilogram; the mass of an object that weighs 1 pound is equal to 0.4536 kilogram.)

SECOND

The official unit of time for both the USCS and the SI is the second. Until 1956, it was defined in terms of the mean solar day, which was divided into 24 hours. Each hour was divided into 60 minutes and each minute into 60 seconds. Thus, there were 86,400 seconds per day, and the second was defined as 1/86,400 of the mean solar day. This proved unsatisfactory because the rate of rotation of the earth is gradually becoming slower. In 1956, the mean solar day of the year 1900 was chosen as the standard on which to base the second. In 1964, the second was officially defined as the time taken by a cesium-133 atom to make 9,192,631,770 vibrations.

NEWTON

One newton is the force required to accelerate 1 kilogram at 1 meter per second per second. This unit is named after Sir Isaac Newton.

JOULE

One joule is equal to the amount of work done by a force of 1 newton acting over a distance of 1 meter. In 1948, the joule was adopted as the unit of energy by the International Conference on Weights and Measures. Therefore, the specific heat of water at 15°C is now given as 4185.5 joules per kilogram Celsius degree. This figure is always associated with the mechanical equivalent of heat—4.1855 joules per calorie.

AMPERE

The ampere is defined as the intensity of the constant electric current that, when maintained in two parallel conductors of infinite length and negligible cross section and placed 1 meter apart in a vacuum, would produce between them a force equal to 2×10^{-7} newton per meter length. In our treatment of electric current in this text, we have used the not-so-official but easier-to-comprehend definition of the ampere as being the rate of flow of 1 coulomb of charge per second, where 1 coulomb is the charge of 6.25×10^{18} electrons.

KELVIN

The fundamental unit of temperature is named after the scientist William Thomson, Lord Kelvin. The kelvin is defined to be 1/273.15 the thermodynamic temperature of the triple point of water (the fixed point at which ice, liquid water, and water vapor coexist in equilibrium). This definition was adopted in 1968 when it was decided to change the name *degree Kelvin* (°K) to *kelvin* (K). The temperature of melting ice at atmospheric pressure is 273.15 K. The temperature at which the vapor pressure of pure water is equal to standard atmospheric pressure is 373.15 K (the temperature of boiling water at standard atmospheric pressure).

AREA

The unit of area is a square that has a standard unit of length as a side. In the USCS, it is a square with sides that are each 1 foot in length, called 1 square foot and written 1 ft^2. In the international system, it is a square with sides that are 1 meter in length, which makes a unit of area of 1 m^2. In the cgs system it is 1 cm^2. The area of a given surface is specified by the number of square feet, square meters, or square centimeters that would fit into it. The area of a rectangle equals the base times the height. The area of a circle is equal to πr^2, where $\pi = 3.14$ and r is the radius of the circle. Formulas for the surface areas of other objects can be found in geometry textbooks.

FIGURE A.2

Unit square.

VOLUME

The volume of an object refers to the space it occupies. The unit of volume is the space taken up by a cube that has a standard unit of length for its edge. In the USCS, one unit of volume is the space occupied by a cube 1 foot on an edge and is called 1 cubic foot, written 1 ft^3. In the metric system it is the space occupied by a cube with sides of 1 meter (SI) or 1 centimeter (cgs). It is written 1 m^3 or 1 cm^3 (or cc). The volume of a given space is specified by the number of cubic feet, cubic meters, or cubic centimeters that will fill it.

In the USCS, volumes can also be measured in quarts, gallons, and cubic inches as well as in cubic feet. There are 1728 (12 × 12 × 12)

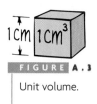

FIGURE A.3

Unit volume.

cubic inches in 1 ft^3. A U.S. gallon is a volume of 231 in^3. Four quarts equal 1 gallon. In the SI volumes are also measured in liters. A liter is equal to 1000 cm^3.

Unit Conversion

Often in science, and especially in a laboratory setting, it is necessary to convert from one unit to another. To do so, you need only multiply the given quantity by the appropriate *conversion factor.*

All conversion factors can be written as ratios in which the numerator and denominator represent the equivalent quantity expressed in different units. Because any quantity divided by itself is equal to 1, all conversion factors are equal to 1. For example, the following two conversion factors are both derived from the relationship 100 centimeters = 1 meter:

$$\frac{100 \text{ centimeters}}{1 \text{ meter}} = 1 \qquad \frac{1 \text{ meter}}{100 \text{ centimeters}} = 1$$

Because all conversion factors are equal to 1, multiplying a quantity by a conversion factor does not change the value of the quantity. What does change are the units. Suppose you measured an item to be 60 centimeters in length. You can convert this measurement to meters by multiplying it by the conversion factor that allows you to cancel centimeters.

EXAMPLE

Convert 60 centimeters to meters.

ANSWER

$$(60 \text{ \sout{centimeters}}) \frac{(1 \text{ meter})}{(100 \text{ \sout{centimeters}})} = 0.6 \text{ meter}$$

↑	↑	↑
quantity in centimeters	conversion factor	quantity in meters

To derive a conversion factor, consult a table that presents unit equalities, such as Table A.2 or on the inside cover of this book. Then multiply the given quantity by the conversion factor, and voilà, the units are converted. Always be careful to write down your units. They are your ultimate guide, telling you what numbers go where and whether you are setting up the equation properly.

STOP AND
CHECK YOURSELF

Multiply each physical quantity by the appropriate conversion factor to find its numerical value in the new unit indicated. You will need paper, pencil, a calculator, and a table of unit equalities.

a. 7320 grams to kilograms

b. 235 kilograms to pounds

c. 2.61 miles to kilometers

d. 100 calories to kilocalories

CHECK YOUR ANSWERS

a. 7.32 kg

b. 518 lb

c. 4.20 km

d. 0.1 kcal

APPENDIX B

LINEAR AND ROTATIONAL MOTION

When we describe the motion of something, we say how it moves relative to something else (Chapter 3). In other words, motion requires a reference frame (an observer, origin, and axes). We are free to choose this frame's location and to have it moving relative to another frame. When our frame of motion has zero acceleration, it is called an *inertial frame*. In an inertial frame, force causes an object to accelerate in accord with Newton's laws. When our frame of reference is accelerated, we observe fictitious forces and motions. Observations from a carousel, for example, are different when it is rotating and when it is at rest. Our description of motion and force depends on our "point of view."

We distinguish between *speed* and *velocity* (Chapter 3). Speed is how fast something moves, or the time rate of change of position (excluding direction): a *scalar* quantity. Velocity includes direction of motion: a *vector* quantity whose magnitude is speed. Objects moving at constant velocity move the same distance in the same time in the same direction.

Another distinction between speed and velocity has to do with the difference between distance and net distance, or *displacement*. Speed is *distance per duration* while velocity is *displacement per duration*. Displacement differs from distance. For example, a commuter who travels 10 kilometers to work and back travels 20 kilometers, but has "gone" nowhere. The distance traveled is 20 kilometers and the displacement is zero. Although the instantaneous speed and instantaneous velocity have the same value at the same instant, the average speed and average velocity can be very different. The average speed of this commuter's round-trip is 20 kilometers divided by the total commute time—a value greater than zero. But the average velocity is zero. In science, displacement is often more important than distance. (To avoid information overload, we have not treated this distinction in the text.)

Acceleration is the rate at which velocity changes. This can be a change in speed only, a change in direction only, or both.

Computing Velocity and Distance Traveled on an Inclined Plane

Recall, from Chapter 3, Galileo's experiments with inclined planes. We considered a plane tilted such that the speed of a rolling ball increases at the rate of 2 meters per second each second—an acceleration of 2 m/s^2.

So at the instant it starts moving its velocity is zero, and 1 second later it is rolling at 2 m/s, at the end of the next second 4 m/s, the end of the next second 6 m/s, and so on. The velocity of the ball at any instant is simply Velocity = acceleration × time. Or, in shorthand notation $v = at$. (It is customary to omit the multiplication sign, ×, when expressing relationships in mathematical form. When two symbols are written together, such as the *at* in this case, it is understood that they are multiplied.)

How fast the ball rolls is one thing; how *far* it rolls is another. To understand the relationship between acceleration and distance traveled, we must first investigate the relationship between instantaneous velocity and *average velocity*. If the ball shown in Figure B.1 starts from rest, it will roll a distance of 1 meter in the first second. What will be its average speed? The answer is 1 m/s (it covered 1 meter in the interval of 1 second). But we have seen that the *instantaneous velocity* at the end of the first second is 2 m/s. Since the acceleration is uniform, the average in any time interval is found the same way we usually find the average of any two numbers: add them and divide by 2. (Be careful not to do this when acceleration is not uniform!) So if we add the initial speed (zero in this case) and the final speed of 2 m/s and then divide by 2, we get 1 m/s for the average velocity.

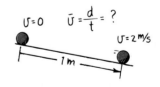

FIGURE B.1

The ball rolls 1 m down the incline in 1 s and reaches a speed of 2 m/s. Its average speed, however, is 1 m/s. Do you see why?

If the ball covers I m during its first second, then in each successive second it will cover the odd-numbered sequence of 3, 5, 7, 9 m, and so on. Note that the total distance covered increases as the square of the total time.

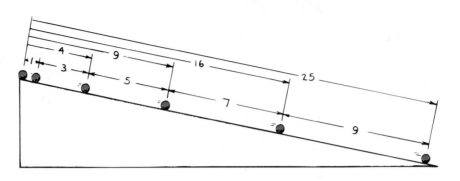

In each succeeding second, we see the ball roll a longer distance down the same slope in Figure B.2. Note the distance covered in the second time interval is 3 meters. This is because the average speed of the ball in this interval is 3 m/s. In the next 1-second interval, the average speed is 5 m/s, so the distance covered is 5 meters. It is interesting to see that successive increments of distance increase as a *sequence of odd numbers*. Nature clearly follows mathematical rules!

Investigate Figure B.2 carefully and note the *total* distance covered as the ball accelerates down the plane. The distances go from zero to 1 meter in 1 second, zero to 4 meters in 2 seconds, zero to 9 meters in 3 seconds, zero to 16 meters in 4 seconds, and so on in succeeding seconds. The sequence for *total distances* covered is of the *squares of the time*. We'll investigate the relationship between distance traveled and the square of the time for constant acceleration more closely in the case of free fall.

STOP AND CHECK YOURSELF

During the span of the second time interval, the ball begins at 2 m/s and ends at 4 m/s. What is the *average speed* of the ball during this 1-s interval? What is its *acceleration*?

CHECK YOUR ANSWERS

$$\text{Average speed} = \frac{\text{beginning} + \text{final speed}}{2}$$

$$= \frac{2\,\text{m/s} + 4\,\text{m/s}}{2} = 3\,\text{m/s}$$

$$\text{Acceleration} = \frac{\text{change in velocity}}{\text{time interval}}$$

$$= \frac{4\,\text{m/s} - 2\,\text{m/s}}{1\,\text{s}} = \frac{2\,\text{m/s}}{1\,\text{s}} = 2\,\text{m/s}^2$$

Computing Distance When Acceleration Is Constant

How far will an object released from rest fall in a given time? To answer this question, let us consider the case in which it falls freely for 3 seconds, starting at rest. Neglecting air resistance, the object will have a constant acceleration of about 10 meters per second each second (actually more like 9.8 m/s², but we want to make the numbers easier to follow).

Velocity at the *beginning* = 0 m/s

Velocity at the *end* of 3 seconds = (10 × 3) m/s

$$\textit{Average velocity} = \frac{1}{2} \text{ the sum of these two speeds}$$

$$= \frac{1}{2} \times (0 + 10 \times 3)\,\text{m/s}$$

$$= \frac{1}{2} \times 10 \times 3 = 15\,\text{m/s}$$

Distance traveled = average velocity × time

$$= \left(\frac{1}{2} \times 10 \times 3\right) \times 3$$

$$= \frac{1}{2} \times 10 \times 3^2 = 45\,\text{m}$$

We can see from the meanings of these numbers that

$$\text{Distance traveled} = \frac{1}{2} \times \text{acceleration} \times \text{square of time}$$

This equation is true for an object falling not only for 3 seconds but for any length of time, as long as the acceleration is constant. If we let d stand for the distance traveled, a for the acceleration, and t for the time, the rule may be written, in shorthand notation,

$$d = \frac{1}{2}at^2$$

This relationship was first deduced by Galileo. He reasoned that if an object falls for, say, twice the time, it will fall with *twice the average speed*. Since it falls for *twice* the time at *twice* the average speed, it will fall *four* times as far. Similarly, if an object falls for *three* times the time, it will have an average speed *three* times as great and will fall *nine* times as far. Galileo reasoned that the total distance fallen should be proportional to the *square* of the time.

In the case of objects in free fall, it is customary to use the letter *g* to represent the acceleration instead of the letter *a* (*g* because acceleration is due to *gravity*). While the value of *g* varies slightly in different parts of the world, it is approximately equal to 9.8 m/s² (32 ft/s²). If we use *g* for the acceleration of a freely falling object (negligible air resistance), the equations for falling objects starting from a rest position become

$$v = gt$$

$$d = \frac{1}{2}gt^2$$

Much of the difficulty in learning physics, like learning any discipline, has to do with learning the language—the many terms and definitions. Speed is somewhat different from velocity, and acceleration is vastly different from speed or velocity. Please be patient with yourself as you find learning the similarities and the differences among physics concepts is not an easy task.

When Chelcie Liu releases both balls simultaneously, he asks, "Which will reach the end of the equal-length tracks first?" (Hint: On which track is the average speed of the ball greater? Then, double hint: Which wins, the fast ball or the slow ball?)

STOP AND CHECK YOURSELF

1. An auto starting from rest has a constant acceleration of 4 m/s². How far will it go in 5 s?

2. How far will an object released from rest fall in 1 s? In this case the acceleration is $g = 9.8$ m/s².

3. If it takes 4 s for an object to freely fall to the water when released from the Golden Gate Bridge, how high is the bridge?

CHECK YOUR ANSWERS

1. Distance $= \frac{1}{2} \times 4\frac{m}{s^2} \times (5s)^2 = 50$ m

2. Distance $= \frac{1}{2} \times 9.8\frac{m}{s^2} \times (1s)^2 = 4.9$ m

3. Distance $= \frac{1}{2} \times 9.8\frac{m}{s^2} \times (4s)^2 = 78$ m

Notice that the units of measurement when multiplied give the proper units of meters for distance:

$$d = \frac{1}{2} \times 9.8\frac{m}{s^2} \times 16s^2 = 78\,m$$

Circular Motion

Linear speed is what we have been calling simply *speed*—the distance traveled in meters or kilometers per unit of time. A point on the perimeter of a merry-go-round or turntable moves a greater distance in one complete rotation than a point nearer the center. Moving a greater distance in the same time means a greater speed. The speed of something moving along a circular path is **tangential speed**, because the direction of motion is tangent to the circle.

The Physics Place
Rotational Motion

The Physics Place
Rotational Speed

Rotational speed (sometimes called angular speed) refers to the number of rotations or revolutions per unit of time. All parts of the rigid merry-go-round

When a phonograph record turns, a ladybug farther from the center travels a longer path in the same time and has a greater tangential speed.

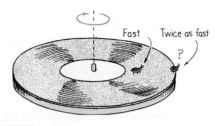

FIGURE B.5

The entire disk rotates at the same rotational speed, but ladybugs at different distances from the center travel at different tangential speeds. A lady-bug twice as far from the center moves twice as fast.

turn about the axis of rotation *in the same amount of time.* All parts share the same rate of rotation, or *number of rotations or revolutions per unit of time.* It is common to express rotational rates in revolutions per minute (rpm).* Phonograph records that were common a few years ago rotate at 33 1/3 rpm. A lady-bug sitting anywhere on the surface of the record revolves at 33 1/3 rpm.

Tangential speed is *directly proportional* to rotational speed (at a fixed radial distance). Unlike rotational speed, tangential speed depends on the distance from the axis (Figure B.5). Something at the center of a rotating platform has no tangential speed at all, and merely rotates. But, approaching the edge of the platform, tangential speed increases. Tangential speed is directly proportional to the distance from the axis (for a given rotational speed). Twice as far from the rotational axis, the speed is twice as great. Three times as far from the rotational axis, there is three times as much tangential speed. When a row of people locked arm in arm at the skating rink makes a turn, the motion of "tail-end Charlie" is evidence of this greater speed. So tangential speed is directly proportional both to rotational speed and to radial distance.**

---

* Physics types usually describe rotational speed in terms of the number of "radians" turned in a unit of time, for which they use the symbol ω (the Greek letter *omega*). There's a little more than 6 radians in a full rotation (2π radians, to be exact).

** When customary units are used for tangential speed v, rotational speed ω, and radial distance r, the direct proportion of v to both r and ω becomes the exact equation $v = r\omega$. So the tangential speed will be directly proportional to r when all parts of a system simultaneously have the same ω, as for a wheel, disk, or rigid wand. (The direct proportionality of v to r is not valid for the planets because planets don't all have the same ω.)

CHECK YOURSELF

On a rotating platform similar to the disk shown in Figure B.5, if you sit halfway between the rotating axis and the outer edge and have a rotational speed of 20 rpm and a tangential speed of 2 m/s, what will be the rotational and tangential speeds of your friend who sits at the outer edge?

CHECK YOUR ANSWER

Since the rotating platform is rigid, all parts have the same rotational speed, so your friend also rotates at 20 rpm. Tangential speed is a different story; since she is twice as far from the axis of rotation, she moves twice as fast—4 m/s.

Torque

Whereas force causes changes in speed, *torque* causes changes in rotation. To understand torque (rhymes with *dork*), hold the end of a meterstick horizontally with your hand. If you dangle a weight from the meterstick near your hand, you can fell the meterstick twist. Now if you slide the weight farther from your hand, the twist you feel is greater, although the weight is the same. The force acting on your hand is the same. What's different is the torque.

🎬 **The Physics Place**
Difference Between Torque and Weight

Torque = lever arm × force

Lever arm is the distance between the point of application of the force and the axis of rotation. It is the shortest distance between the applied force and the rotational axis. Torques are intuitively familiar to youngsters playing on a seesaw. Kids can balance a seesaw even when their weights are unequal. Weight alone doesn't produce rotation. Torque does, and children soon learn that the distance they sit from the pivot point is every bit as important as weight (Figure B.7). When the

FIGURE B.6

If you move the weight away from your hand, you will feel the difference between force and torque.

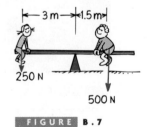

FIGURE B.7

No rotation is produced when the torques balance each other.

torques are equal, making the net torque zero, no rotation is produced.

Recall the equilibrium rule in Chapter 3—that the sum of the forces acting on a body or any system must equal zero for mechanical equilibrium. That is, $\Sigma F = 0$. We now see an additional condition. The *net torque* on a body or on a system must also be zero for mechanical equilibrium. Anything in mechanical equilibrium doesn't accelerate—neither linearly nor rotationally.

Suppose that the seesaw is arranged so that the half-as-heavy girl is suspended from a 4-meter rope hanging from her end of the seesaw (Figure B.8). She is now 5 meters from the fulcrum, and the seesaw is still balanced. We see that the lever-arm distance is 3 meters, not 5 meters. The lever arm about any axis of rotation is the perpendicular distance from the axis to the line along which the force acts. This will always be the shortest distance between the axis of rotation and the line along which the force acts.

This is why the stubborn bolt shown in Figure B.9 is turned more easily when the applied force is perpendicular to the handle, rather than at an oblique angle, as shown in the first figure. In the first figure, the lever arm is shown by the dashed line and is less than the length of the wrench handle. In the second figure, the lever arm is equal to the length of the wrench handle. In the third figure, the lever arm is extended with a pipe to provide more leverage and a greater torque.

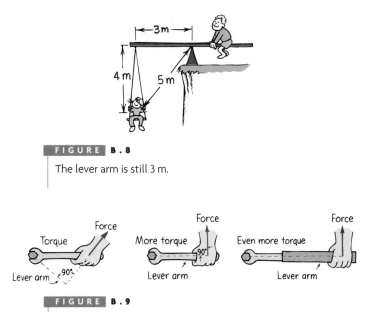

FIGURE B.8

The lever arm is still 3 m.

FIGURE B.9

Although the magnitudes of the force are the same in each case, the torques are different.

STOP AND CHECK YOURSELF

1. If a pipe effectively extends a wrench handle to three times its length, by how much will the torque increase for the same applied force?
2. Consider the balanced seesaw in Figure B.7. Suppose the girl on the left suddenly gains 50 N, such as by being handed a bag of apples. Where should she sit in order to balance, assuming the heavier boy remains in place?

CHECK YOUR ANSWERS

1. Three times more leverage for the same force gives three times more torque. (This method of increasing torque sometimes results in shearing off the bolt!)
2. She should sit $\frac{1}{2}$ m closer to the center. Then her lever arm is 2.5 m. This checks:

$$300 \text{ N} \times 2.5 \text{ m} = 500 \text{ N} \times 1.5 \text{ m}.$$

Angular Momentum

Things that rotate, whether a cylinder rolling down an incline or an acrobat doing a somersault, keep on rotating until something stops them. A rotating object has an "inertia of rotation." Recall, from Chapter 3, that all moving objects have "inertia of motion" or *momentum*—the product of mass and velocity. This kind of momentum is **linear momentum**. Similarly, the "inertia of rotation" of rotating objects is called **angular momentum**.

For the case of an object that is small compared with the radial distance to its axis of rotation, like a tetherball swinging from a long string or a planet orbiting around the sun, the angular momentum can be expressed as the magnitude of its linear momentum, mv, multiplied by the radial distance, γ (Figure B.10).* In shorthand notation, Angular momentum = mvr. Like linear momentum, angular momentum is a vector quantity and has direction as well as magnitude. In this appendix, we won't treat the vector nature of angular momentum (or even of torque, which also is a vector).

* For rotating bodies that are large compared with radial distance—for example, a planet rotating about its own axis—the concept of *rotational inertia* must be introduced. Then angular momentum is rotational inertia × rotational speed. See any of Hewitt's *Conceptual Physics* textbooks for more information.

FIGURE B.10

A small object of mass *m* whirling in a circular path of radius *r* with a speed *v* has angular momentum *mvr*.

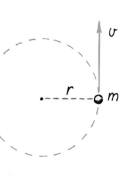

FIGURE B.11
INTERACTIVE FIGURE

Conservation of angular momentum. When the man pulls his arms and the whirling weights inward, he decreases the radial distance between the weights and the axis of rotation, and the rotational speed increases correspondingly.

Just as an external net force is required to change the linear momentum of an object, an external net torque is required to change the angular momentum of an object. We can state a rotational version of Newton's first law (the law of inertia):

An object or system of objects will maintain its angular momentum unless acted upon by an unbalanced external torque.

We see application of this rule when we look at a spinning top. If friction is low and torque also low, the top tends to remain spinning. The earth and planets spin in torque-free regions, and once they are spinning, they remain so.

Conservation of Angular Momentum

Just as the linear momentum of any system is conserved if no net forces are acting on the system, angular momentum is conserved if no net torque acts on the system. In the absence of an unbalanced external torque, the angular momentum of that system is constant. This means that its angular momentum at any one time will be the same as at any other time.

Conservation of angular momentum is shown in Figure B.11. The man stands on a low-friction turntable with weights extended. To simplify, consider only the weights in his hands. When he is slowly turning with his arms extended, much of the angular momentum is due to the distance between

The Physics Place

Conservation of Angular Momentum Using a Rotating Platform

the weights and the rotational axis. When he pulls the weights inward, the distance is considerably reduced. What is the result? His rotational speed increases!* This example is best appreciated by the turning person, who feels changes in rotational speed that seem to be mysterious. But it's straight physics! This procedure is used by a figure skater who starts to whirl with her arms and perhaps a leg extended and then draws her arms and leg in to obtain a greater rotational speed. Whenever a rotating body contracts, its rotational speed increases.

The law of angular momentum conservation is seen in the motions of the planets and the shape of the galaxies. When a slowly rotating ball of gas in space gravitationally contracts, the result is an increase in its rate of rotation. The conservation of angular momentum is far-reaching.

---

* When a direction is assigned to rotational speed, we call it *rotational velocity* (often called *angular velocity*). By convention, the rotational velocity vector and the angular momentum vector have the same direction and lie along the axis of rotation.

VECTORS

Vectors and Scalars

A *vector* quantity is a directed quantity—one that must be specified not only by magnitude (size) but

FIGURE C.1

by direction as well. Recall, from Chapter 4, that velocity is a vector quantity. Other examples are force, acceleration, and momentum. In contrast, a *scalar* quantity can be specified by magnitude alone. Some examples of scalar quantities are speed, time, temperature, and energy.

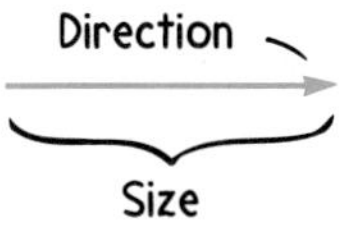

Vector quantities may be represented by arrows. The length of the arrow tells you the magnitude of the vector quantity, and the arrowhead tells you the direction of the vector quantity. Such an arrow drawn to scale and pointing appropriately is called a *vector*.

Adding Vectors

Vectors that add together are called *component vectors*. The sum of component vectors is called a *resultant*.

To add two vectors, make a parallelogram with two component vectors acting as two of the adjacent sides (Figure C.2). (Here our parallelogram is a rectangle.) Then draw a diagonal from the origin of the vector pair; this is the resultant (Figure C.3).

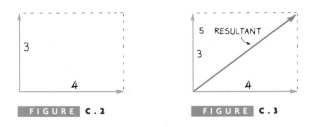

FIGURE C.2 **FIGURE C.3**

Caution: Do not try to mix vectors! We cannot add apples and oranges, so velocity vectors combine only with velocity vectors, force vectors combine only with force vectors, and acceleration vectors combine only with acceleration vectors—each on its own vector diagram. If you ever show different kinds of vectors on the same diagram, use different colors or some other method of distinguishing the different kinds of vectors.

Finding Components of Vectors

Recall, from Chapter 4, that to find a pair of perpendicular components for a vector, first draw a dashed line through the tail of the vectors (in the direction of one of the desired components). Second, draw another dashed line through the tail end of the vector at right angles to the first dashed line. Third, make a rectangle whose diagonal is the given vector. Draw in the two components. Here we let **F** stand for "total force," **U** stand for "upward force," and **S** stand for "sideways force."

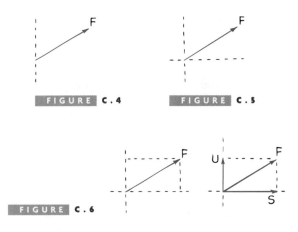

FIGURE C.4 **FIGURE C.5**

FIGURE C.6

Examples

1. Ernie Brown pushes a lawnmower and applies a force that pushes it forward and also against the ground. In Figure C.7, F represents the force applied by the man. We can separate this force into two components. The vector D represents the downward component, and S is the sideways component, the force that moves the lawnmower forward. If we know the magnitude and direction of the vector F, we can estimate the magnitude of the components from the vector diagram.

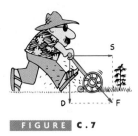

FIGURE C.7

2. Would it be easier to push or pull a wheelbarrow over a step? Figure C.8 shows the force at the wheel's center. When you push a wheelbarrow, part of the force is directed downward, which makes it harder to get over the step. When

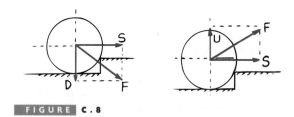

FIGURE C.8

you pull, however, part of the pulling force is directed upward, which helps to lift the wheel over the step. Note that the vector diagram suggests that pushing the wheelbarrow may not get it over the step at all. Do you see that the height of the step, the radius of the wheel, and the angle of the applied force determine whether the wheelbarrow can be pushed over the step? We see how vectors help us analyze a situation so that we can see just what the problem is!

3. If we consider the components of the weight of an object rolling down an incline, we can see why its speed depends on the angle. Note that the steeper the incline, the greater the component **S** becomes and the faster the object rolls. When the incline is vertical, **S** becomes equal to the weight, and the object attains maximum acceleration, 9.8 m/s^2. There are two more force vectors that are not shown: the normal force **N**, which is equal and oppositely directed to **D**, and the friction force **f**, acting at the barrel-plane contact.

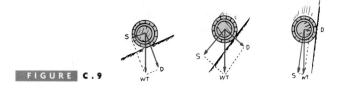

FIGURE C.9

4. When moving air strikes the underside of an airplane wing, the force of air impact against the wing may be represented by a single vector perpendicular to the plane of the wing (Figure C.10). We represent the force vector as acting midway along the lower wing surface, where the dot is, and pointing above the wing to show the direction of the resulting wind impact force. This force can be broken up into two components, one sideways and the other up. The upward component, U, is called *lift*. The sideways component, S, is called *drag*. If the aircraft is

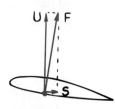

FIGURE C.10

to fly at constant velocity at constant altitude, then lift must equal the weight of the aircraft and the thrust of the plane's engines must equal drag. The magnitude of lift (and drag) can be altered by changing the speed of the airplane or by changing the angle (called *angle of attack*) between the wing and the horizontal.

5. Consider the satellite moving clockwise in Figure C.11. Everywhere in its orbital path, gravitational force **F** pulls it toward the center of the host planet. At position A we see **F** separated into two components: **f**, which is tangent to the path of the projectile, and **f′**, which is perpendicular to the path. The relative magnitudes of these components in comparison to the magnitude of **F** can be seen in the imaginary rectangle they compose; **f** and **f′** are the sides, and **F** is the diagonal. We see that

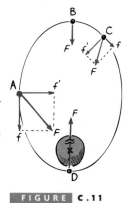

FIGURE C.11

component **f** is along the orbital path but against the direction of motion of the satellite. This force component reduces the speed of the satellite. The other component, **f′**, changes the direction of the satellite's motion and pulls it away from its tendency to go in a straight line. So the path of the satellite curves. The satellite loses speed until it reaches position B. At this farthest point from the planet (apogee), the gravitational force is somewhat weaker but perpendicular to the satellite's motion, and component **f** has reduced to zero. Component **f′**, on the other hand, has increased and is now fully merged to become **F**. Speed at this point is not enough for circular orbit, and the satellite begins to fall toward the planet. It picks up speed because the component **f** reappears and is in the direction of motion as shown in position C. The satellite picks up speed until it whips around to position D (perigee), where once again the direction of motion is perpendicular to the gravitational force, **f′** blends to full **F**, and **f** is nonexistent. The speed is in excess of that needed for circular orbit at this distance, and it overshoots to repeat the cycle. Its loss in speed in going from D to B equals its gain in speed from B to D. Kepler discovered that planetary paths are elliptical, but never knew why. Do you?

6. Refer to the Polaroids held by Ludmila back in Chapter 14, in Figure 12.46. In the first picture (a), we see that light is transmitted through the pair of Polaroids because their axes are aligned. The emerging light can be represented as a vector aligned with the polarization axes of the Polaroids. When the Polaroids are crossed (b), no light emerges because light passing through the first Polaroid is perpendicular to the polarization axes of the second Polaroid, with no components along its axis. In the third picture (c), we see that light is transmitted when a third Polaroid is sandwiched at an angle between the crossed Polaroids. The explanation for this is shown in Figure C.12.

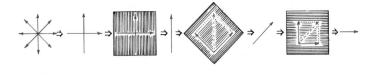

FIGURE C.12

Sailboats

Sailors have always known that a sailboat can sail downwind, in the direction of the wind. Sailors have not always known, however, that a sailboat can sail upwind, against the wind. One reason for this has to do with a feature that is common only to recent sailboats—a fin-like keel that extends deep beneath the bottom of the boat to ensure that the boat will knife through the water only in a forward (or backward) direction. Without a keel, a sailboat could be blown sideways.

Figure C.13 shows a sailboat sailing directly downwind. The force of wind impact against the sail accelerates the boat. Even if the drag of the water and all other resistance forces are negligible, the maximum speed of the boat is the wind speed. This is because the wind will not make impact against the sail if the boat is moving as fast as the wind. The wind would have no speed relative to the boat and the sail would simply sag. With no force, there is no acceleration. The force vector in Figure C.13 *decreases* as the boat travels faster. The force vector is maximum when the boat is at rest and the full impact of the wind fills the sail, and is minimum when the boat travels as fast as the wind. If

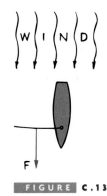

FIGURE C.13

the boat is somehow propelled to a speed faster than the wind (by way of a motor, for example), then air resistance against the front side of the sail will produce an oppositely directed force vector. This will slow the boat down. Hence, the boat when driven only by the wind cannot exceed wind speed.

If the sail is oriented at an angle, as shown in Figure C.14, the boat will move forward, but with less acceleration. There are two reasons for this:

1. The force on the sail is less because the sail does not intercept as much wind in this angular position.

2. The direction of the wind impact force on the sail is not in the direction of the boat's motion but is perpendicular to the surface of the sail. Generally speaking, whenever any fluid (liquid or gas) interacts with a smooth surface, the force of interaction is perpendicular to the smooth surface.* The boat does not move in the same direction as the perpendicular force on the sail, but is constrained to move in a forward (or backward) direction by its keel.

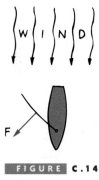

FIGURE C.14

We can better understand the motion of the boat by resolving the force of wind impact, **F**, into perpendicular components. The important component is that which is parallel to the keel, which we label **K**, and the other component is perpendicular to the keel, which we label **T**. It is the component **K**, as shown in Figure C.15, that is responsible for the forward motion of the boat. Component **T** is a useless force that tends to tip the boat

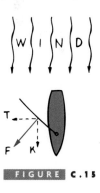

FIGURE C.15

* You can do a simple exercise to see that this is so. Try bouncing a coin off another on a smooth surface, as shown. Note that the struck

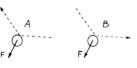

coin moves at right angles (perpendicular) to the contact edge. Note also that it makes no difference whether the projected coin moves along path A or path B. See your instructor for a more rigorous explanation, which involves momentum conservation.

over and move it sideways. This component force is offset by the deep keel. Again, maximum speed of the boat can be no greater than wind speed.

Many sailboats sailing in directions other than exactly downwind (Figure C.16) with their sails properly oriented can exceed wind speed. In the case of a sailboat cutting across the wind, the wind may continue to make impact with the sail even after the boat exceeds wind speed. A surfer, in a similar way, exceeds the velocity of the propelling wave by angling his surfboard across the wave. Greater angles to the propelling medium (wind for the boat, water wave for the surfboard) result in greater speeds. A sailcraft can sail faster cutting across the wind than it can sailing downwind.

FIGURE C.16

As strange as it may seem, maximum speed for most sailcraft is attained by cutting into (against) the wind, that is, by angling the sailcraft in a direction upwind! Although a sailboat cannot sail directly upwind, it can reach a destination upwind by angling back and forth in a zigzag fashion. This is called *tacking*. Suppose the boat and sail are as shown in Figure C.17. Component K will push the boat along in a forward direction, angling into the wind. In the position shown,

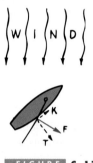

FIGURE C.17

the boat can sail faster than the speed of the wind. This is because as the boat travels faster, the impact of wind is increased. This is similar to running in a rain that comes down at an angle. When you run into the direction of the downpour, the drops strike you harder and more frequently, but when you run away from the direction of the downpour, the drops don't strike you as hard or as frequently. In the same way, a boat sailing upwind experiences greater wind impact force, while a boat sailing downwind experiences a decreased wind impact force. In any case, the boat reaches its terminal speed when opposing forces cancel the force of wind impact. The opposing forces consist mainly of water resistance against the hull of the boat. The hulls of racing boats are shaped to minimize this resistive force, which is the principal deterrent to high speeds.

Iceboats (sailcraft equipped with runners for traveling on ice) encounter no water resistance and can travel at several times the speed of the wind when they tack upwind. Although ice friction is nearly absent, an iceboat does not accelerate without limits. The terminal velocity of a sailcraft is determined not only by opposing friction forces but also by the change in relative wind direction. When the boat's orientation and speed are such that the wind seems to shift in direction, so the wind moves parallel to the sail rather than into it, forward acceleration ceases—at least in the case of a flat sail. In practice, sails are curved and produce an airfoil that is as important to sailcraft as it is to aircraft, as discussed in Chapter 7.

EXPONENTIAL GROWTH AND DOUBLING TIME*

One of the most important things we seem unable to perceive is the process of exponential growth. We think we understand how compound interest works, but we can't get it through our heads that a fine piece of tissue paper folded upon itself 50 times (if that were possible) would be more than 20 million kilometers thick. If we could, we could "see" why our income buys only half of what it did 4 years ago, why the price of everything has doubled in the same time, why populations and pollution proliferate out of control.**

When a quantity such as money in the bank, population, or the rate of consumption of a resource steadily grows at a fixed percent per year, we say the growth is exponential. Money in the bank may grow at 4 percent per year; electric power generating capacity in the United States grew at about 7 percent per year for the first three-quarters of the 20th century. The important thing about exponential growth is that the time required for the growing quantity to double in size (increase by 100 percent) is also constant. For example, if the population of a growing city takes 12 years to double from 10,000 to 20,000 inhabitants and its growth remains steady, in the next 12 years the population will double to 40,000, and in the next 10 years to 80,000, and so on.

There is an important relationship between the percent growth rate and its *doubling time*, the time it takes to double a quantity:[†]

$$\text{Doubling time} = \frac{69.3}{\text{percent growth per unit time}}$$

$$\approx \frac{70}{\%}$$

---

* This appendix is drawn from material by University of Colorado physics professor Albert A. Bartlett, who strongly asserts, "The greatest shortcoming of the human race is man's inability to understand the exponential function." See Professor Bartlett's still-timely article, "Forgotten Fundamentals in the Energy Crisis" (*American Journal of Physics*, September 1978) or his revised version (*Journal of Geological Education*, January 1980).

** K. C. Cole, *Sympathetic Vibrations* (New York: Morrow, 1984).

† For exponential decay we speak about half-life, the time required for a quantity to reduce to half its value. This case is treated in Chapter 16.

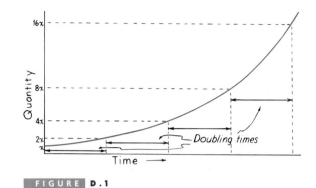

FIGURE D.1

An exponential curve. Notice that each of the successive equal time intervals noted on the horizontal scale corresponds to a doubling of the quantity indicated on the vertical scale. Such an interval is called the doubling time.

So to estimate the doubling time for a steadily growing quantity, we simply divide the number 70 by the percentage growth rate. For example, the 7 percent growth rate of electric power generating capacity in the United States means that in the past the capacity had doubled every 10 years [70%/(7%/year) = 10 years]. A 2 percent growth rate for world population means the population of the world doubles every 35 years [70%/(2%/year) = 35 years]. A city planning commission that accepts what seems like a modest 3.5 percent growth rate may not realize that this means that doubling will occur in 70/3.5 or 20 years; that's double capacity for such things as water supply, sewage-treatment plants, and other municipal services every 20 years.

What happens when you put steady growth in a finite environment? Consider the growth of bacteria that grow by division, so that one bacterium becomes two, the two divide to become four, the four divide to become eight, and so on. Suppose the division time for a certain strain of bacteria is 1 minute. This is then steady growth—the number of bacteria grows exponentially with a doubling time of 1 minute. Further, suppose that one bacterium is put in a bottle at 11:00 A.M. and that growth continues steadily until the bottle becomes full of bacteria at 12 noon. Consider seriously the following question.

FIGURE D.2

CHECK YOURSELF

When was the bottle half-full?

CHECK YOUR ANSWER

11:59 A.M.; the bacteria will double in number every minute!

It is startling to note that at 2 minutes before noon the bottle was only 1/4 full. Table D.1 summarizes the amount of space left in the bottle in the last few minutes before noon. If you were an average bacterium in the bottle, at which time would you first realize that you were running out of space? For example, would you sense there was a serious problem at 11:55 A.M., when the bottle was only 3% filled (1/32) and had 97% of open space (just yearning for development)? The point here is that there isn't much time between the moment that the effects of growth become noticeable and the time when they become overwhelming.

Suppose that at 11:58 A.M. some farsighted bacteria see that they are running out of space and launch a full-scale search for new bottles. Luckily, at 11:59 A.M. they discover three new empty bottles, three times as much space as they had ever known. This quadruples the total resource space ever known to the bacteria, for they now have a total of four bottles, whereas before the discovery they had only one. Further suppose that, thanks to their technological proficiency, they are able to migrate to their new habitats without difficulty. Surely, it seems to most of the bacteria that their problem is solved—and just in time.

CHECK YOURSELF

If the bacteria growth continues at the unchanged rate, what time will it be when the three new bottles are filled to capacity?

CHECK YOUR ANSWER

12:02 P.M.!

TABLE D.2

Effects of the Discovery of Three New Bottles

Time	Effect
11:58 A.M.	Bottle 1 is 1/4 full
11:59 A.M.	Bottle 1 is 1/2 full
12:00 noon	Bottle 1 is full
12:01 P.M.	Bottles 1 and 2 are both full
12:02 P.M.	Bottles 1, 2, 3, and 4 are all full

We see from Table D.2 that quadrupling the resource extends the life of the resource by only two doubling times. In our example the resource is space—but it could as well be coal, oil, uranium, or any nonrenewable resource.

Continued growth and continued doubling lead to enormous numbers. In two doubling times, a quantity will double twice ($2^2 = 4$; quadruple) in size; in three doubling times, its size will increase eightfold ($2^3 = 8$); in four doubling times, it will increase sixteenfold ($2^4 = 16$); and so on.

This is best illustrated by the story of the court mathematician in India who years ago invented the game of chess for his king. The king was so pleased with the game that he offered to repay the mathematician, whose request seemed modest enough. The mathematician requested a single grain of wheat on the first square of the chessboard, two grains on the second square, four on the third square, and so on, doubling the number of grains on each succeeding square until all squares had been used. At this rate there would be 2^{63} grains of wheat on the 64th square. The king soon saw that he could not fill this "modest" request, which amounted to more wheat than had been harvested in the entire history of the earth!

It is interesting and important to note that the number of grains on any square is one grain more than the total of all grains on the preceding squares. This is true anywhere on the board. Note from

---

TABLE D.1

The Last Minutes in the Bottle

Time	Part Full (%)	Part Empty
11:54 A.M.	1/64 (1.5%)	63/64
11:55 A.M.	1/32 (3%)	31/32
11:56 A.M.	1/16 (6%)	15/16
11:57 A.M.	1/8 (12%)	7/8
11:58 A.M.	1/4 (25%)	3/4
11:59 A.M.	1/2 (50%)	1/2
12:00 noon	full (100%)	none

FIGURE D.3

A single grain of wheat placed on the first square of the chessboard is doubled on the second square, this number is doubled on the third, and so on, presumably for all 64 squares. Note that each square contains one more grain than all the preceding squares combined. Does enough wheat exist in the world to fill all 64 squares in this manner?

TABLE D.3

Filling the Squares on the Chessboard

Square Number	Grains on Square	Total Grains Thus Far
1	1	1
2	2	3
3	4	7
4	8	15
5	16	31
6	32	63
7	64	127
.	.	.
.	.	.
.	.	.
64	2^{63}	$2^{64} - 1$

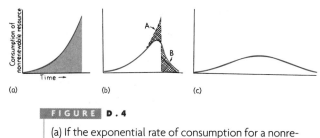

FIGURE D.4

(a) If the exponential rate of consumption for a nonrenewable resource continues until it is depleted, consumption falls abruptly to zero. The shaded area under this curve represents the total supply of the resource. (b) In practice, the rate of consumption levels off and then falls less abruptly to zero. Note that the crosshatched area A is equal to the crosshatched area B. Why? (c) At lower consumption rates, the same resource lasts a longer time.

Table D.3 that when eight grains are placed on the fourth square, the eight is one more than the total of seven grains that were already on the board. Or the 32 grains placed on the sixth square is one more than the total of 31 grains that were already on the board. We see that in one doubling time we use more than all that had been used in all the preceding growth!

So if we speak of doubling energy consumption in the next however many years, bear in mind that this means in these years we will consume more energy than has heretofore been consumed during the entire preceding period of steady growth. And if power generation continues to use predominantly fossil fuels, then except for some improvements in efficiency, we would burn up in the next doubling time a greater amount of coal, oil, and natural gas than has already been consumed by previous power generation, and except for improvements in pollution control, we can expect to discharge even more toxic wastes into the environment than the millions upon millions of tons already discharged over all the previous years of industrial civilization. We would also expect more human-made calories of heat to be absorbed by Earth's ecosystem than have been absorbed in the entire past! At the previous 7% annual growth rate in energy production, all this would occur in one doubling time of a single decade. If over the coming years the annual growth rate remains at half this value, 3.5 percent, then all this would take place in a doubling time of two decades. Clearly this cannot continue!

The consumption of a nonrenewable resource cannot grow exponentially for an indefinite period, because the resource is finite and its supply finally expires. The most drastic way this could happen is

shown in Figure D.4(a), where the rate of consumption, such as barrels of oil per year, is plotted against time, say in years. In such a graph the area under the curve represents the supply of the resource. We see that when the supply is exhausted, the consumption ceases altogether. This sudden change is rarely the case, for the rate of extracting the supply falls as it becomes more scarce. This is shown in Figure D.4(b). Note that the area under the curve is equal to the area under the curve in (a). Why? Because the total supply is the same in both cases. The principal difference is the time taken to finally extinguish the supply. History shows that the rate of production of a nonrenewable resource rises and falls in a nearly symmetric manner, as shown in (c). The time during which production rates rise is approximately equal to the time during which these rates fall to zero or near zero.

Production rates for all nonrenewable resources decrease sooner or later. Only production rates for renewable resources, such as agriculture or forest products, can be maintained at steady levels for long periods of time (Figure D.5), provided such production does not depend on waning nonrenewable resources such as petroleum. Much of today's

FIGURE D.5

A curve showing the rate of consumption of a renewable resource such as agricultural or forest products, where a steady rate of production and consumption can be maintained for a long period, provided this production is not dependent upon the use of a nonrenewable resource that is waning in supply.

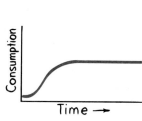

agriculture is so petroleum-dependent that it can be said that modern agriculture is simply the process whereby land is used to convert petroleum into food. The implications of petroleum scarcity go far beyond rationing of gasoline for cars or fuel oil for home heating.

The consequences of unchecked exponential growth are staggering. It is important to ask: Is growth really good? In answering this question, bear in mind that human growth is an early phase of life that continues normally through adolescence. Physical growth stops when physical maturity is reached. What do we say of growth that continues in the period of physical maturity? We say that such growth is obesity—or worse, cancer.

QUESTIONS TO PONDER

1. According to a French riddle, a lily pond starts with a single leaf. Each day the number of leaves doubles, until the pond is completely covered by leaves on the 30th day. On what day was the pond half covered? One-quarter covered?

2. In an economy that has a steady inflation rate of 7% per year, in how many years does a dollar lose half its value?

3. At a steady inflation rate of 7%, what will be the price every 10 years for the next 50 years for a theater ticket that now costs $20? For a coat that now costs $200? For a car that now costs $20,000? For a home that now costs $200,000?

4. If the sewage treatment plant of a city is just adequate for the city's current population, how many sewage treatment plants will be necessary 42 years later if the city grows steadily at 5% annually?

5. If world population doubles in 40 years and world food production also doubles in 40 years, how many people then will be starving each year compared to now?

6. Suppose you get a prospective employer to agree to hire your services for wages of a single penny for the first day, 2 pennies for the second day, and double each day thereafter providing the employer keeps to the agreement for a month. What will be your total wages for the month?

7. In the preceding exercise, how will your wages for only the 30th day compare to your total wages for the previous 29 days?

8. If fusion power were harnessed today, the abundant energy resulting would probably sustain and even further encourage our present appetite for continued growth and in a relatively few doubling times produce an appreciable fraction of the solar power input to the earth. Make an argument that the current delay in harnessing fusion is a blessing for the human race.

Some Significant Dates in the History of Physics

ca. 320 BC	**Aristotle** describes motion in terms of natural tendencies.
ca. 250 BC	**Archimedes** discovers the principle of buoyancy.
ca. AD 150	**Ptolemy** refines the Earth-centered system of the world.
1543	**Copernicus** publishes his Sun-centered system of the world.
1575–1596	**Brahe** measures precise positions of the planets in the sky.
1609	**Galileo** first uses a telescope as an astronomical tool.
1609/1619	**Kepler** publishes three laws of planetary motion.
1634	**Galileo** advances understanding of accelerated motion.
1661	**Boyle** relates pressure and volume of gases at constant temperature.
1676	**Roemer** demonstrates that light has finite speed.
1678	**Huygens** develops a wave theory of light.
1687	**Newton** presents the theory of mechanics in his *Principia*.
1738	**Bernoulli** explains the behavior of gases in terms of molecular motions.
1747	**Franklin** suggests the conservation of electrical "fire" (charge).
1780	**Galvani** discovers "animal electricity."
1785	**Coulomb** precisely determines the law of electric force.
1795	**Cavendish** measures the gravitational constant *G*.
1798	**Rumford** argues that heat is a form of motion.
1800	**Volta** invents the battery.
1802	**Young** uses wave theory to account for interference.
1811	**Avogadro** suggests that, at equal temperature and pressure, all gases have equal numbers of molecules per unit volume.
1815–1820	**Young** and others provide evidence for the wave nature of light.
1820	**Oersted** discovers the magnetic effect of an electric current.
1820	**Ampère** establishes the law of force between current-carrying wires.
1821	**Fraunhofer** invents the diffraction grating.
1824	**Carnot** states that heat cannot be transformed wholly to work.
1831	**Faraday** and **Henry** discover electromagnetic induction.
1842–1843	**Mayer** and **Joule** suggest a general law of energy conservation.
1846	**Adams** and **Leverrier** predict the existence of the planet Neptune.
1865	**Maxwell** gives the electromagnetic theory of light.
1869	**Mendeleev** organizes the elements into a periodic table.
1877	**Boltzmann** relates entropy to probability.
1885	**Balmer** finds numerical regularity in the spectrum of hydrogen.
1887	**Michelson** and **Morley** fail to detect the ether.
1888	**Hertz** generates and detects radio waves.
1895	**Roentgen** discovers X-rays.
1896	**Bequerel** discovers radioactivity.
1897	**Thomson** identifies cathode rays as negative corpuscles (electrons).
1900	**Planck** introduces the quantum idea.
1905	**Einstein** introduces the light corpuscle (photon) concept.
1905	**Einstein** advances the special theory of relativity.
1911	**Rutherford** reveals the nuclear atom.

1913	**Bohr** gives a quantum theory of the hydrogen atom.
1915	**Einstein** advances the general theory of relativity.
1923	**Compton's** experiments confirm the existence of the photon.
1924	**de Broglie** advances the wave theory of matter.
1925	**Goudsmit** and **Uhlenbeck** establish the spin of the electron.
1925	**Pauli** states the exclusion principle.
1926	**Schrödinger** develops the wave theory of quantum mechanics.
1927	**Davisson** and **Germer** and **Thomson** verify the wave nature of electrons.
1927	**Heisenberg** proposes the uncertainty principle.
1928	**Dirac** blends relativity and quantum mechanics in a theory of the electron.
1929	**Hubble** discovers the expanding universe.
1932	**Anderson** discovers antimatter in the form of the positron.
1932	**Chadwick** discovers the neutron.
1932	**Heisenberg** gives the neutron-proton explanation of nuclear structure.
1934	**Fermi** proposes a theory of the annihilation and creation of matter.
1938	**Meitner** and **Frisch** interpret results of **Hahn** and **Strassmann** as nuclear fission.
1939	**Bohr** and **Wheeler** give a detailed theory of nuclear fission.
1942	**Fermi** builds and operates the first nuclear reactor.
1945	**Oppenheimer's** Los Alamos team creates a nuclear explosion.
1947	**Bardeen, Brattain,** and **Shockley** develop the transistor.
1956	**Reines** and **Cowan** identify the antineutrino.
1957	**Feynman** and **Gell-Mann** account for all weak interactions with a "left-handed" neutrino.
1960	**Maiman** invents the laser.
1965	**Penzias** and **Wilson** discover background radiation in the universe left over from the Big Bang.
1967	**Bell** and **Hewish** discover pulsars, which are neutron stars.
1968	**Wheeler** names black holes.
1969	**Gell-Mann** suggests quarks as the building blocks of nucleons.
1977	**Lederman** and his team discover the bottom quark.
1981	**Binning** and **Rohrer** invent the scanning tunneling microscope.
1987	**Bednorz** and **Müller** discover high-temperature superconductivity.
1995	**Cornell** and **Wieman** create a "Bose–Einstein condensate" at 20 billionths of a degree.
2000	**Pogge** and **Martini** provide evidence for supermassive black holes in other galaxies.
2001	**David Smith** and colleagues create materials with negative index of refraction.
2002	**Rolf Landau** and colleagues create anti-hydrogen atoms.
2003	**Charles Bennett** and colleagues establish the age of the universe as 13.7 billion years and the fraction of all energy in the form of ordinary matter as only 4 percent.
2006	**Angelika Drees** and colleagues find evidence for a quark-gluon "liquid."

A (a) Abbreviation for *ampere*. (b) When in lowercase italic *a*, the symbol for *acceleration*.

aberration Distortion in an image produced by a lens or mirror, caused by limitations inherent to some degree in all optical systems. See *spherical aberration* and *chromatic aberration*.

absolute zero Lowest possible temperature that any substance can have; the temperature at which the atoms of a substance have their minimum kinetic energy. The temperature of absolute zero is $-273.15°C$, which is $-459.7°F$ and 0 kelvin.

absorption lines Dark lines that appear in an absorption spectrum. The pattern of lines is unique for each element.

absorption spectrum Continuous spectrum, like that generated by white light, interrupted by dark lines or bands that result from the absorption of light of certain frequencies by a substance through which the light passes.

ac Abbreviation for *alternating current*.

acceleration (a) Rate at which an object's velocity changes with time; the change in velocity may be in magnitude (speed), or direction, or both.

$$\text{acceleration} = \frac{\text{change of velocity}}{\text{time interval}}$$

acceleration due to gravity (g) Acceleration of a freely falling object. Its value near Earth's surface is about 9.8 meters per second each second.

achromatic lenses See *chromatic aberration*.

acoustics Study of the properties of sound, especially its transmission.

action force One of the pair of forces described in Newton's third law. See also *Newton's laws of motion, Law 3*.

additive primary colors Three colors of light—red, blue, and green—that when added in certain proportions will produce any color of the spectrum.

adhesion Molecular attraction between two surfaces making contact.

adiabatic Term applied to expansion or compression of a gas occurring without gain or loss of heat.

adiabatic process Process, often of fast expansion or compression, wherein no heat enters or leaves a system. As a result, a liquid or gas undergoing an expansion will cool, or undergoing a compression will warm.

air resistance Friction, or drag, that acts on something moving through air.

alchemist Practitioner of the early form of chemistry called alchemy, which was associated with magic. The goal of alchemy was to change base metals to gold and to discover a potion that could produce eternal youth.

alloy Solid mixture composed of two or more metals or of a metal and a nonmetal.

alpha particle Nucleus of a helium atom, which consists of two neutrons and two protons, ejected by certain radioactive nuclei.

alpha ray Stream of alpha particles (helium nuclei) ejected by certain radioactive nuclei.

alternating current (ac) Electric current that rapidly reverses in direction. The electric charges vibrate about relatively fixed positions, usually at the rate of 60 hertz.

AM Abbreviation for *amplitude modulation*.

ammeter A device that measures current. See *galvanometer*.

ampere (A) SI unit of electric current. One ampere is a flow of one coulomb of charge per second—6.25×10^{18} electrons (or protons) per second.

amplitude For a wave or vibration, the maximum displacement on either side of the equilibrium (midpoint) position.

amplitude modulation (AM) Type of modulation in which the amplitude of the carrier wave is varied above and below its normal value by an amount proportional to the amplitude of the impressed signal.

amu Abbreviation for *atomic mass unit*.

analog signal Signal based on a continuous variable, as opposed to a digital signal made up of discrete quantities.

aneroid barometer Instrument used to measure atmospheric pressure; based on the movement of the lid of a metal box, rather than on the movement of a liquid.

angle of incidence Angle between an incident ray and the normal to the surface it encounters.

angle of reflection Angle between a reflected ray and the normal to the surface of reflection.

angle of refraction Angle between a refracted ray and the normal to the surface at which it is refracted.

angular momentum Product of a body's rotational inertia and rotational velocity about a particular axis. For an object that is small compared with the radial distance, it is the product of mass, speed, and radial distance of rotation.

$$\text{angular momentum} = mvr$$

antimatter Matter composed of atoms with negative nuclei and positive electrons.

antinode Any part of a standing wave with maximum displacement and maximum energy.

antiparticle Particle having the same mass as a normal particle, but a charge of the opposite sign. The antiparticle of an electron is a positron.

antiproton Antiparticle of a proton; a negatively charged proton.

apogee Point in an elliptical orbit farthest from the focus around which orbiting takes place. See also *perigee*.

Archimedes' principle Relationship between buoyancy and displaced fluid: An immersed object is buoyed up by a force equal to the weight of the fluid it displaces.

armature Part of an electric motor or generator where an electromotive force is produced. Usually the rotating part.

astigmatism Defect of the eye caused when the cornea is curved more in one direction than in another.

atmospheric pressure Pressure exerted against bodies immersed in the atmosphere resulting from the weight of air pressing down from above. At sea level, atmospheric pressure is about 101 kPa.

atom Smallest particle of an element that has all the element's chemical properties. Consists of protons and neutrons in a nucleus surrounded by electrons.

atomic bonding Linking together of atoms to form larger structures, such as molecules and solids.

atomic mass number Number associated with an atom, equal to the number of nucleons (protons plus neutrons) in the nucleus.

atomic mass unit (amu) Standard unit of atomic mass. It is based on the mass of the common carbon atom, which is arbitrarily given the value of exactly 12. An amu of one is one-twelfth the mass of this common carbon atom.

atomic number Number associated with an atom, equal to the number of protons in the nucleus, or, equivalently, to the number of electrons in the electron cloud of a neutral atom.

average speed Path distance divided by time interval.

$$\text{average speed} = \frac{\text{total distance covered}}{\text{time interval}}$$

Avogadro's number 6.02×10^{23} molecules.

Avogadro's principle Equal volumes of all gases at the same temperature and pressure contain the same number of molecules, 6.02×10^{23} in one mole (a mass in grams equal to the molecular mass of the substance in atomic mass units).

axis (pl. axes) (a) Straight line about which rotation takes place. (b) Straight lines for reference in a graph, usually the x-axis for measuring horizontal displacement and the y-axis for measuring vertical displacement.

barometer Device used to measure the pressure of the atmosphere.

beats Sequence of alternating reinforcement and cancellation of two sets of superimposed waves differing in frequency, heard as a throbbing sound.

Bernoulli's principle Pressure in a fluid decreases as the speed of the fluid increases.

beta particle Electron (or positron) emitted during the radioactive decay of certain nuclei.

beta ray Stream of beta particles (electrons or positrons) emitted by certain radioactive nuclei.

Big Bang Primordial explosion that is thought to have resulted in the creation of our expanding universe.

bimetallic strip Two strips of different metals welded or riveted together. Because the two substances expand at different rates when heated or cooled, the strip bends; used in thermostats.

bioluminescence Light emitted from certain living things that have the ability to chemically excite molecules in their bodies; these excited molecules then give off visible light.

biomagnetism Magnetic material located in living organisms that may help them navigate, locate food, and affect other behaviors.

black hole Concentration of mass resulting from gravitational collapse, near which gravity is so intense that not even light can escape.

blind spot Area of the retina where all the nerves carrying visual information exit the eye and go to the brain; this is a region of no vision.

blue shift Increase in the measured frequency of light from an approaching source; called the blue shift because the apparent increase is toward the high-frequency, or blue, end of the color spectrum. Also occurs when an observer approaches a source. See also *Doppler effect*.

boiling Change from liquid to gas occurring beneath the surface of the liquid; rapid vaporization. The liquid loses energy, the gas gains it.

bow wave V-shaped wave produced by an object moving on a liquid surface faster than the wave speed.

Boyle's law The product of pressure and volume is a constant for a given mass of confined gas regardless of changes in either pressure or volume individually, as long as temperature remains unchanged.

$$P_1V_1 = P_2V_2$$

breeder reactor Nuclear fission reactor that not only produces power but produces more nuclear fuel than it consumes by converting a nonfissionable uranium isotope into a fissionable plutonium isotope. See also *nuclear reactor*.

British thermal unit (BTU) Amount of heat required to change the temperature of 1 pound of water by 1 Fahrenheit degree.

Brownian motion Haphazard movement of tiny particles suspended in a gas or liquid resulting from bombardment by the fast-moving molecules of the gas or liquid.

BTU Abbreviation for *British thermal unit*.

buoyancy Apparent loss of weight of an object immersed or submerged in a fluid.

buoyant force Net upward force exerted by a fluid on a submerged or immersed object.

C Abbreviation for *coulomb*.

cal Abbreviation for *calorie*.

calorie (cal) Unit of heat. One calorie is the heat required to raise the temperature of one gram of water

1 Celsius degree. One Calorie (with a capital *C*) is equal to one thousand calories and is the unit used in describing the energy available from food; also called a kilocalorie (kcal).

$$1 \text{ cal} = 4.18 \text{ J} \text{ or } 1 \text{ J} = 0.24 \text{ cal}$$

capacitor   Device used to store charge in a circuit.

carbon dating   Process of determining the time that has elapsed since death by measuring the radioactivity of the remaining carbon-14 isotopes.

carrier wave   High-frequency radio wave modified by a lower-frequency wave.

Celsius scale   Temperature scale that assigns 0 to the melt-freeze point for water and 100 to the boil-condense point of water at standard pressure (one atmosphere at sea level).

center of gravity (CG)   Point at the center of an object's weight distribution, where the force of gravity can be considered to act.

center of mass   Point at the center of an object's mass distribution, where all its mass can be considered to be concentrated. For everyday conditions, it is the same as the center of gravity.

centrifugal force   Apparent outward force on a rotating or revolving body.

centripetal force   Center-directed force that causes an object to follow a curved or circular path.

CG   Abbreviation for *center of gravity*.

chain reaction   Self-sustaining reaction that, once started, steadily provides the energy and matter necessary to continue the reaction.

charge   See *electric charge*.

charging by contact   Transfer of electric charge between objects by rubbing or simple touching.

charging by induction   Redistribution of electric charges in and on objects caused by the electrical influence of a charged object close by but not in contact.

chemical formula   Description that uses numbers and symbols of elements to describe the proportions of elements in a compound or reaction.

chemical reaction   Process of rearrangement of atoms that transforms one molecule into another.

chinook   Warm, dry wind that blows down from the eastern side of the Rocky Mountains across the Great Plains.

chromatic aberration   Distortion of an image caused when light of different colors (and thus different speeds and refractions) focuses at different points when passing through a lens. Achromatic lenses correct this defect by combining simple lenses made of different kinds of glass.

circuit   Any complete path along which electric charge can flow. See also *series circuit* and *parallel circuit*.

circuit breaker   Device in an electric circuit that breaks the circuit when the current gets high enough to risk causing a fire.

coherent light   Light of a single frequency with all photons exactly in phase and moving in the same direction. Lasers produce coherent light. See also *incoherent light* and *laser*.

complementarity   Principle enunciated by Niels Bohr stating that the wave and particle aspects of both matter and radiation are necessary, complementary parts of the whole. Which part is emphasized depends on what experiment is conducted (i.e., on what questions one puts to nature.)

complementary colors   Any two colors of light that, when added, produce white light.

component   Parts into which a vector can be separated and that act in different directions from the vector. See *resultant*.

compound   Chemical substance made of atoms of two or more different elements combined in a fixed proportion.

compression   (a) In mechanics, the act of squeezing material and reducing its volume. (b) In sound, the region of increased pressure in a longitudinal wave.

concave mirror   Mirror that curves inward like a "cave."

condensation   Change of phase of a gas into a liquid; the opposite of evaporation.

conduction   (a) In heat, energy transfer from particle to particle within certain materials, or from one material to another when the two are in direct contact. (b) In electricity, the flow of electric charge through a conductor.

conduction electrons   Electrons in a metal that move freely and carry electric charge.

conductor   (a) Material through which heat can be transferred. (b) Material, usually a metal, through which electric charge can flow. Good conductors of heat are generally good electric charge conductors.

cones   See *retina*.

conservation of angular momentum   When no external torque acts on an object or a system of objects, no change of angular momentum takes place. Hence, the angular momentum before an event involving only internal torques is equal to the angular momentum after the event.

conservation of charge   Principle that net electric charge is neither created nor destroyed but is transferable from one material to another.

conservation of energy   Principle that energy cannot be created or destroyed. It may be transformed from one form into another, but the total amount of energy never changes.

conservation of energy for machines   Work output of any machine cannot exceed the work input.

conservation of momentum   In the absence of a net external force, the momentum of an object or system of objects is unchanged.

$$mv_{\text{(before event)}} = mv_{\text{(after event)}}$$

conserved Term applied to a physical quantity, such as momentum, energy, or electric charge, that remains unchanged during interactions.

constructive interference Combination of waves so that two or more waves overlap to produce a resulting wave of increased amplitude. See also *interference*.

convection Means of heat transfer by movement of the heated substance itself, such as by currents in a fluid.

converging lens Lens that is thicker in the middle than at the edges and refracts parallel rays of light passing through it to a focus. See also *diverging lens*.

convex mirror Mirror that curves outward. The virtual image formed is smaller and closer to the mirror than the object. See also *concave mirror*.

cornea Transparent covering over the eyeball, which helps focus the incoming light.

correspondence principle If a new theory is valid, it must account for the verified results of the old theory in the region where both theories apply.

cosmic ray One of various high-speed particles that travel throughout the universe and originate in violent events in stars.

cosmology Study of the origin and development of the entire universe.

coulomb (C) SI unit of electrical charge. One coulomb is equal to the total charge of 6.25×10^{18} electrons.

Coulomb's law Relationship among electrical force, charges, and distance: The electrical force between two charges varies directly as the product of the charges (q) and inversely as the square of the distance between them. (k is the proportionality constant 9×10^9 N·m^2/C^2) If the charges are alike in sign, the force is repulsive; if the charges are unlike, the force is attractive.

$$F = k\frac{q_1 q_2}{d^2}$$

crest One of the places in a wave where the wave is highest or the disturbance is greatest in the opposite direction from a trough. See also *trough*.

critical angle Minimum angle of incidence for which a light ray is totally reflected within a medium.

critical mass Minimum mass of fissionable material in a nuclear reactor or nuclear bomb that will sustain a chain reaction. A subcritical mass is one in which the chain reaction dies out. A supercritical mass is one in which the chain reaction builds up explosively.

crystal Regular geometric shape found in a solid in which the component particles are arranged in an orderly, three-dimensional, repeating pattern.

current See *electric current*.

cyclotron Particle accelerator that imparts high energy to charged particles such as protons, deuterons, and helium ions.

dark matter Unseen and unidentified matter that is evident by its gravitational pull on stars in the galaxies—comprising perhaps 90% of the matter of the universe.

dc Abbreviation for *direct current*.

de Broglie matter waves All particles have wave properties; in de Broglie's equation, the product of momentum and wavelength equals Planck's constant.

de-excitation See *excitation*.

density Mass of a substance per unit volume. Weight density is weight per unit volume. In general, any item per space element (e.g., number of dots per area).

$$\text{density} = \frac{\text{mass}}{\text{volume}}$$

$$\text{weight density} = \frac{\text{weight}}{\text{volume}}$$

destructive interference Combination of waves so that crest parts of one wave overlap trough parts of another, resulting in a wave of decreased amplitude. See also *interference*.

deuterium Isotope of hydrogen whose atom has a proton, a neutron, and an electron. The common isotope of hydrogen has only a proton and an electron; therefore, deuterium has more mass.

deuteron Nucleus of a deuterium atom; it has one proton and one neutron.

dichroic crystal Crystal that divides unpolarized light into two internal beams polarized at right angles and strongly absorbs one beam while transmitting the other.

diffraction Bending of light that passes around an obstacle or through a narrow slit, causing the light to spread and to produce light and dark fringes.

diffraction grating Series of closely spaced parallel slits or grooves that are used to separate colors of light by interference.

diffuse reflection Reflection of waves in many directions from a rough surface. See also *polished*.

digital audio Audio reproduction system that uses binary code to record and reproduce sound.

digital signal Signal made up of discrete quantities or signals, as opposed to an analog signal which is based on a continuous signal.

diode Electronic device that restricts current to a single direction in an electric circuit; a device that changes alternating current to direct current.

dipole See *electric dipole*.

direct current (dc) Electric current whose flow of charge is always in one direction.

dispersion Separation of light into colors arranged according to their frequency, for example by interaction with a prism or a diffraction grating.

diverging lens Lens that is thinner in the middle than at the edges, causing parallel rays of light passing through it to diverge as if from a point. See also *converging lens*.

Doppler effect   Change in frequency of a wave of sound or light due to the motion of the source or the receiver. See also *red shift* and *blue shift*.

echo   Reflection of sound.

eddy   Changing, curling paths in turbulent flow of a fluid.

efficiency   In a machine, the ratio of useful energy output to total energy input, or the percentage of the work input that is converted to work output.

$$\text{Efficiency} = \frac{\text{useful energy output}}{\text{total energy input}}$$

elastic collision   Collision in which colliding objects rebound without lasting deformation or heat generation.

elastic limit   Distance of stretching or compressing beyond which an elastic material will not return to its original state.

elasticity   Property of a solid wherein a change in shape is experienced when a deforming force acts on it, with a return to its original shape when the deforming force is removed.

electric charge   Fundamental electrical property to which the mutual attractions or repulsions between electrons or protons are attributed.

electric current   Flow of electric charge that transports energy from one place to another. Measured in amperes, where one ampere is the flow of 6.25×10^{18} electrons (or protons) per second.

electric dipole   Molecule in which the distribution of charge is uneven, resulting in slightly opposite charges on opposite sides of the molecule.

electric field   Force field that fills the space around every electric charge or group of charges. Measured by force per charge (newtons/coulomb).

electric potential   Electric potential energy (in joules) per unit of charge (in coulombs) at a location in an electric field; measured in volts and often called voltage.

$$\text{voltage} = \frac{\text{electrical energy}}{\text{charge}} = \frac{\text{joules}}{\text{coulomb}}$$

electric potential energy   Energy that a charge has due to its location in an electric field.

electric power   Rate of electrical energy transfer or the rate of doing work, which can be measured by the product of current and voltage.

$$\text{power} = \text{current} \times \text{voltage}$$

electrical force   Force that one charge exerts on another. When the charges are the same sign, they repel; when the charges are opposite, they attract.

electrical resistance   Resistance of a material to the flow of electric charge through it; measured in ohms (symbol Ω).

electrically polarized   Term applied to an atom or molecule in which the charges are aligned so that one side is slightly more positive or negative than the opposite side.

electricity   General term for electrical phenomena, much like gravity has to do with gravitational phenomena, or sociology with social phenomena.

electrode   Terminal, for example of a battery, through which electric current can pass.

electromagnet   Magnet whose magnetic properties are produced by electric current.

electromagnetic induction   Phenomenon of inducing a voltage in a conductor by changing the magnetic field near the conductor. If the magnetic field within a closed loop changes in any way, a voltage is induced in the loop. The induction of voltage is actually the result of a more fundamental phenomenon: the induction of an electric field. See also *Faraday's law*.

electromagnetic radiation   Transfer of energy by the rapid oscillations of electromagnetic fields, which travel in the form of waves called electromagnetic waves.

electromagnetic spectrum   Range of frequencies over which electromagnetic radiation can be propagated. The lowest frequencies are associated with radio waves; microwaves have a higher frequency, and then infrared waves, light, ultraviolet radiation, X-rays, and gamma rays in sequence.

electromagnetic wave   Energy-carrying wave emitted by vibrating charges (often electrons) that is composed of oscillating electric and magnetic fields that regenerate one another. Radio waves, microwaves, infrared radiation, visible light, ultraviolet radiation, X-rays, and gamma rays are all composed of electromagnetic waves.

electromotive force (emf)   Any voltage that gives rise to an electric current. A battery or a generator is a source of emf.

electron   Negative particle in the shell of an atom.

electron volt (eV)   Amount of energy equal to that an electron acquires in accelerating through a potential difference of 1 volt.

electrostatics   Study of electric charges at rest, as opposed to electrodynamics.

element   Substance composed of atoms that all have the same atomic number and, therefore, the same chemical properties.

elementary particles   Subatomic particles. The basic building blocks of all matter, consisting of two classes of particles, the quarks and the leptons.

ellipse   Closed curve of oval shape wherein the sum of the distances from any point on the curve to two internal focal points is a constant.

emf   Abbreviation for *electromotive force*.

emission spectrum   Distribution of wavelengths in the light from a luminous source.

energy   Commonly defined as the ability to do work, it comes in many forms and is conserved (its total amount never changes). It is *not* a material substance.

entropy   A measure of the disorder of a system. Whenever energy freely transforms from one form to another, the direction of transformation is toward a state of greater disorder and therefore toward one of greater entropy.

equilibrium   In general, a state of balance. For mechanical equilibrium, the state in which no net forces and no net torques act. In liquids, the state in which evaporation equals condensation. More generally, the state in which no net change of energy occurs.

equilibrium rule $\sum F = 0$. On an object or system of objects in mechanical equilibrium, the sum of forces equals zero. Also, $\sum \tau = 0$; the sum of the torques equal zero.

escape velocity   Velocity that a projectile, space probe, etc., must reach to escape the gravitational influence of Earth or a celestial body to which it is attracted.

ether   Hypothetical invisible medium that was formerly thought to be required for the propagation of electromagnetic waves, and thought to fill space throughout the universe.

eV   Abbreviation for *electron volt*.

evaporation   Change of phase from liquid to gas that takes place at the surface of a liquid. The opposite of condensation.

excitation   Process of boosting one or more electrons in an atom or molecule from a lower to a higher energy level. An atom in an excited state will usually decay (de-excite) rapidly to a lower state by the emission of radiation. The frequency and energy of emitted radiation are related by

$$E = hf$$

excited   See *excitation*.

eyepiece   Lens of a telescope closest to the eye; which enlarges the real image formed by the first lens.

fact   Close agreement by competent observers of a series of observations of the same phenomena.

Fahrenheit scale   Temperature scale in common use in the United States. The number 32 is assigned to the melt-freeze point of water, and the number 212 to the boil-condense point of water at standard pressure (1 atmosphere, at sea level).

Faraday's law   Induced voltage in a coil is proportional to the product of the number of loops and the rate at which the magnetic field changes within those loops. In general, an electric field is induced in any region of space in which a magnetic field is changing with time. The magnitude of the induced electric field is proportional to the rate at which the magnetic field changes. See also *Maxwell's counterpart to Faraday's law*.

voltage induced $\sim$ number of loops
$$\times \frac{\text{magnetic field change}}{\text{change in time}}$$

field   See *force field*.

field lines   See *magnetic field lines*.

flotation   See *principle of flotation*.

fluid   Anything that flows; in particular, any liquid or gas.

fluorescence   Property of certain substances to absorb radiation of one frequency and to re-emit radiation of a lower frequency.

FM   Abbreviation for *frequency modulation*.

focal length   Distance between the center of a lens and either focal point; the distance from a mirror to its focal point.

focal plane   Plane, perpendicular to the principal axis, that passes through a focal point of a lens or mirror. For a converging lens or a concave mirror, any incident parallel rays of light converge to a point somewhere on a focal plane. For a diverging lens or a convex mirror, the rays appear to come from a point on a focal plane.

focal point   For a converging lens or a concave mirror, the point at which rays of light parallel to the principal axis converge. For a diverging lens or a convex mirror, the point from which such rays appear to come.

focus   (pl. foci) (a) For an ellipse, one of the two points for which the sum of the distances to any point on the ellipse is a constant. A satellite orbiting Earth moves in an ellipse that has Earth at one focus. (b) For optics, a focal point.

force   Any influence that tends to accelerate an object; a push or pull; measured in newtons. Force is a vector quantity.

force field   That which exists in the space surrounding a mass, electric charge, or magnet, so that another mass, electric charge, or magnet introduced into this region will experience a force. Examples of force fields are gravitational fields, electric fields, and magnetic fields.

forced vibration   Vibration of an object caused by the vibrations of a nearby object. The sounding board in a musical instrument amplifies the sound through forced vibration.

Fourier analysis   Mathematical method that disassembles any periodic wave form into a combination of simple sine waves.

fovea   Area of the retina that is in the center of the field of view; region of most distinct vision.

frame of reference   Vantage point (usually a set of coordinate axes) with respect to which position and motion may be described.

Fraunhofer lines   Dark lines visible in the spectrum of the Sun or a star.

free fall   Motion under the influence of gravity only.

free radical   Unbonded, electrically neutral, very chemically active atom or molecular fragment.

freezing   Change in phase from liquid to solid; the opposite of melting.

frequency   For a vibrating body or medium, the number of vibrations per unit time. For a wave, the number of crests that pass a particular point per unit time. Frequency is measured in hertz.

frequency modulation (FM)   Type of modulation in which the frequency of the carrier wave is varied above and below its normal frequency by an amount that is proportional to the amplitude of the impressed signal. In this case, the amplitude of the modulated carrier wave remains constant.

friction   Force that acts to resist the relative motion (or attempted motion) of objects or materials that are in contact.

fuel cell   A device that converts chemical energy to electrical energy, but unlike a battery is continually fed with fuel, usually hydrogen.

fulcrum   Pivot point of a lever.

fundamental frequency   See *partial tone.*

fuse   Device in an electric circuit that breaks the circuit when the current gets high enough to risk causing a fire.

g   (a) Abbreviation for *gram.* (b) When in lowercase italic *g,* the symbol for the acceleration due to gravity (at Earth's surface, 9.8 m/s^2). (c) When in lowercase bold **g** the gravitational field vector (at Earth's surface, 9.8 N/kg). (d) When in uppercase italic *G,* the symbol for the *universal gravitation constant* (6.67×10^{-11} N · m^2/kg^2).

galvanometer   Instrument used to detect electric current. With the proper combination of resistors, it can be converted to an ammeter or a voltmeter. An ammeter is calibrated to measure electric current. A voltmeter is calibrated to measure electric potential.

gamma ray   High-frequency electromagnetic radiation emitted by atomic nuclei.

gas   Phase of matter beyond the liquid phase, wherein molecules fill whatever space is available to them, taking no definite shape.

general theory of relativity   Einstein's generalization of special relativity, which deals with accelerated motion and features a geometric theory of gravitation.

generator   Machine that produces electric current, usually by rotating a coil within a stationary magnetic field.

geodesic   Shortest path between points on any surface.

geosynchronous orbit   A satellite orbit in which the satellite orbits Earth once each day. When moving westward, the satellite remains at a fixed point (about 42,000 km) above Earth's surface.

global warming   See *greenhouse effect.*

gram (g)   A metric unit of mass. It is one thousandth of a kilogram.

gravitation   Attraction between objects due to mass. See also *law of universal gravitation* and *universal gravitational constant.*

gravitational field   Force field that exists in the space around every mass or group of masses in which other bodies experience gravitational attraction; measured in newtons per kilogram.

gravitational potential energy   Energy that a body possesses because of its position in a gravitational field. On Earth, potential energy (PE) equals mass (*m*) times the acceleration due to gravity (*g*) times height (*h*) from a reference level such as Earth's surface.

$$PE = mgh$$

gravitational red shift   Shift in wavelength toward the red end of the spectrum experienced by light leaving the surface of a massive object, as predicted by the general theory of relativity.

gravitational wave   Gravitational disturbance that propagates through spacetime made by a moving mass. (Undetected at this writing.)

greenhouse effect   Warming effect caused by short-wavelength radiant energy from the Sun that easily enters the atmosphere and is absorbed by Earth, but when radiated at longer wavelengths cannot easily escape Earth's atmosphere.

grounding   Allowing charges to move freely along a connection from a conductor to the ground.

group   Elements in the same column of the periodic table.

h   (a) Abbreviation for hour (though hr. is often used). (b) When in italic *h,* the symbol for *Planck's constant.*

hadron   Elementary particle that can participate in strong nuclear force interactions.

half-life   Time required for half the atoms of a radioactive isotope of an element to decay. This term is also used to describe decay processes in general.

harmonic   See *partial tone.*

heat   The energy that flows from one object to another by virtue of a difference in temperature. Measured in *calories* or *joules.*

heat capacity   See *specific heat capacity.*

heat engine   A device that uses heat as input and supplies mechanical work as output, or that uses work as input and moves heat "uphill" from a cooler to a warmer place.

heat of fusion   Amount of energy that must be added to a kilogram of a solid (already at its melting point) to melt it.

heat of vaporization   Amount of energy that must be added to a kilogram of a liquid (already at its boiling point) to vaporize it.

heat pump   A device that transfers heat out of a cool environment and into a warm environment.

heat waves   See *infrared waves.*

heavy water   Water (H_2O) that contains the heavy hydrogen isotope deuterium.

hertz (Hz)   SI unit of frequency. One hertz is one vibration per second.

hologram   Two-dimensional microscopic interference pattern that shows three-dimensional optical images.

Hooke's law   Distance of stretch or squeeze (extension or compression) of an elastic material is directly proportional to the applied force. Where *x* is the change in length and *k* is the spring constant,

$$F = k\Delta x$$

humidity   Measure of the amount of water vapor in the air. Absolute humidity is the mass of water per volume of air. Relative humidity is absolute humidity at that temperature divided by the maximum possible, usually given as a percent.

Huygens' principle   Light waves spreading out from a light source can be regarded as a superposition of tiny secondary wavelets.

hypothesis   Educated guess; a reasonable explanation of an observation or experimental result that is not fully accepted as factual until tested over and over again by experiment.

Hz   Abbreviation for *hertz.*

ideal efficiency   Upper limit of efficiency for all heat engines; it depends on the temperature difference between input and exhaust.

$$\text{ideal efficiency} = \frac{T_{\text{hot}} T_{\text{cold}}}{T_{\text{hot}}}$$

impulse   Product of force and the time interval during which the force acts. Impulse produces change in momentum.

$$\text{impulse} = Ft = \Delta mv$$

incandescence   State of glowing while at a high temperature, caused by electrons bouncing around over dimensions larger than the size of an atom, emitting radiant energy in the process. The peak frequency of radiant energy is proportional to the absolute temperature of the heated substance:

$$\bar{f} \sim T$$

incoherent light   Light containing waves with a jumble of frequencies, phases, and possibly directions. See also *coherent light* and *laser.*

index of refraction (n)   Ratio of the speed of light in a vacuum to the speed of light in another material.

$$n = \frac{\text{speed of light in vaccum}}{\text{speed of light in material}}$$

induced   (a) Term applied to electric charge that has been redistributed on an object due to the presence of a charged object nearby. (b) Term applied to a voltage, electric field, or magnetic field that is created due to a change in or motion through a magnetic field or electric field.

induction   Charging of an object without direct contact. See also *electromagnetic induction.*

inelastic   Term applied to a material that does not return to its original shape after it has been stretched or compressed.

inelastic collision   Collision in which the colliding objects become distorted and/or generate heat during the collision, and possibly stick together.

inertia   Sluggishness or apparent resistance of an object to change its state of motion. Mass is the measure of inertia.

inertial frame of reference   Unaccelerated vantage point in which Newton's laws hold exactly.

infrared   Electromagnetic waves of frequencies lower than the red of visible light.

infrared waves   Electromagnetic waves that have a lower frequency than visible red light.

infrasonic   Term applied to sound frequencies below 20 hertz, the normal lower limit of human hearing.

in parallel   Term applied to portions of an electric circuit that are connected at two points and provide alternative paths for the current between those two points.

in phase   Term applied to two or more waves whose crests (and troughs) arrive at a place at the same time, so that their effects reinforce each other.

in series   Term applied to portions of an electric circuit that are connected in a row so that the current that goes through one must go through all of them.

instantaneous speed   Speed at any instant.

insulator   (a) Material that is a poor conductor of heat and that delays the transfer of heat. (b) Material that is a poor conductor of electricity.

intensity   Power per square meter carried by a sound wave, often measured in decibels.

interaction   Mutual action between objects where each object exerts an equal and opposite force on the other.

interference   Result of superposing different waves, often of the same wavelength. Constructive interference results from crest-to-crest reinforcement; destructive interference results from crest-to-trough cancellation. The interference of selected wavelengths of light produces colors known as interference colors. See also *constructive interference, destructive interference, interference pattern,* and *standing wave.*

interference pattern   Pattern formed by the overlapping of two or more waves that arrive in a region at the same time.

internal energy   The total energy stored in the atoms and molecules within a substance. Changes in internal energy are of principal concern in thermodynamics.

inverse-square law   Law relating the intensity of an effect to the inverse square of the distance from the cause. Gravity, electric, magnetic, light, sound, and radiation phenomena follow the inverse-square law.

$$\text{intensity} \sim \frac{1}{\text{distance}^2}$$

inversely   When two values change in opposite directions, so that if one increases and the other decreases by the same amount, they are said to be inversely proportional to each other.

ion   Atom (or group of atoms bound together) with a net electric charge, which is due to the loss or gain of electrons. A positive ion has a net positive charge. A negative ion has a net negative charge.

ionization   Process of adding or removing electrons to or from the atomic nucleus.

iridescence   Phenomenon whereby interference of light waves of mixed frequencies reflected from the top and bottom of thin films produces an assortment of colors.

iris   Colored part of the eye that surrounds the black opening through which light passes. The iris regulates the amount of light entering the eye.

isotopes   Atoms whose nuclei have the same number of protons but different numbers of neutrons.

J   Abbreviation for *joule.*

joule (J)   SI unit of work and of all other forms of energy. One joule of work is done when a force of one newton is exerted on an object moved one meter in the direction of the force.

K   (a) Abbreviation for *kelvin.* (b) When in lowercase k, the abbreviation for the prefix *kilo-.* (c) When in lowercase italic *k,* the symbol for the electrical proportionality constant in *Coulomb's law.* It is approximately $9 \times 10^9 \, \text{N} \cdot \text{m}^2/\text{C}^2$.(d) When in lowercase italic *k,* the symbol for the spring constant in *Hooke's law.*

kcal   Abbreviation for *kilocalorie.*

KE   Abbreviation for *kinetic energy.*

kelvin   SI unit of temperature. A temperature measured in kelvins (symbol K) indicates the number of units above absolute zero. Divisions on the Kelvin scale and Celsius scale are the same size, so a change in temperature of one kelvin equals a change in temperature of one Celsius degree.

Kelvin scale   Temperature scale, measured in kelvins K, whose zero (called absolute zero) is the temperature at which it is impossible to extract any more internal energy from a material. $0 \, \text{K} = -273.15°\text{C}$. There are no negative temperatures on the Kelvin scale.

Kepler's laws
Law 1: The path of each planet around the Sun is an ellipse with the Sun at one focus.

Law 2: The line from the Sun to any planet sweeps out equal areas of space in equal time intervals.

Law 3: The square of the orbital period of a planet is directly proportional to the cube of the average distance of the planet from the Sun ($T^2 \sim r^3$ for all planets).

kg   Abbreviation for *kilogram.*

kilo-   Prefix that means thousand, as in kilowatt or kilogram.

kilocalorie (kcal)   Unit of heat. One kilocalorie equals 1000 calories, or the amount of heat required to raise the temperature of one kilogram of water by 1°C. Equal to one food Calorie.

kilogram (kg)   Fundamental SI unit of mass. It is equal to 1000 grams. One kilogram is very nearly the amount of mass in one liter of water at 4°C.

kilometer (km)   One thousand *meters.*

kilowatt (kW)   One thousand *watts.*

kilowatt-hour (kWh)   Amount of energy consumed in 1 hour at the rate of 1 kilowatt.

kinetic energy (KE)   Energy of motion, equal (nonrelativistically) to half the mass multiplied by the speed squared.

$$\text{KE} = \tfrac{1}{2}mv^2$$

km   Abbreviation for *kilometer.*

kPa   Abbreviation for *kilopascal.* See *pascal.*

kWh   Abbreviation for *kilowatt-hour.*

L   Abbreviation for *liter.* (In some textbooks, lowercase l is used.)

laser   Optical instrument that produces a beam of coherent light—that is, light with all waves of the same frequency, phase, and direction. The word is an acronym for light amplification by stimulated emission of radiation.

latent heat of fusion   The amount of energy required to change a unit mass of a substance from solid to liquid (and vice versa).

latent heat of vaporization   The amount of energy required to change a unit mass of a substance from liquid to gas (and vice versa).

law   General hypothesis or statement about the relationship of natural quantities that has been tested over and over again and has not been contradicted. Also known as a *principle.*

law of conservation of momentum   In the absence of an external force, the momentum of a system remains unchanged. Hence, the momentum before an event involving only internal forces is equal to the momentum after the event:

$$mv_{(\text{before event})} = mv_{(\text{after event})}$$

law of inertia   See *Newton's laws of motion, Law 1.*

law of reflection   The angle of incidence for a wave that strikes a surface is equal to the angle of reflection. This is true for both partially and totally reflected waves. See also *angle of incidence* and *angle of reflection.*

law of universal gravitation   For any pair of particles, each particle attracts the other particle with a force that is directly proportional to the product of the masses of the particles, and inversely proportional to the square of the distance between them (or their centers of mass if spherical objects), where *F* is the force, *m* is the mass, *d* is distance, and *G* is the gravitation constant:

$$F \sim \frac{m_1 m_2}{d^2} \quad or \quad F = G\frac{m_1 m_2}{d^2}$$

lens   Piece of glass or other transparent material that can bring light to a focus.

lepton   Class of elementary particles that are not involved with the nuclear force. It includes the electron and its neutrino, the muon and its neutrino, and the tau and its neutrino.

lever Simple machine made of a bar that turns about a fixed point called the fulcrum.

lever arm Perpendicular distance between an axis and the line of action of a force that tends to produce rotation about that axis.

lift In application of Bernoulli's principle, the net upward force produced by the difference between upward and downward pressures. When lift equals weight, horizontal flight is possible.

light Visible part of the electromagnetic spectrum.

light-year The distance light travels in a vacuum in one year: 9.46×10^{12} km.

line spectrum Pattern of distinct lines of color, corresponding to particular wavelengths, that are seen in a spectroscope when a hot gas is viewed. Each element has a unique pattern of lines.

linear momentum Product of the mass and the velocity of an object. Also called momentum. (This definition applies at speeds much less than the speed of light.)

linear motion Motion along a straight-line path.

linear speed Path distance moved per unit of time. Also called simply speed.

liquid Phase of matter between the solid and gaseous phases in which the matter possesses a definite volume but no definite shape: it takes on the shape of its container.

liter (L) Metric unit of volume. A liter is equal to 1000 cm^3.

logarithmic Exponential.

longitudinal wave Wave in which the individual particles of a medium vibrate back and forth in the direction in which the wave travels—for example, sound.

loudness Physiological sensation directly related to sound intensity or volume. Relative loudness, or sound level, is mea-sured in decibels.

lunar eclipse Event wherein the full Moon passes into the shadow of Earth.

m (a) Abbreviation for *meter*. (b) When an italic *m*, the abbreviation for *mass*.

Mach number Ratio of the speed of an object to the speed of sound. For example, an aircraft traveling *at* the speed of sound is rated Mach 1.0; traveling at *twice* the speed of sound, Mach 2.0.

machine Device for increasing (or decreasing) a force or simply changing the direction of a force.

magnet Any object that has magnetic properties, that is the ability to attract objects made of iron or other magnetic substances. See also *electromagnetism* and *magnetic force*.

magnetic declination Discrepancy between the orientation of a compass pointing toward magnetic north and the true geographic north.

magnetic domain Microscopic cluster of atoms with their magnetic fields aligned.

magnetic field Region of magnetic influence around a magnetic pole or a moving charged particle.

magnetic field lines Lines showing the shape of a magnetic field. A compass placed on such a line will turn so that the needle is aligned with it.

magnetic force (a) Between magnets, it is the attraction of unlike magnetic poles for each other and the repulsion between like magnetic poles. (b) Between a magnetic field and a moving charged particle, it is a deflecting force due to the motion of the particle: The deflecting force is perpendicular to the magnetic field lines and the direction of motion. This force is greatest when the charged particle moves perpendicular to the field lines and is smallest (zero) when it moves parallel to the field lines.

magnetic monopole Hypothetical particle having a single north or a single south magnetic pole, analogous to a positive or negative electric charge.

magnetic pole One of the regions on a magnet that produces magnetic forces.

magnetic pole reversal When the magnetic field of an astronomical body reverses its poles, that is, the location where the north magnetic pole existed becomes the south magnetic pole, and the south magnetic pole becomes the north magnetic pole.

magnetism Property of being able to attract objects made of iron, steel, or magnetite. See also *electromagnetism* and *magnetic force*.

magnetohydrodynamic (MDH) generator Device generating electric power by interaction of a plasma and a magnetic field.

mass (*m*) Quantity of matter in an object; the measurement of the inertia or sluggishness that an object exhibits in response to any effort made to start it, stop it, or change in any way its state of motion; a form of energy.

mass spectrometer Device that magnetically separates charged ions according to their mass.

mass-energy equivalence Relationship between mass and energy as given by the equation

$$E = mc^2$$

where *c* is the speed of light.

matter waves See *de Broglie matter waves*.

Maxwell's counterpart to Faraday's law A magnetic field is created in any region of space in which an electric field is changing with time. The magnitude of the induced magnetic field is proportional to the rate at which the electric field changes. The direction of the induced magnetic field is at right angles to the changing electric field.

mechanical advantage Ratio of output force to input force for a machine.

mechanical energy Energy due to the position or the movement of something; potential or kinetic energy (or a combination of both).

mechanical equilibrium   State of an object or system of objects for which any impressed forces cancel to zero and no acceleration occurs and when no net torque exists. That is, $\sum F = 0$, and $\sum \tau = 0$.

mega-   Prefix that means million, as in megahertz or megajoule.

melting   Change in phase from solid to liquid; the opposite of freezing. Melting is a different process from dissolving, in which an added solid mixes with a liquid and the solid dissociates.

meson   Elementary particle with an atomic weight of zero; can participate in the strong interaction.

metastable state   State of excitation of an atom that is characterized by a prolonged delay before de-excitation.

meter (m)   Standard SI unit of length (3.28 feet).

MeV   Abbreviation for million *electron volts,* a unit of energy, or equivalently a unit of mass.

MHD   Abbreviation for *magnetohydrodynamic.*

mi   Abbreviation for mile.

microscope   Optical instrument that forms enlarged images of very small objects.

microwaves   Electromagnetic waves with frequencies greater than radio waves but less than infrared waves.

min   Abbreviation for minute.

mirage   False image that appears in the distance and is due to the refraction of light in Earth's atmosphere.

mixture   Substances mixed together without combining chemically.

MJ   Abbreviation for megajoules, million *joules.*

model   Representation of an idea created to make the idea more understandable.

modulation   Impressing a signal wave system on a higher frequency carrier wave, amplitude modulation (AM) for amplitude signals and frequency modulation (FM) for frequency signals.

molecule   Two or more atoms of the same or different elements bonded to form a larger particle.

momentum   Inertia in motion. The product of the mass and the velocity of an object (provided the speed is much less than the speed of light). Has magnitude and direction and therefore is a vector quantity. Also called linear momentum, and abbreviated *p.*

$$p = mv$$

monochromatic light   Light made of only one color and therefore waves of only one wavelength and frequency.

muon   Elementary particle in the class of elementary particles called leptons. It is short-lived with a mass that is 207 times that of an electron; may be positively or negatively charged.

music   Scientifically speaking, sound with periodic tones, which appear on an oscilloscope as a regular pattern.

N   Abbreviation for *newton.*

nanometer   Metric unit of length that is 10^{-9} meter (one billionth of a meter).

natural frequency   Frequency at which an elastic object naturally tends to vibrate if it is disturbed and the disturbing force is removed.

neap tide   Tide that occurs when the Moon is halfway between a new Moon and a full Moon, in either direction. The tides due to the Sun and the Moon partly cancel, so that the high tides are lower than average and the low tides are not as low as average. See also *spring tide.*

net force   Combination of all the forces that act on an object.

neutrino   Elementary particle in the class of elementary particles called leptons. It is uncharged and almost massless; three kinds—electron, muon, and tau neutrinos, are the most common high-speed particles in the universe; more than a billion pass unhindered through each person every second.

neutron   Electrically neutral particle that is one of the two kinds of nucleons that compose an atomic nucleus.

neutron star   Star that has undergone a gravitational collapse in which electrons are compressed into protons to form neutrons.

newton (N)   SI unit of force. One newton is the force applied to a one-kilogram mass that will produce an acceleration of one meter per second per second.

Newton's law of cooling   The rate of cooling of an object—whether by conduction, convection, or radiation—is approximately proportional to the temperature difference between the object and its surroundings.

Newton's laws of motion
Law 1: Every body continues in its state of rest or of uniform speed in a straight line unless acted on by a nonzero net force. Also known as the law of inertia.

Law 2: The acceleration produced by a net force on a body is directly proportional to the magnitude of the net force, is in the same direction as the net force, and is inversely proportional to the mass of the body.

Law 3: Whenever one body exerts a force on a second body, the second body exerts an equal and opposite force on the first.

node   Any part of a standing wave that remains stationary; a region of minimal or zero energy.

noise   Scientifically speaking, sound that corresponds to an irregular vibration of the eardrum produced by some irregular vibration, which appears on an oscilloscope as an irregular pattern.

nonlinear motion   Any motion not along a straight-line path.

normal   At right angles to, or perpendicular to. A normal force acts at right angles to the surface on which it acts. In optics, a normal defines the line perpendicular to a surface about which angles of light rays are measured.

normal force   Component of support force perpendicular to a supporting surface. For an object resting on a horizontal surface, it is the upward force that balances the weight of the object.

nuclear fission   Splitting of an atomic nucleus, particularly that of a heavy element such as uranium-235, into two lighter elements, accompanied by the release of much energy.

nuclear force   Attractive force within a nucleus that holds neutrons and protons together. Part of the nuclear force is called the strong interaction. The strong interaction is an attractive force that acts between protons, neutrons, and mesons (another nuclear particle); however, it acts only over very short distances (10^{-15} meter). The weak interaction is the nuclear force responsible for beta (electron) emission.

nuclear fusion   Combining of nuclei of light atoms, such as hydrogen, into heavier nuclei, accompanied by the release of much energy. See also *thermonuclear fusion*.

nuclear reactor   Apparatus in which controlled nuclear fission or nuclear fusion reactions take place.

nucleon   Principal building block of the nucleus. A neutron or a proton; the collective name for either or both.

nucleus   (pl. nuclei) Positively charged center of an atom, which contains protons and neutrons and has almost all the mass of the entire atom but only a tiny fraction of the volume.

objective lens   In an optical device using compound lenses, the lens closest to the object observed.

octave   In music, the eighth full tone above or below a given tone. The tone an octave above has twice as many vibrations per second as the original tone; the tone an octave below has half as many vibrations per second.

ohm (Ω)   SI unit of electrical resistance. One ohm is the resistance of a device that draws a current of one ampere when a voltage of one volt is impressed across it.

Ohm's law   Current in a circuit is directly proportional to the voltage impressed across the circuit, and is inversely proportional to the resistance of the circuit.

$$\text{Currency} = \frac{\text{voltage}}{\text{resistance}}$$

opaque   Term applied to materials that absorb light without re-emission, and consequently do not allow light through them.

optical fiber   Transparent fiber, usually of glass or plastic, that can transmit light down its length by means of total internal reflection.

oscillation   Same as vibration: a repeating to-and-fro motion about an equilibrium position. Both oscillation and vibration refer to periodic motion, that is, motion that repeats.

oscillatory motion   To-and-fro vibratory motion, such as that of a pendulum.

out of phase   Term applied to two waves for which the crest of one wave arrives coincident with a trough of the second wave. Their effects tend to cancel each other.

overtone   Musical term where the first overtone is the second harmnic. See also *partial tone*.

oxidize   Chemical process in which an element or molecule loses one or more electrons.

ozone   Gas, found in a thin layer in the upper atmosphere, composed of molecules of three oxygen atoms. Atmospheric oxygen gas is composed of molecules of two oxygen atoms.

Pa   Abbreviation for the SI unit *pascal*.

parabola   Curved path followed by a projectile acting under the influence of gravity only.

parallax   Apparent displacement of an object when viewed by an observer from two different positions; often used to calculate the distance of stars.

parallel circuit   Electric circuit with two or more devices connected in such a way that the same voltage acts across each one and any single one completes the circuit independently of the others. See also *in parallel*.

partial tone   One of the many tones that make up one musical sound. Each partial tone (or partial) has only one frequency. The lowest partial of a musical sound is called the fundamental frequency. Any partial whose frequency is a multiple of the fundamental frequency is called a harmonic. The fundamental frequency is also called the first harmonic. The second harmonic has twice the frequency of the fundamental; the third harmonic, three times the frequency, and so on.

pascal (Pa)   SI unit of pressure. One pascal of pressure exerts a normal force of one newton per square meter. A kilopascal (kPa) is 1000 pascals.

Pascal's principle   Changes in pressure at any point in an enclosed fluid at rest are transmitted undiminished to all points in the fluid and act in all directions.

PE   Abbreviation for *potential energy*.

penumbra   Partial shadow that appears where some of the light is blocked and other light can fall. See also *umbra*.

percussion   In musical instruments, the striking of one object against another.

perigee   Point in an elliptical orbit closest to the focus about which orbiting takes place. See also *apogee*.

period   In general, the time required to complete a single cycle. (a) For orbital motion, the time required for a complete orbit. (b) For vibrations or waves, the time required for one complete cycle, equal to 1/frequency.

periodic table   Chart that lists elements by atomic number and by electron arrangements, so that elements with similar chemical properties are in the same column (group). See Figure 2.9, page 19.

perturbation   Deviation of an orbiting object (e.g., a planet) from its path around a center of force (e.g., the Sun) caused by the action of an additional center of force (e.g., another planet).

phase   (a) One of the four main forms of matter: solid, liquid, gas, and plasma. Often called *state*. (b) The part of a cycle that a wave has advanced at any moment. See also *in phase* and *out of phase*.

phosphor   Powdery material, such as that used on the inner surface of a fluorescent light tube, that absorbs ultraviolet photons, then gives off visible light.

phosphorescence   Type of light emission that is the same as fluorescence except for a delay between excitation and de-excitation, which provides an afterglow. The delay is caused by atoms being excited to energy levels that do not decay rapidly. The afterglow may last from fractions of a second to hours, or even days, depending on such factors as the type of material and temperature.

photoelectric effect   Ejection of electrons from certain metals when exposed to certain frequencies of light.

photon   Localized corpuscle of electromagnetic radiation whose energy is proportional to its radiation frequency: $E \sim f$, or $E = hf$, where h is Planck's constant.

pigment   Fine particles that selectively absorb light of certain frequencies and selectively transmit others.

pitch   Term that refers to our subjective impression about the "highness" or "lowness" of a tone, which is related to the frequency of the tone. A high-frequency vibrating source produces a sound of high pitch; a low-frequency vibrating source produces a sound of low pitch.

Planck's constant (h)   Fundamental constant of quantum theory that determines the scale of the small-scale world. Planck's constant multiplied by the frequency of radiation gives the energy of a photon of that radiation.

$$E = hf, \quad h = 6.6 \times 10^{-34} \text{ joule-second}$$

plane mirror   Flat-surfaced mirror.

plane-polarized wave   A wave confined to a single plane.

plasma   Fourth phase of matter, in addition to solid, liquid, and gas. In the plasma phase, existing mainly at high temperatures, matter consists of positively charged ions and free electrons.

polarization   Aligning of vibrations in a transverse wave, usually by filtering out waves of other directions. See also *plane-polarized wave* and *dichroic crystal*.

polished   Describes a surface that is so smooth that the distances between successive elevations of the surface are less than about one-eighth the wavelength of the light or other incident wave of interest. The result is very little diffuse reflection.

positron   Antiparticle of an electron; a positively charged electron.

potential difference   Difference in electric potential (voltage) between two points. Free charge flows when there is a difference, and will continue until both points reach a common potential.

potential energy (PE)   Energy of position, usually related to the relative position of two things, such as a stone and Earth (gravitational PE), or an electron and a nucleus (electric PE).

power   Rate at which work is done or energy is transformed, equal to the work done or energy transformed divided by time; measured in watts.

$$\text{power} = \frac{\text{work}}{\text{time}}$$

pressure   Force per surface area where the force is normal (perpendicular) to the surface; measured in pascals. See also *atmospheric pressure*.

$$\text{pressure} = \frac{\text{force}}{\text{area}}$$

primary colors   See a*dditive primary colors* and *subtractive primary colors*.

principal axis   Line joining the centers of curvature of the surfaces of a lens. Line joining the center of curvature and the focus of a mirror.

principle   General hypothesis or statement about the relationship of natural quantities that has been tested over and over again and has not been contradicted; also known as a law.

principle of flotation   A floating object displaces a weight of fluid equal to its own weight.

prism   Triangular solid of a transparent material such as glass, that separates incident light by refraction into its component colors. These component colors are often called the spectrum.

projectile   Any object that moves through the air or through space, acted on only by gravity (and air resistance, if any).

proton   Positively charged particle that is one of the two kinds of nucleons in the nucleus of an atom.

pseudoscience   Fake science that pretends to be real science.

pulley   Wheel that acts as a lever used to change the direction of a force. A pulley or system of pulleys can also multiply forces.

pupil   Opening in the eyeball through which light passes.

quality   Characteristic timbre of a musical sound, governed by the number and relative intensities of partial tones.

quantum   (pl. quanta) From the Latin word *quantus*, meaning "how much," a quantum is the smallest elemental unit of a quantity, the smallest discrete amount of something. One quantum of electromagnetic energy is called a photon. See also *quantum mechanics* and *quantum theory*.

quantum mechanics   Branch of physics concerned with the atomic microworld based on wave functions and probabilities, introduced by Max Planck (1900) and developed by Werner Heisenberg (1925), Erwin Schrödinger (1926), and others.

quantum physics Branch of physics that is the general study of the microworld of photons, atoms, and nuclei.

quantum theory Theory that describes the microworld, where many quantities are granular (in units called quanta), rather than continuous, and where particles of light (photons) and particles of matter (such as electrons) exhibit wave as well as particle properties.

quark One of the two classes of elementary particles. (The other is the lepton.) Two of the six quarks (up and down) are the fundamental building blocks of nucleons (protons and neutrons).

rad Unit used to measure a dose of radiation; the amount of energy (in centijoules) absorbed from ionizing radiation per kilogram of exposed material.

radiant energy Any energy, including heat, light, and X-rays, that is transmitted by radiation. It occurs in the form of electromagnetic waves.

radiation (a) Energy transmitted by electromagnetic waves. (b) The particles given off by radioactive atoms such as uranium. Do not confuse radiation with radioactivity.

radiation curve of sunlight See *solar radiation curve.*

radio waves Electromagnetic waves of the longest frequency.

radioactive Term applied to an atom having an unstable nucleus that can spontaneously emit a particle and become the nucleus of another element.

radioactivity Process of the atomic nucleus that results in the emission of energetic particles. See *radiation.*

radiotherapy Use of radiation as a treatment to kill cancer cells.

rarefaction Region of reduced pressure in a longitudinal wave.

rate How fast something happens or how much something changes per unit of time; a change in a quantity divided by the time it takes for the change to occur.

ray Thin beam of light. Also lines drawn to show light paths in optical ray diagrams.

reaction force Force that is equal in strength and opposite in direction to the action force, and one that acts simultaneously on whatever is exerting the action force. See also *Newton's third law.*

real image Image formed by light rays that converge at the location of the image. A real image, unlike a virtual image, can be displayed on a screen.

red shift Decrease in the measured frequency of light (or other radiation) from a receding source; called the *red shift* because the decrease is toward the low-frequency, or red, end of the color spectrum. See also *Doppler effect.*

reflection Return of light rays from a surface in such a way that the angle at which a given ray is returned is equal to the angle at which it strikes the surface. When the reflecting surface is irregular, the light is returned in irregular directions; this is *diffuse reflection.* In general, the bouncing back of a particle or wave that strikes the boundary between two media.

refraction Bending of an oblique ray of light when it passes from one transparent medium to another. This is caused by a difference in the speed of light in the transparent media. In general, the change in direction of a wave as it crosses the boundary between two media in which the wave travels at different speeds.

regelation Process of melting under pressure and the subsequent refreezing when the pressure is removed.

relationship of impulse and momentum Impulse is equal to the change in the momentum of the object that the impulse acts upon. In symbol notation,

$$Ft = \Delta mv$$

relative Regarded in relation to something else, depending on point of view, or frame of reference. Sometimes referred to as "with respect to."

relative humidity Ratio between how much water vapor is in the air and the maximum amount of water vapor that could be in the air at the same temperature.

rem Acronym of **r**oentgen **e**quivalent **m**an, it is a unit used to measure the effect of ionizing radiation on human beings.

resistance See *electrical resistance.*

resistor Device in an electric circuit designed to resist the flow of charge.

resolution (a) Method of separating a vector into its component parts. (b) Ability of an optical system to make clear or to separate the components of an object viewed.

resonance Phenomenon that occurs when the frequency of forced vibrations on an object matches the object's natural frequency, producing a dramatic increase in amplitude.

rest energy The "energy of being" given by the equation $E = mc^2$.

resultant Net result of a combination of two or more vectors.

retina Layer of light-sensitive tissue at the back of the eye, composed of tiny light-sensitive antennae called rods and cones. Rods sense light and darkness. Cones sense color.

reverberation Persistence of a sound, as in an echo, due to multiple reflections.

revolution Motion of an object turning around an axis that lies outside the object.

Ritz combination principle For an element, the frequencies of some spectral lines are either the sum or the difference of the frequencies of two other lines in that element's spectrum.

rods See *retina.*

rotation Spinning motion that occurs when an object rotates about an axis located within the object (usually an axis through its center of mass).

rotational inertia Reluctance or apparent resistance of an object to change its state of rotation, determined by the distribution of the mass of the object and the location of the axis of rotation or revolution.

rotational speed Number of rotations or revolutions per unit of time; often measured in rotations or revolutions per second or minute.

rotational velocity Rotational speed together with a direction for the axis of rotation or revolution.

RPM Abbreviation for rotations or revolutions per minute.

s Abbreviation for second.

satellite Projectile or smaller celestial body that orbits a larger celestial body.

saturated Term applied to a substance, such as air, that contains the maximum amount of another substance, such as water vapor, at a given temperature and pressure.

scalar quantity Quantity in physics, such as mass, volume, and time, that can be completely specified by its magnitude, and has no direction.

scale In music, a succession of notes or frequencies that are in simple ratios to one another.

scaling Study of how size affects the relationship among weight, strength, and surface area.

scatter To absorb sound or light and re-emit it in all directions.

scattering Emission in random directions of light that encounters particles that are small compared to the wavelength of light; more often at short wavelengths (blue) than at long wavelengths (red).

Schrödinger's wave equation Fundamental equation of quantum mechanics, which interprets the wave nature of material particles in terms of probability wave amplitudes. It is as basic to quantum mechanics as Newton's laws of motion are to classical mechanics.

scientific method Orderly method for gaining, organizing, and applying new knowledge.

self-induction Induction of an electric field within a single coil, caused by the interaction of the loops within the same coil. This self-induced voltage is always in a direction opposing the changing voltage that produces it, and is commonly called back electromotive force or back emf.

semiconductor Device made of material not only with properties that fall between a conductor and an insulator but with resistance that changes abruptly when other conditions change, such as temperature, voltage, and electric or magnetic field.

series circuit Electric circuit with devices connected in such a way that the electric current through each of them is the same. See also *in series.*

shadow Shaded region that appears where light rays are blocked by an object.

shell model of the atom Model in which the electrons of an atom are pictured as grouped in concentric shells around the nucleus.

shock wave Cone-shaped wave produced by an object moving at supersonic speed through a fluid.

short circuit Disruption in an electric circuit, caused by the flow of charge along a low-resistance path between two points that should not be directly connected, thus deflecting the current from its proper path; an effective "shortening of the circuit."

SI Abbreviation for Système International, an international system of units of metric measure accepted and used by scientists throughout the world. See Appendix A for more details.

simple harmonic motion Vibratory or periodic motion, like that of a pendulum, in which the force acting on the vibrating body is proportional to its displacement from its central equilibrium position and acts toward that position.

sine curve Curve whose shape represents the crests and troughs of a wave, as traced out by a pendulum that drops a trail of sand while swinging at right angles to and over a moving conveyor belt.

sine wave The simplest of waves with only one frequency and the shape of a sine curve.

sliding friction Contact force produced by the rubbing together of the surface of a moving object with the material over which it slides.

solar constant 1400 J/m^2 received from the Sun each second at the top of Earth's atmosphere, expressed in terms of power, 1.4 kW/m^2.

solar eclipse Event wherein the Moon blocks light from the Sun and the Moon's shadow falls on part of Earth.

solar power Energy per unit time derived from the Sun. See also *solar constant.*

solar radiation curve Graph of brightness versus frequency (or wavelength) of sunlight.

solid Phase of matter characterized by definite volume and shape.

solidify To become solid, as in freezing or the setting of concrete.

sonic boom Loud sound resulting from the incidence of a shock wave.

sound Longitudinal wave phenomenon that consists of successive compressions and rarefactions of the medium through which the wave travels.

sound barrier The pile up of sound waves in front of an aircraft approaching or reaching the speed of sound, believed in the early days of jet aircraft to create a barrier of sound that a plane would have to break through in order to go faster than the speed of sound. The sound barrier does not exist.

spacetime   Four-dimensional continuum in which all events take place and all things exist: Three dimensions are the coordinates of space and the fourth is of time.

specific heat capacity   Quantity of heat required to raise the temperature of a unit mass of a substance by one degree Celsius (or equivalently, by one kelvin). Often simply called specific heat.

spectral lines   Colored lines that form when light is passed through a slit and then through a prism or diffraction grating, usually in a spectroscope. The pattern of lines is unique for each element.

spectrometer   See *spectroscope.*

spectroscope   An optical instrument that separates light into its constituent frequencies or wavelengths in the form of spectral lines. A *spectrometer* is an instrument that can also measure the frequencies or wavelengths.

spectrum   (pl. spectra) For sunlight and other white light, the spread of colors seen when the light is passed through a prism or diffraction grating. The colors of the spectrum, in order from lowest frequency (longest wavelength) to highest frequency (shortest wavelength) are red, orange, yellow, green, blue, indigo, violet. See also *absorption spectrum, electromagnetic spectrum, emission spectrum,* and *prism.*

speed   How fast something moves; the distance an object travels per unit of time; the magnitude of velocity. See also *average speed, linear speed, rotational speed,* and *tangential speed.*

$$\text{speed} = \frac{\text{distance}}{\text{time}}$$

spherical aberration   Distortion of an image caused when the light that passes through the edges of a lens focuses at slightly different points from the point where the light passing through the center of the lens focuses. It also occurs with spherical mirrors.

spring tide   High or low tide that occurs when the Sun, Earth, and Moon are all lined up so that the tides due to the Sun and Moon coincide, making the high tides higher than average and the low tides lower than average. See also *neap tide.*

stable equilibrium   State of an object balanced so that any small displacement or rotation raises its center of gravity.

standing wave   Stationary wave pattern formed in a medium when two sets of identical waves pass through the medium in opposite directions. The wave appears not to be traveling.

static friction   Force between two objects at relative rest by virtue of contact that tends to oppose sliding.

streamline   Smooth path of a small region of fluid in steady flow.

strong force   Force that attracts nucleons to each other within the nucleus; a force that is very strong at close distances but decreases rapidly as the distance increases. Also called strong interaction. See also *nuclear force.*

strong interaction   See *strong force.*

subcritical mass   See *critical mass.*

sublimation   Direct conversion of a substance from the solid to the vapor phase without passing through the liquid phase.

subtractive primary colors   The three colors of light-absorbing pigments—magenta, yellow, and cyan—that when mixed in certain proportions will reflect any color in the spectrum.

superconductor   Material that is a perfect conductor with zero resistance to the flow of electric charge.

supercritical mass   See *critical mass.*

superposition principle   In a situation where more than one wave occupies the same space at the same time, the displacements add at every point.

supersonic   Traveling faster than the speed of sound.

support force   Upward force that balances the weight of an object on a surface.

surface tension   Tendency of the surface of a liquid to contract in area and thus behave like a stretched elastic membrane.

tangent   Line that touches a curve in one place only and is parallel to the curve at that point.

tangential speed   Linear speed along a curved path.

tangential velocity   Component of velocity tangent to the trajectory of a projectile.

technology   Method and means of solving practical problems by implementing the findings of science.

telescope   Optical instrument that forms images of very distant objects.

temperature   Measure of the average translational kinetic energy per molecule of a substance, measured in degrees Celsius or Fahrenheit, or in kelvins.

temperature inversion   Condition wherein upward convection of air is stopped, sometimes because an upper region of the atmosphere is warmer than the region below it.

terminal speed   Speed attained by an object wherein the resistive forces, often air resistance, counterbalance the driving forces, so motion is without acceleration.

terminal velocity   Terminal speed together with the direction of motion (down for falling objects).

terrestrial radiation   Radiant energy emitted from Earth.

theory   Synthesis of a large body of information that encompasses well-tested and verified hypotheses about aspects of the natural world.

thermal contact   State of two or more objects or substances in contact such that heat can flow from one object or substance to the other.

thermal equilibrium   State of two or more objects or substances in thermal contact when they have reached a common temperature.

thermal pollution   Undesirable heat expelled from a heat engine or other source.

thermodynamics   Study of heat and its transformation to mechanical energy, characterized by two principal laws:

First Law: A restatement of the law of conservation of energy as it applies to systems involving changes in temperature: Whenever heat is added to a system, it transforms to an equal amount of some other form of energy.

*Second Law: H*eat cannot be transferred from a colder body to a hotter body without work being done by an outside agent.

thermometer   Device used to measure temperature, usually in degrees Celsius, degrees Fahrenheit, or kelvins.

thermonuclear fusion   Nuclear fusion brought about by extremely high temperatures; in other words, the welding together of atomic nuclei by high temperature.

thermostat   Type of valve or switch that responds to changes in temperature and that is used to control the temperature of something.

torque   Product of force and lever-arm distance, which tends to produce rotational acceleration.

$$\text{torque} = \text{lever-arm distance} \times \text{force}$$

total internal reflection   The 100% reflection (with no transmission) of light that strikes the boundary between two media at an angle greater than the critical angle.

transformer   Device for increasing or decreasing voltage or transferring electric power from one coil of wire to another by means of electromagnetic induction.

transistor   See *semiconductor.*

transmutation   Conversion of an atomic nucleus of one element into an atomic nucleus of another element through a loss or gain in the number of protons.

transparent   Term applied to materials that allow light to pass through them in straight lines.

transuranic element   Element with an atomic number above 92, beyond uranium in the periodic table.

transverse wave   Wave with vibration at right angles to the direction the wave is traveling. Light consists of transverse waves.

tritium   Unstable, radioactive isotope of hydrogen whose atom has a proton, two neutrons, and an electron.

trough   One of the places in a wave where the wave is lowest or the disturbance is greatest in the opposite direction from a crest. See also *crest.*

turbine   Paddle wheel driven by steam, water, etc., that is used to do work.

turbogenerator   Generator that is powered by a turbine.

ultrasonic   Term applied to sound frequencies above 20,000 hertz, the normal upper limit of human hearing.

ultraviolet (UV)   Electromagnetic waves of frequencies higher than those of violet light.

umbra   Darker part of a shadow where all the light is blocked. See also *penumbra.*

uncertainty principle   The principle formulated by Heisenberg, stating that Planck's constant, *h,* sets a limit on the accuracy of measurement at the atomic level. According to the uncertainty principle, it is not possible to measure exactly both the position and the momentum of a particle at the same time, nor the energy and the time associated with a particle simultaneously.

universal gravitational constant   The proportionality constant *G* that measures the strength of gravity in the equation for Newton's law of universal gravitation.

$$F = G\frac{m_1 m_2}{d^2}$$

unstable equilibrium   State of an object balanced so that any small displacement or rotation lowers its center of gravity.

UV   Abbreviation for *ultraviolet.*

V   (a) In lowercase italic *v,* the symbol for *speed* or *velocity.* (b) In uppercase V, the abbreviation for *voltage.*

vacuum   Absence of matter; void.

Van Allen radiation belts   Two donut-shaped belts of radiation that surround Earth.

vaporization   The process of a phase change from liquid to vapor; evaporation.

vector   Arrow whose length represents the magnitude of a quantity and whose direction represents the direction of the quantity.

vector quantity   Quantity in physics that has both magnitude and direction. Examples are force, velocity, acceleration, torque, and electric and magnetic fields.

velocity   Speed of an object and its direction of motion; a vector quantity.

vibration   Oscillation; a repeating to-and-fro motion about an equilibrium position—a "wiggle in time."

virtual image   Image formed by light rays that do not converge at the location of the image. Mirrors, converging lenses used as magnifying glasses, and diverging lenses all produce virtual images. The image can be seen by an observer, but cannot be projected onto a screen.

visible light   Part of the electromagnetic spectrum that the human eye can see.

visible spectrum   See *electromagnetic spectrum.*

volt (V)   SI unit of electric potential. One volt is the electric potential difference across which one coulomb of charge gains or loses one joule of energy. 1 V = 1 J/C

voltage   Electrical "pressure" or a measure of electrical potential difference.

$$\text{voltage} = \frac{\text{electric potential energy}}{\text{unit of charge}}$$

voltage source   Device, such as a dry cell, battery, or generator, that provides a potential difference.

voltmeter   See *galvanometer*.

volume   Quantity of space an object occupies.

W   (a) Abbreviation for *watt*. (b) When in uppercase italic *W*, the abbreviation for *work*.

watt   SI unit of power. One watt is expended when one joule of work is done in one second. 1 W = 1 J/s

wave   A "wiggle in space and time"; a disturbance that repeats regularly in space and time and that is transmitted progressively from one place to the next with no net transport of matter.

wave front   Crest, trough, or any continuous portion of a two-dimensional or three-dimensional wave in which the vibrations are all the same way at the same time.

wave speed   Speed with which waves pass a particular point.

$$\text{wave speed} = \text{wavelength} \times \text{frequency}$$

wave velocity   Wave speed stated with the direction of travel.

wavelength   Distance between successive crests, troughs, or identical parts of a wave.

weak force   Also called weak interaction. The force within a nucleus that is responsible for beta (electron) emission. See *nuclear force*.

weak interaction   See *nuclear force* and *weak force*.

weight   The force that an object exerts on a supporting surface (or, if suspended, in a supporting string)—often, but not always, due to the force of gravity.

weight density   See *density*.

weightlessness   Condition of free fall toward or around Earth, in which an object experiences no support force (and exerts no force on a scale).

white light   Light, such as sunlight, that is a combination of all the colors. Under white light, white objects appear white and colored objects appear in their individual colors.

work (*W*)   Product of the force on an object and the distance through which the object is moved (when force is constant and motion is in a straight line in the direction of the force); measured in joules.

$$\text{work} = \text{force} \times \text{distance}$$

work-energy theorem   Work done on an object is equal to the kinetic energy gained by the object.

$$\text{work} = \text{change in energy or } W = \Delta\text{KE}$$

X-ray   Electromagnetic radiation, higher in frequency than ultraviolet, emitted by atoms when the innermost orbital electrons undergo excitation.

zero-point energy   Extremely small amount of kinetic energy that molecules or atoms have even at absolute zero.

PHOTO CREDITS

Frontmatter: p xv Lillian Lee Hewitt

Part Openers: Part 1 p 29 John Suchocki; **Part 2 p 159** Paul G. Hewitt; **Part 3 p 199** Paul G. Hewitt; **Part 4 p 245** Paul G. Hewitt; **Part 5 p 319** Paul G. Hewitt

Chapter 1: Opening photo Paul G. Hewitt; **1.2** Corbis Los Angeles; **1.3** Jay M. Pasachoff

Chapter 2: Opening photo Paul G. Hewitt; **2.4** The Enrico Fermi Institute; **2.5** IBM Corporate Archives; **2.8** Paul G. Hewitt

Chapter 3: Opening photo Paul G. Hewitt; **History of Science p 31** Corbis Los Angeles; **History of Science p 33** Art Resource, N.Y.; **3.11** Paul G. Hewitt; **3.17** Animals Animals/Earth Scenes; **3.18** Addison Wesley Longman, Inc./San Francisco; **Everyday Applications p 46** Getty Images

Chapter 4: Opening photo Exploratorium; **p 53** Art Resource, N.Y.; **4.14** Photo Researchers, Inc.; **4.15** Fundamental Photographs, NYC; **4.30** Animals Animals/Earth Scenes; **4.31** Paul G. Hewitt; **History of Science p 71** Giraudon/Art Resource. Artist: Godfrey Kneller; **p 75** Paul G. Hewitt

Chapter 5: Opening photo Paul G. Hewitt; **5.3** The Harold D. Edgerton Trust/Palm Press; **5.7** Paul G. Hewitt; **5.8** Paul G. Hewitt; **5.14** Paul G. Hewitt; **5.15** AP Wide World Photos; **5.20** Paul G. Hewitt; **5.23** Paul G. Hewitt; **5.24** Paul G. Hewitt; **5.26** NASA/Goddard Institute for Space Studies; **p 102** Collection of Paul G. Hewitt

Chapter 6: Opening photo Paul G. Hewitt; **6.11** NASA Earth Observing System; **6.16** Fundamental Photographs, NYC; **6.23** Getty Images; **Everyday Application p 118** Getty Images **6.36** Fundamental Photographs, NYC; **6.43** NASA/Goddard Space Flight Center; **6.44** NASA Earth Observing System

Chapter 7: Opening photo Paul G. Hewitt; **7.3 (top, bottom)** Paul G. Hewitt; **7.4** Paul G. Hewitt; **7.5** Paul G. Hewitt; **7.18** Paul G. Hewitt; **7.21** The Granger Collection; **7.28** Paul G. Hewitt; **7.34** Construction Photography.com; **7.39** Paul G. Hewitt; **7.42** Corbis Los Angeles; **p 155** Paul G. Hewitt

Chapter 8: Opening photo Paul G. Hewitt; **8.2** Kasai Werel; **8.9** Paul G. Hewitt; **8.10** Paul G. Hewitt; **8.14** AP/Wide World Photos; **8.15** Paul G. Hewitt; **8.16** Meidor Hu; **8.20** Nuridsany et Perennov/Photo Researchers, Inc; **p 175 (top)** Ed Young/Photo Researchers, Inc.; **p 175 (bottom)** Paul G. Hewitt

Chapter 9: Opening photo Tracy Suchocki; **9.3** Paul G. Hewitt; **9.4** Don Hynek/Wisconsin Division of Labor; **9.6** Nancy Rogers; **9.7** Paul G. Hewitt; **9.8** Paul G. Hewitt; **9.17 (left, right)** Robert D. Carey; **9.20** Paul G. Hewitt; **9.21** Tammy Tunison; **9.22** Lillian Lee Hewitt; **9.24** Paul G. Hewitt; **9.27** Dennis Wong; **9.32** Nicole Minor/Exploratorium; **Everyday**

Applications p 189 (left, right, middle) Paul G. Hewitt; **9.35** Lillian Lee Hewitt; **9.36** Paul G. Hewitt

Chapter 10: Opening photo Howard Lukefahr; **10.10** Princeton University, Palmer Physical Laboratory; **10.11** Evan Jones; **10.16** Paul G. Hewitt; **10.18** Animals Animals/Earth Scenes; **10.19** Addison Wesley Longman, Inc./San Francisco; **10.24** Paul G. Hewitt; **10.27** Paul G. Hewitt; **10.28** Addison Wesley Longman, Inc./San Francisco; **10.29** Addison Wesley Longman, Inc./San Francisco; **10.30** Paul G. Hewitt; **10.33** Paul G. Hewitt; **10.34** Addison Wesley Longman, Inc./San Francisco; **10.35** Katia Chrchourova

Chapter 11: Opening photo Fred Myers; **11.3** Richard Megna/Fundamental Photographs, NYC; **11.5a** Fundamental Photographs, NYC; **11.5b** Richard Megna; **11.9** Paul G. Hewitt; **p 229** Fundamental Photographs, NYC; **11.12a,b** Fundamental Photographs, NYC; **11.12c** Peter Arnold, Inc; **11.13** AP Wide World Photos; **11.14** John Suchocki; **11.20 (left, right)** Addison Wesley Longman, Inc./San Francisco; **11.26** Paul G. Hewitt; **11.27** Paul G. Hewitt; **11.33** Lillian Lee Hewitt; **11.34** Lillian Lee Hewitt; **11.36** Lillian Lee Hewitt; **p 243** Lillian Lee Hewitt;

Chapter 12: Opening photo Dave Eddy; **12.3** Corbis Los Angeles; **12.10** Paul G. Hewitt; **12.12** Terrence McCarthy/San Francisco Symphony; **12.14** Getty Images Inc./ Stone Allstock; **12.15** Laura Pike & Steve Eggen; **12.17** Paul G. Hewitt; **12.18 (left, right)** AP Wide World Photos; **12.18 (middle)** Corbis Los Angeles; **12.20a** Paul G. Hewitt; **12.20b** Fundamental Photographs, NYC; **12.24** Paul G. Hewitt; **12.32** U.S. Navy News Photo; **12.35** The Harold E. Edgerton Trust/Palm Press; **12.39** Paul G. Hewitt; **12.45** Meidor Hu; **p 269** Paul G. Hewitt

Chapter 13: Opening photo Udo Von Mulert; **13.7** Paul G. Hewitt; **13.9** Paul G. Hewitt; **13.17** Paul G. Hewitt; **13.18** Paul G. Hewitt; **13.19a-f** Paul G. Hewitt; **13.21** Paul G. Hewitt; **13.23** Meidor Hu; **13.25** Getty Images/Retrofile; **13.26** Don King/Getty Images Inc.-Image Bank; **13.28a,b,c** Education Development Center, Inc.; **13.30** Ken Kay/Fundamental Photographs, NYC; **13.38** Paul G. Hewitt; **p 292** Paul G. Hewitt

Chapter 14: Opening photo Paul G. Hewitt; **14.3** Paul G. Hewitt; **14.6** David Nunek/Photo Researchers, Inc; **14.7** Institute of Paper Science & Technology; **14.14** Ted Mathieu; **14.16** Robert Greenler; **14.22** Paul G. Hewitt; **14.31** Photo Researchers, Inc.; **14.40** Paul G. Hewitt; **14.48** Fundamental Photographs, NYC/Diane Schiumo; **14.50a,b,c** Paul G. Hewitt; **p 315 (top)** Armstrong Roberts; **p 315 (bottom)** Barbara Thomas; **p 316** Milo Patterson

Chapter 15: Opening photo Mary Murphy Waldorf; **p 321** Corbis Los Angeles; **15.3** Neil Chapman, Collection of Paul G. Hewitt; **15.5** Lillian Lee Hewitt;

411

Page numbers followed by "n" indicate a footnote reference.